STUDENT COMPANION

Akif Uzman
University of Houston-Downtown

Joseph Eichberg
University of Houston

William Widger
University of Houston

Donald Voet
University of Pennsylvania

Judith G. Voet
Swarthmore College

Charlotte W. Pratt
Seattle, Washington

to accompany

FUNDAMENTALS OF BIOCHEMISTRY

Donald Voet Judith G. Voet Charlotte W. Pratt

John Wiley & Sons, Inc.

New York / Chichester / Weinheim / Brisbane / Singapore / Toronto

The front cover shows some of the molecular assemblies that form the circle of life:
DNA makes RNA makes protein makes DNA.

The images are (*clockwise from the top*):

1. B-DNA, *based on an X-ray structure by Richard Dickerson and Horace Drew.*
2. The nucleosome, *courtesy of Timothy Richmond.*
3. Model of the *lac* repressor in complex with DNA and CAP protein, *courtesy of Ponzy Lu and Mitchell Lewis.*
4. Ribozyme RNA, *based on an X-ray structure by Jennifer Doudna.*
5. The ribosome in complex with tRNAs, *courtesy of Joachim Frank.*
6. DNA polymerase in complex with DNA, *courtesy of Tom Ellenberger.*

The central image is based on Leonardo da Vinci's drawing *Study of Proportions.*
It represents for us the never ending human quest for understanding. (© G. Bartholomew/ Westlight)

To order books or for customer service call 1-800-CALL-WILEY (225-5945).

ISBN 0-471-17046-1

Printed in the United States of America

10 9 8 7 6 5 4 3 2 1

Printed and bound by Bradford & Bigelow, Inc.

Preface

Welcome to biochemistry! You are about to become acquainted with one of the most exciting scientific disciplines. The biotechnology industry, with its roots in molecular genetics, is one of the most visible manifestations of the explosion of biochemical knowledge that has occurred during your lifetime. Drug design and novel approaches such as gene therapy rely on fundamental knowledge of the chemistry of biological molecules, particularly proteins. Our most common diseases (e.g., diabetes and heart disease) have pleiotropic multifaceted physiological effects that are best understood in terms of biochemistry. You will soon discover that biochemistry's impact on our lives cannot be over-emphasized. We are excited to bring you an ever-expanding understanding of this magnificent subject!

Learning biochemistry is not easy but it can be fun! Most students discover that biochemistry is a synthetic science, merging knowledge of general chemistry, organic chemistry, and biology. Hence, a more mature and creative kind of thinking is required to gain deep understanding of biochemistry. In addition to a solid foundation in chemistry and biology, you will need to recognize and assimilate some general principles from other disciplines, including physiology, genetics, and cell biology. In this respect, biochemistry is not all that different from nonscientific pursuits that require some degree of "lateral thinking."

This Student Companion accompanies *Fundamentals of Biochemistry* by Donald Voet, Judith G. Voet, and Charlotte W. Pratt. It is designed to help you master the basic concepts and exercise your analytic skills as you work your way through the textbook. The Student Companion is divided into four parts. Each chapter begins with a general summary reminding you of the topics covered in that chapter. This is followed by a section called Essential Concepts, which provides an overview of the main facts and ideas that are essential for your understanding of biochemistry. This can be regarded as a set of brief notes for each chapter, alerting you to the key facts you need to commit to memory and to the concepts you need to master. You will soon notice that biochemical knowledge is cumulative: New concepts often rely on a solid understanding of previously presented concepts. Hence, one of the key goals of this Companion is to help you gain this understanding. The third section is the Guide to Study Exercises, a compilation of abbreviated answers to the Study Exercises found at the end of each textbook chapter. The last part is the Questions. These are organized in a manner to help you gain a firm understanding of each section of a chapter. Some questions ask you to recall essential facts while others exercise your problem-solving skills. Answers to all the questions are provided in the Companion. However, you do yourself a great disservice by turning to them too soon. Don't know the answer right away? Keep trying! Use the answers to check yours, not to fill in a temporary void in your understanding.

"How should I study biochemistry beyond reading the textbook and working in this Companion?" is a likely question from students. The phrase "if you don't use it, you lose it" applies here. It is pointless to simply read your biochemistry textbook over and over. Unless you are actively engaged in working with the material, you become a passive reader. Active engagement includes using your hands to work problems, drawing pictures, or writing an outline. As you read, ask yourself questions and seek answers from the text and your instructor. By doing these things, you use the material, and it becomes incorporated into your long-term memory.

"How often should I study biochemistry?" is also a common question. Most instructors agree that frequent short study sessions—even daily—will pay far greater dividends than a single long session every week or so. Because short-term memory lasts just a few minutes, take a few minutes after class to review your notes. Talk biochemistry with anyone who will listen. Form a study group with your classmates and quiz each other. Use the study group to enrich your knowledge and test your memory. All these activities allow you to transfer knowledge from short-term memory to long-term memory. In other words, the more you use biochemistry, the better you know it and the more fun you will have with it. Good luck!

We have many people to thank for helping us get this Companion to you. First and foremost, we would like to acknowledge our ever-patient editors, Jennifer Yee and Cliff Mills, at John Wiley & Sons. Drs. Elaine Brown of Prose-Professional Science Editing, Caroline Breitenberger of the Ohio State University, and Laura Mitchell of St. Joseph's University provided much appreciated reviews of our original draft. We would also like to thank our students at Swarthmore College, the University of Houston, and the University of Houston-Downtown for pointing out errors and ambiguities in earlier drafts of this work. It is still possible that errors persist, and so we would greatly appreciate being alerted to them. Please forward your comments to:

Cliff Mills
Chemistry Editor, College Division
John Wiley and Sons
605 Third Avenue
New York, NY 10158

Akif Uzman

Joseph Eichberg

William Widger

Charlotte W. Pratt

Donald Voet

Judith G. Voet

Table of Contents

Part I Student Companion

1	Life	1
2	Water	11
3	Nucleotides and Nucleic Acids	23
4	Amino Acids	35
5	Proteins: Primary Structure	51
6	Proteins: Three-Dimensional Structure	65
7	Protein Function	79
8	Carbohydrates	91
9	Lipids	103
10	Biological Membranes	113
11	Enzymatic Catalysis	125
12	Enzyme Kinetics, Inhibition, and Regulation	137
13	Introduction to Metabolism	149
14	Glucose Catabolism	163
15	Glycogen Metabolism and Gluconeogenesis	177
16	Citric Acid Cycle	191
17	Electron Transport and Oxidative Phosphorylation	201
18	Photosynthesis	219
19	Lipid Metabolism	231
20	Amino Acid Metabolism	249
21	Mammalian Fuel Metabolism: Integration and Regulation	265
22	Nucleotide Metabolism	279
23	Nucleic Acid Structure	289
24	DNA Replication, Repair, and Recombination	301

25	Transcription and RNA Processing	315
26	Translation	327
27	Regulation of Gene Expression	339
PART II	**Kinemage Exercises**	**1**

Chapter 1 Life

This chapter introduces you to life at the biochemical and cellular level. It begins with a discussion of the chemical origins of life and its early evolution. This discussion continues into ideas and theories about the evolution of organisms, followed by a brief introduction to taxonomy and phylogeny viewed from a molecular perspective. The chapter concludes with an introduction to the basic concepts of thermodynamics and its application to living systems. Biochemistry, like all other sciences, is based on the measurement of observable phenomena. Hence, it is important to become familiar with the conventions used in the measurement of energy and mass. Box 1-1 presents the essential biochemical conventions that we will encounter throughout *Fundamentals of Biochemistry*.

At the end of each chapter is a list of *Key Terms* used in the chapter. These terms will become our working vocabulary, so it is important that you understand their meanings. The *Study Exercises* alert you to some of the major ideas developed in the chapter; however, they are not exhaustive. In this *Student Companion* we will review and explore these ideas more deeply.

Essential Concepts

Origin of Life

1. Living matter consists of a relatively small number of elements, of which C, N, O, H, Ca, P, K, and S account for ~98% of the dry weight of most organisms (which are 70% water).

2. The current model for the origin of life proposes that organisms arose from the polymerization of simple organic molecules to form more complex molecules, some of which were capable of self-replication.

3. Most polymerization reactions involving the building of small organic molecules into larger more complex ones occur by the formation of water. This is called a condensation reaction.

4. A key development in the origin of life was the formation of a membrane that could separate the critical molecules required for replication and energy capture from a potentially degradative environment.

5. Modern cells can be classified as either prokaryotic or eukaryotic. Eukaryotic cells are distinguished by a variety of membrane-bounded organelles and an extensive cytoskeleton.

Organismal Evolution

6. Phylogenetic evidence based on comparisons of ribosomal RNA genes have been used by Woese and colleagues to group all organisms into three domains: archaea, bacteria, and eukarya.

7. The evolution of sexual reproduction marked an important step of the evolution of organisms, because it allowed for genetic exchanges that lead to an increase in the adaptability of a population of organisms to changing environments.

8. Modern eukaryotes may harbor ancient symbionts in the form of mitochondria and chloroplasts.

9. Biological evolution is not goal-directed, requires some built-in sloppiness, is constrained by its past, and is ongoing.

10. Natural selection directs the evolution of species.

Thermodynamics

11. The first law of thermodynamics states that energy (U) is conserved; it can neither be created nor destroyed.

12. Enthalpy is a thermodynamic function that is the sum of the energy of the system and the product of the pressure and the volume (PV). Since biochemical processes occur at constant pressure and have negligible changes in volume, the change in energy of the system is nearly equivalent to the change in enthalpy ($\Delta U = \Delta H$).

13. The second law of thermodynamics states that spontaneous processes are characterized by an increase in the entropy of the universe, that is, by the conversion of order to disorder.

14. The spontaneity of a process *is determined by its free energy change* ($\Delta G = \Delta H - T\Delta S$). Spontaneous reactions have $\Delta G < 0$ (exergonic) and nonspontaneous reactions have $\Delta G > 0$ (endergonic).

15. For any process at equilibrium, the rate of the forward reaction is equal to the rate of the reverse reaction, and $\Delta G = 0$.

16. Energy, enthalpy, entropy, and free energy are state functions; that is, they depend only on the state of the system, not its history. Hence, they can be measured by considering only the initial and final states of the system and ignoring the path by which the system reached its final state.

17. The entropy of a solute varies with its concentration; therefore, so does its free energy. The free energy change of a chemical reaction depends on the concentration of both its reactants and its products.

18. For the general chemical reaction

$$a\text{A} + b\text{B} \rightleftharpoons c\text{C} + d\text{D}$$

the free energy change is given by the following relationship:

$$\Delta G = \Delta G^\circ + RT \ln\left(\frac{[\text{C}]^c\,[\text{D}]^d}{[\text{A}]^a\,[\text{B}]^b}\right)$$

19. The equilibrium constant of a chemical reaction is related to the standard free energy of the reaction when the reaction is at equilibrium ($\Delta G = 0$), so that we have

$$\Delta G^\circ = -RT \ln K_{eq}$$

where K_{eq} is the equilibrium constant of the reaction:

$$K_{eq} = \frac{[C]^c{}_{eq}\,[D]^d{}_{eq}}{[A]^a{}_{eq}\,[B]^b{}_{eq}}$$

The equilibrium constant can therefore be calculated from standard free energy data and *vice versa.*

20. The equilibrium constant varies with temperature by the relation

$$\ln K_{eq} = \frac{-\Delta H^\circ}{R}\left(\frac{1}{T}\right) + \frac{\Delta S^\circ}{R}$$

where ΔH° and ΔS° represent enthalpy and entropy in the standard state. A plot of K_{eq} versus $1/T$, known as a van't Hoff plot, permits the values of ΔH° and ΔS° (and therefore ΔG° at any temperature) to be determined from measurements of K_{eq} at two (or more) temperatures.

21. The biochemical standard state is defined as follows: The temperature is 25°C, the pH is 7.0, and the pressure is 1 atm. The activities of reactants and products are taken to be the total activities of all their ionic species, except for water, which is assigned an activity of 1. $[H^+]$ is also assigned an activity of 1 at the physiologically relevant pH of 7. These conditions are different than the chemical standard state, so that the biochemical standard free energy is designated as $\Delta G^{\circ\prime}$ and the chemical standard state is ΔG°. We assume that activity equals molarity for dilute solutions.

22. An isolated system cannot exchange matter or energy with its surroundings. A closed system can exchange only energy with its surroundings. A closed system inevitably reaches equilibrium. Open systems exchange both matter and energy with their surroundings and therefore cannot be at equilibrium. Living organisms must exchange both matter and energy with their surroundings and are thus open systems. Living organisms tend to maintain a constant flow of matter and energy, referred to as the steady state.

23. Living systems can respond to slight perturbations from the steady state to restore the system back to the steady state. This process underlies the physiological concept of homeostasis.

24. The recovery of free energy from a biochemical process is never total and some energy is lost to the surroundings as heat. Hence, while the system becomes more ordered, the surroundings experience an increase in entropy.

25. Enzymes accelerate the rate at which biochemical processes reach equilibrium. They accomplish this by interacting with reactants and products to provide a more energetically favorable pathway for the biochemical process to take place.

Key Equations

Be sure to know the conditions for which the following thermodynamic equations apply and be able to interpret their meaning.

1. $\Delta U = q - w$
2. $H = U + PV$
3. $\Delta H = q_p - w + P\Delta V$
4. $S = k_B \ln W$
5. $\Delta S \geq \frac{q}{T} = \frac{\Delta H}{T}$
6. $\Delta G = \Delta H - T\Delta S$
7. $\Delta G = \Delta G° + RT \ln\left(\frac{[C]^c [D]^d}{[A]^a [B]^b}\right)$
8. $\Delta G° = -RT \ln K_{eq}$
9. $K_{eq} = \frac{[C]^c_{eq} [D]^d_{eq}}{[A]^a_{eq} [B]^b_{eq}}$
10. $\ln K_{eq} = \frac{-\Delta H°}{R}\left(\frac{1}{T}\right) + \frac{\Delta S°}{R}$

Guide to Study Exercises (text p. 21)

1. The major stages of chemical and organismal evolution are:
   Formation of simple organic compounds.
   Polymerization of small molecules to form more complex molecules.
   Replication, the self-directed synthesis of additional molecules.
   Concentration of molecules and compartmentation of chemical reactions.
   Development of systems for synthesizing precursors and generating energy.
   Organization of simple cells.
   Emergence of sexual reproduction.
   Cooperation among cells in multicellular organisms. (Sections 1-1 to 1-3)

2. Evolution occurs when the makeup of a population is altered as specific variants are passed from individuals to their offspring. The variants that come to predominate are those that increase the ability of individuals to survive and reproduce under the prevailing conditions; that is, those that are "selected" by nature. These principles of evolution by natural selection apply to living organisms or any self-replicating system. (Sections 1-1B and 1-3C)

3. The archaea include methanogens, halobacteria, and some thermophiles. The bacteria include organisms such as *E. coli* and cyanobacteria. The eukarya include protozoans, fungi, plants, and animals. (Section 1-3A)

4. The first law of thermodynamics states that energy is conserved and can be neither created nor destroyed. It may, however, be transferred, for example, in the form of heat or work, between a closed or open system and its surroundings. The second law of thermodynamics states that spontaneous processes are characterized by increasing disorder (increasing entropy). These laws are sometimes restated as "You can't win" and "You can't even break even." (Sections 1-4A and B)

5. The overall free energy change of a process (ΔG) varies with the difference between the enthalpy change (ΔH) and the product of the temperature and the entropy change ($T\Delta S$): $\Delta G = \Delta H - T\Delta S$. Therefore, ΔG varies not only with the relative magnitudes of ΔH and ΔS but also with the temperature. (Section 1-4C)

6. In the biochemical standard state (which is indicated by a degree symbol and a prime following the symbol for the state function, e.g., $\Delta G^{\circ\prime}$), the temperature is 25°C, the pressure is 1 atm, the activity of each solute is equivalent to its total molar activity (assumed to equal molarity for dilute solutions; water has an activity of 1), and the pH is 7.0 ($[H^+] = 10^{-7}$ M). (Section 1-4D)

7. While they are alive, organisms resist the second law of thermodynamics by maintaining a steady state in which energy and matter constantly flow through the organism, far from equilibrium. The high degree of organization of a living thing requires free energy; without this free energy (i.e., when $\Delta G = 0$), the organism dies and, thermodynamically speaking, comes to equilibrium. (Section 1-4E)

Questions

The Origin of Life

1. What are the elements that account for 98% of the dry weight of living cells?

2. In what critical ways was the atmosphere of the primitive earth different from the earth's current atmosphere?

3. What are the rationale and significance of the Miller and Urey experiments?

4. What was, and continues to be, the source of atmospheric oxygen?

5. Examine the reaction shown on the next page for the condensation and hydrolysis of lactose (two covalently linked sugars, a disaccharide). Circle the functional groups that form water during condensation.

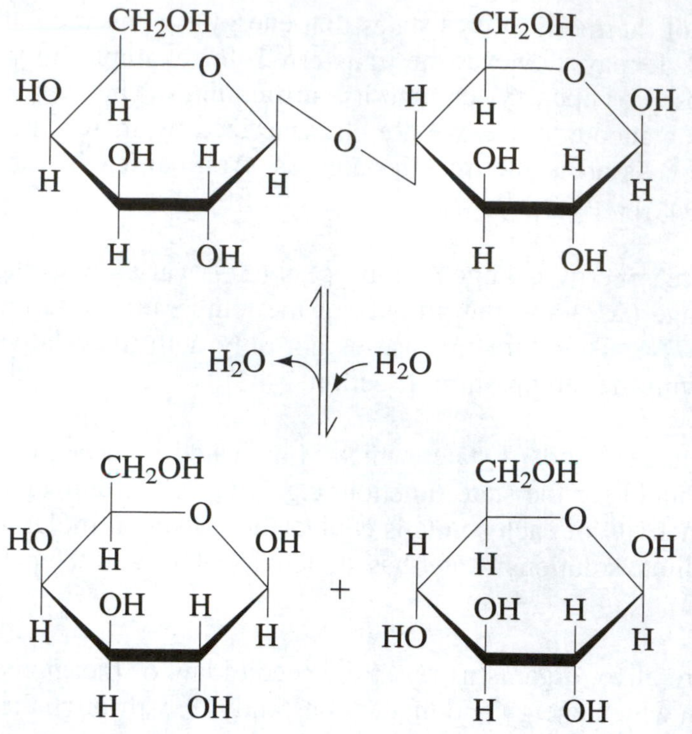

6. The condensation of two functional groups can result in the formation of another common functional group, which can be referred to as a compound functional group. Examine the functional groups in Figure 1-2. Which compound functional groups are the combination of two other functional groups found in that figure? Show how these compound functional groups form.

Cellular Architecture

7. Draw a schematic diagram of a eukaryotic cell showing its principal organelles.

8. Give the principal distinguishing feature(s) of each pair of terms:
   (a) Prokaryote and eukaryote
   (b) Cytosol and cytoplasm
   (c) Endoplasmic reticulum and cytoskeleton

Organismal Evolution

9. Living organisms are classified into three major domains: ______________________,
______________________, and ______________________.

10. What is the most compelling evidence that mitochondria and chloroplasts represent descendants of symbiotic bacteria that lived inside of ancient eukaryotic cells?

11. What is the relationship between mutation and genetic variation in a population of organisms? Of what significance is it to evolution?

Thermodynamics

12. Distinguish between enthalpy and energy. Under what conditions are their changes equivalent?

13. What does it mean when q and w are positive?

14. When crystalline urea is dissolved in water, the temperature of the solution drops precipitously. Does the enthalpy of the system increase or decrease? Explain.

15. List and define the four major thermodynamic state functions.

16. Which of the following pairs of states has higher entropy?
    (a) Two separate beakers of NaCl and KCH_3COO in solution and a beaker containing a solution of both salts.
    (b) A set of dice in which all the dice show 6 dots on the top side and a set of dice in which the 6's show up on one of the side faces.
    (c) A small symmetric molecule that can form a polymer through reaction at either end and a small asymmetric molecule that can polymerize from only one end.

17. Rationalize the temperature dependence of Gibbs free energy changes when both the enthalpy and entropy terms are positive values and when they are both negative values.

18. Hydrogen gas combines spontaneously with oxygen gas to form water

$$2\ H_2 + O_2 \rightarrow 2\ H_2O$$

Which term, enthalpy or entropy, predominates in the equation for the Gibbs free energy? How are the surroundings affected by this reaction?

19. Evaluate the following statement: An enzyme accelerates the rate of a reaction by increasing the spontaneity of the reaction.

20. Based on your reading of this chapter, suggest simple criteria for a reasonable definition of life.

Answers to Questions

1. Carbon, nitrogen, oxygen, hydrogen, phosphorus, calcium, potassium, and sulfur account for 98% of the dry weight of living cells.

2. The primitive earth's atmosphere was a relatively reducing atmosphere lacking appreciable amounts of O_2. Much recent controversy revolves around just how much NH_3 and CH_4 existed in the young earth's atmosphere. Relatively low amounts would indicate that the atmosphere was both nonreducing and nonoxidizing. In either case, the absence of significant levels of O_2 indicates that the atmosphere was nonoxidizing compared to the present atmosphere.

3. The Miller and Urey experiments were designed to ask whether biological molecules could be generated by mimicking the atmospheric conditions of the early earth. The early atmos-

phere was thought then to consist of H_2O, CH_4, NH_3, SO_2, and possibly H_2. Miller and Urey subjected these molecules, except for SO_2, to electrical discharges that were meant to simulate discharges believed to be prevalent in the earth's early atmosphere. The generation of diverse organic acids showed that the precursors for larger biological molecules could be spontaneously produced in the early atmosphere, paving the way for subsequent chemical evolution.

4. Photosynthesis.

5. Shown below is the condensation of galactose and glucose to form lactose. The atoms involved in the elimination of water are circled.

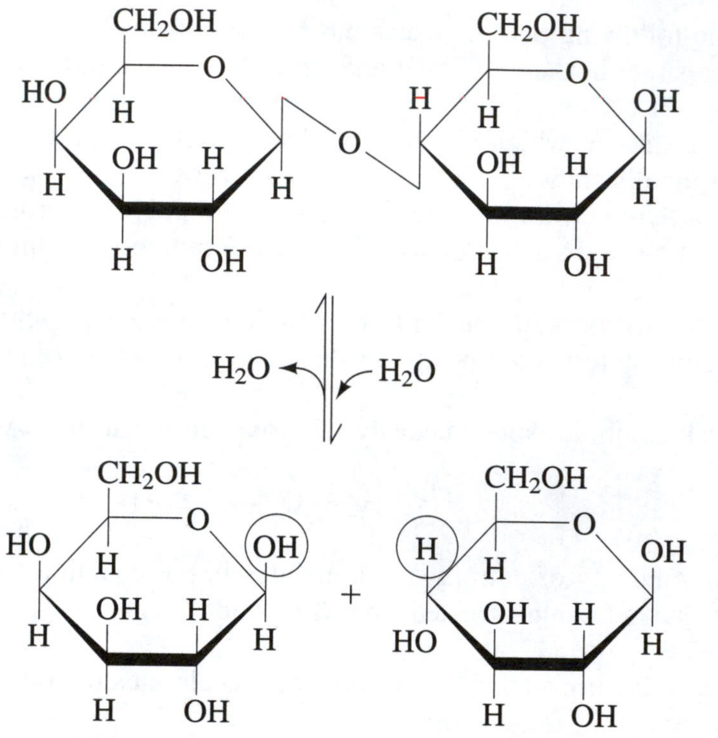

6. Shown below are condensation reactions of two pairs of functional groups found in Figure 1-2 to form two compound functional groups. The atoms involved in the elimination of water are circled.

$$R'\text{—}\overset{\displaystyle O}{\overset{\|}{C}}\text{—}OH + R''\text{—}O\text{—}H \xrightarrow{-H_2O} R'\text{—}\overset{\displaystyle O}{\overset{\|}{C}}\text{—}OR''$$

carboxyl alcohol ester

$$R'\text{—}\overset{\displaystyle O}{\overset{\|}{C}}\text{—}OH + R''\text{—}\overset{\displaystyle H}{\overset{|}{N}}\text{—}H \xrightarrow{-H_2O} R'\text{—}\overset{\displaystyle O}{\overset{\|}{C}}\text{—}NH\text{—}R''$$

carboxyl amine amide

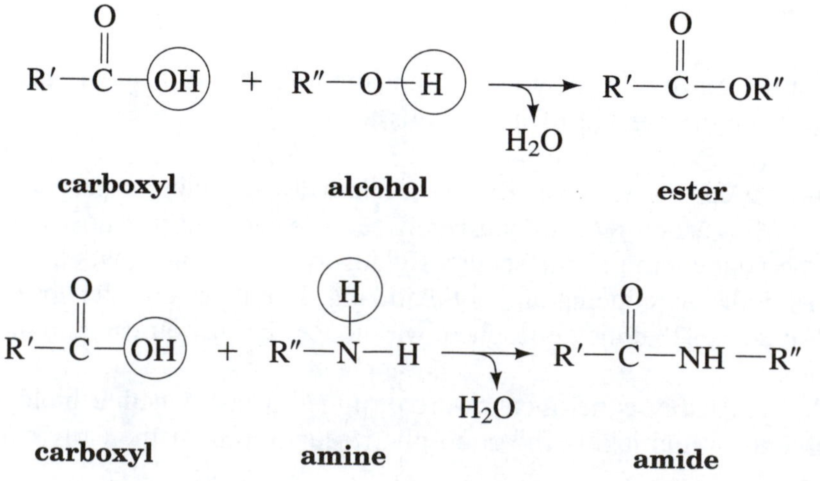

7. See Figure 1-8.

8. (a) Eukaryotes are defined by their elaborate internal membrane systems and the enclosure of genomic DNA inside a double-membrane compartment, the nucleus, which is lacking in prokaryotes.
   (b) The cytoplasm is the contents of a prokaryotic cell and also refers to the cellular contents outside of the nucleus in eukaryotes. The cytosol is the cellular contents minus all of the membranous compartments of the cell (e.g., mitochondria, endoplasmic reticulum, Golgi apparatus, chloroplasts, and vacuoles).
   (c) The endoplasmic reticulum is an extensive network of internal membranes in eukaryotic cells, which is topologically continuous with the nuclear membrane. The cytoskeleton is the extensive network of protein filaments in the cytosol of eukaryotic cells.

9. All living organisms can be classified into three domains: archaea (archaebacteria), bacteria (eubacteria), and eukarya (eukaryotes).

10. The most compelling evidence that these organelles were once symbiotic bacteria is the presence of distinct genetic material and protein synthesis machinery inside of these organelles. The RNA and proteins that make up the protein synthesis machinery of these organelles is much more similar to that of bacteria than that of eukaryotes (see Chapter 26).

11. Genetic variation results from mutations that persist in a population and have not been eliminated by natural selection because they did not significantly decrease fitness. Evolution of a population occurs when variations that increase an individual's chances for survival and reproduction spread throughout the population.

12. Enthalpy is defined as $H = U + PV$. Hence $\Delta H = \Delta U + P\Delta V$. The enthalpy of a system is equal to its energy change when $P\Delta V = 0$, which occurs at the constant pressure and volume conditions typical of living things.

13. If q is positive, then heat has been transferred *to* the system, increasing its internal energy. If w is positive, then work has been done *by* the system, decreasing its internal energy.

14. In this example, urea and the water are the system, and the vessel and beyond are the surroundings. As urea dissolves, the enthalpy increases (an endothermic process). Heat is absorbed into the interactions between the urea and water, making the solution cooler.

15. **Energy** is measured as the heat absorbed by a system minus the work done by the system on its surroundings. **Enthalpy,** or heat content, is the amount of heat generated or absorbed by a system when a process occurs at constant pressure, as in biological systems, and no work is done other than the work of expansion or contraction (ΔV) of the system. **Entropy** is a measure of the heat absorbed or generated by a system at constant temperature and reflects the number of equivalent ways of arranging a system with no change in its internal energy. **Gibbs free energy** is the energy available to do work; it is a combination of enthalpy and entropy ($\Delta H - T\Delta S$) and an indicator of the spontaneity of a process at constant pressure and temperature.

16. (a) The beaker containing the solution with both salts has higher entropy. The molecules in this solution can be arranged in many more orientations with respect to each other than each salt solution alone.
    (b) The set of dice with the 6's showing on the side faces has more entropy since there are many more ways to obtain this configuration.
    (c) The symmetric molecule has greater entropy, since it can polymerize by joining reactions involving either end. Note that the information content of the asymmetric molecule is higher, however. All biological polymers are formed from asymmetric subunits.

17. When enthalpy and entropy are both positive, ΔG decreases with increasing temperature, and the temperature at which the reaction occurs spontaneously must be high enough that the $T\Delta S$ term is larger than the ΔH term in the equation $\Delta G = \Delta H - T\Delta S$. For instance, dissolving crystalline urea in water is endothermic, however the process is spontaneous when it is carried out at room temperature. When enthalpy and entropy are both negative, ΔG decreases with decreasing temperature, and for the reaction to be spontaneous, the temperature must be low enough that the $T\Delta S$ term is not more negative than the ΔH term.

18. The expression for the Gibbs free energy is $\Delta G = \Delta H - T\Delta S$. In this reaction the entropy decreases since the number of molecules decreases and the product is therefore more ordered than the reactants. Hence, in order for this reaction to be spontaneous the enthalpy term (ΔH) must be more negative than the entropy term ($T\Delta S$). In many cases such as this, the enthalpy of the reaction is negative and the reaction releases heat to the surroundings.

19. This statement is incorrect. An enzyme does not alter the spontaneity of a reaction; rather, it increases the rate at which a reaction reaches equilibrium.

20. Catalysis, replication, and mutability have been argued to be the minimum criteria for life. In addition, an organism must be able to maintain a novel chemical environment that is not in equilibrium with its surroundings and to resist the environmental fluctuations that might disturb its ability to carry out the other three essential functions of living systems—this steady state condition is called homeostasis.

Chapter 2 Water

This chapter introduces you to the unique properties of water and to acid–base reactions. The discussion of water begins with a look at its structure and how its polarity provides a basis for understanding its powers as a solvent. You are then introduced to the hydrophobic effect, osmosis, and diffusion. The chemical properties of water are then described, beginning with the ionization of water, which sets the stage for a discussion of acid–base chemistry and the behavior of weak acids and buffers. This discussion includes the Brønsted–Lowry definition for acids and bases, the definition of pH, and the derivation of the Henderson–Hasselbalch equation. As we shall see in subsequent chapters, a solid understanding of acid–base equilibria is fundamental to understanding key aspects of amino acid biochemistry, protein structure, enzyme catalysis, transport across membranes, energy metabolism, and other metabolic transformations.

Essential Concepts

1. Water is essential to biochemistry because:
   (a) Biological macromolecules assume specific shapes in response to the chemical and physical properties of water.
   (b) Biological molecules undergo chemical reactions in an aqueous environment.
   (c) Water is a key reactant in many reactions, usually in the form of H^+ and OH^- ions.
   (d) Water is oxidized in photosynthesis to produce molecular oxygen, O_2, as part of the process that converts the sun's energy into usable chemical form. Expenditure of that energy under aerobic conditions leads to the reduction of O_2 back to water.

Physical Properties of Water

2. The structure of water closely approximates a tetrahedron with its two hydrogen atoms and the two lone pairs of electrons of its oxygen atom "occupying" the vertices of the tetrahedron.

3. The high electronegativity of oxygen relative to hydrogen results in the establishment of a permanent dipole in water molecules.

4. The polar nature of water results in negative portions of the molecule being attracted to the positive portions of neighboring water molecules by a largely electrostatic interaction known as the hydrogen bond.

5. Hydrogen bonds are represented as $D{-}H \cdots A$, where $D{-}H$ is a weakly acidic compound so that the hydrogen atom (H) has a partial positive charge, and A is a weakly basic group that bears lone pairs of electrons. A is often an oxygen atom or a nitrogen atom (occasionally sulfur).

6. Water is strongly hydrogen bonded, with each water molecule participating in four hydrogen bonds with its neighbors: two in which it donates and two in which it accepts. Hydrogen bonds commonly form between water molecules and the polar functional groups of biomolecules, or between the polar functional groups themselves.

7. The strongly hydrogen bonded character of water is responsible for many of its characteristic properties, most notably:
   (a) A high heat of fusion, which allows water to act as a heat sink, such that greater heat loss is necessary for the freezing of water compared to other substances of similar molecular mass.
   (b) A high heat of vaporization, such that relatively more heat must be input to vaporize water compared to other substances of similar molecular mass.
   (c) An ability to dissolve most polar compounds.
   (d) An open structure makes ice less dense than liquid water, thereby making ice float, insulating the water beneath it, and inhibiting total freezing of large bodies of water.

8. A variety of weak electrostatic interactions are critical to the structure and reactivity of biological molecules. These interactions include, in order of increasing strength, London dispersion forces, dipole–dipole interactions, hydrogen bonds, and ionic interactions (see Table 2-1).

9. Water is an excellent solvent of polar and ionic substances due to its property of surrounding polar molecules and ions with oriented shells of water, thereby attenuating the electrostatic forces between these molecules and ions.

10. The tendency of water to minimize its contact with nonpolar (hydrophobic) molecules is called the hydrophobic effect. This effect is largely driven by the increase in entropy caused by the necessity for water to order itself around nonpolar molecules. This causes the nonpolar molecules to aggregate, thereby reducing the surface area that water must order itself about. Consequently, nonpolar substances are poorly soluble in aqueous solution.

11. Many biological molecules have both polar (or charged) and nonpolar functional groups and are therefore simultaneously hydrophilic and hydrophobic. These molecules are said to be amphiphilic or amphipathic.

12. Osmosis is the movement of solvent across a semipermeable membrane from a region of lower concentration of solute to a region of higher concentration of solute. Osmotic pressure of a solution is the pressure that must be applied to the solution to prevent an inflow of solvent. Hence, an increase in solute concentration results in an increase in osmotic pressure.

13. Diffusion is the random movement of molecules in solution (or in the gas phase). It is responsible for the movement of solutes from a region of high concentration to a region of low concentration.

Chemical Properties of Water

14. Water is a neutral, polar molecule that has a slight tendency to ionize into H^+ and OH^-. However, the proton is never free and binds to a water molecule to form H_3O^+ (hydronium ion).

15. The ionization of water is described as an equilibrium between the unionized water (reactant) and its ionized species (products)

$$H_2O \rightleftharpoons H^+ + OH^-$$

In which

$$K = \frac{[H^+][OH^-]}{[H_2O]}$$

Since in dilute aqueous solution, the concentration of water is essentially constant (55.5 M), the concentration of H_2O is incorporated into the value of K, which is referred to as K_w, the ionization constant of water.

$$K_w = [H^+][OH^-]$$

16. The values of both H^+ and K are inconveniently small; hence, their values are more conveniently expressed as negative logarithms, so that

$$\text{pH} = -\log [H^+]$$

$$\text{p}K = -\log K$$

17. According to the Brønsted–Lowry definition, an acid is a substance that can donate a proton, and a base is a substance that can accept a proton. The strength of a weak acid is proportional to its dissociation constant, which is expressed as

$$K = \frac{[H^+][A^-]}{[HA]}$$

18. The pH of a solution of a weak acid is determined by the relative concentrations of the acid and its conjugate base. The equilibrium expression for the dissociation of a weak acid can be rearranged to

$$\text{pH} = \text{p}K + \log \frac{[A^-]}{[HA]}$$

This relationship is known as the Henderson–Hasselbalch equation. When the concentration of a weak acid is equal to the concentration of its conjugate base, pH = pK. Hence, the stronger the acid, the lower its pK (see Table 2-5).

19. Solutions of a weak acid at pH's near its pK resist large changes in pH as OH^- or H^+ is added. Added protons react with the weak acid's conjugate base to reform the weak acid; whereas added OH^- combines with the acid to form its conjugate base and water. A solution of a weak acid and its conjugate base (in the form of a salt) is referred to as a buffer. Buffers are effective within 1 pH unit of the pK of the component acid.

Key Equation

Know how to use the Henderson–Hasselbalch equation:

$$\text{pH} = \text{p}K + \log \frac{[\text{A}^-]}{[\text{HA}]}$$

Guide to Study Exercises (text p. 38)

1. A water molecule in ice is linked to its four nearest neighbors to form a three-dimensional lattice in which the hydrogen-bonded atoms assume a tetrahedral arrangement (see Fig. 2-3). In liquid water, there are only slightly fewer hydrogen bonds between water molecules, but the bonds are no longer strictly tetrahedral, so that irregular groupings of hydrogen bonded molecules occur. (Section 2-1A)

2. Polar substances dissolve in water because the polar water molecules interact with the dipoles of the solute, thereby weakening the attractive forces between solute molecules. Nonpolar substances do not dissolve in water because they lack the bond dipoles that can interact with polar water molecules. (Section 2-1B)

3. The dispersion of a nonpolar substance in an aqueous medium is accompanied by a decrease in entropy due to the ordering of polar water molecules in cages around the nonpolar molecules. The entropy of this process is minimized by the aggregation of the nonpolar molecules, which reduces the area of the water cages and thus the number of ordered water molecules. The loss of entropy of the aggregated nonpolar substance is more than offset by the collective increase in the entropy of the water molecules. The aggregation and exclusion from aqueous solvent of the nonpolar solute is known as the hydrophobic effect. (Section 2-1C)

4. Amphiphiles form micelles in water in order that their nonpolar portions can aggregate by exclusion from water while their polar portions can interact with the polar water molecules. (Section 2-1C)

5. Osmosis is the movement of molecules across a semipermeable membrane from a region of relatively high concentration (e.g., pure water, which can pass through a membrane) to a region of relatively low concentration (e.g., water containing a dissolved solute that cannot pass through the membrane). Diffusion is the random movement of molecules in solution (or in a gas) from a region of higher solute concentration to a region of lower solute concentration. (Section 2-1D)

6. In both the Arrhenius and Brønsted–Lowry definitions, an acid is a substance that can donate a proton. The Arrhenius definition of a base is somewhat limited: an Arrhenius base is a substance that can donate a hydroxide ion. The Brønsted–Lowry definition of a base is more general: a Brønsted–Lowry base is a substance that can accept a proton and hence includes substances that don't contain hydroxyl groups. (Section 2-2B)

7. Because HCl is a strong acid, it dissociates completely in solution. Therefore, the $[H^+]$ of 1 M HCl is 1 M. Since pH = $-\log [H^+]$, the pH of a 1 M solution of HCl is 0. (Section 2-2B)

Questions

Physical Properties of Water

1. Draw a 3-dimensional structure of a water molecule, including any lone pairs of electrons, and indicate the dipole moment. What is the name of the geometrical figure you have drawn?

2. Which gas in each of the following pairs would you expect to be more soluble in water? Why?
   (a) oxygen and carbon dioxide
   (b) nitrogen and ammonia
   (c) methane and hydrogen sulfide

3. Why do bottles of beer break when placed in a freezer?

4. Distinguish between hydrophilic, hydrophobic, and amphipathic substances.

5. Mixing olive oil with vinegar creates a salad dressing that is an emulsion, a mixture of vinegar with many tiny oil droplets. However, in a few minutes, the olive oil separates entirely from the vinegar. Describe the changes in entropy that occur during the initial mixing and the subsequent separation of the olive oil and vinegar.

6. Shown below is a beaker that contains a solution of 0.1 M NaCl. Floating inside is a cellulose bag that is permeable to water and small ions like Na^+ and Cl^- but is impermeable to protein. The bag contains 1 M NaCl and 0.1 M protein. Describe the movement of solutes and solvent.

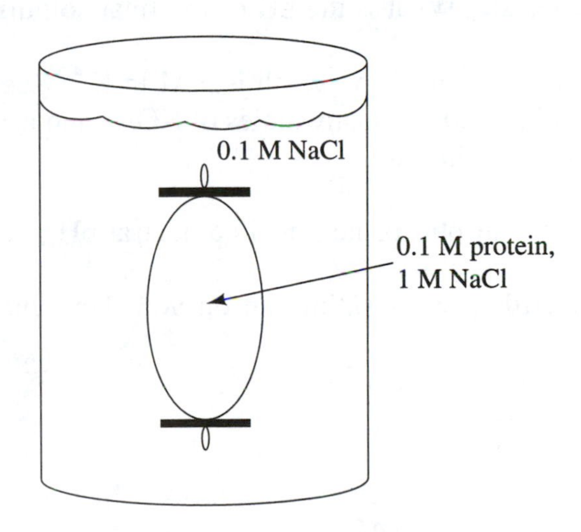

Acids, Bases, and Buffers

7. In the most general definition of an acid and a base, a Lewis acid is a compound that can accept an electron pair, and a Lewis base is a compound that can donate an electron pair. Which of the following compounds can be classified as Brønsted–Lowry acids and bases and which as Lewis acids and bases?

$$BF_3,\quad NH_3,\quad Zn^{2+},\quad HCOOH,\quad NH_4^+,\quad OH^-,\quad Cl^-,\quad H_2O$$

8. Define pK and pH, and write the Henderson–Hasselbalch equation that relates the two.

9. The pK's of trichloroacetic acid and acetic acid are 0.7 and 4.76, respectively. Which is the stronger acid? What is the dissociation constant of each?

10. What is the $[OH^-]$ in a 0.05 M HCl solution? What is the pH?

11. A solution of 0.1 M HCl has a pH of 1. A solution of 0.1 M acetic acid has a pH of 2.8. How much 1 M NaOH is needed to titrate a 100 mL sample of each acid to its respective equivalence point? *Hint:* It may be useful to review Le Chatelier's principle.

12. A 0.01 M solution of a weak acid in water is 0.05% ionized at 25°C. What is its pK?

13. What is a buffer? How does it work? What compounds act as buffers in cells?

14. What is the pH of a 0.1 M solution of cacodylic acid (pK = 6.27)?

15. How would the pH of a 0.1 M solution of acetic acid be affected by the addition of a 4.5 M solution of sodium acetate (NaOAc)? NaOAc is a relatively strong base (weak acids have strong conjugate bases). *Hint:* Consider Le Chatelier's principle.

16. A 0.02 M solution of lactic acid (pK = 3.86) is mixed with an equal volume of a 0.05 M solution of sodium lactate. What is the pH of the final solution?

17. You need a KOAc solution at pH 5, which is 3 M in K^+. Such a solution is used in bacterial plasmid DNA isolation. How many moles of KOAc and acetic acid (HOAc) do you need to make 500 mL of this solution?

18. What are the predominant phosphate ions in a neutral pH phosphate buffer?

19. A beaker of pure distilled water sitting out on your lab bench is slightly acidic. Explain. *Hint:* See Box 2-2.

Answers to Questions

1.

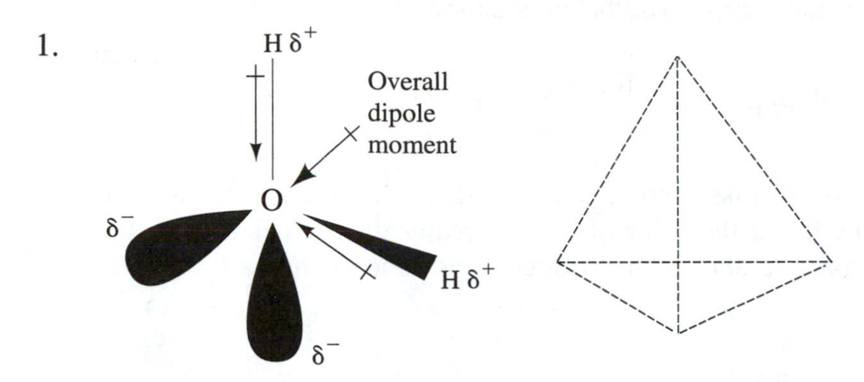

The molecule forms a tetrahedron.

2. (a) Carbon dioxide (O═C═O), which is more polarizable than oxygen (O═O).
   (b) Ammonia (NH_3), which is more polar than nitrogen (N≡N).
   (c) Hydrogen sulfide (HS), which is more polar than methane (CH_4).

3. Water expands upon freezing. Full bottles of beer have little room for expansion and the frozen liquid expands beyond the boundaries of the glass and cap. Similarly, bottles of wine lose their corks in the freezer.

4. Hydrophilic substances are polar compounds that are readily soluble in water. Hydrophobic substances are nonpolar compounds that are insoluble in water. Amphipathic substances have both polar and nonpolar segments; they form micelles or bilayers in aqueous solutions in which the polar groups face the water and the nonpolar groups exclude water and face each other (see Figure 2-10).

5. When the olive oil and vinegar (which is an aqueous solution) are mixed, the entropy of the solution decreases (the action of mixing is an input of energy that drives this process). Mechanical mixing results in an ordering of water around the thousands of oil droplets. Once the mixing has stopped, the system moves toward equilibrium, which maximizes entropy. As the oil droplets fuse, the surface-to-volume ratio of all of the oil decreases, thereby decreasing the amount of ordered water. Consequently, the entropy of the solution increases.

6. Na^+ and Cl^- ions diffuse out of the bag down their concentration gradient across the cellulose membrane. Water moves into the bag via osmosis since its concentration inside the bag is relatively lower due to the presence of the protein that cannot traverse the membrane (and initially due to the higher [NaCl] inside the bag—which eventually equalizes with the [NaCl] outside the bag). The bag expands as water enters. This setup is called dialysis and is used to remove or add salts to solutions of macromolecules.

7. BF_3, Zn^{2+}, HCOOH, NH_4^+, and H_2O are Lewis acids; however, only HCOOH, NH_4^+, and H_2O are Brønsted–Lowry acids. NH_3, OH^-, and Cl^- are considered bases under both definitions.

8. pK is the negative logarithm of the dissociation constant of a weak acid. pH is the negative logarithm of the $[H^+]$. The Henderson–Hasselbalch equation is

$$\text{pH} = \text{p}K + \log \frac{[\text{conjugate base}]}{[\text{acid}]}$$

9. pK is the negative logarithm of the dissociation constant, K, so the smaller the value of pK, the larger the value of K. The larger the value of K, the greater the dissociation of the acid in water. Therefore, trichloroacetic acid is the stronger acid since it has a lower pK than acetic acid.

For trichloroacetic acid, p$K = -\log K$

Therefore, $K = 10^{-0.7} = 0.2$

Similarly for acetic acid,

$$\log K = -\,4.76$$
$$K = 10^{-4.76} = 1.74 \times 10^{-5}$$

10. HCl is a strong acid and completely dissociates in water, so $[\text{HCl}] = [H^+] = 0.05$ M. The pH is given by

$$\begin{aligned} \text{pH} &= -\log\,[H^+] \\ &= -\log 0.05 \\ &= 1.3 \end{aligned}$$

Since $[H^+][OH^-] = K_w$,

$$\begin{aligned} [OH^-] &= \text{p}K_w\,/\,[H^+] \\ &= 10^{-14}\,/\,0.05 \\ &= 2 \times 10^{-13}\ \text{M} \end{aligned}$$

11. By Le Chatelier's principle, the equilibrium for the dissociation of a weak acid is shifted toward the dissociation of the acid by removal of H^+. Therefore, as H^+ is removed by titration with OH^-, more acid dissociates, until there is no acid remaining. Hence, the amount of OH^- needed to titrate equivalent amounts of all monovalent acids is the same. In this case, 10 mL of 1 M NaOH will bring the titration of each acid to its end point. (*Note:* 10 mL of 1 M NaOH provides 0.01 moles of OH^- to neutralize the 0.01 moles of H^+.)

12. The equilibrium constant for a weak acid is

$$K = \frac{[H^+][A^-]}{[\text{HA}]}$$

The concentration of each of the dissociated ions is 0.05% of 0.01 M, or 0.0005×0.01 M = 5×10^{-6} M. Because only 0.05% of the acid is ionized, we can assume that [HA] does not change. We then calculate

$$K = \frac{(5 \times 10^{-6})(5 \times 10^{-6})}{0.01} = 2.5 \times 10^{-9}$$

$$\text{p}K = -\log(2.5 \times 10^{-9}) = 8.6$$

13. A buffer is a solution of a weak acid and its conjugate base. A buffer works by "absorbing" base or acid equivalents within about one pH unit of its pK. For example, when a small amount of OH^- is added, it reacts with HA to form water and A^-, with little change in pH. Similarly, a small amount of H^+ reacts with A^- to form HA. In biological systems, protons produced during catabolic reactions must be buffered by the cell. Phosphate and carbonate ions, as well as proteins, nucleic acids, and fatty acids, serve as buffers in cells.

14. Since pK = 6.27, $K = 10^{-6.27} = 5.37 \times 10^{-7}$. For the dissociation of cacodylic acid

$$\text{cacodylic acid} \rightleftharpoons H^+ + \text{cacodylate}^-$$

we can approximate $[H^+] = [\text{cacodylate}^-] = x$.

We approximate [cacodylic acid] by $0.1 - x$.

Therefore,

$$K = \frac{[H^+][\text{cacodylate}^-]}{[\text{cacodylic acid}]}$$

$$5.37 \times 10^{-7} = \frac{x^2}{0.1 - x}$$

$$0 = x^2 + (5.37 \times 10^{-7})x - (5.37 \times 10^{-8})$$

Solve for x using the quadratic equation

$$x = \frac{-b \pm \sqrt{b^2 - 4ac}}{2a}$$

$$x = \frac{-(5.37 \times 10^{-7}) \pm \sqrt{(5.37 \times 10^{-7})^2 + 4(5.37 \times 10^{-8})}}{2}$$

$$x = -2.32 \times 10^{-4} \text{ or } 2.31 \times 10^{-4}$$

The result can only be 2.31×10^{-4} since there is no such thing as a negative concentration.

Therefore, $x = [H^+] = 2.31 \times 10^{-4}$ M and

$$pH = 3.64$$

Note that for weak acids with p$K > 5$, one can approximate the pH by ignoring the change in acid concentration due to its dissociation. Thus, [cacodylic acid] = 0.1 M and

$$5.37 \times 10^{-7} = \frac{x^2}{0.1}$$

Therefore, $x = 2.32 \times 10^{-4}$ M

15. By Le Chatelier's principle, the added acetate anion drives the reaction toward the formation of undissociated acid, thereby causing a decrease in $[H^+]$ and an increase in pH.

16. The final concentrations are [lactic acid] = 0.01 M and [lactate] = 0.025 M. Use the Henderson–Hasselbalch equation to calculate the pH:

$$\begin{aligned} pH &= pK + \log \frac{[\text{lactate}]}{[\text{lactic acid}]} \\ &= 3.86 + \log 2.5 \\ &= 3.86 + 0.398 \\ &= 4.26 \end{aligned}$$

17. Use the Henderson–Hasselbalch equation to calculate the concentration of acetic acid (x) necessary to obtain a solution of pH 5 that contains 3 M K^+ (3 M KOAc). The pK of acetic acid is 4.76 (Table 2-5).

$$pH = pK + \log \frac{[\text{KOAc}]}{[\text{HOAc}]}$$

$$5 = 4.76 + \log \frac{3}{x}$$

$$\begin{aligned} 0.24 &= \log 3 - \log x \\ 0.24 - 0.48 &= -\log x \\ 0.24 &= \log x \\ x &= 1.73 \text{ M} \end{aligned}$$

Therefore, for a 500 mL solution, the amount of acetic acid added should be 1.73/2 or 0.86 moles. The amount of KOAc should be 3/2 or 1.5 moles.

18. At pH 7, $H_2PO_4^-$ is in equilibrium with HPO_4^{2-} at nearly a 1:1 ratio since pH 7 is near pK_2 (6.82). See Figure 2-16 and Table 2-5.

19. Atmospheric carbon dioxide reacts with the distilled water in the beaker to form carbonic acid. The carbonic acid dissociates into protons and bicarbonate with a p*K* of about 6. These protons make the distilled water slightly acidic.

$$CO_2 + H_2O \rightleftharpoons H_2CO_3$$

$$H_2CO_3 \rightleftharpoons H^+ + HCO_3^-$$

Chapter 3 Nucleotides and Nucleic Acids

This chapter introduces you to the structure and function of nucleotides and their polymers, ribonucleic acid (RNA) and deoxyribonucleic acid (DNA). The chapter begins with a discussion of the various kinds of nucleotides and the large variety of their functions in cellular processes. The nucleic acid polymers, RNA and DNA, are the primary players in the storage, transmission, and decoding of the genetic material. Scientists use a variety of powerful techniques to characterize and manipulate DNA from any organism. This chapter discusses the sequence-specific cleavage of DNA by restriction endonucleases; DNA sequencing; amplification of DNA by cloning in unicellular organisms such as bacteria and yeast; and the *in vitro* amplification of DNA by the polymerase chain reaction.

Essential Concepts

Nucleotide Structure and Function

1. Of the four major classes of biological molecules (amino acids, sugars, lipids, and nucleotides), nucleotides are the most functionally diverse. They are involved in energy transfer, catalysis, and signaling within and between cells, and are essential for the storage, decoding, and transmission of genetic information.

2. Nucleotides are composed of a nitrogenous base linked to a ribose sugar to which at least one phosphate group is attached. The eight common nucleotides, which are the monomeric units of RNA and DNA, contain the bases adenine, guanine, cytosine, thymine, and uracil.

3. The best known nucleotide is the energy transmitter adenosine triphosphate (ATP), which is synthesized from adenosine diphosphate (ADP). Transfer of one or two of the phosphoryl groups of ATP is an exergonic process whose free energy can be used to drive an otherwise nonspontaneous process.

4. Nucleotide derivatives such as flavin adenine dinucleotide (FAD) and nicotinamide adenine dinucleotide (NAD^+) undergo reversible oxidation–reduction reactions in cells. Neither the heterocyclic ring system of FAD nor the nicotinamide base of NAD^+ can be synthesized by mammals and are obtained from the diet in the form of riboflavin (vitamin B_2) and niacin, respectively. Nicotinamide adenine dinucleotide phosphate ($NADP^+$) is identical to NAD^+ except for an additional phosphate group at the 2′ carbon of ribose. In general, NAD^+ is used in energy-capturing reactions such as in glycolysis, whereas reduced $NADP^+$ (NADPH) is used in biosynthetic reactions.

5. The nucleotide derivative coenzyme A (CoA) contains the vitamin pantothenic acid (vitamin B_3). CoA functions metabolically as a carrier of acetyl and other acyl groups.

Nucleic Acid Structure

6. Nucleic acids are polymers of nucleotides in which phosphate groups link the 3′ and 5′ positions of neighboring ribose residues. This linkage is called a phosphodiester bond because

the phosphate is esterified to the two ribose groups. The phosphates are acidic at biological pH and so the polynucleotide is a polyanion.

7. Nucleic acids are inherently asymmetric, so that one end (with a 5′ phosphate) is different from the other end (with a 3′ hydroxyl). This asymmetry, or polarity, is critical for the information storage function of nucleic acids. In fact, all linear biological molecules show this kind of polarity.

8. The structure of DNA was determined by Francis Crick and James Watson in 1953. Key information used to build their model included the following:
   (a) DNA has equal numbers of adenine and thymine residues and equal numbers of cytosine and guanine residues (called Chargaff's rules).
   (b) X-Ray diffraction studies (principally by Rosalind Franklin) revealed that the polymer is most likely helical with a uniform width.
   (c) Structural studies had indicated that the nitrogenous bases should assume the keto tautomeric form.
   (d) Chemical evidence indicated that the polymer was linked by phosphodiester bonds between the 3′ and the 5′ carbons of adjoining ribose units.

9. The Watson–Crick model has the following major features:
   (a) Two polynucleotide chains wind around a central axis to form a right-handed helix.
   (b) The polynucleotide chains are antiparallel to each other.
   (c) The nitrogenous bases occupy the core of the double helix, while the sugar–phosphate chains are the backbones of DNA, running along the outside of the helical structure.
   (d) Adenine is hydrogen bonded to thymine to form a planar base pair (like the rung of a ladder). In a similar fashion, cytosine is hydrogen bonded to guanine.

10. The complementary strands of DNA immediately suggest that each strand of DNA can act as a template for the synthesis of its complementary strand and so transmit hereditary information across generations.

11. The DNA of an organism (its genome) is unique to each organism and, in general, increases in amount in rough proportion to the complexity of the organism. In eukaryotes, genomic DNA occurs in discrete linear pieces called chromosomes.

12. While single-stranded DNA is uncommon in cells, RNA (ribonucleic acid) is usually single-stranded. However, by complementary base pairing, RNA can form intrastrand double helical sections, which bend and fold these molecules into unique three-dimensional shapes.

Overview of Nucleic Acid Function

13. The chemical link between DNA and proteins is RNA. DNA replication, its transcription into RNA, and the translation of messenger RNA into a protein constitute the central dogma of molecular biology: DNA makes RNA makes protein.

14. RNA has many functional forms. The product of DNA transcription that encodes a protein is called messenger RNA (mRNA) and is translated in the ribosome, an organelle containing ribosomal RNA (rRNA). In the ribosome, each set of three nucleotides in mRNA base

pairs with a transfer RNA (tRNA) molecule that is covalently linked to a specific amino acid. The order of tRNA binding to the mRNA determines the sequence of amino acids in the protein. It follows that any alteration to the genetic material (mutation) manifests itself in a new nucleotide sequence in mRNA, which often results in the pairing of a different RNA and hence a new amino acid sequence.

15. Some RNAs have catalytic capabilities. In cells, the joining of two amino acids in the ribosome appears to be catalyzed by rRNA.

Nucleic Acid Sequencing

16. Nucleic acid sequences provide the following:
    (a) Information on the probable amino acid sequence of a protein (but not on any post-translational modifications to these amino acids).
    (b) Clues about protein structure and function (based on comparisons of known proteins).
    (c) Information about the regulation of transcription (DNA sequences adjoining amino acid coding sequences are usually involved in regulation of transcription).
    (d) Discoveries of new genes and regulatory elements in DNA.

17. The overall strategy for sequencing DNA or any other polymer is
    (a) Cleave the polymer into specific fragments that are small enough to be fully sequenced.
    (b) Sequence the residues in each segment.
    (c) Determine the order of the fragments in the original polymer by repeating the preceding steps using a different set of fragments obtained by a different degradation procedure.

18. Nucleic acid sequencing became relatively easy with the following:
    (a) The development of recombinant DNA techniques, which allows the isolation of large amounts of specific segments of DNA.
    (b) The discovery of restriction endonucleases, which cleave DNA at specific sequences.
    (c) The development of the chain-termination method of sequencing.

19. Individuality in humans and other organisms derives from their high degree of genetic polymorphism. For example, homologous human chromosomes differ in sequence about every 200 to 500 base pairs. These differences create or eliminate sites at which restriction endonucleases can cleave DNA. Hence, DNA fragments generated by a specific restriction endonuclease vary in size among individuals. These differences are called restriction fragment length polymorphisms (RFLPs). If a particular RFLP is closely linked to a known gene, it can be used as a diagnostic tool to evaluate inheritance patterns of that gene.

20. Gel electrophoresis is a technique in which charged molecules move through a gel-like matrix under the influence of an electric field. The rate a molecule moves is proportional to its charge density, its size, and its shape. Nucleic acids have a relatively uniform shape and charge density, so their movement through the gel is largely a function of their size.

21. The most common method used to sequence DNA is called the chain-terminator or dideoxy method. This technique relies on the interruption of DNA replication *in vitro* by small amounts of dideoxynucleotides. These lack the 3′-hydroxyl group to which another nu-

cleotide would normally be added during polymerization. The DNA fragments, each ending at a position corresponding to one of the bases, are separated according to size by electrophoresis, thereby revealing the sequence of the DNA.

22. Phylogenetic relationships can be evaluated by comparing sequences of similar genes in different organisms. The number of nucleotide differences roughly corresponds to the extent of evolutionary divergence between the organisms.

Recombinant DNA Technology

23. Recombinant DNA technology makes it possible to amplify and purify specific DNA sequences that may represent as little as one part per million of an organism's DNA.

24. Cloning refers to the production of multiple genetically identical organisms from a single ancestor. DNA cloning is the production of large amounts of a particular DNA segment through the cloning of cells harboring a vector containing this DNA. Cloned DNA can be purified, after which it can be sequenced. If the DNA of interest is flanked by the appropriate regulatory sequences (which themselves can be introduced into the vector), then the cells harboring the cloned DNA will transcribe and translate the cloned DNA to produce large amounts of protein. The protein can be purified and used for a variety of analytical or biomedical purposes.

25. Organisms such as *E. coli* and yeast are commonly used to carry cloning vectors, which are small, autonomously replicating DNA molecules. The most frequently used vectors are circular DNA molecules called plasmids, which occur as one or hundreds of copies in the host cell. Other important vectors for DNA cloning are bacteriophage λ in *E. coli* and yeast artificial chromosomes (YACs) in yeast. YACs are linear pieces of DNA that contain all the chromosomal structures needed for normal chromosome replication and segregation during cell division in yeast. Although their copy number is not as high as that of plasmids, they can accommodate much larger pieces of foreign DNA than can plasmids.

26. The following strategy is typically used to clone a segment of DNA:
    (a) A fragment of DNA is obtained using restriction endonucleases that generate sticky ends. The fragment is then isolated for subsequent ligation to a vector that has been cut with the same restriction endonuclease. This restriction site may be in a selectable gene, so that cells containing recombinant plasmids can be distinguished from cells containing nonrecombinant vectors.
    (b) The fragment is ligated to the vector using an enzyme called DNA ligase, which catalyzes the formation of phosphodiester bonds.
    (c) The recombinant DNA molecule (vector plus the DNA insert) is introduced into cells by a method called transformation. Cells containing the recombinant DNA are usually selected for by their ability to resist specific antibiotics.
    (d) Cells containing the desired DNA are isolated and stored or allowed to grow in order to obtain large quantities of the recombinant DNA.

27. In most cases, the desired DNA exists in just a few copies per cell, so all the cell's DNA is cloned at once by a method called shotgun cloning to produce a genomic library. The challenge is then to use a screening technique to find the desired clone among thousands to mil-

lions in the library. Colony hybridization is a common screening technique in which cells harboring recombinant DNA are transferred from a master plate to a filter that traps the DNA after the cells have been lysed. The filters are then probed using a gene fragment or mRNA that binds specifically to the DNA of interest.

28. If some sequence information is available, a specific DNA molecule can be amplified using a powerful technique called the polymerase chain reaction (PCR). In PCR, a DNA sample (often quite impure) is denatured into two strands and incubated with DNA polymerase, deoxynucleotides, and two oligonucleotides flanking the DNA of interest that serve as primers for DNA replication by DNA polymerase. Twenty rounds of DNA replication and DNA denaturation (all of which are automated in simple computerized incubators) can increase the amount of the target sequence by a million-fold.

29. DNA cloning can be used to obtain large amounts of protein, referred to as recombinant protein, in bacteria or other organisms (although posttranslational modifications unique to eukaryotes cannot be obtained in bacteria). Another important application is the ability to mutate the amino acid sequence in the DNA of interest at specific sites by a technique called site-directed mutagenesis. Another exciting application of DNA cloning is the ability to introduce modified or novel genes into an intact animal or plant, creating a transgenic organism. The transplanted gene is called the transgene. When this technique produces a therapeutic effect, it is referred to as gene therapy. Gene therapy also refers to the introduction of recombinant DNA into select cells to obtain a therapeutic effect.

30. Recombinant DNA technology raises ethical considerations that must be evaluated before its use in medicine, forensics, and agriculture.

Key Equation

$$P = 1 - (1 - f)^N$$

Guide to Study Exercises (text p. 75)

1. See Table 3-1.

2. See Fig. 3-1.

3. DNA contains the pentose 2′-deoxyribose; RNA contains ribose. DNA contains the bases adenine, guanine, cytosine, and thymine; RNA contains adenine, guanine, cytosine, and uracil. DNA is (usually) double-stranded and hence contains equal amounts of A and T and of C and G, due to the requirements of base pairing. RNA is (usually) single-stranded and therefore has no such constraints. (Sections 3-1 and 3-2)

4. According to the central dogma, DNA directs its own replication as well as its transcription to RNA. The RNA is then translated into protein. (Section 3-3A)

5. The Watson–Crick model of DNA is a molecule containing two antiparallel strands of DNA

that wind around a common axis to form a double helix. The strands associate through hydrogen bonding between bases on opposite chains: A pairs with T and G pairs with C. The paired bases occupy the core of the helix and the sugar–phosphate backbone runs along the periphery, defining the major and minor grooves on the surface of the helix. (Section 3-2B)

6. In the restriction–modification system, a bacterium modifies (by addition of a methyl group) certain nucleotides in specific sequences of its own DNA. Foreign DNA, whose sequences are not methylated in the same sequence-specific manner, is susceptible to cleavage by the host bacterium's restriction endonuclease, which recognizes the same nucleotide sequence as does the host's modification methylase. This system destroys foreign DNA. (Section 3-4A)

7. The chain-terminator (dideoxy) sequencing method uses a DNA polymerase to synthesize DNA strands complementary to the DNA strand of interest. The new strands, which are chemically tagged for easy detection, terminate prematurely at positions where a 2′,3′-dideoxy nucleotide has been incorporated, because further polymerization requires a 3′-OH group. When the products of four reaction mixtures, each containing a different dideoxynucleotide, are electrophoresed in parallel lanes, the lengths of the truncated chains indicate the positions of the corresponding nucleotides. (Section 3-4C)

Questions

Nucleotide Structure and Function

1. Indicate which of the following are purine or pyrimidine bases, nucleosides, or nucleotides:
   (a) uracil
   (b) deoxythymidine
   (c) guanosine monophosphate
   (d) adenosine
   (e) cytosine
   (f) guanylic acid
   (g) UMP
   (h) guanine

Nucleic Acid Structure

2. What key sets of data were used to build the Watson–Crick model of DNA structure?

3. Define or explain the following terms: (a) antiparallel; (b) complementary base pairing.

4. The base compositions of samples of genomic DNA from several different animals are given below. Which samples are likely to come from the same species?
   (a) 27.3% T
   (b) 29.5% G
   (c) 13.1% C
   (d) 36.9% A
   (e) 22.7% C
   (f) 19.9% T

Nucleic Acid Sequencing

5. A certain gene in one strain of *E. coli* can be cleaved into fragments of 3 kb and 4 kb using the restriction endonuclease *Pst*I, but it is unaffected by *Pvu*II. In another strain of bacteria, the same gene cannot be cleaved with *Pst*I, but treatment with *Pvu*II yields fragments of 3 kb and 4 kb. Explain the probable genetic difference between the two strains.

6. You have inserted a 5-kb piece of eukaryotic DNA into a 3.5-kb plasmid at the *Bam*HI site. When you digest the recombinant DNA with various restriction enzymes, you obtain the fragments listed below. Draw a circular map of the recombinant plasmid showing all the restriction sites. Is it possible to determine the orientation of the insert in the plasmid? If so, what information reveals this? If not, what restriction digest would you perform to determine the orientation of the insert?

Restriction enzyme	***Size of DNA fragments (kb)***
*Eco*RI and *Bam*HI	1, 1.25, 2.75, 3.5
*Pst*I and *Bam*HI	1.75, 3.25, 3.5
*Eco*RI, *Pst*I, and *Bam*HI	0.75, 1, 1.25, 2, 3.5
*Sal*I	8.5
*Sal*I and *Bam*HI	0.5, 3, 5
*Bam*HI	3.5, 5

7. In order to separate different-sized fragments of purified DNA by agarose gel electrophoresis at pH 7.5, should one allow the DNA to migrate from the cathode (the − electrode) or from the anode (the + electrode)? Explain.

8. T7 DNA polymerase is less sensitive to dideoxynucleotides than the *E. coli* DNA polymerase I fragment described in Section 3-4C.
   (a) For a given concentration of dideoxynucleotides, which enzyme is more likely to give a longer "ladder" on the sequencing gel?
   (b) Which enzyme is best used for sequencing DNA close to the primer?

9. You are interested in the sequence of a 15-base segment of a gene. Using the chain-terminator procedure, you obtain the following results after gel electrophoresis. What is the sequence of the gene segment?

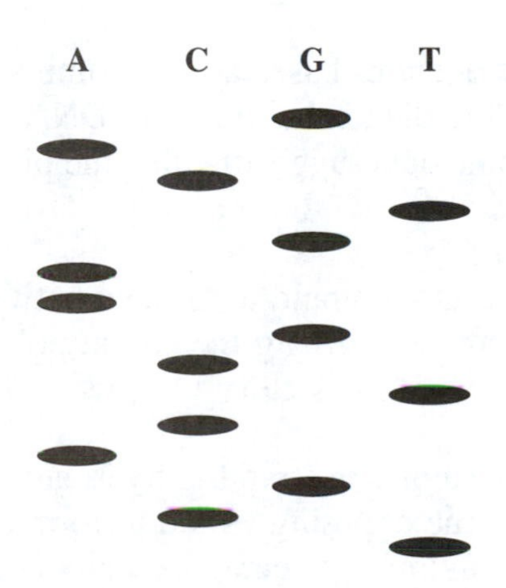

Recombinant DNA Technology

10. Outline a procedure for identifying *E. coli* cells that contain a recombinant pBR322 plasmid in which a foreign gene has been inserted at the *Pvu*I site.

11. The size of the human haploid genome is about 3×10^6 kb. Using human sperm DNA and bacteriophage λ as a cloning vector, you have made a human genome library containing 10^5 clones. Each clone has a 10-kb insert on average. What is the probability that a particular 10-kb single-copy gene is present in your library? Are you confident that you can find your gene with this library? Does a library with 10 times more clones significantly improve your chances? If not, why not?

12. The amplification of DNA by PCR is exponential for many rounds and then becomes linear. What factors are likely to become rate limiting so that rate of DNA accumulation becomes linear?

13. Many eukaryotic peptide hormones cloned in and isolated from bacteria have no biological activity. Why?

14. DNA synthesized using RNA as a template is known as cDNA. In the technique called subtractive hybridization cloning, a small amount of cDNA from all the mRNA of tissue A is hybridized to all the RNA from tissue B. The unhybridized cDNA is then cloned. What has been cloned?

Answers to Questions

1. (a) Pyrimidine base
   (b) Pyrimidine nucleoside
   (c) Purine nucleotide
   (d) Purine nucleoside
   (e) Pyrimidine base
   (f) Purine nucleotide
   (g) Pyrimidine nucleotide
   (h) Purine base

2. (a) Chargaff's rules
   (b) Knowledge that nitrogenous bases are predominantly in the keto tautomeric form
   (c) X-ray crystallographic data indicating that DNA has a double-helical structure
   (d) X-ray crystallographic data suggesting that the planar aromatic bases form a stack parallel to the fiber axis

3. (a) The nucleotides of a DNA strand are linked between the 3′ and 5′ ribose carbons by a phosphate group, thereby giving the strand polarity. The two strands of DNA form a duplex in which the strands run in opposite directions, that is, in an antiparallel arrangement.
   (b) Each nitrogenous base of one strand is hydrogen bonded to a different specific nitrogenous base on the opposing strand to form a planar base pair. This qualitatively specific hydrogen bonding between base pairs is called complementary base pairing.

In the Watson–Crick structure, adenine specifically pairs with thymine, and guanine pairs with cytosine.

4. Samples (a) and (e), since by Chargaff's rules, 27.3% T implies 27.3% A, so that G + C = 45.4% and C = 22.7%. Samples (c) and (d), since 13.1% C implies 13.1% G and 36.9% each of A and T.

5. The gene in the first strain contains the sequence CTGCAG, which is recognized and cleaved by *Pst*I but not by *Pvu*II. In the second strain, the site has mutated so that it cannot be recognized by *Pst*I. Its sequence must be CAGCTG, since it is recognized and cleaved by *Pvu*II.

6. Shown below is one of two possible restriction maps of the recombinant plasmid (in the other possible map, the *Sal*I site would be 0.5 kb from the other *Bam*HI site). To determine the orientation of the insert, it would be necessary to perform a restriction enzyme digest with *Sal*I and *Pst*I (or *Sal*I and *Eco*RI) to reveal the location of the *Sal*I site. For the map shown below, the *Sal*I/*Pst*I digestion products would be 2.25 kb and 6.25 kb.

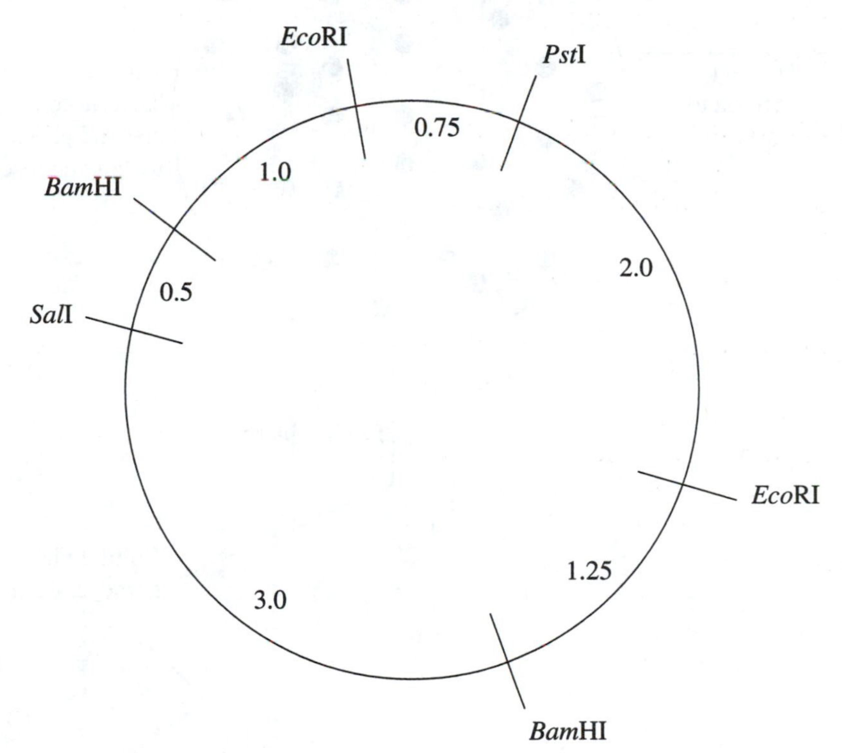

7. At pH 7.5, purified DNA is negatively charged; hence, the DNA fragments will migrate from the cathode toward the anode.

8. (a) T7 DNA polymerase is likely to give a longer ladder, since it is less likely to terminate DNA chain elongation.
   (b) The *E. coli* polymerase I fragment is better suited for obtaining sequences close to the primer, since it terminates DNA chain elongation more often and hence produces more short fragments than does T7 DNA polymerase.

9. The bands on the gel identify the 3′ ends of DNA fragments synthesized by the DNA polymerase in the presence of dideoxynucleotides. Their sequence, read from bottom to top, is

5′-T-C-G-A-C-T-C-G-A-A-G-T-C-A-G-3′

This is complementary to the DNA of interest, which has the sequence

5′-C-T-G-A-C-T-T-C-G-A-G-T-C-G-A-3′

10. Bacteria containing the plasmids will all be resistant to tetracycline. The *Pvu*I site is located in the ampicillin-resistance gene, making bacteria containing recombinant plasmids sensitive to ampicillin. Shown below is a strategy for selecting bacteria containing the recombinant plasmid. Note that one identifies the recombinant plasmids by the absence of specific bacterial colonies in the culture plate containing both tetracycline and ampicillin. Hence, one must recover recombinant plasmids from either the first (master) plate or the tetracycline-containing plate.

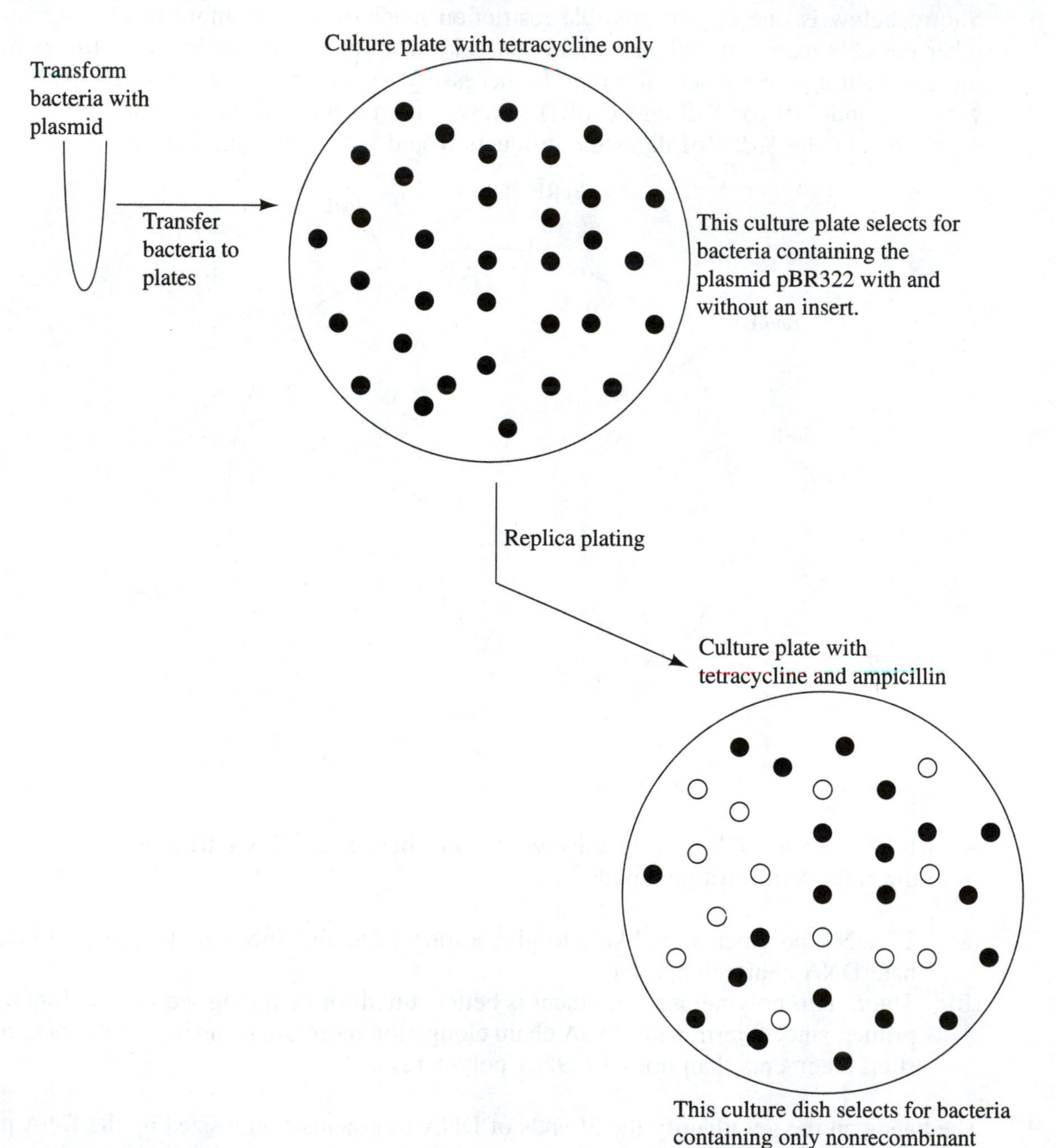
Transform bacteria with plasmid
Transfer bacteria to plates
Culture plate with tetracycline only
This culture plate selects for bacteria containing the plasmid pBR322 with and without an insert.
Replica plating
Culture plate with tetracycline and ampicillin
This culture dish selects for bacteria containing only nonrecombinant plasmids.

11. The frequency (f) of the gene of interest is $10/(3 \times 10^6) = 3.33 \times 10^{-6}$. The probability ($P$) of finding this gene is

$$P = 1 - (1 - f)^N$$

where N is the number of clones. Hence, $P = 1 - (1 - 3.33 \times 10^{-6})^{100{,}000} = 1 - 0.72 = 0.28$. This low probability inspires no confidence that a single-copy gene can be easily found. The probability may be even lower, since not all genomic DNA fragments will be incorporated into the cloning vector with equivalent efficiency. Using the relationship above, the probability increases to 0.96 when the library contains 10^6 clones. Most investigators will shoot for a probability of 0.999 when constructing a recombinant DNA library to ensure the best chance of identifying a single-copy gene.

12. The finite amounts of nucleotides, primers, and enzyme will become rate-limiting. In the early rounds of PCR, these elements are at or near saturating levels (that is, their concentrations are much higher than that of the DNA template). Consequently, the rate of DNA production is directly proportional to the amount of template, which accumulates in an exponential fashion. But as the template DNA accumulates, its concentration approaches that of the enzyme, primers, and nucleotides. This slows the rate at which these elements come together to generate additional DNA, so the rate of DNA accumulation becomes linear.

13. These hormones require posttranslational processing that cannot be carried out in the bacteria.

14. What has been cloned is the cDNA corresponding to the mRNA sequences of tissue A that are absent from tissue B. The hybridization reaction eliminates the mRNA sequences shared by the tissues, including mRNAs encoding proteins required for common metabolic activities. The cloned cDNA therefore represents the genes that are active (i.e., that are transcribed into mRNA) in tissue A but not tissue B.

Chapter 4 Amino Acids

This chapter introduces you to the structure and chemistry of amino acids. The chapter begins by discussing the zwitterionic character of amino acids at physiological pH, followed by a brief introduction to the amide linkage known as the peptide bond. Amino acids are categorized by the chemistry of their unique R groups and by their mode of biosynthesis. The standard amino acids are specified by the genetic code; nonstandard amino acids include the D-stereoisomers and modified forms of the standard amino acids. This chapter also introduces you to the conventions used to describe stereoisomers, namely the Fischer convention and the *RS* system. Amino acids contain ionizable groups, so this chapter discusses the acid–base properties of amino acids and introduces you to the concept of the isoelectric point. A clear understanding of the acid–base properties of amino acids is critical for appreciating the structural and catalytic behavior of proteins. The chapter closes with a brief discussion of the biological significance of the nonstandard amino acids.

Essential Concepts

Amino Acid Structure

1. All proteins are composed of 20 standard amino acids, which are specified by the genetic code.

2. The standard amino acids are called α-amino acids because they have a primary amino group and a carboxyl group bound to the same carbon atom (the α carbon). Only proline has a secondary amino group attached to the α carbon, but it is still commonly referred to as an α-amino acid.

3. The generic structure of an amino acid at pH 7 is shown below.

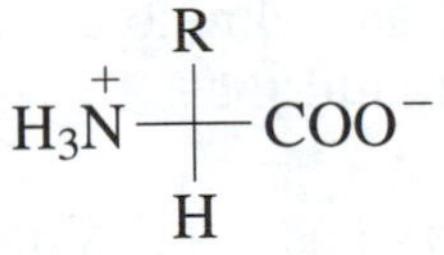

$$H_3\overset{+}{N}-\underset{H}{\overset{R}{\overset{|}{\underset{|}{C}}}}-COO^-$$

   At pH 7, the amino acid is a zwitterion, or dipolar ion. A unique side chain, or R group, characterizes each amino acid.

4. Amino acids are polymerized by condensation reactions to form a chain called a polypeptide. Each polypeptide is polarized: One end has a free amino group and the other end has a free carboxyl group, referred to as the N-terminus and the C-terminus, respectively.

5. The R groups of the standard 20 amino acids are classified into three categories based on their polarities and charge at pH 7: the nonpolar amino acids, the polar uncharged amino acids, and the charged amino acids (see Table 4-1).

6. Among the nonpolar amino acids, glycine (shorthand Gly or G) has a hydrogen atom as its R group. Alanine (Ala; A), valine (Val; V), leucine (Leu; L), isoleucine (Ile; I), and methionine (Met; M) have aliphatic chains as R groups (Met has a sulfur rather than a methylene group). Tryptophan (Trp; W) and phenylalanine (Phe; F) contain bulky indole and phenyl groups, respectively.

7. The polar uncharged amino acids include asparagine (Asn; N), glutamine (Gln; Q), serine (Ser; S), threonine (Thr; T), tyrosine (Tyr; Y), and cysteine (Cys; C). Amide functional groups occur in Gln and Asn. Alcoholic functional groups occur in Ser and Thr. Tyrosine and cysteine are characterized by a phenolic group and a thiol group, respectively.

8. Among the charged amino acids, aspartate (Asp; D) and glutamate (Glu; E) contain carboxylic groups in their R groups. Lysine (Lys; K), arginine (Arg; R), and histidine (His; H) contain a butylammonium group, a guanidino group, and an imidazole group, respectively.

9. The p*K* values of ionizable groups depend on the electrostatic influences of nearby groups. Inside of proteins, the p*K* values of ionizable R groups may shift by several pH units from their values in the free amino acids.

Stereochemistry

10. Except for glycine, the standard amino acids have asymmetric structures and rotate the plane of polarized light; thus, they are optically active. These molecules cannot be superimposed on their mirror images. Such nonsuperimposable pairs of molecules are called enantiomers. The asymmetric atom of an optically active molecule is called the chiral center and the molecule is said to have the property of chirality.

11. Fischer projections are used to represent the absolute configuration of substituents around a chiral center. In the Fischer convention, horizontal lines extend above the plane of the paper, while vertical lines extend below the surface of the paper. The α-amino acid shown above is a Fischer projection of an L-amino acid. The specific arrangement of substituents around the chiral carbon is related to that of L-glyceraldehyde.

12. A molecule with n chiral centers has 2^n different possible stereoisomers. Molecules with two or more chiral centers are better described by the *RS* system, in which each substituent bound to the chiral center is prioritized according to its atomic number. Hence, the exact molecular arrangement of a molecule can be unambiguously described.

13. Biochemical reactions almost invariably produce pure stereoisomers, in large part due to the precise arrangement of chiral groups inside of enzymes, which restricts the geometry of the reactants.

Nonstandard Amino Acids

14. Nonstandard amino acids in polypeptides arise from posttranslational modifications of the R groups of amino acid residues of the polypeptide. These modifications have critical roles in the structure and function of proteins. Many unpolymerized nonstandard amino acids are synthesized by chemical modifications of one of the standard amino acids. Cells use many of these amino acids as signaling molecules, particularly in the central nervous system. Among the nonstandard amino acids are also the D isomers of the standard amino acids; many of these occur in bacterial cell walls and bacterially produced antibiotics.

Key Equation

$$pI = \frac{1}{2}(pK_i + pK_j)$$

Guide to Study Exercises (text p. 92)

1. See Table 4-1.

2. *Polarity*. The nonpolar amino acids are alanine, glycine, isoleucine, leucine, methionine, phenylalanine, proline, tryptophan, and valine. Six of the polar amino acids are uncharged: these are asparagine, cysteine, glutamine, serine, threonine, and tyrosine. Five polar amino acids are charged; these are arginine, aspartate, histidine, glutamate, and lysine.

   Structure. There are many ways to classify amino acids based on the structures of their side chains. Glycine has the smallest side chain, just an H atom. Amino acids with aliphatic side chains are alanine, isoleucine, leucine, valine, and proline (which is really a cyclic imino acid). The sulfur-containing amino acids are cysteine and methionine; the aromatic amino acids are phenylalanine, tryptophan, and tyrosine. Serine and threonine have simple alcoholic side chains. Similarly, aspartate and glutamate have small carboxylic acid side chains; their amide forms are asparagine and glutamine. Arginine, histidine, and lysine have basic side chains with nitrogenous groups.

   Type of functional group. Eight of the amino acids have side chains without functional groups: alanine, glycine, isoleucine, leucine, phenylalanine, proline, tryptophan, and valine. Methionine, whose thiomethyl group is unreactive, also belongs to this class of amino acids. Three amino acids are alcohols: serine, threonine, and tyrosine. Cysteine, with its thiol group, could also be classified with the alcohols. Three of the amino acids have basic functional groups: arginine, histidine, and lysine. Aspartate and glutamate are acidic amino acids. Their counterparts asparagine and glutamine are amides.

   Acid–base properties. All of the amino acids have an acidic group (COOH) and a basic group (NH_2) attached to the α carbon. Two of the amino acids have very acidic side chains: aspartate and glutamate. In certain cases, the side chains of cysteine and tyrosine can also ionize. Three of the amino acids have basic side chains: arginine, histidine, and lysine. (Section 4-1)

3. Five amino acids have side chains that are typically ionized at physiological pH (the side chains of cysteine and tyrosine sometimes ionize). Of these five, two are negatively charged: The β-COOH group of aspartic acid has a p*K* of 3.90, and the γ-COOH group of glutamic acid has a p*K* of 4.07. Therefore, at pH 7, these groups are entirely in the COO^- form. Two amino acids are positively charged at physiological pH since their p*K*'s are much greater than the physiological pH of 7: The guanidino group of arginine has a p*K* of 12.48, and the ε-amino group of lysine has a p*K* of 10.54. Only one amino acid, histidine, whose imidazole group has a p*K* of 6.04, exists in both the neutral and protonated forms at physiological pH. Keep in mind that the p*K* of an ionizable group is influenced by its microenvironment, so the p*K* values of amino acids may vary by several pH units when the amino acid is part of a folded polypeptide chain. (Sections 4-1C and D)

4. The Fischer convention describes the absolute configuration of a chiral molecule by referring it to one of the two stereoisomers of glyceraldehyde. The comparison is made by mentally replacing the H, OH, CHO, and CH_2OH groups of D- or L-glyceraldehyde with the groups around the chiral center of the other molecule. (Section 4-2)

5. Amino acids in proteins may be covalently modified by hydroxylation, methylation, acetylation, carboxylation, and phosphorylation. In addition, larger groups such as lipids and carbohydrates may be attached. (Section 4-3A)

Questions

Amino Acid Structure

1. Without consulting the text, draw a generic amino acid using the Fischer convention.

2. Examine the amino acids below. Assume that the p*K* of the carboxyl group attached to the α carbon is 2.0 and that of the primary amine attached to the α carbon is 9.5 The p*K*'s of ionizable R groups are shown below each structure. (a) Categorize these amino acids as nonpolar, uncharged polar, or charged polar at pH 7. (b) Which of the structures cannot exist as shown at any pH in aqueous solution? (c) Name each of the amino acids.

A
COOH
$H_3\overset{+}{N}$—C—H
CH_2
NH

B
COO^-
$H_3\overset{+}{N}$—C—H
$(CH_2)_4$
NH_2
p*K* = 10.54

C
COO^-
$H_3\overset{+}{N}$—C—H
CH_2
OH
p*K* = 10.46

D
COO^-
H_2N—C—H
$(CH_2)_2$
COOH
p*K* = 4.07

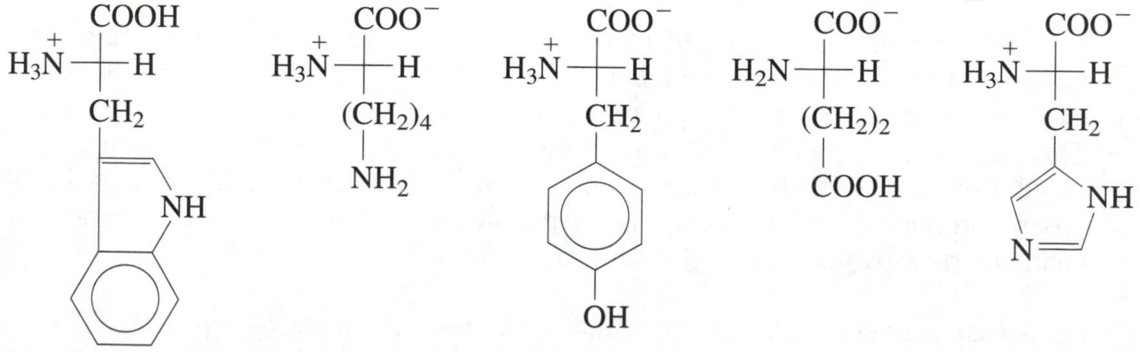

E
COO^-
$H_3\overset{+}{N}$—C—H
CH_2
NH
N
p*K* = 6.04

3. Which of the amino acids in Table 4-1 can be converted to another amino acid by mild hydrolysis that liberates ammonia?

4. Which of the amino acids in Table 4-1 can generate a new amino acid by the addition of a hydroxyl group?

5. The ionic characters of which amino acids are likely to be sensitive to pH changes in the physiological range? Explain.

6. Circle the functional groups that are eliminated in the formation of a peptide bond between the amino acids shown below. Draw the structure of the dipeptide.

$$^{-}O{-}C(=O){-}C(R_1)(H){-}N^{+}H_3 \qquad ^{-}O{-}C(=O){-}C(R_2)(H){-}N^{+}H_3$$

7. What percentage of the histidine imidazole group is protonated at pH 7.2?

8. What percentage of the cysteine sulfhydryl group is deprotonated at pH 7.6?

9. In proteins, the imidazole groups of histidine play key roles in catalysis involving reversible protonation/deprotonation events. The p*K* of the imidazole group is influenced by surrounding amino acids and in many enzymes is near 7.
   (a) What is the significance of this apparent p*K* in catalysis?
   (b) Some unprotonated imidazole groups participate in hydrogen bonding a substrate to an enzyme. Do you expect the p*K* of these imidazole groups also to be near 7?

10. Using the p*K* values in Table 4-1, identify the amino acids that have the titration curves shown below.

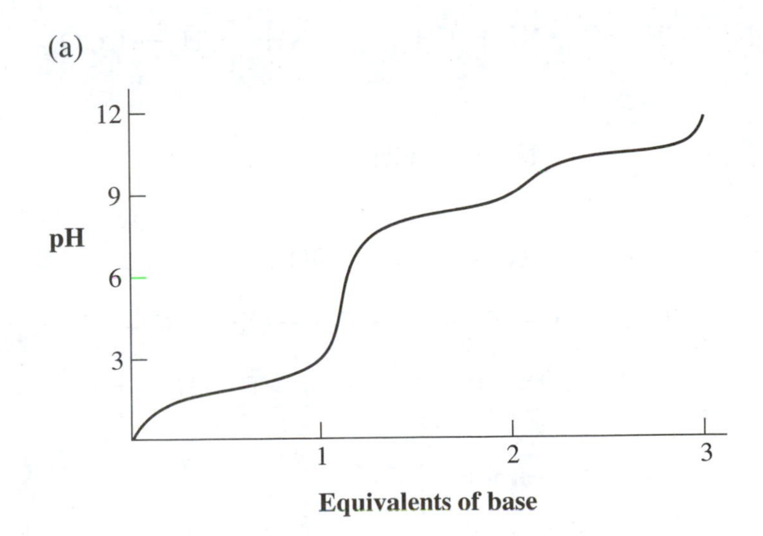

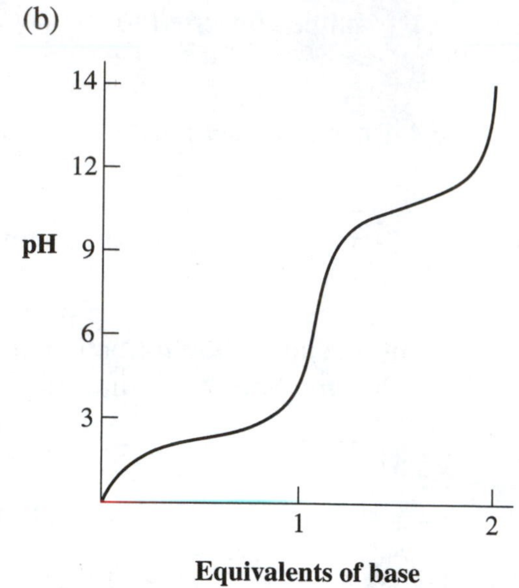

11. Which of the following tripeptides contain peptide bonds NOT commonly found in proteins?

A

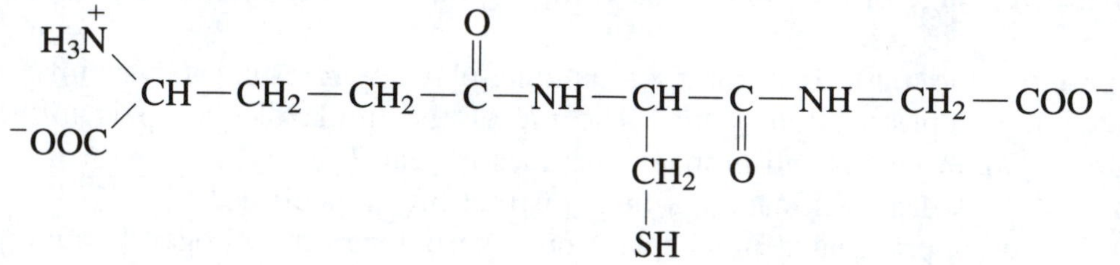

$$\overset{+}{H_3N}(^{-}OOC)CH-CH_2-CH_2-\overset{O}{\overset{\|}{C}}-NH-\underset{\underset{SH}{|}}{\underset{CH_2}{|}}{CH}-\underset{\underset{O}{\|}}{C}-NH-CH_2-COO^-$$

glutathione

B

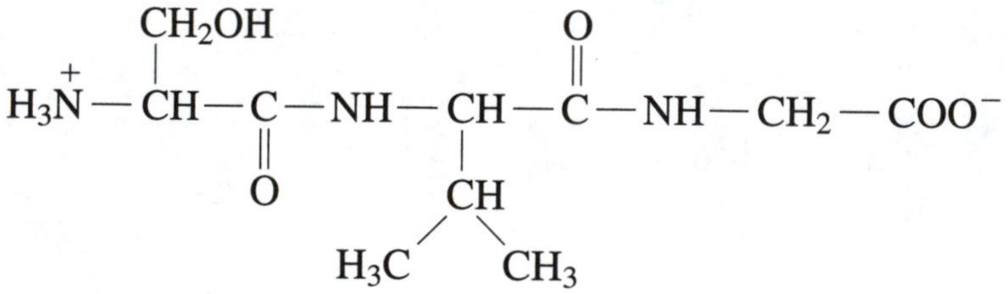

$$H_3\overset{+}{N}-\overset{CH_2OH}{\overset{|}{CH}}-\underset{\underset{O}{\|}}{C}-NH-\underset{\underset{H_3C\quad CH_3}{CH}}{\underset{|}{CH}}-\overset{O}{\overset{\|}{C}}-NH-CH_2-COO^-$$

C

$$H_3\overset{+}{N}-CH_2-CH_2-\underset{\underset{O}{\|}}{C}-NH-\overset{COO^-}{\overset{|}{CH}}-CH_2-\underset{HN}{\underset{|}{C}}=\underset{N}{\underset{|}{CH}}\quad \underset{CH}{}$$

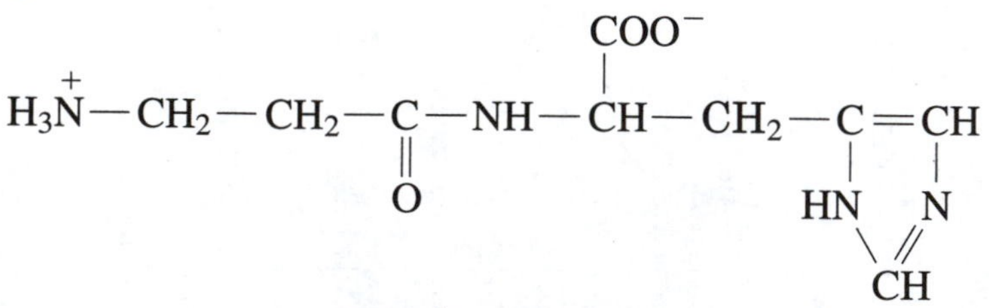

carnosine

D

$$
\begin{array}{l}
\overset{+}{H_3N}-\underset{\substack{|\\ COO^-}}{CH}-CH_2-CH_2-CH_2-NH-\underset{\substack{|\\ C=O\\ |\\ NH\\ |\\ CH_2\\ |\\ COO^-}}{CH}-CH_2-SH
\end{array}
$$

12. Why is the p*K* of the carboxyl group of glycine (p*K* = 2.3) less than that for acetic acid (p*K* = 4.76)?

13. Calculate the p*I*'s of aspartate, lysine and serine.

14. In proteins, the p*K* of the C-terminus is about 3.8, while that of the N-terminus is about 7.8. Rationalize these differences from the p*K* values of the α-carboxyl and the α-amino groups in free amino acids.

15. Is a protein as good a *cellular* buffer at physiological pH as its constituent amino acids would be if they were present as free amino acids in proportional concentrations in the cell? Explain.

16. 100 mg of anhydrous powder of lysine is not completely soluble in 10 mL of water but dissolves completely when base is added. Explain.

17. Shown below is a titration curve for glutamic acid. Draw the structure of the species that predominates at each labeled point.

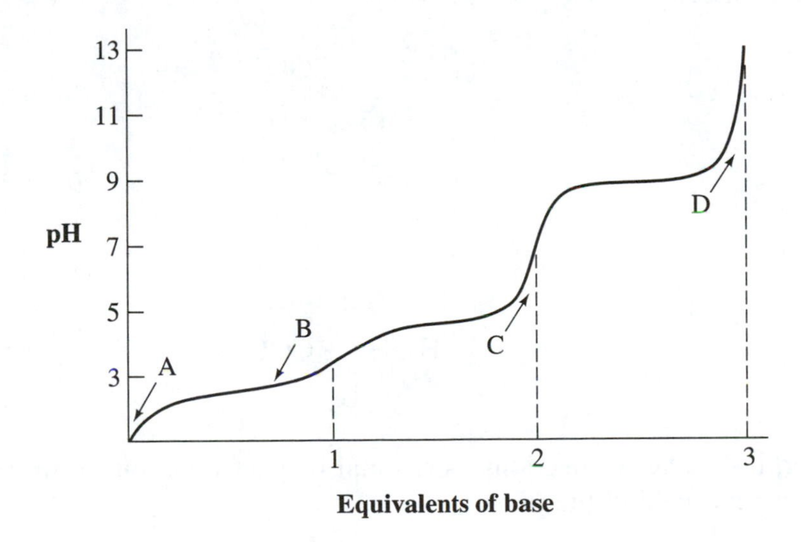

Stereochemistry

18. Which amino acid found in proteins has no optical activity?

19. Draw Fischer projections of L-aspartate and L-cysteine and show why L-Asp is (*S*)-Asp and L-Cys is (*R*)-Cys. (Refer to Box 4-1)

20. Which amino acids in Table 4-1 have two or more prochiral centers? A prochiral center can be made chiral by substituting a different group for one of the two identical groups attached to it.

Nonstandard Amino Acids

21. Which of the nonstandard amino acids below cannot occur in the interior of a polypeptide chain? Explain.

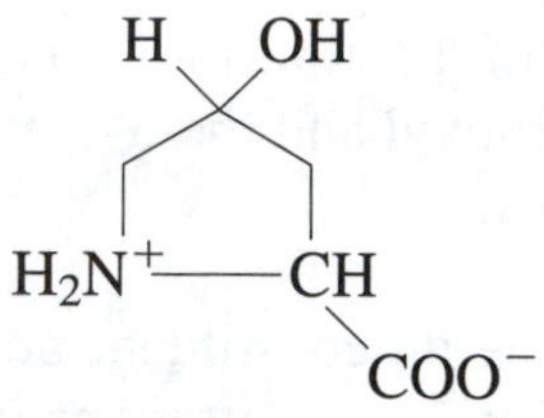

H OH (ring C), H_2N^+—CH—COO^-

4-hydroxyproline

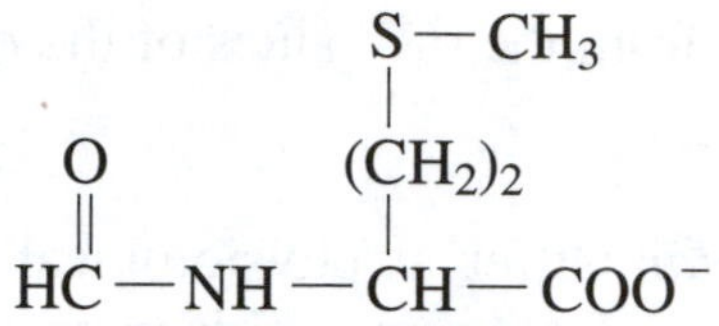

$S-CH_3$ — $(CH_2)_2$ — HC(=O)—NH—CH—COO^-

***N*-formylmethionine**

$(CH_3)_3\overset{+}{N}$—CH(CH_3)—COO^-

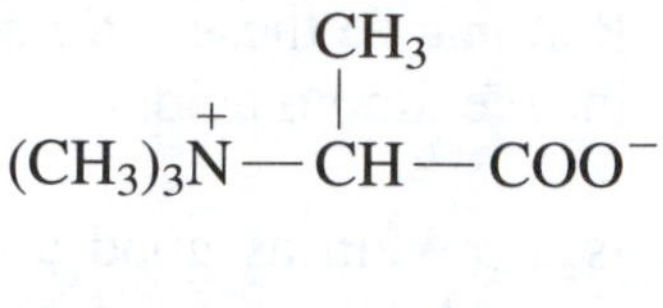

***N,N,N*-trimethylalanine**

22. From which of the standard amino acids are the following physiologically active amines derived? What modifications gave rise to these products? (See Figure 4-15 for structures.)
(a) GABA (b) histamine (c) thyroxine (d) dopamine

Answers to Questions

1.

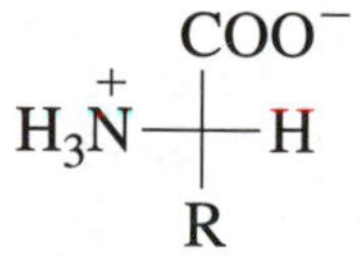

COO^- (top), $H_3\overset{+}{N}$ (left), H (right), R (bottom)

or

R (top), $H_3\overset{+}{N}$ (left), COO^- (right), H (bottom)

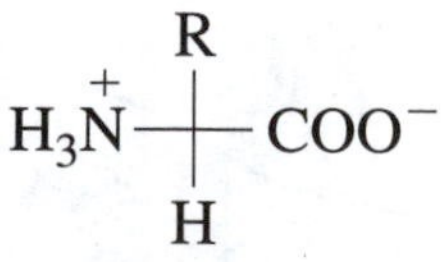

Remember, in Fischer projections, horizontal bonds extend out from the paper while vertical bonds extend behind the paper.

2. (a) Nonpolar: **A**; uncharged, polar: **C** and **E** (because **E** has a p*K* of 6.0, it is largely uncharged at pH 7); charged, polar: **B** and **D** (at pH 7, the ε-amino group of **B** is positively charged, and the γ-carboxyl group of **D** is negatively charged).

(b) **B** and **D** cannot exist at any pH in aqueous solution. In **B**, the ε-amino group will become deprotonated at lower pH values than the α-amino group since its p*K* is lower. Similarly, the γ-carboxyl group of **D** is more acidic than the α-amino group.

(c) **A** is tryptophan; **B** is lysine; **C** is tyrosine; **D** is glutamate; and **E** is histidine.

3. Glutamine and asparagine can be converted to glutamate and aspartate, respectively.

4. Alanine gives rise to serine, and phenylalanine gives rise to tyrosine.

5. Histidine and cysteine are the most likely to be sensitive in the physiological pH range since the p*K* values of their R groups are 6.04 and 8.37, respectively.

6.

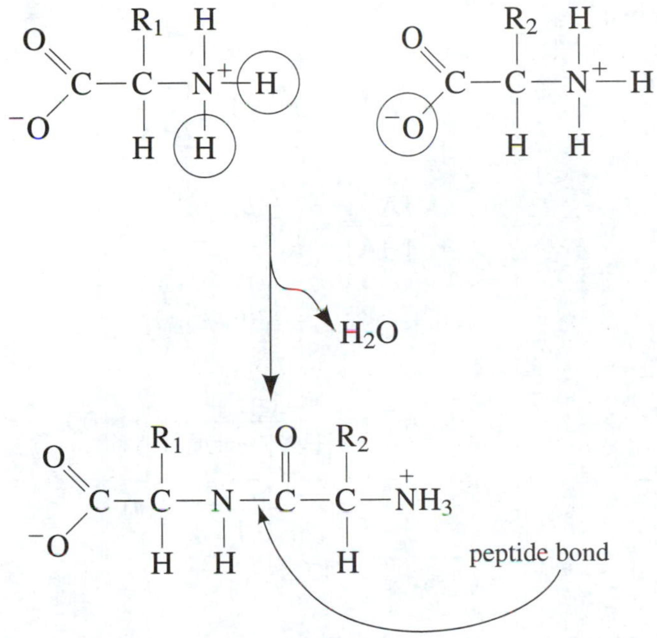

H_2O

peptide bond

7. Use the Henderson–Hasselbalch equation and rearrange terms to find the percentage of protonated species:

$$\mathrm{pH} = \mathrm{p}K + \log\frac{[\mathrm{A}^-]}{[\mathrm{HA}]}$$

$$\frac{[\mathrm{A}^-]}{[\mathrm{HA}]} = 10^{(\mathrm{pH}-\mathrm{p}K)}$$

$$= 10^{(7.2-6.04)}$$

$$= 10^{1.16} = 14.45$$

To obtain the percentage of the protonated species,

$$\% \text{ HA} = \frac{[\text{HA}]}{[\text{HA}] + [\text{A}^-]} \times 100$$
$$= \frac{1}{1 + 14.45} \times 100$$
$$= 6.47\%$$

Hence, 6.47% is in the protonated form.

8. Use the Henderson–Hasselbalch equation and rearrange terms to find the percentage of deprotonated species:

$$\text{pH} = \text{p}K + \log\frac{[\text{A}^-]}{[\text{HA}]}$$

$$\frac{[\text{A}^-]}{[\text{HA}]} = 10^{(\text{pH}-\text{p}K)}$$
$$= 10^{(7.6-8.37)}$$
$$= 10^{-0.77} = 0.17$$

$$\% \text{ A}^- = \frac{[\text{A}^-]}{[\text{HA}] + [\text{A}^-]} \times 100$$
$$= \frac{0.17}{1 + 0.17} \times 100$$
$$= 14.53\%$$

Therefore, about 14.53% of the sulfhydryl groups are deprotonated.

9. (a) The intracellular pH is also near 7, so about half the imidazole groups will be protonated. In other words, the imidazole groups can easily abstract a proton or give one up, as is required for catalysis, and remain unchanged at the end of the reaction.
   (b) It is more likely that the pK of these groups is near the pK of free histidine (6.04) so that they are available to form hydrogen bonds with potential substrates.

10. (a) Lysine; tyrosine is similar but its pK_2 is closer to 9.2.
    (b) Proline; it has only two pK's, one at about 2 and the other well above 10. The only other amino acid with a pK_2 above 10 is cysteine, but it has three pK's.

(a)

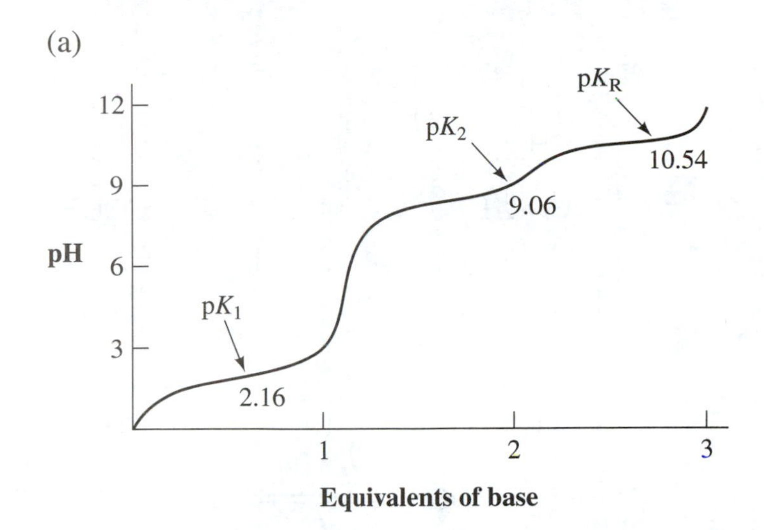

(b)

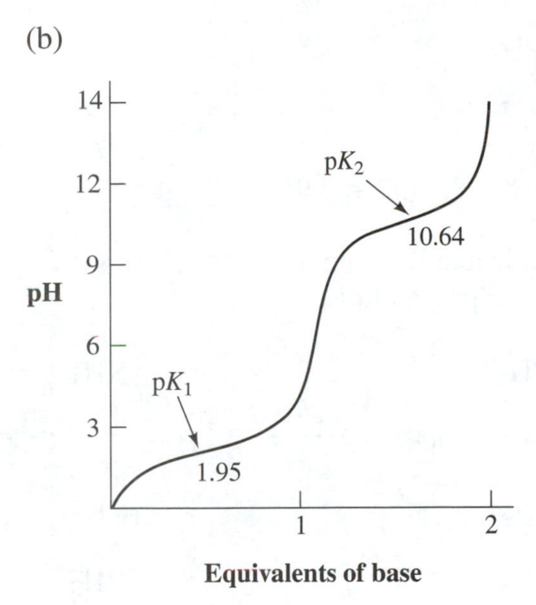

11. **A**, **C**, and **D** contain peptide bonds not commonly found in proteins. In **A**, the γ-carboxylate group of the first amino acid participates in an amide linkage (the second and third amino acids are linked by a conventional peptide bond). **C** contains a linkage between the carboxylate of a β-amino acid and the amino group of an α-amino acid. In **D**, the side chain of the first amino acid is linked to the amino group of the second.

12. The positively charged amino group of glycine electrostatically stabilizes the nearby carboxylate ion. Hence, the carboxylic acid group of glycine dissociates more readily to form the carboxylate ion than does the carboxylic acid group of acetic acid.

13. The pI is approximately midway between the pK's of the two ionizations involving the neutral species. In amino acids with ionizable side chains, the relevant pK's may be pK_1 and

pK_R or pK_R and pK_2. For aspartate, the neutral species results from ionization of the α-COOH (pK_1) and the β-COOH (pK_R):

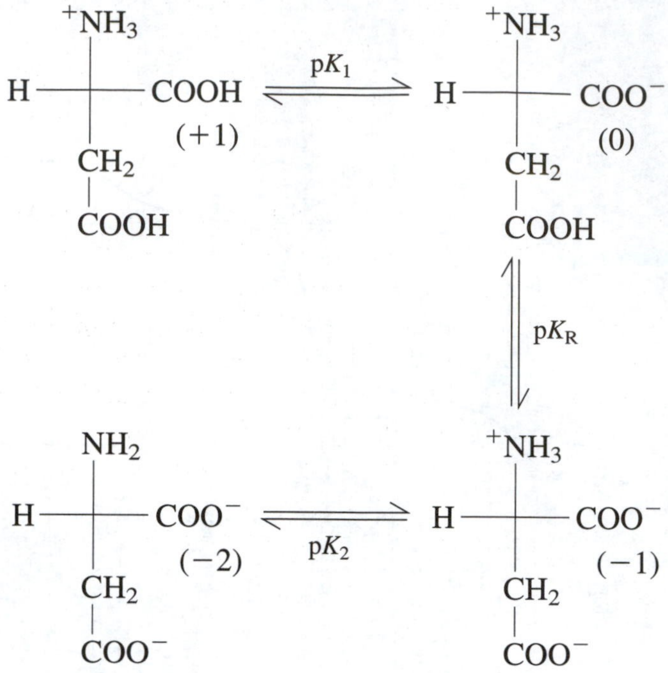

$^{+}NH_3$
H—C—COOH (+1)
CH_2
COOH

$\overset{pK_1}{\rightleftharpoons}$

$^{+}NH_3$
H—C—COO^- (0)
CH_2
COOH

$\Updownarrow pK_R$

$^{+}NH_3$
H—C—COO^- (−1)
CH_2
COO^-

$\underset{pK_2}{\rightleftharpoons}$

NH_2
H—C—COO^- (−2)
CH_2
COO^-

$pI = (pK_1 + pK_R)/2 = (1.99 + 3.90)/2 = 2.95.$

For serine, which has no ionizable R group, $pI = (pK_1 + pK_2)/2 = (2.19 + 9.21)/2 = 5.7$. For lysine, the p$I$ lies between pK_2 and pK_R:

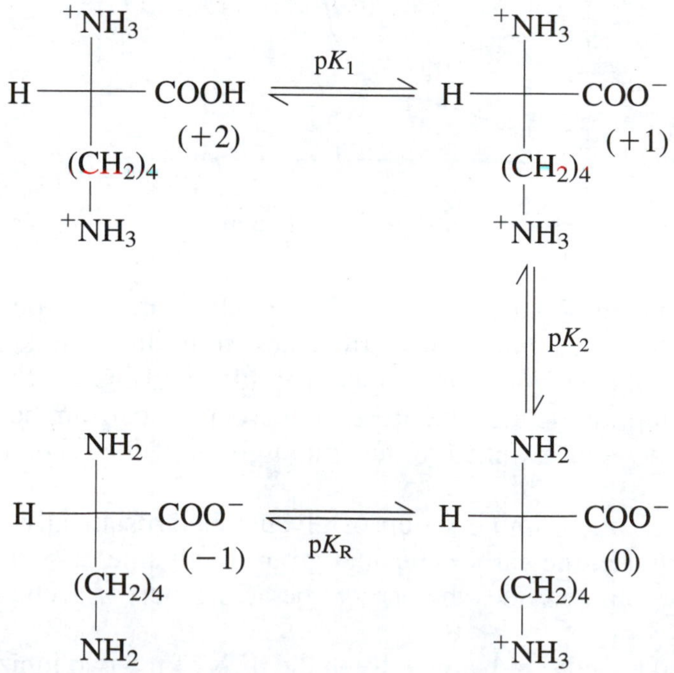

$^{+}NH_3$
H—C—COOH (+2)
$(CH_2)_4$
$^{+}NH_3$

$\overset{pK_1}{\rightleftharpoons}$

$^{+}NH_3$
H—C—COO^- (+1)
$(CH_2)_4$
$^{+}NH_3$

$\Updownarrow pK_2$

NH_2
H—C—COO^- (0)
$(CH_2)_4$
$^{+}NH_3$

$\underset{pK_R}{\rightleftharpoons}$

NH_2
H—C—COO^- (−1)
$(CH_2)_4$
NH_2

Hence, $pI = (pK_2 + pK_R)/2 = (9.06 + 10.54)/2 = 9.8.$

14. In free amino acids, the neighboring α-carboxyl and α-amino groups affect each other's p*K* values. For instance, the negatively charged carboxyl group stabilizes the protonated amino group, making it a weaker acid and thus raising its p*K* value. Similarly, the positively charged amino group stabilizes the anionic carboxyl group, making it a stronger acid and thereby lowering its p*K* value. In proteins, the free amino group and the free carboxyl group are separated by at least ~40 amino acid linkages, and hence have little or no effect on each other's p*K* values.

15. Yes. The formation of the peptide bond between each amino acid in the intact protein eliminates two ionizable groups. However, at physiological pH (pH 5–9), the α-carboxyl and the α-amino groups of amino acids are poor buffers, because their respective p*K* values are far from the physiological pH range. However, at physiological pH, a protein containing a number of histidines and cysteines, for instance, may have a buffering capacity similar to that of an equimolar solution of its constituent amino acids; the side groups of these amino acids are free to ionize regardless of whether they are free or in a protein. Other ionizable R groups have p*K*'s too far from physiological pH to be effective buffers.

16. Lysine solid is of necessity neutral and therefore has the structure

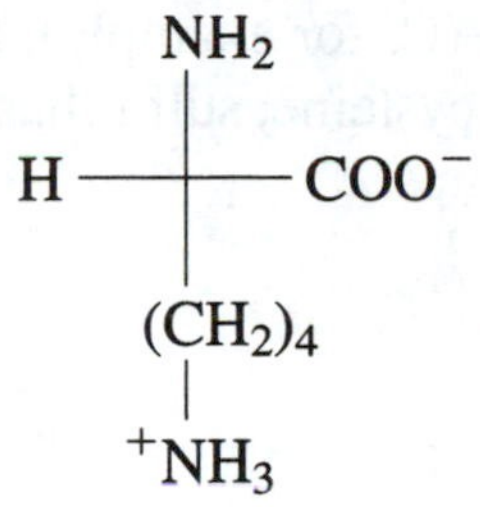

$$H - \underset{\underset{^{+}NH_3}{|}}{\underset{(CH_2)_4}{|}}{\overset{NH_2}{C}} - COO^-$$

Neutral species are not as soluble as charged species. Adding base causes lysine's ε-NH_3^+ group to be converted to the neutral ε-NH_2, giving the molecule a net negative charge and higher solubility.

17.

A

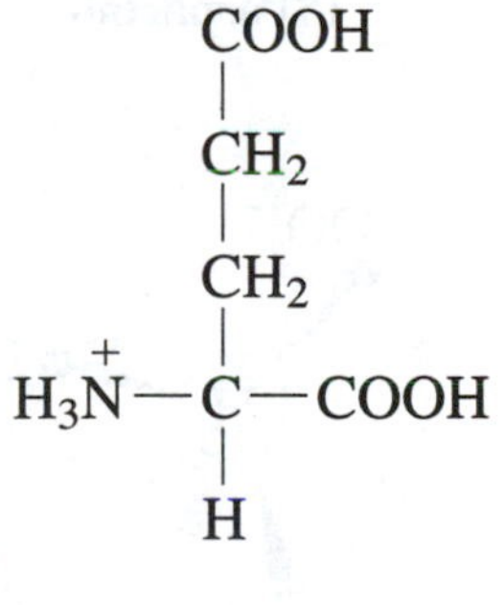

$$HOOC-CH_2-CH_2-\underset{H}{\overset{|}{C}}(-\overset{+}{N}H_3)-COOH$$

(COOH–CH₂–CH₂– attached to C; $H_3\overset{+}{N}-C-COOH$, with H below C)

B

$$HOOC-CH_2-CH_2-\underset{H}{\overset{|}{C}}(-\overset{+}{N}H_3)-COO^-$$

(COOH–CH₂–CH₂– attached to C; $H_3\overset{+}{N}-C-COO^-$, with H below C)

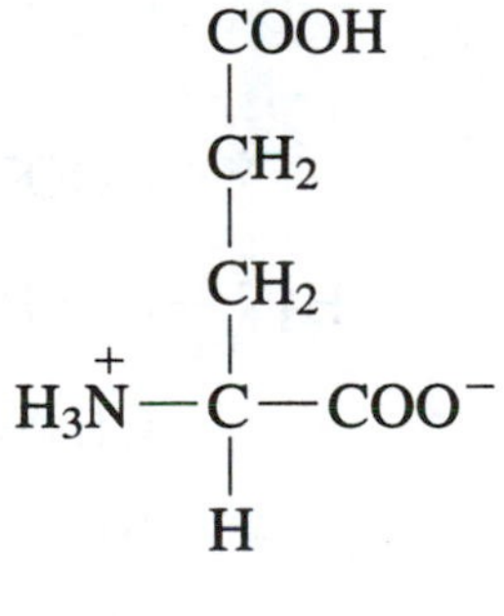

C

$$\begin{array}{c} COO^- \\ | \\ CH_2 \\ | \\ CH_2 \\ | \\ H_3\overset{+}{N}-C-COO^- \\ | \\ H \end{array}$$

D

$$\begin{array}{c} COO^- \\ | \\ CH_2 \\ | \\ CH_2 \\ | \\ H_2N-C-COO^- \\ | \\ H \end{array}$$

18. Glycine; its C_α has two identical substituents, H.

19. In a Fischer projection, the vertical bonds by convention point into the paper while the horizontal bonds point out from the paper. When the carbon groups are aligned vertically as drawn below, the amino group is on the left in an L-amino acid; it is on the right in a D-amino acid. To identify *R* or *S* configurations, swing the α-carbon hydrogen behind the α carbon. Then assign a priority number based on the atomic mass of the atom or group bound to the α carbon. The configuration is *R* or *S* according to whether the groups with decreasing priority have a clockwise or counterclockwise orientation, respectively. If two substituent atoms are the same (C, for example), the other atoms attached to them are used to assign priority. Note that for cysteine, sulfur has a higher atomic mass than oxygen, which reverses the priority of the groups.

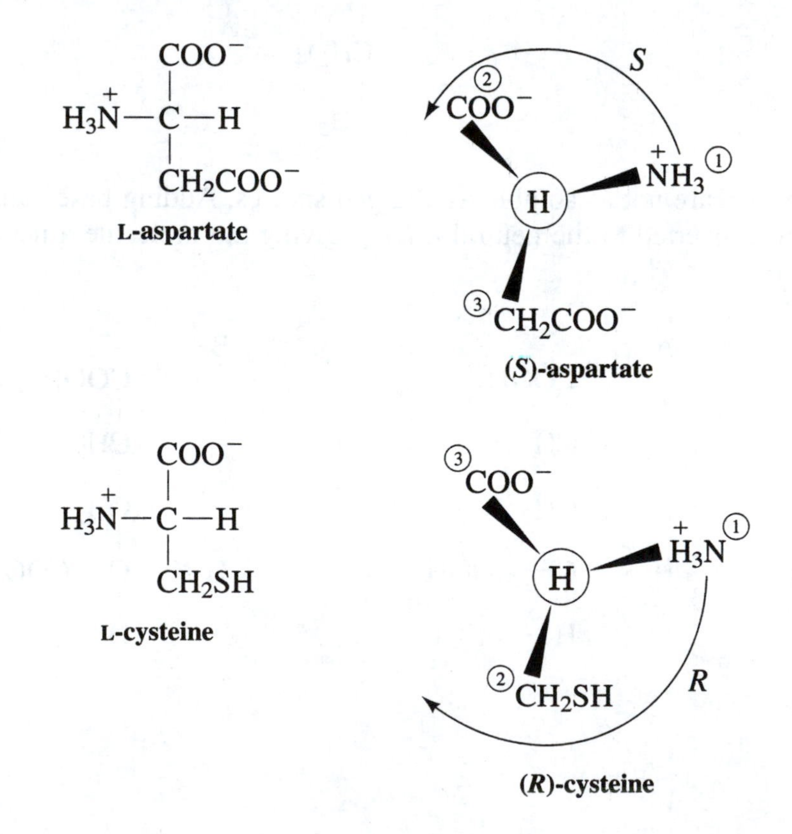

20. The amino acids with two or more prochiral centers are Arg, Gln, Glu, Leu, Lys, Met, and Pro.

21. *N*-formylmethionine and *N,N,N*-trimethylalanine cannot occur in the interior of a polypeptide, because they lack free amino groups to form peptide bonds.

22. (a) GABA is derived from glutamate via decarboxylation of the α-carboxyl group.
    (b) Histamine is derived from histidine also by decarboxylation of the α-carboxyl group.
    (c) Thyroxine is derived from tyrosine by addition of a phenyl group and iodination.
    (d) Dopamine is derived from tyrosine via hydroxylation of the phenyl group and decarboxylation of the α-carboxyl group.

Chapter 5 Proteins: Primary Structure

This chapter covers protein purification and primary structure. In this chapter you will learn how a protein's size, charge, and general shape can be analyzed and used to develop procedures to purify the protein. Many of the principles you learned in the preceding chapters about thermodynamics, aqueous solutions, and acid–base chemistry can be applied to the isolation of proteins. This chapter also includes a discussion of the strategies and chemical methods for determining the primary structure of proteins, focusing on Edman degradation. The chapter concludes with a section discussing protein evolution. Here you will see how comparisons of primary structures have led biochemists to categorize proteins into families and identify specific modules or motifs involved in specific functions.

Essential Concepts

Polypeptide Diversity

1. The primary structure of a protein is the amino acid sequence of its polypeptide chain (or chains if the protein contains more than one polypeptide).

2. Proteins are synthesized in cells by the stepwise polymerization of amino acids in the order specified by the sequence of nucleotides in its gene. The 5′ end of the messenger RNA corresponds to the amino terminus of the polypeptide, which is the end that contains a free amino group bound to the α carbon.

3. Although the theoretical possibilities for polypeptide composition and length are unlimited, polypeptides found in nature are limited somewhat in size and composition. Most polypeptides contain between 100 and 1000 residues and do not necessarily include all 20 genetically encoded amino acids. Leu, Ala, Gly, Ser, Val, and Glu are the most abundant amino acids in proteins, while Trp, Cys, Met, and His are the least common.

Protein Purification

4. A general approach to protein purification requires the following:
   (a) A rapid and efficient method to disrupt cells so that the contents of the lysed cells (lysate) can be quickly stabilized in a buffer of appropriate pH and ionic strength.
   (b) Consideration of factors that affect the stability of the desired protein, such as pH, temperature, presence of degradative enzymes, adsorption to surfaces, and solvent conditions for long term storage.
   (c) Appropriate separation techniques to effectively purify the desired protein from the total proteins of the lysate.
   (d) A test or assay to easily assess the activity of the desired protein at each step in the purification.

5. A protein contains multiple charged groups, so its solubility in aqueous solution varies with the concentrations of dissolved salts (ionic strength), pH, and temperature. When salt is added to a protein at low ionic strength, the protein's solubility increases with increasing

ionic strength (called salting in). At some point, the solubility of the protein decreases (called salting out). This variable solubility can be exploited to selectively precipitate desired or undesired proteins.

6. Chromatography, which is used for analysis as well as purification, relies on the ability of a substance dissolved in a solvent (mobile phase) to interact with a solid matrix (stationary phase) as the solvent percolates through it. The chromatographic matrix is often in a column so that the solvent that elutes (exits) from it can be collected and assayed for the desired substance.

7. Ion exchange chromatography takes advantage of a molecule's ionizable groups. The binding affinity of a substance for a cation or anion exchanger depends on its net charge and the pH and ionic strength of the solvent. The bound substance is eluted by increasing the salt concentration or changing the pH of the solvent.

8. Hydrophobic interaction chromatography, which uses a matrix substituted with nonpolar groups, takes advantage of the exposed hydrophobic regions of a protein. The bound protein is eluted with a solvent that weakens the hydrophobic effects that cause the protein to bind to the matrix.

9. Gel filtration chromatography separates molecules according to their size and shape. The stationary phase consists of beads containing pores that span a relatively narrow size range. Smaller molecules spend more time inside the beads than larger molecules and therefore elute later (after a larger volume of mobile phase has passed through the column). Within the molecular mass range that is fractionated by the specific stationary phase used, there is a linear relationship between a substance's elution time and the logarithm of its molecular mass.

10. Affinity chromatography exploits a protein's specific ligand-binding behavior. The protein binds to a ligand that is covalently attached to the stationary phase and can be eluted with a solution containing a high salt concentration or excess ligand, which competes with the bead-bound ligand for binding sites on the protein.

11. Electrophoresis separates molecules according to size and net charge. The molecules move through a polyacrylamide or agarose gel under the influence of an electric field. Proteins can be visualized by soaking the gel in a dye that binds to proteins or by electroblotting the proteins to nitrocellulose paper and probing with an antibody that is specific for the desired protein (Western blotting).

12. In SDS polyacrylamide gel electrophoresis (SDS-PAGE), proteins are denatured with the detergent sodium dodecyl sulfate (SDS), which binds to the protein and imparts a large negative charge by virtue of its negatively charged sulfate group. The net charge of an SDS-polypeptide is proportional to its length, giving all proteins the same charge density, so SDS-PAGE separates proteins almost entirely by sieving effects. Smaller proteins migrate faster since they are less impeded by the cross-linked polyacrylamide.

13. The ultracentrifuge, which can generate centrifugal forces of over 600,000*g*, has been used to determine protein molecular mass and subunit composition. The rate at which a mole-

cule sediments in an ultracentrifuge depends largely on its mass and its shape. Polypeptides have sedimentation coefficients ranging from 1S to 50S.

14. Sedimentation in a density gradient improves the resolving power of the ultracentrifuge. In zonal ultracentrifugation, the sample is added on top of a preformed gradient of an inert substance such as sucrose. During centrifugation, macromolecules move through the gradient according to their sedimentation coefficient. In equilibrium density gradient ultracentrifugation, the sample is dissolved in a concentrated solution of a fast-diffusing substance such as CsCl, which forms a density gradient during centrifugation. The macromolecules in the sample form bands at positions where the density of the solution is equivalent to their own. This technique is widely used to analyze and isolate nucleic acids.

Protein Sequencing

15. Knowledge of a protein's amino acid sequence (its primary structure) is important for:
    (a) Determining its three-dimensional structure and elucidating its molecular mechanism of action.
    (b) Comparing sequences of analogous proteins from different species to gain insights into protein function and evolutionary relationships among the proteins and the organisms that produce them.
    (c) Developing diagnostic tests and effective therapies for inherited diseases that are caused by single amino acid changes in a protein.

16. Determining a protein's sequence may require several steps:
    (a) End group analysis reveals the number of different polypeptides or subunits in a protein. Either dansyl chloride or phenylisothiocyanate (Edman degradation) can be used to identify the N-terminal residue of a polypeptide. The C-terminus can be identified using carboxypeptidases, enzymes that catalyze hydrolysis of a C-terminal amino acid.
    (b) Disulfide bonds within and between polypeptide chains are cleaved by reducing agents such as 2-mercaptoethanol, and the free sulfhydryl groups are then alkylated, e.g., with iodoacetate, to prevent re-formation of the disulfide bonds.
    (c) The different polypeptides of a multisubunit protein are separated so that each can be sequenced.
    (d) The amino acid composition of a polypeptide can be determined by the complete hydrolysis of the polypeptide followed by separation and analysis of the liberated amino acids using HPLC.
    (e) Polypeptides longer than 100 residues cannot be sequenced directly and must therefore be cleaved into smaller fragments using endopeptidases such as trypsin or chymotrypsin or chemical agents such as cyanogen bromide.
    (f) Each fragment is isolated and then sequenced using repeated cycles of the Edman degradation.
    (g) The order of sequenced fragments in the intact polypeptide is established by cleaving the polypeptide with a different reagent to obtain a second set of fragments whose sequences overlap those of the first.
    (h) The positions of disulfide bonds are identified by cleaving the polypeptide before it has been reduced. Each fragment containing a disulfide bond is reduced, and the resulting two peptides are separated and sequenced.

17. In Edman degradation, phenylisothiocyanate (PITC) reacts exclusively with the N-terminal amino group of a polypeptide chain. Treating the resulting PTC polypeptide with trifluoroacetic acid releases the N-terminal residue as a thiazolinone derivative, which is converted to a phenylthiohydantoin–amino acid that can be identified by chromatography. The procedure is then repeated for the newly exposed N-terminal residue of the polypeptide. This procedure has been automated and refined to the point where up to 100 residues can be sequenced with as little as 5 to 10 picomoles of protein (<0.1 μg)!

Protein Evolution

18. Mutations that alter a protein's primary structure may or may not affect the protein's function. Mutations that are deleterious to the organism are less likely to be passed on to the next generation. Mutations that enhance an organism's ability to survive and reproduce tend to propagate quickly. This is the essence of Darwinian evolution.

19. Because related species have evolved from a common ancestor, the genes specifying each of their proteins have also evolved from a corresponding gene in that ancestor. Hence, the more closely related the two species, the more similar are the primary structures of their proteins. A phylogenetic tree can therefore be constructed by comparing the sequence divergence of proteins in different organisms.

20. Comparisons of the primary structure of evolutionarily related proteins (homologous proteins) may indicate which amino acid residues are essential to the protein's function: Essential positions tend to contain invariant residues; less essential positions tend to contain chemically similar (conserved) residues; and positions that do not significantly affect protein structure or function can accommodate a variety of residues.

21. The rates of evolution of proteins vary considerably. This reflects the ability of a given protein to accept amino acid changes without compromising its function.

22. Large families of proteins such as the immunoglobulins or globins probably arose by gene duplication and subsequent primary structure divergence among the copies. Hence, the new copies are capable of evolving new functions.

23. Many proteins contain amino acid sequence modules of about 40–100 residues that may have unique functions. Some modules are repeated many times in a single polypeptide. Hence, shuffling of modules is potentially a mechanism for rapidly generating new proteins with new functions.

24. Most primary structures of proteins are known from sequencing the genes that encode them. However, DNA sequencing cannot provide information about the locations of disulfide bonds or posttranslational modifications of amino acids. In addition, some newly synthesized polypeptides are trimmed by site-specific cleavage to generate the mature functional protein. In practice, protein sequences are used to confirm nucleic acid sequences and *vice versa.*

Guide to Study Exercises (text p. 122)

1.

Leu	9.1%	Pro	5.2%
Ala	7.8%	Arg	5.1%
Gly	7.2%	Asn	4.3%
Ser	6.8%	Gln	4.3%
Val	6.6%	Phe	3.9%
Glu	6.3%	Tyr	3.2%
Lys	5.9%	His	2.3%
Thr	5.9%	Met	2.2%
Asp	5.3%	Cys	1.9%
Ile	5.3%	Trp	1.4%

2. The stability of purified proteins depends on the pH, the temperature, the absence of degradative enzymes, minimal contact with protein-adsorbing surfaces, the absence of oxidizing agents, and the prevention of microbial growth. (Section 5-2A)

3. The charge of a molecule is the basis for separation by ion exchange chromatography and electrophoresis; polarity is the basis for separation by hydrophobic interaction chromatography; size is the basis for separation by gel filtration chromatography, SDS-PAGE, and ultracentrifugation; and binding specificity for small molecules is the basis for separation by affinity chromatography. (Section 5-2)

4. Ion exchange chromatography takes advantage of the ability of a macromolecule's charged surface residues to bind to oppositely charged groups that have been immobilized on a chromatographic matrix. Elution is accomplished by disrupting these nonspecific charge–charge interactions by increasing the salt concentration or changing the pH. Affinity chromatography relies on the ability of a macromolecule to bind specifically to an immobilized ligand. Elution is accomplished by changing the pH or the salt concentration or by adding an excess of soluble ligand. Affinity chromatography is a much more selective technique than ion exchange chromatography because it takes advantage of a molecule's unique biochemical properties rather than a general attribute such as overall charge. (Section 5-2C)

5. Ultracentrifugation provides information about the molecular mass of a macromolecule from its rate of sedimentation. For a molecule of known molecular mass, slower than expected sedimentation may indicate an elongated shape or aggregation of the molecule. Similarly, faster than expected sedimentation may indicate dissociation of a molecular aggregate into smaller units. (Section 5-2E)

6. The N-terminal residue of a polypeptide can be identified in several ways: (a) by reacting the primary amino group with dansyl chloride, which produces a fluorescent amino acid derivative after acid hydrolysis of the polypeptide; (b) by reacting the N-terminal amino group with phenylisothiocyanate followed by trifluoroacetic acid, which produces a free PTH-amino acid that can be identified by chromatography; or (c) by digesting the polypeptide with an aminopeptidase and identifying the first amino acid released. The C-terminal residue of a polypeptide can be identified by digesting the polypeptide with a carboxypeptidase and identifying the first residue released. These techniques do have limitations. For example, the chemical derivatization methods cannot be used on a polypeptide with a chemically modified (e.g.,

acetylated) N-terminus. Exopeptidases may cleave certain residues slowly or not at all or may quickly cleave off several residues. Of course, if the protein consists of more than one polypeptide chain, multiple N- or C-terminal residues will be present. (Section 5-3A)

7. Disulfide bonds must be cleaved so that the sequencing reactions can use linear, reagent-accessible polypeptide chains as substrates. (Section 5-3A)

8. Information about a protein's amino acid composition may be used to confirm unexpected results from amino acid sequencing, for example, if the protein contains a high proportion of rare amino acids or modified amino acids. The amino acid composition may also provide guidance about which enzymatic or chemical cleavage methods might generate peptide fragments for sequencing. Amino acid compositions may also be used to quickly compare proteins, since proteins that are closely linked by evolution have similar amino acid compositions. (Section 5-3A)

9. Edman degradation begins with the reaction of phenylisothiocyanate with the N-terminal amino group of a polypeptide. The phenylthiocarbamyl product is treated with trifluoroacetic acid to cleave off the N-terminal residue as a thiazolinone derivative, which is removed and converted to a phenylthiohydantoin-amino acid by acid treatment. The PTH product is identified by chromatography. The entire reaction process can be repeated, up to ~100 times, in order to successively liberate and identify a polypeptide's residues starting from the N-terminus. (Section 5-3C)

10. Automated procedures, which can be used for fractionating macromolecules by chromatography and electrophoresis, for performing sequencing reactions, and for performing biochemical assays, offer the advantages of speed and reproducibility. Automation minimizes the opportunities for human error and helps ensure consistent conditions for many runs. Automation is virtually a necessity when many samples must be analyzed. In many cases, automated procedures save time even when analyzing a single sample. (Sections 5-2 and 5-3)

11. Comparing the sequences of proteins from different species reveals evolutionary relationships, because the number of amino acid differences between the proteins indicates the time since the divergence of the species that produce the proteins. Sequence comparisons also provide information about the structure and function of the protein, since residues that are highly conserved often have essential structural or functional roles, whereas more variable residues are often found on the protein surface or in areas that are not essential for its structure or function. (Section 5-4A)

12. A phylogenetic tree is constructed by counting the amino acid differences between homologous proteins from different species. These are used to draw a tree whose branch points represent a putative common ancestor for the species above it. The lengths of the branches correlate with the number of amino acid differences between the proteins. (Section 5-4A)

Questions

Protein Purification

1. Discuss the disadvantages of each of the following in the handling of proteins:
   (a) Low concentrations of protein
   (b) Sudsing of protein solutions
   (c) Lack of sterile conditions
   (d) Absence of protease inhibitors
   (e) Warm environments
   (f) pH's far from neutrality

2. List possible ways to track the presence of a protein during its purification.

3. You have several cell lysates and want to determine which of them contain protein X in its phosphorylated form (with phosphorylation occurring on a tyrosine residue). You have two antibodies: One recognizes phosphotyrosine and the other recognizes protein X. Describe how you would analyze the cell lysates for the presence of phosphorylated protein X using just these two antibodies.

4. The salting out procedure outlined in Fig. 5-4 results in significant protein purification by eliminating many less- and more-soluble proteins. What is its other advantage?

5. You wish to separate two proteins in solution by selectively precipitating one of them. What method could you use other than salting out?

6. The p*I* of pepsin is <1.0, whereas that of lysozyme is 11.0. What amino acids must predominate in each protein to generate such p*I*s?

7. Evaluate the solubility of the following peptides at the given pH. For each pair, which peptide is more soluble and why? *Hint:* Evaluate the ionization states of the amino acid functional groups.
   (a) pH 7: Gly_{20} and $(\text{Glu-Asp})_{10}$
   (b) pH 5: $(\text{Cys-Ser-Ala})_5$ and $(\text{Pro-Ile-Leu})_5$
   (c) pH 7: Leu_3 and Leu_{10}

8. Would you use an anion or cation exchange column to purify bovine histone at pH 7.0? (See Table 5-2.)

9. Why is a DEAE ion exchange column ineffective above pH 9?

10. When immunoaffinity chromatography is used to purify a protein, the cell lysate is often subjected to one or more purification steps before the material is applied to the immunoaffinity column. Why is this necessary?

11. List three methods for determining the molecular mass of a polypeptide. Which is the most accurate and why?

12. Protein A has a molecular mass of 100 kD, while protein B has a molecular mass of 70 kD. However, the sedimentation coefficients of the two proteins are 2.65S and 5.15S, respectively. Explain.

13. Evaluate the protein separation capabilities of zonal versus equilibrium density gradient centrifugation.

14. Shown below is the biochemical fractionation and purification of an enzyme. Fill in the table with the values for specific activity and fold purification, and answer the accompanying questions. *Hint:* Use Table 5-3 as a guide.
    (a) Which step provides the greatest purification?
    (b) Is the enzyme pure? How can purity be assessed?

	Total Protein (mg)	*Activity (nkat)*	*Specific Activity (nkat/mg)*	*Fold Purification*
Crude extract	50,000	10,000,000		
Ammonium sulfate precipitation	5,000	3,750,000		
DEAE-cellulose, stepwise KCl gradient	500	500,000		
DEAE-cellulose, KCl linear gradient	250	500,000		
Gel filtration	25	250,000		
Substrate affinity chromatography	1	100,000		

Protein Sequencing

15. Match the terms below with the accompanying statements. (More than one term may apply to a statement.)

A. dansyl chloride
B. phenylisothiocyanate
C. endopeptidase
D. 2-mercaptoethanol
E. cyanogen bromide
F. iodoacetate

______ An alkylating agent that reacts with Cys residues.
______ A reducing agent that carries out reductive cleavage of disulfide bonds.
______ A reagent used to identify N-terminal residues.
______ A reagent that cleaves polypeptides into smaller fragments.
______ A reagent used in sequencing polypeptides from the N-terminus.

16. Could carboxypeptidase A have been used to identify the C-terminal residue of chicken cytochrome *c* (see Table 5-6)?

17. How many fragments can be obtained by treating bovine insulin (Fig. 5-1) with endopeptidase V8? Does this number depend on whether the protein is first reduced and alkylated?

18. Describe the limitations of the following methods for determining amino acid composition:
    (a) acid hydrolysis
    (b) base hydrolysis
    (c) enzymatic hydrolysis

19. You have a protein whose molecular mass is about 6.8 kD according to SDS-PAGE. However, SDS-PAGE of the protein in the presence of 2-mercaptoethanol reveals two bands of 2.3 kD and 4.5 kD. Are these polypeptides too large for direct sequencing by an automated sequenator using Edman degradation? What steps must you take to determine the complete sequence of this protein?

20. You have a polypeptide that has been degraded with cyanogen bromide and trypsin. The sequences of the fragments are listed below (using one-letter codes). Is there enough information to derive the complete amino acid sequence of the polypeptide? What further treatments might support the deduced sequence?

Cyanogen bromide treatment:	***Trypsin treatment:***
(1) **K**	(5) **FMK**
(2) **RFM**	(6) **GMNIK**
(3) **MLYCRGM**	(7) **GLMR**
(4) **NIKGLM**	(8) **MLYCR**

21. You want to know the amino acid sequence of a purified polypeptide (1,341 D). Your new technician has performed the analyses outlined below but forgot to obtain the total amino acid composition of the polypeptide. You forgive your technician because you feel you may have enough information. Do you have enough data to determine the amino acid sequence?
    (a) Partial acid hydrolysis yields tripeptides with the following amino acid sequences as determined by manual Edman degradation: **HSE, EGT, DYS, TSD, FTS, YSK.**
    (b) Dansyl chloride treatment yields **H** and **K.**
    (c) Brief carboxypeptidase A treatment yields one **K** per mole peptide.
    (d) Chymotrypsin treatment yields 2 peptides (A and B) with molecular masses of 659 D and 467 D, respectively, and a dipeptide composed of **S** and **K.**
        Peptide A
        (1) Dansyl chloride treatment yields **H.**
        (2) Brief carboxypeptidase A treatment yields **F** and **T.**
        Peptide B
        (3) Dansyl chloride treatment yields **T.**
        (4) Brief carboxypeptidase A treatment yields **Y** and **D.**
    (e) Trypsin treatment results in no fragmentation of the polypeptide.

Protein Evolution

22. Explain why single-nucleotide mutations in DNA occur at a constant rate, but the rate of protein evolution varies.

23. Rank the following residue positions in cytochrome *c* in order from least to most conserved: 56, 61, 74, 85, and 89.

24. Which protein(s) shown in Figure 5-19 would be most useful for assessing the phylogenetic relationships between mammalian species (note that mammals diverged from reptiles about 300 million years ago)? Explain.

Answers to Questions

1. (a) Proteins at low concentrations can be adsorbed by glass or some plastic surfaces, thus significantly decreasing their recovery.
   (b) Sudsing of protein solutions indicates increased contact with the air–water interface, which can denature the proteins.
   (c) Proteins can be eaten or degraded by contaminating bacteria and fungi.
   (d) Protease inhibitors are essential for limiting protein degradation by proteases released from lysosomes during cell lysis.
   (e) Many proteins are only marginally stable and slowly denature over time at temperatures above 25°C.
   (f) Protein structure as well as enzymatic activity are strictly dependent on pH.

2. (a) If the protein is an enzyme, incubate it with a substrate that generates an easily detected (e.g., colored) product or a product that can be converted to a detectable product by a second enzyme.
   (b) If an antibody to the protein is available, the protein can be detected using an immunoassay such as an RIA or ELISA.

3. Immunoaffinity chromatography, immunoassay, and Western blotting could work. One protocol would be to coat a surface with antibodies to protein X, then incubate it with a cell lysate to allow any protein X present to bind to the immobilized antibodies. Wash away unbound proteins, then incubate the surface-bound antibody–protein X complex with the anti-phosphotyrosine antibody to which a readily assayed enzyme has been attached. Wash away unbound antibody and assay the enzyme. The appearance of the enzyme's reaction product indicates the presence of phosphorylated protein X. Unphosphorylated protein X will not bind the second antibody, and lysates that contain no protein X will not cause the second antibody to bind.

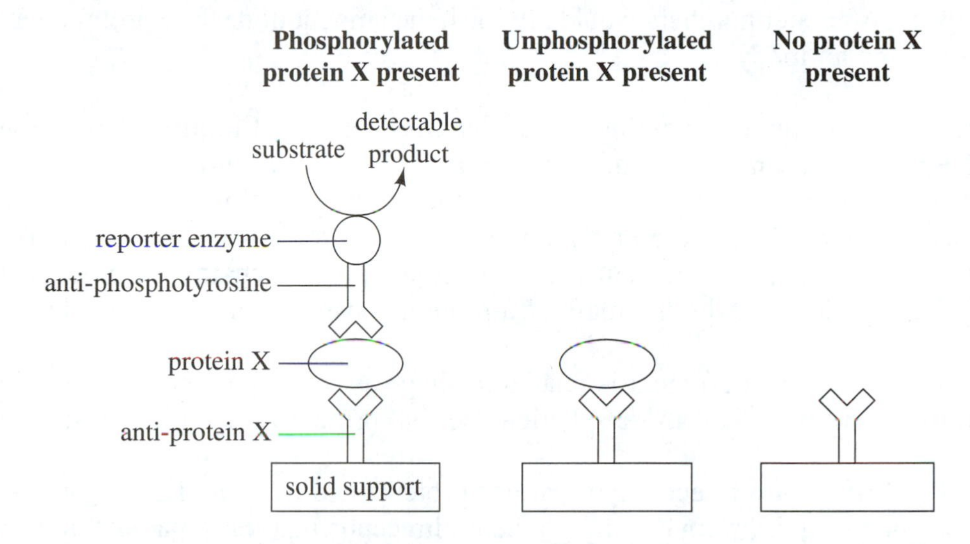

4. Following salting out, the precipitated protein can be redissolved in a very small volume, thus concentrating it. A highly concentrated protein solution may be necessary for a subsequent gel filtration step or other analytical procedure. Concentrated protein solutions are also easier to handle, and they minimize protein loss by adsorption to surfaces.

5. Adjust the pH of the solution so that it approaches the p*I* of one of the proteins. As that protein's net charge approaches 0, it will become less soluble and may precipitate, whereupon it can be eliminated by centrifugation. Alternatively, add a water-soluble organic solvent to selectively precipitate one of the proteins.

6. In pepsin, aspartate and glutamate must predominate because they have acidic R groups. Lysine, arginine, and tyrosine, all with basic R groups, must predominate in lysozyme.

7. (a) At pH 7.0, the side chains of Asp and Glu are ionized. Although both peptides contain ionized N- and C-terminal groups, $(\text{Glu-Asp})_{10}$ contains more total ionizable groups than Gly_{20} and is therefore the more soluble peptide.
   (b) Although Cys would be ionized at higher pH, neither peptide has ionized R groups at pH 5.0. However, $(\text{Cys-Ser-Ala})_5$ has more polar groups and is therefore more soluble than $(\text{Pro-Ile-Leu})_5$.
   (c) Leu_3 is less hydrophobic since it has two charged groups (its N- and C-termini) per three residues, compared to two charged groups per 10 residues in Leu_{10}. The shorter peptide is more soluble since it requires less disruption of water structure to dissolve.

8. Bovine histone is highly basic (pI = 10.8) and therefore positively charged at pH 7.0. It would therefore bind to the anionic matrix of a cation exchanger (e.g., carboxymethyl cellulose).

9. Above pH 9, the amino groups of the DEAE matrix lose their protons and are therefore unable to bind anionic substances.

10. Some pre-purification before immunoaffinity chromatography may be necessary because the cell lysate may contain other proteins that compete with the desired protein for binding to the immobilized antibody. Although such proteins would bind with lower affinity to the antibody, their greater numbers would diminish the amount of desired protein that binds specifically to the antibody.

11. Three methods for determining molecular mass are gel filtration chromatography, SDS-PAGE, and ultracentrifugation. SDS-PAGE is the most accurate. Gel filtration and ultracentrifugation are sensitive to the native shape of a protein, whereas SDS-PAGE is performed on an SDS-denatured polypeptide so that its folded shape is not an issue. Note that SDS-PAGE reveals the molecular mass of a polypeptide chain; the other methods are appropriate for determining the mass of a protein consisting of multiple subunits.

12. The sedimentation coefficient is related to shape as well as molecular mass. Protein A may be rodlike and protein B more spherical, so that protein B sediments faster.

13. Zonal centrifugation effectively separates proteins and other molecules on the basis of molecular mass. Equilibrium density gradient ultracentrifugation separates molecules based on their density. Since most proteins have similar densities, this is not a useful technique for protein separation. However, it is useful for separating nucleic acids from proteins and other cellular molecules.

14.

	Total Protein (mg)	*Activity (nkat)*	*Specific Activity (nkat/mg)*	*Fold Purification*
Crude extract	50,000	10,000,000	200	1
Ammonium sulfate precipitation	5,000	3,750,000	750	3.75
DEAE-cellulose, stepwise KCl gradient	500	500,000	1,000	5
DEAE-cellulose, KCl linear gradient	250	500,000	2,000	10
Gel filtration	25	250,000	10,000	50
Substrate affinity chromatography	1	100,000	100,000	500

(a) The step that provides the greatest purification is the step that gives the largest fold purification from the previous step, or the greatest increase in specific activity from the previous step. In this example, the step is substrate affinity chromatography, with a 10-fold purification from the previous step.

(b) The enzyme may be pure, or there may yet be other proteins present. SDS-PAGE provides the most powerful technique for ascertaining whether the final protein preparation contains a single polypeptide. However, if SDS-PAGE reveals equivalent amounts of more than one polypeptide, it is possible that the native enzyme is a multimer of nonidentical subunits.

15. __F__ An alkylating agent that reacts with Cys residues.
__D__ A reducing agent that carries out reductive cleavage of disulfide bonds.
__A, B__ A reagent used to identify N-terminal residues.
__C, E__ A reagent that cleaves polypeptides into smaller fragments.
__B__ A reagent used in sequencing polypeptides from the N-terminus.

16. No, because the C-terminal residue is Lys, which cannot be cleaved by carboxypeptidase A.

17. If the insulin is reduced and alkylated to eliminate disulfide bonding between Cys residues, then endopeptidase V8, which cleaves after Glu residues, will generate six peptides. If the disulfide bonds between the two chains of insulin are intact, the peptidase will generate four fragments.

18. (a) Acid hydrolysis degrades Ser, Thr, Tyr, and Trp and converts Asn and Gln to Asp and Glu.
(b) Base hydrolysis degrades Arg, Cys, Ser, and Thr.
(c) Protease digestion may be incomplete, and digestion of the protease itself may contribute amino acids to the sample.

19. Since the average mass of an amino acid residue is ~110 D, a peptide with a mass of 4500 D contains ~44 residues. Thus, both peptides are small enough to be directly sequenced. The two peptides are linked by a disulfide bond, so determining the protein's complete sequence requires the following steps: (a) reduce the disulfide bond; (b) alkylate the free thiol groups; (c) separate the peptides; and (d) sequence each peptide. If either peptide contains more than one Cys residue, additional fragmentation and sequencing steps would be necessary to identify the position of the disulfide bond in the intact protein.

20. Yes, there is enough information. The overlap of the relevant fragments is

```
MLYCRGM
     GMNIK
       NIKGLM
          GLMR
             RFM
              FMK
```

Thus, the peptide sequence is **MLYCRGMNIKGLMRFMK.** To confirm the deduced sequence, perform one round of Edman degradation to verify that the N-terminus is methionine, and perform a brief carboxypeptidase A digestion to verify that lysine is at the C-terminus.

21. Since there are no cysteines, there is only one N-terminus in this polypeptide. Therefore, treatment (b) reveals that His or Lys must be at the N-terminus (the ε-amino group of Lys can also react with dansyl chloride). Treatment (c) indicates that there is a Lys at the C-terminus, and since trypsin treatment did not cleave the polypeptide, the **YSK** peptide must represent the C-terminal peptide. Since peptide A contains **F** and must end in **F** due to the specificity of chymotrypsin, **T** must precede **F.** We can now deduce the following:

```
HSE
  EGT...

              FTS
                TSD
                   DYS
                     YSK
```

From the masses of the two peptides produced by chymotrypsin treatment, we know that peptide A is about 6 amino acids long (~110 D per residue) and peptide B is 4 amino acids long. Hence, the sequence of peptide A must be **HSEGTF** and that for peptide B **TSDY.** Since **YSK** is the C-terminal peptide, we can deduce that the sequence of the polypeptide is **HSEGTFTSDYSK.**

In some cases, analysis of the amino acid composition can confirm whether all the amino acids of the peptide have been accounted for in the sequence, particularly if partial acid hydrolysis, which may release some free amino acids, is used. In the example above, the sum of the masses of the residues of the sequence match the mass of the intact peptide, so no residues were missed in the sequencing.

22. The rate of protein evolution depends on the ability of a protein to accept mutations without significantly altering its function (neutral drift), which varies among proteins. Hence, proteins with slow rates of evolution cannot accept many amino acid substitutions without loss of function, thereby eliminating mutant forms by natural selection.

23. From inspection of the last row of Table 5-6, it is seen that residue 56 has three different amino acids in the table, 61 has four, 74 has one, 85 has two, and 89 has nine. Hence their order, from least to most conserved is: 89, 61, 56, 85, 74.

24. The fibrinopeptides are evolving too rapidly to show the more ancient relationships among mammals, since over periods greater than 100 million years, many residues will have mutated two or more times. Histone H4 has evolved too slowly to reveal variations in mammalian species (they all have nearly identical histone H4's). However, cytochrome *c* and hemoglobin evolve at rates that would be useful for assessing phylogenetic relationships among mammalian species, with hemoglobin, the faster-evolving protein of the two, providing a more sensitive measure.

Chapter 6 Proteins: Three-Dimensional Structure

This chapter introduces you to an area of biochemistry that has shown an incredible growth of information in recent years: the three-dimensional structures of proteins. The chapter first discusses in some detail the geometric and physical properties of the planar peptide group, which contains the peptide bond. Here you see that the number of polypeptide conformations, while quite large, is finite due to restrictions imposed by colliding van der Waals spheres of the atoms of the polypeptide. The chapter then examines the major forms of secondary or local three-dimensional structure in polypeptides, in particular, the α helix and the β sheet. Keratin, silk fibroin, and collagen are the most common and best-characterized fibrous proteins; each is largely composed of one type of secondary structure. The structural properties and biology of these proteins are discussed at this point.

The chapter then turns its attention to the methods by which overall three-dimensional structure, or tertiary structure, is determined: X-ray diffraction of protein crystals and nuclear magnetic resonance (NMR) of proteins in solution. You are then introduced to the various kinds of supersecondary structures (motifs) and domains that have been identified in many proteins. This section concludes with a brief discussion of the symmetries observed in the quaternary structures of proteins that consist of more than one subunit.

The factors that affect protein stability are then explored, followed by a discussion of how newly synthesized proteins fold. In this presentation, you are introduced to proteins that facilitate protein folding *in vivo:* protein disulfide isomerase and the molecular chaperones (the Hsp70 family and the chaperonins). The chapter closes with a brief discussion of protein dynamics and their biological significance. In Boxes 6-1 and 6-4, you will see the importance of protein structure in diseases that arise from altered tertiary and quaternary structures. Boxes 6-2 and 6-3 discuss recent progress in predicting the structures of proteins and designing proteins, and the use of NMR in protein structure determination, respectively.

Essential Concepts

1. There are four levels of organization in protein structure:
   (a) Primary structure, which is the amino acid sequence.
   (b) Secondary structure, which is the local three-dimensional structure of a polypeptide without regard to the conformations of its side chains.
   (c) Tertiary structure, which refers to the overall three-dimensional structure of an entire polypeptide.
   (d) Quaternary structure, which refers to the three-dimensional arrangement of polypeptides in a protein composed of multiple polypeptides.

Secondary Structure

2. Amino acids are joined by the condensation of the amino group of one amino acid with the carboxyl group of another, which results in a linkage called a peptide bond. The peptide

group has a resonance structure such that the bond connecting the carboxyl carbon and the amino nitrogen has ~40% double-bond character.

With rare exceptions, the *trans* conformation shown above occurs in proteins.

3. The backbone of a polypeptide chain can be viewed as a linked series of planar peptide groups that can rotate about the single covalent bonds involving the α carbon. When the peptide groups are fully extended and all lie in a plane, the torsion angles of the C_α—N bond (ϕ) and the C_α—C bond (ψ) are defined as 180° (Figure 6-4). As these planar groups rotate, the van der Waals spheres of their various atoms collide; hence the rotational freedom of the bonds is restricted. The allowed conformations, given as ϕ and ψ values, are summarized in the Ramachandran diagram (Figure 6-6).

4. In regular secondary structures, successive residues have similar backbone conformations, that is, repeating values of ϕ and ψ (e.g., the α helix and the β sheet).

5. The α helix is a right-handed helix with 3.6 residues per turn and a pitch of 5.4 Å. Hydrogen bonds connect the peptide C=O group of the *n*th residue with the peptide N—H group of the $(n + 4)$th residue. The core of the helix is tightly packed, and the amino acid side chains project outward and downward from the helix.

6. β sheets result from hydrogen bonding between segments of extended polypeptide chains. Antiparallel β sheets consist of polypeptide strands that extend in alternate directions. Parallel β sheets consist of polypeptide chains that all extend in the same direction. Parallel β sheets are slightly less stable than antiparallel sheets because the hydrogen bonds connecting the polypeptides are distorted. Mixed β sheets, consisting of both parallel and antiparallel polypeptide chains, are common.

7. Proteins are typically classified as fibrous or globular. A single type of secondary structure dominates fibrous proteins. The three best-characterized fibrous proteins are α keratin, silk fibroin, and collagen, all of which form higher order structures that are insoluble in water.
   (a) α Keratin consists of two α-helical chains that wrap around each other in a left-handed coiled coil. Each polypeptide chain has a 7-residue pseudorepeat, *a-b-c-d-e-f-g,* such that hydrophobic residues at positions *a* and *d* form the contacts between the helices. This dimer is assembled further into higher order structures that are less well characterized.
   (b) Silk fibroin contains extensive regions of antiparallel β sheets whose chains extend parallel to the fiber axis. The silk fibroin made by the silkworm, *Bombyx mori,* contains long stretches of the six-residue repeat $(\text{Gly-Ser-Gly-Ala-Gly-Ala})_n$ in which the Gly and Ala (or Ser) side chains extend to opposite sides of the β sheet.

(c) Collagen contains long stretches of the three-residue repeating unit Gly-X-Y, in which X is often Pro, and Y is often 4-hydroxyPro (Hyp; hydroxylation of Pro residues requires ascorbic acid). Each polypeptide strand forms a left-handed helix with around three residues per turn. Three such parallel chains then wrap around each other in a gentle right-handed coil (Figure 6-17). Interchain hydrogen bonds connect all three strands, which are staggered so that a Gly residue occurs at every position along the triple helix. Covalent cross-links between modified Lys residues (allysine) and His residues bind together aggregates of collagen fibers.

8. Irregular secondary structures include β bulges, reverse turns (β bends), Ω loops, and other structures whose residues have nonrepeating backbone conformations.

9. Certain amino acid residues occur more often in α helices or β sheets. Moreover, certain residues tend to disrupt or break secondary structures. These tendencies provide a foundation for predicting protein structure from amino acid sequences and for designing proteins with particular structures.

Tertiary Structure

10. The tertiary structure of a protein can be determined by X-ray diffraction or nuclear magnetic resonance (NMR).
    (a) The interaction of a beam of X-rays with the electrons of a crystallized protein yields a diffraction pattern that can be mathematically analyzed to reconstruct a three-dimensional electron density map of the atoms of the protein. Knowledge of the protein's primary structure is necessary to identify the amino acid residues in a three-dimensional model of the protein.
    (b) NMR spectroscopy can be used to obtain information about the tertiary structure of a protein in solution. This technique is limited to proteins with $<$250 residues.

11. In globular proteins, nonpolar residues occur most often in the protein interior. Charged residues most commonly occur on the protein surface. Polar residues that are buried in the protein interior are almost always "neutralized" by forming hydrogen bonds to other internal groups.

12. α Helices and β sheets can be combined in a variety of ways to form motifs (e.g., a βαβ motif or a β barrel). Larger globular domains form in polypeptides containing more than ~200 amino acids. Domains are often associated with a biological function, such as binding a small molecule.

13. Although over 7000 protein structures are known, a few dozen folding patterns account for nearly half of all known protein structures. This may reflect evolutionary constraints based on the ability of these domains to
    (a) Form stable folding patterns.
    (b) Tolerate amino acid deletions, substitutions, and insertions without significant loss of biological activity.
    (c) Support essential biological functions.

Quaternary Structure and Symmetry

14. Most proteins with molecular masses >100 kD consist of more than one polypeptide chain. The arrangement of the chains is the protein's quaternary structure. The contact regions between subunits of a protein resemble the interior of a polypeptide.

15. The advantages of multisubunit proteins include the following:
    (a) A defective region of a protein (e.g., caused by mistranslation or improper folding) can be easily repaired by replacing the defective subunit.
    (b) The only genetic information necessary to specify a large protein is that specifying its few different self-assembling subunits.
    (c) It provides a structural basis for the regulation of enzymatic activity.

16. Proteins with more than one subunit are called oligomers, and their identical subunits (which may contain more than one polypeptide) are called protomers. In most proteins, protomers occupy geometrically equivalent positions, displaying rotational symmetry. The most common are cyclic symmetry (usually involving two protomers) and dihedral symmetry (in which an n-fold rotation axis intersects a two-fold rotation axis at right angles).

Protein Folding and Stability

17. Native proteins are only marginally stable under physiological conditions. The hydrophobic effect is the principal force that stabilizes protein structures. Electrostatic interactions, especially hydrogen bonds and ion pairs, contribute little to the overall stability of tertiary structures because water interacts similarly with fully unfolded (denatured) proteins. However, these weak interactions play important roles in aligning residues in specific secondary structures and domains.

18. Disulfide bonds cross-link structures in extracellular proteins, which occur in a relatively oxidizing environment. Disulfide bonds are rare in intracellular proteins, presumably because the reducing environment of the cytosol weakens them. In some proteins, metal ions such as Zn^{2+} cross-link small structures that would otherwise be unstable.

19. Most proteins unfold or denature at temperatures well below 100°C, with a sharp transition indicating that denaturation is a cooperative process. Conditions or substances that denature proteins include heat, variations in pH, detergents, and chaotropic agents (e.g., guanidinium ion and urea).

20. Under proper conditions, most unfolded proteins will renature spontaneously, thereby indicating that tertiary structure is dictated by primary structure.

21. Protein folding is an ordered process rather than a random search for a stable conformation. During folding, secondary structures form first, followed by motifs and domains. A key driving force may be a hydrophobic collapse that buries hydrophobic regions in the interior of the protein, out of contact with the aqueous medium (to yield a state referred to as a molten globule).

22. Certain proteins facilitate protein folding *in vivo,* which is often considerably faster than folding *in vitro.* Protein disulfide isomerase mediates the formation of disulfide bonds. Molecular chaperones prevent improper associations between exposed hydrophobic segments of unfolded polypeptides that could lead to non-native folding as well as nonspecific aggregation of the hydrophobic regions of different unfolded polypeptides. The Hsp70 chaperones are a family of 70 kD proteins that prevent premature folding of polypeptides as they are being synthesized in the ribosome. The chaperonins are very large multimeric complexes that consist of two types of proteins: (1) the Hsp60 proteins (GroEL in *E. coli*) and the Hsp10 proteins (GroES in *E. coli*). The GroEL–GroES complex is a barrel-shaped structure that encloses a polypeptide and facilitates its folding in a protected environment in an ATP-dependent manner.

23. Several neurological diseases are characterized by the intracellular accumulation of fibrous proteins in amyloid plaques. One set of diseases, which includes scrapie and "mad cow disease," appears to be caused by small proteins called prions. In disease states, an α-helical prion changes its conformation to a mixture of α helices and β sheets and forms insoluble fibers. It has been proposed that prions act as infectious agents by catalyzing the misfolding of other prion proteins.

24. Proteins are flexible molecules whose conformations rapidly fluctuate. This conformational flexibility is important in that it allows the diffusion of small molecules (e.g., ligands or substrates) into the interior of the protein.

Exercise

Use a molecular model set to satisfy yourself that glycine can assume conformations not shaded in the Ramachandran diagram (i.e., ϕ values between 60° and 180° and all ψ values except those between −45° and 45°).

Guide to Study Exercises (text p. 159)

1. The conformational freedom of a peptide bond is limited by its partial double-bond character. The peptide group (consisting of the C and O atoms of the carbonyl group and the N and H atoms of the amino group) forms a rigid plane to maximize π-bonding overlap. Thus, rotation is possible only for the bonds involving the α carbon. (Section 6-1A)

2. Regular secondary structures are characterized by sequences of residues having the same (or closely similar) ϕ and ψ values, for example, α helices and β sheets. Irregular structures are characterized by ϕ and ψ values that vary from one residue to the next. (Sections 6-1B and D)

3. In an α helix, each peptide C=O group hydrogen bonds to the backbone N—H group four residues further on. Consequently, the four residues at each end of the helix are incompletely hydrogen bonded. (Section 6-1B)

4. β Sheets are pleated, rather than fully extended, to optimize hydrogen bonding between neighboring polypeptide chains. (Section 6-1B)

5. Fibrous proteins, which are elongated molecules that often have one dominant type of secondary structure, are responsible for the mechanical properties (e.g., high tensile strength) of structures such as hair, horns, bones, and tendons. (Section 6-1C)

6. Turns and loops usually occur on the protein surface because they link the largely straight elements of secondary structure (e.g., α helices and β sheets) that comprise the core of the protein. (Section 6-1D)

7. Nonpolar residues tend to occupy the core of a native protein due to hydrophobic effects. Hence the aliphatic side chains of Ile, Leu, Met, and Val and the aromatic side chains of Phe and Trp are usually found in the protein interior, out of contact with the aqueous solvent. More hydrophilic side chains, particularly the charged groups of Arg, His, Lys, Asp, and Glu, tend to occur at the protein surface, where they can interact with water. (Section 6-2B)

8. The number of possible protein sequences (20^n for an n-residue polypeptide) far exceeds the number of possible protein structures, even for a relatively small protein, because a given protein structural element may accommodate many different sequences (that is, many sequences have essentially the same structure). Natural selection favors structures that are both stable and able to tolerate some amino acid changes. (Section 6-2C)

9. Multiple subunits offer the following advantages:
   (a) A single relatively small gene can direct the synthesis of the multiple identical subunits that self-assemble (in contrast to a large gene that would direct the synthesis of a very large polypeptide).
   (b) A defect in transcription or translation affects only a single subunit, not the entire assembly.
   (c) The subunits can be transported to and assembled at a location that would be inaccessible to a large protein (e.g., the extracellular space).
   (d) If each subunit contains an active site, the assembly would have much greater activity than a similarly sized monomer with only one active site.
   (e) A multisubunit structure provides the structural basis for the regulation of an enzyme's activity. (Section 6-3)

10. Proteins are synthesized exclusively from L-amino acids and therefore have inherent chirality. The mirror image of a protein would have the opposite chirality, which would require that it be made from D-amino acids. Consequently, only rotational symmetry is possible in proteins. (Section 6-3)

11. Proteins are stabilized by:
    (a) The hydrophobic effect, which causes nonpolar side chains to aggregate in the protein interior, out of contact with the aqueous medium.
    (b) van der Waals forces that act over short distances in the protein interior.
    (c) Other electrostatic interactions, including ion pairing and hydrogen bonding, which have a relatively small effect.

(d) Cross-links such as disulfide bonds or the liganding of protein groups to metal ions, which stabilize structural motifs that may be too small to maintain a stable structure on their own. (Section 6-4A)

12. As a protein folds, its free energy decreases until it reaches a minimum. The protein's entropy (degree of disorder) simultaneously decreases as portions of the unfolded, fluctuating polypeptide chain adopt a fixed structure. Some elements of secondary structure coalesce into larger elements, leading to discrete tertiary structures. Thus, the unfolded protein has high free energy and a high entropy, whereas the fully folded protein has low free energy and a low entropy. (Section 6-4C)

13. Protein renaturation *in vitro* starts with a full-length unfolded (denatured) polypeptide. In contrast, protein folding *in vivo* occurs as the polypeptide is being synthesized so that substantial lengths of polypeptide rarely exist in an unfolded state. The progressive nature of folding in a newly synthesized polypeptide may permit formation of structures that are not easily attained by a denatured polypeptide that refolds all at once. Furthermore, protein folding *in vivo* may be facilitated by cellular factors, such as molecular chaperones, that are usually absent *in vitro*. (Sections 6-4B and C)

Questions

Secondary Structure

1. Why does the peptide bond result in a planar configuration for its adjoining functional groups?

2. In the dipeptide below, indicate which bonds are described by ϕ and ψ.

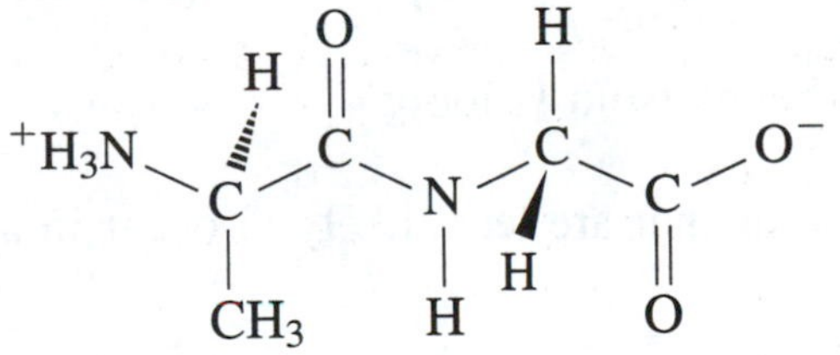

3. Shown below is a portion of poly-β-alanine$_{10}$. Comment on the secondary structure of this peptide.

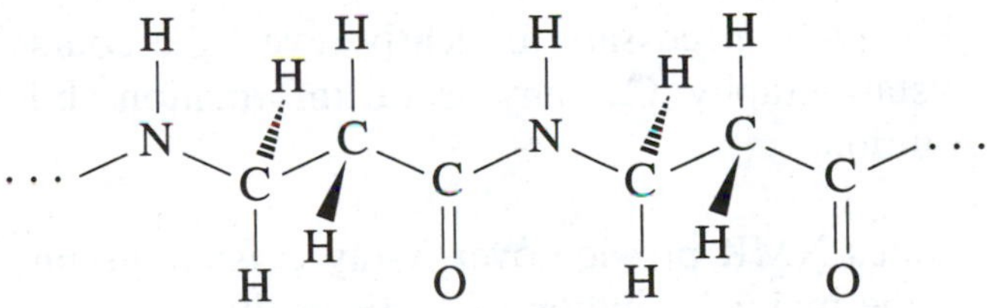

4. When $\phi = 180°$ and $\psi = 0°$, what groups come into closer-than-van-der-Waals contact? (See Figure 6-4.)

5. How do R groups constrain the potential conformations of a protein? (See Figure 6-4 and Kinemage 3-1.)

6. Study Figure 6-8 and describe the orientation of the groups that contribute to the dipole moment of the α helix and the overall direction of this dipole moment. *Hint:* The sum of local dipole moments determines the overall dipole moment of the helix.

7. Why are antiparallel β sheets more stable than parallel β sheets?

8. How does the 7-residue repeat of α keratin promote formation of a coiled coil?

9. Wool from sheep is steamed to generate long fibers for knitting. Wool sweaters shrink after subsequent drying with heat. Explain what is happening at the molecular level.

10. Would fabric made from silk containing more bulky residues be more or less likely to shrink after a hot-water wash and hot-air dry regimen? Explain.

11. What is the repeating sequence of collagen and how is it essential to the structure of the triple coils of collagen fibers?

12. Scurvy is caused by a deficiency of _______________, which is necessary for the activity of _______________.

13. Organic compounds containing radioactive atoms can be used to follow the biosynthesis of molecules in cells and organisms (see Section 13-4A). In these experiments, the amount of radioisotope in proteins derived from the cells or organisms is monitored. The administration of ^{14}C-labeled 4-hydroxyproline to mice results in no appearance of ^{14}C in the animals' collagen fibers. However, the administration of ^{14}C-labeled proline does result in the appearance of ^{14}C in the collagen fibers of mice. Explain these results.

14. What distinguishes β bends from Ω loops?

15. List the three amino acids that are least likely to occur in a β sheet.

Tertiary Structure

16. Examine Table 4-1 (pp. 80–81) and determine which amino acids might exhibit identical electron densities at 2.0 Å resolution.

17. A resolution of ~3.5 Å is necessary to clearly reveal the course of the polypeptide backbone in X-ray crystallography. Can any useful information about protein structure be obtained at 6 Å resolution?

18. What advantages does NMR provide over X-ray crystallography in characterizing protein structure? What is the major limitation of NMR analysis?

19. What specific role does knowledge of the primary amino acid sequence play in the determination of tertiary structure by X-ray crystallography?

20. How does the polarity or charge of an amino acid affect its likely location within a protein?

21. How does the ratio of hydrophilic to hydrophobic amino acid residues change in a series of globular proteins ranging from 10 to 50 kD?

Quaternary Structure and Symmetry

22. What are the possible symmetries of (a) a tetrameric protein and (b) a pentameric protein? Assume all the subunits are identical.

23. The subunit composition of an oligomeric protein can be determined by treating the protein with a cross-linking agent (a bifunctional molecule that reacts with and links groups in two different portions of the polypeptide), denaturing the protein, and analyzing the products by SDS-PAGE (Section 5-2D).
    (a) An oligomer analyzed by SDS-PAGE shows a single 40 kD polypeptide. Brief treatment with a cross-linking agent yields the SDS-PAGE banding pattern shown below. What is the protein's probable subunit composition?

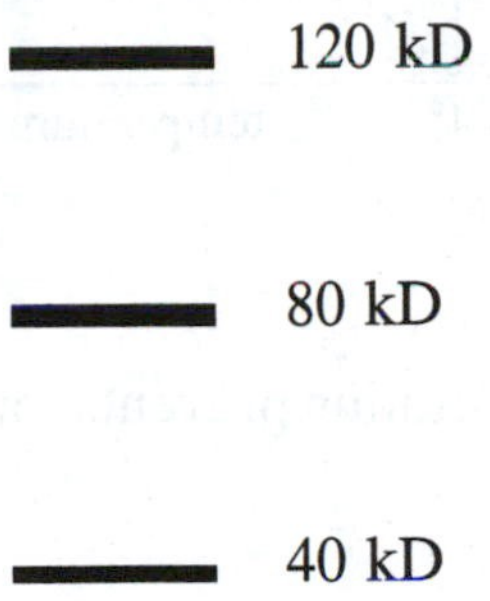

120 kD

80 kD

40 kD

    (b) Not shown in the electrophoretogram above are faint bands at 240 kD and 360 kD. What do these bands represent?
    (c) Another protein analyzed by SDS-PAGE shows two polypeptides at 20 kD and 50 kD. Chemical cross-linking of the native protein yields the results shown below in SDS-PAGE. Not shown is a very faint band at 540 kD. What is the subunit composition of this protein?

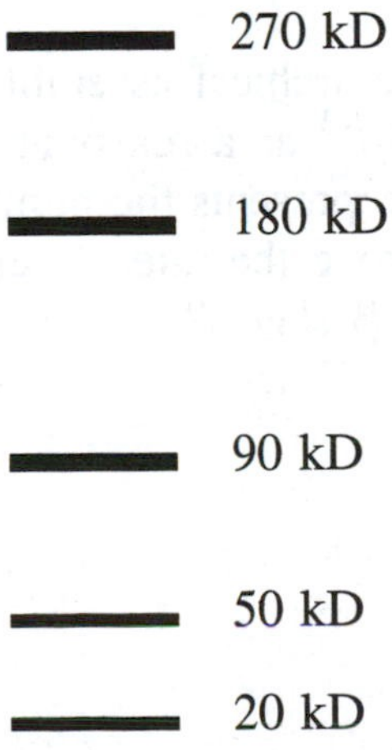

270 kD

180 kD

90 kD

50 kD

20 kD

24. Consider the equilibrium of an oligomer:

$$2 \text{ monomers} \rightleftharpoons \text{dimer}$$

The $\Delta H°$ is positive and varies little between 4°C and 45°C. $T\Delta S°$, in contrast, varies widely over the same temperature range. The behaviors of $\Delta H°$ and $T\Delta S°$ are shown in the graph below. How does the position of equilibrium of this oligomer differ at 4°C versus 45°C?

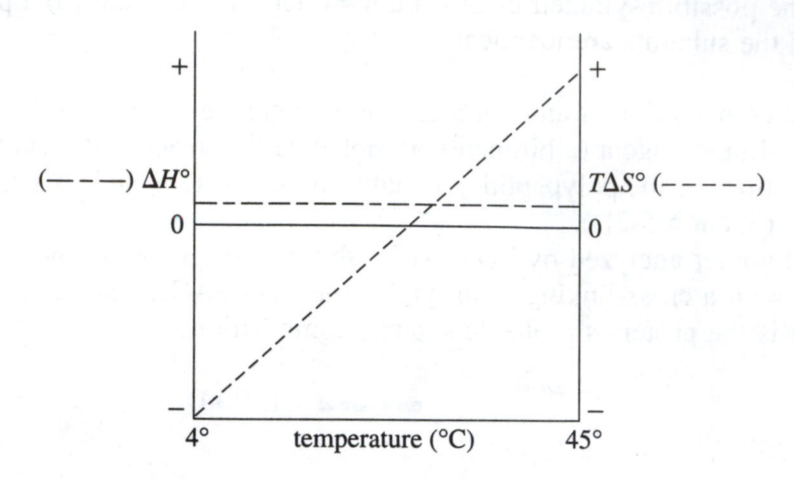

Protein Folding and Stability

25. What thermodynamic consideration prevents ion pairs from contributing significantly to protein stability?

26. List four denaturants of proteins.

27. What is the probability of recovering the native conformation (as judged by biological activity) of a dimeric protein, each of whose subunits contains three disulfide bonds? Assume that the disulfide linkages form randomly.

28. Calculate the maximum length of a polypeptide that could recover its native conformation in about one day if the polypeptide explored all possible conformations (see page 153).

29. To appreciate the role of hierarchical assembly in the rapid renaturation of proteins, consider the equation $t = 10^n/10^{13}$ as a reasonable approximation of renaturation time for a globular protein in which n represents the number of nucleating structures (α helices and β sheets). How does this improve the rate of renaturation of a 200-residue globular protein with eight α helices and six β sheets?

Answers to Questions

1. The C—N bond has a partial double-bond character due to resonance interactions. This limits rotation around the C—N bond, constraining the carbonyl and amide groups to lie in the same plane.

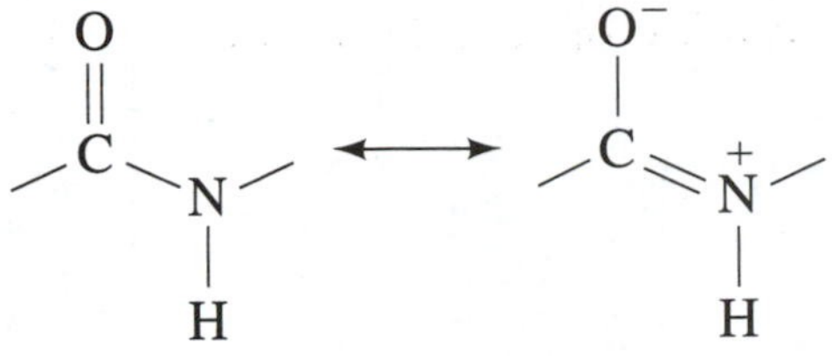

2. 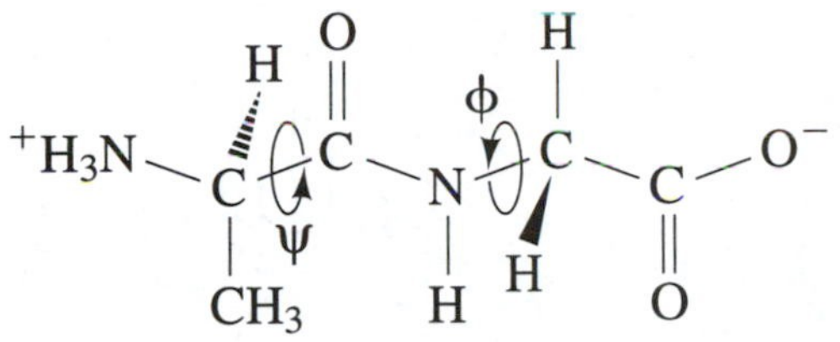 Note that both ϕ and ψ increase clockwise when viewed from C_α.

3. In the structure shown below, the thicker lines represent the peptide bonds. The polypeptide backbone contains three single bonds capable of free rotation between each peptide bond (versus two in a standard polypeptide). This greater flexibility allows for more possible conformations of the peptide chain, so that the polymer is more likely to assume a random, fluctuating conformation rather than a regular secondary structure.

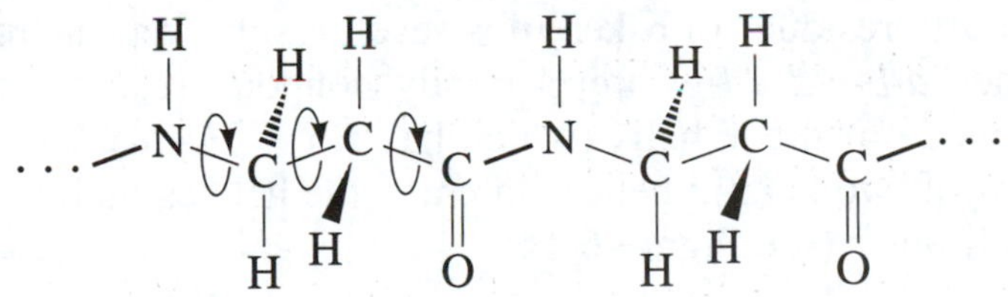

4. When $\phi = 180°$ and $\psi = 0°$, the van der Waals spheres of the amide hydrogens overlap.

5. Rotation around ϕ will cause the van der Waals sphere of the carbonyl oxygen of residue n to collide with the van der Waals sphere of the C_β of residue $n + 1$. This can be clearly seen in Kinemage 3-1 on the CD that accompanies the text. Turn on AAs 1–3, PH1, planes, O1, and CB. Then try rotating around ϕ to observe collision of van der Waals spheres.

6. Shown below is the dipole of a peptide unit. In an α helix, all the hydrogen bonds point in the same direction, so the dipoles of each peptide unit sum up to an overall dipole moment for the entire helix. Since the carbonyl groups all point from the N-terminus toward the C-terminus of the helix, the N-terminus of the helix has a partial positive charge and the C-terminus has a partial negative charge.

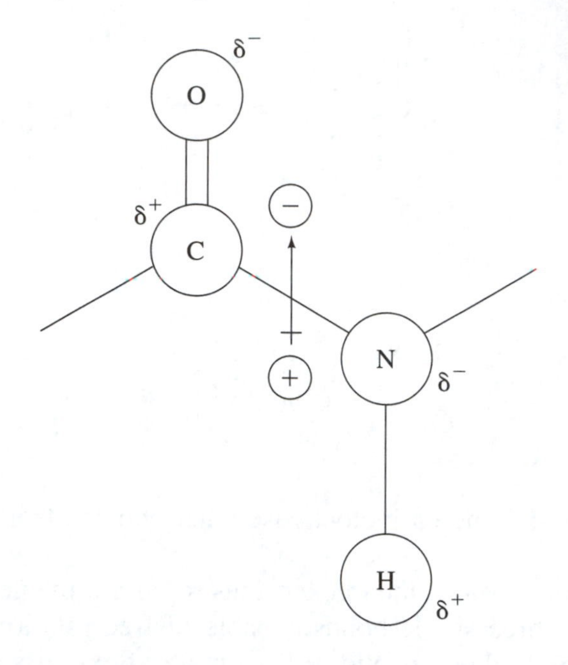

7. Antiparallel β sheets are more stable because the hydrogen bonds connecting adjacent polypeptide strands are not distorted, as they are in parallel β sheets (see Figure 6-9).

8. The first and fourth residues of α keratin's seven-residue repeat (residues *a* and *d* of the repeating sequence *a-b-c-d-e-f-g*) are primarily nonpolar residues that form a hydrophobic strip along one side of the α helix (which has $3.6 \approx 7/2$ residues per turn). When the hydrophobic strips of two keratin helices interact, the helices incline slightly, causing them to coil around each other (see Figure 6-14).

9. The steaming process stretches α keratin into a pleated β sheet conformation. When subsequently heated, the keratin in the wool is converted back to an α-helical conformation, which is more compact.

10. Silk with bulky side chains is more likely to shrink or lose its shape, because the β sheets are less ordered and are therefore less elastic and less likely to recover their original shape after heating.

11. The repeating sequence is Gly-X-Y, where X is often Pro and Y is often Hyp. The glycine residues occur in the interior of the triple helix, where there is no room to accommodate any side chain larger than a hydrogen atom. The highly constrained Pro and Hyp side chains confer rigidity on the helical structure.

12. Scurvy is caused by a deficiency of ascorbic acid (vitamin C), which is necessary for the activity of prolyl hydroxylase.

13. Hydroxyproline is not one of the 20 standard amino acids used in the synthesis of proteins from mRNA. Specific proline residues in collagen are oxidized by prolyl hydroxylase to form 4-hydroxyproline only after peptide synthesis. Hence, only the administration of ^{14}C-labeled proline can yield radioactively labeled collagen fibers.

14. β bends are 4-residue structures associated with reversals of direction between two successive polypeptide segments. Ω loops are larger neck-shaped structures containing 6 to 16 residues that are located on protein surfaces.

15. Glu, Asp, and Pro are the least likely to occur in a β sheet as judged by their low propensity for appearing in the β sheets of known proteins (see Table 6-1 on p. 139).

16. Several pairs of amino acids give indistinguishable electron density maps: Asp and Asn, Glu and Gln, Thr and Val. Other amino acids that may have similar shapes are Ser and Cys, although the S atom of Cys has a much greater electron density than the O atom of Ser.

17. At 6 Å resolution, larger elements of protein structure can be identified, most notably helices, which have a diameter of several Å and are visible as rodlike shapes.

18. The advantages of NMR include (a) structural information from proteins that fail to crystallize and (b) information about protein folding and dynamics since protein movements can be traced over relatively long time scales. The primary disadvantage of NMR is that the protein must be no larger than ~250 residues.

19. When interpreting an electron density map, knowledge of the primary structure allows the identification of specific amino acids along the polypeptide chain. This would otherwise be quite difficult if not impossible for R groups with similar shapes. (See Question 16.)

20. Charged or polar amino acids are commonly on the outside of the molecule, where they are exposed to the aqueous solvent, whereas uncharged or nonpolar amino acids are often buried in the interior of the protein, out of contact with the aqueous solvent, due to hydrophobic effects.

21. As protein size increases, the surface-to-volume ratio decreases. Recall that the surface of a sphere increases as the square of the radius (r^2), while the volume of a sphere increases as the cube of the radius (r^3). Hence, the ratio of hydrophilic to hydrophobic residues decreases as the molecular mass increases since the interior (volume) of the protein increases more rapidly than its surface area.

22. A tetrameric protein can have C_4 or D_2 symmetry. A pentameric protein can have only C_5 symmetry.

23. (a) The protein is probably a trimer of 40 kD subunits. The gel shows cross-linked dimers and trimers in addition to free monomers.
    (b) The apparent masses of the faint bands are multiples of 120 kD and therefore probably represent protein trimers that have been chemically cross-linked to other trimers.

(c) The protein is a trimer of two 20 kD subunits and one 50 kD subunit. Some of the 90-kD oligomers have been cross-linked to form larger structures of 180 kD (a dimer), 270 kD (a trimer), and 540 kD (a hexamer).

24. Recall that $\Delta G° = \Delta H° - T\Delta S°$. At 4°C, $T\Delta S°$ is negative, so $\Delta G°$ is positive, indicating that the equilibrium favors monomer formation (see Table 1-3, p. 16). At 45°C, $T\Delta S°$ is positive, so $\Delta G°$ is negative, indicating that the equilibrium now favors the formation of dimers. The cytoskeletal elements called microtubules behave in such a manner.

25. The favorable free energy of formation of an ion pair is nearly equivalent to the loss of solvation free energy that occurs when the charged groups form the ion pair.

26. Heat, pH changes, detergents, and chaotropic agents.

27. The probability for each monomer is $1/5 \times 1/3 \times 1/1 = 1/15 = 0.067$ or 6.7%. The probability that a native dimer can form is $0.067 \times 0.067 = 0.0044$ or 0.44%.

28. $t = 10^n/10^{13}$, where t is the time in seconds, and n is the number of amino acid residues.

Rearranging terms,

$$n = \log 10^{13} + \log t$$

Here, $t = (60 \text{ s/min})(60 \text{ min/h})(24 \text{ h/day}) = 86{,}400 \text{ s}$

so that,

$$\begin{aligned} n &= 13 + \log 86{,}400 \\ &= 13 + 4.9 \\ &= 17.9 \end{aligned}$$

Therefore, the peptide could have no more than 18 residues.

29. For a 200-residue protein, $t = 10^n/10^{13} = 10^{200}/10^{13} = 10^{187}$ s or about 3×10^{179} years. If the α helices and β sheets act as nucleating elements for the formation of tertiary structure, $n = 14$ and $t = 10^{14}/10^{13} = 10$ s.

Chapter 7 Protein Function

The proteins responsible for oxygen binding and transport, muscle contraction, and the immune response are discussed in this chapter. These examples are chosen because of the depth of information known about them as well as their importance to human health. The function of a protein is dictated by its amino acid sequence and is revealed in its secondary, tertiary, and quaternary structure. Myoglobin and hemoglobin are globular soluble proteins that bind oxygen. Muscle proteins, on the other hand, form long filamentous bundles that contract upon the hydrolysis of ATP. Antibodies are proteins that have common overall structures but variable sequences, which thereby form billions of possible proteins, many of which specifically bind foreign macromolecules. Many amino acid residues are critical for these functions, so as you read and study this chapter, try to glean the importance of structure–function relationships in each example.

Myoglobin and hemoglobin are involved in oxygen storage and transport. Myoglobin, whose primary role is to facilitate oxygen transport in muscle (and, in aquatic mammals, to store oxygen), binds oxygen with a simple equilibrium constant. Hemoglobin transports oxygen, first binding oxygen in the capillaries of the lungs (or gills or skin), where oxygen concentrations are high, and then releasing the oxygen in the tissues, where the oxygen concentrations are lower. The binding of oxygen to hemoglobin is not as simple as that for myoglobin. Several factors influence oxygen binding to hemoglobin, including the partial pressure of oxygen itself, pH, the concentration of CO_2, and 2,3-bisphosphoglycerate (BPG). The study of mutant hemoglobin molecules with amino acid substitutions that affect hemoglobin function provides important insights into the roles of individual amino acids.

Muscle fibers consist of bundles of myofibrils, which are striated due to the presence of repeating protein assemblies. The myofibrils consist of interdigitated thick and thin filaments. The thick filaments are made of myosin, whose two heavy chains and four light chains form a long rodlike segment with two globular heads. Hundreds of myosin molecules aggregate in a thick bundle. Thin filaments contain three proteins: actin, tropomyosin, and troponin. Interactions between the thick and thin filaments allow the proteins to move past each other in a process that is driven by the hydrolysis of ATP.

Antibodies are the first line of defense against disease-causing pathogens such as microorganisms and viruses. Cellular immunity is mediated by *T* lymphocytes, which are formed in the thymus. Humoral immunity is mediated by antibodies (immunoglobulins) that are produced by *B* lymphocytes, which mature in the bone marrow. A *B* cell makes only one type of antibody. Antigen binding to a specific antibody located on the surface of a *B* cell triggers an immune response in which the *B* cells secreting that antibody proliferate. *B* cells usually live for a few days; however, memory *B* cells that recognize specific antigens remain and proliferate when more antigen is present. The specificity of an antibody–antigen interaction arises from variable amino sequences in the immunoglobulin, which create a unique antigen-binding site.

Essential Concepts

Myoglobin

1. Myoglobin is a single polypeptide containing a heme group, which consists of a porphyrin ring whose coordinated Fe(II) atom binds molecular oxygen. The protein prevents oxidation of the heme iron to Fe(III), which does not bind oxygen. Oxidized myoglobin is called metmyoglobin.

2. Myoglobin facilitates oxygen transport in muscle, where the solubility of oxygen is low, and acts as an oxygen-storage protein in aquatic mammals. A simple equilibrium equation describes O_2 binding to myoglobin (Mb): $Mb + O_2 \rightleftharpoons MbO_2$. The fractional saturation of myoglobin is defined as $Y_{O_2} = pO_2/(K + pO_2)$, where pO_2 is the partial pressure of oxygen and K is the dissociation constant. A plot of Y_{O_2} versus pO_2 is a simple hyperbolic binding curve. It is convenient to define K as p_{50}, the partial pressure of O_2 at which 50% of myoglobin has bound oxygen. The p_{50} for myoglobin is 2.8 torr, which is much lower than the pO_2 in venous blood (30 torr), so that myoglobin is nearly saturated with oxygen under physiological conditions.

Hemoglobin

3. Hemoglobin binds oxygen in the lungs (pO_2 = 100 torr) and releases it in the capillaries (pO_2 = 30 torr). The efficiency of oxygen transport is greater than expected if oxygen binding were hyperbolic, as in myoglobin. Hemoglobin, which is an $\alpha_2\beta_2$ tetramer, each of whose subunits contains a heme group, binds O_2 cooperatively and thus has a sigmoidal oxygen binding curve. Deoxyhemoglobin is bluish (the color of venous blood), whereas oxyhemoglobin is bright red (the color of arterial blood).

4. The p_{50} of hemoglobin is about 26 torr, which is nearly 10 times greater than that of myoglobin. Because hemoglobin exhibits a sigmoidal oxygen-binding curve, it releases a much greater fraction of its bound O_2 in passing from the lungs to the tissues than would myoglobin. The Hill equation describes the cooperative nature of oxygen binding, and the Hill constant, *n,* describes the degree of cooperativity. The Hill constant, which is not necessarily an integer, is obtained experimentally. The binding of O_2 to hemoglobin is said to be cooperative because the binding of O_2 to one subunit increases the O_2 affinity of the other subunits. The fourth oxygen to bind to hemoglobin does so with a 100-fold greater affinity than the first.

5. Hemoglobin has only two stable conformational states, the T state (the conformation of deoxyhemoglobin) and the R state (the conformation of oxyhemoglobin). Oxygen binding causes the T state to shift to the R state, which has greater affinity for oxygen. The T to R shift is triggered by oxygen binding to the heme iron, which pulls the heme iron atom into the heme plane. This movement is transmitted to the F helix through His F8, which ligands the iron atom. Conformational changes in one subunit are transmitted across the α_1–β_2 and α_2–β_1 interfaces. Due to the conformational constraints at these interfaces, the conformational shift of one subunit must be accompanied by the conformational shift of all subunits, thereby increasing the oxygen affinity of the unoccupied subunits.

6. Decreases in pH promote the release of oxygen from hemoglobin. This effect, called the Bohr effect, is driven by dissolved carbon dioxide, which forms bicarbonate ion and a hydrogen ion. The hydrogen ion protonates hemoglobin, thereby stabilizing its T (deoxy) state. In the lungs, the reaction is reversed: Oxygen binding causes a switch to the R (oxy) state. The released hydrogen ions shift the equilibrium between bicarbonate and carbon dioxide, thereby forming carbon dioxide for expulsion from the lungs. Due to the Bohr effect, the low pH in highly active muscles causes the amount of oxygen delivered by hemoglobin to increase by nearly 10%. Carbon dioxide also binds preferentially to the N-terminal amino

groups of T-state hemoglobin as carbamates and hence is released by R-state hemoglobin. This accounts for about half of the carbon dioxide released from the blood in the lungs.

7. Hemoglobin stripped of 2,3-bisphosphoglycerate (BPG) binds oxygen more tightly than hemoglobin in the blood. BPG binds to the T state but not the R state, thereby decreasing hemoglobin's oxygen affinity. This allows nearly 40% of the oxygen to be unloaded in venous blood. Fetal hemoglobin does not bind BPG as tightly and therefore has a higher affinity for oxygen than adult hemoglobin.

8. The analysis of mutant hemoglobins with altered functions has conveyed considerable knowledge about protein structure–function relationships. One variant, hemoglobin S, has a substitution of valine for glutamic acid in the β subunit. In the deoxy state, hemoglobin S aggregates, causing erythrocytes to sickle and block the capillaries. Individuals who are homozygous for the gene specifying hemoglobin S have sickle cell anemia, a debilitating and often fatal disease.

9. Allosteric proteins are oligomers with multiple ligand-binding sites, in which ligand binding at one site alters the protein's binding affinity for a ligand at another site. In the symmetry model of allosterism, the oligomer has only two binding states, T and R, and ligands bind preferentially to one state. In the sequential model, ligand binding progressively induces conformational changes in the subunits of an oligomer.

Myosin and Actin

10. Striated muscle is made of parallel bundles called myofibrils. As seen in the electron microscope, I bands of lesser electron density alternate with A bands of greater density. The repeating unit, the sarcomere, is bounded by Z disks at the centers of adjacent I bands and includes the A band, which is centered on the M disk. Within the sarcomere, thick filaments are linked to thin filaments by cross-bridges. Muscle contraction occurs when the filaments slide past each other, bringing the Z disks closer together.

11. Thick filaments are made of myosin, which consists of six subunits: two heavy chains, two essential light chains (ELC), and two regulatory light chains (RLC). A heavy chain consists of a globular head and a long α-helical tail. Two such tails associate in a 1600-Å-long left-handed coiled coil, yielding a rodlike molecule that has two globular heads. One subunit each of the RLC and ELC bind to each globular head. Thin filaments are composed of actin, tropomyosin, and troponin. Actin, which has two globular domains, polymerizes to form the core of the thin filament. Tropomyosin is a coiled coil that winds in the actin polymer's helical grove so as to contact seven successive actin monomers. Each troponin molecule binds to a tropomyosin molecule.

12. The myosin heads of the thick filaments bind to the thin filaments. When ATP binds to the myosin head, myosin releases the actin. Subsequent ATP hydrolysis cocks the head of myosin and allows it to rebind weakly to actin. The release of inorganic phosphate increases the strength of binding and causes the head of myosin to snap back in the power stroke. This pulls the filaments past each other. Each myosin head acts in this manner to cause muscle contraction.

13. Ca^{2+}, which binds to troponin C, triggers muscle contraction. Nerve impulses induce the release of Ca^{2+}, which binds to troponin and causes a conformational change that exposes additional myosin binding sites on the thin filament. At lower Ca^{2+} concentrations, the myosin head is blocked from binding actin and the muscle is relaxed.

14. Actin also occurs in nonmuscle cells, where its assembly and disassembly motivates such cellular processes as ameboid locomotion, cytokinesis, cytoplasmic streaming, and the extension and retraction of various cellular protuberances.

Antibodies

15. Antibodies (immunoglobulins) are proteins produced by the immune system of higher organisms to protect them against pathogens such as viruses and bacteria. Antibodies are produced by *B* lymphocytes, which recognize foreign macromolecules (antigens). The primary response to an antigen requires several days for *B* cells to generate the required antibodies. If the organism subsequently encounters the same antigen, a secondary response results, in which large amounts of the antibody are produced more rapidly.

16. Antibodies contain at least four subunits, two identical light chains and two identical heavy chains, which are held together in part by interchain disulfide bonds to form a Y-shaped molecule. Of the five classes of antibodies, IgG is the most abundant. The classes are distinguished by the type of heavy chain (α, δ, ε, γ, and μ). There are two types of light chain (κ and λ). Heavy chains each consist of three domains of constant sequence, designated C_H, and one variable domain, designated V_H. The light chains each consist of one constant domain, C_L, and one variable domain, V_L. The V_H and V_L are located at the two ends of the Y-shaped protein.

17. The variable regions contain the antigen-binding sites in which hypervariable sequences determine the exquisite specificity of antibody–antigen interactions.

18. A *B* cell produces one kind of antibody. An antibody-producing *B* cell can be immortalized by fusing it with a myeloma cell. Cloning of the resulting hybridoma cell yields a colony of cells that produce a single type of antibody, called a monoclonal antibody.

19. The binding of an antibody to its antigen is highly specific and has a dissociation constant ranging from 10^{-4} to 10^{-10} M. Because antibodies have two antigen-binding sites, a population of antibodies can form large antigen–antibody aggregates that hasten the removal of the antigen and induce *B* cell proliferation.

20. Antigens stimulate the proliferation of a population of pre-existing *B* cells whose diversity arises from genetic changes that occur in *B* cell development. Sequence variation allows the synthesis of potentially billions of immunoglobulins with different antigen-binding specificities.

21. In autoimmune diseases, the organism loses its self-tolerance and produces antibodies against its own tissues. This process is sometimes triggered by trauma or infection and may result from the resemblance of a self-antigen to some foreign antigen. Autoimmune diseases have variable symptoms ranging from mild to lethal.

Key Equations

$$Y_{O_2} = \frac{pO_2}{K + pO_2}$$

$$Y_{O_2} = \frac{(pO_2)^n}{(p_{50})^n + (pO_2)^n}$$

Guide to Study Exercises (text p. 194)

1. The structures of myoglobin and hemoglobin differ in two ways: (1) Myoglobin contains a single globin polypeptide, whereas hemoglobin is an $\alpha_2\beta_2$ tetramer (a dimer of $\alpha\beta$ dimers) with four globin chains. (2) Myoglobin does not undergo a significant conformational change on binding O_2, whereas hemoglobin undergoes a dramatic change in conformation when O_2 is bound to at least one subunit in each $\alpha\beta$ dimer.

   Myoglobin facilitates O_2 transport in muscle under conditions of high exertion, whereas hemoglobin functions to transport O_2 in blood. These different functions of myoglobin and hemoglobin are reflected in their structural differences. Myoglobin, which has a relatively high affinity for O_2 (p_{50} = 2.8 torr), has a simple O_2-binding behavior so that its binding curve is hyperbolic. Hemoglobin has a more modest affinity for O_2 (p_{50} = 26 torr) and its four subunits bind O_2 cooperatively so that its binding curve is sigmoidal. (Sections 7-1 and 7-2A and B)

2. The keys to the hemoglobin–myoglobin O_2-delivery system are the high affinity of myoglobin for O_2 and the cooperative binding of O_2 to hemoglobin. The p_{50} of hemoglobin is 26 torr, so that, in the lungs, hemoglobin becomes almost fully saturated (arterial pO_2 is ~100 torr). At low pO_2, the affinity of hemoglobin for O_2 falls off rapidly because of its sigmoidal (cooperative) binding behavior, so at venous pO_2 (~30 torr), hemoglobin is only about half-saturated with O_2. The O_2 released from hemoglobin is efficiently bound by myoglobin. Even in the venous circulation, the pO_2 is still much greater than the p_{50} of myoglobin (2.8 torr), so the myoglobin is essentially saturated with O_2. This, in effect, increases the solubility of O_2 in the tissues and thereby increases the rate at which O_2 can diffuse from the capillaries to the tissues, where it is consumed. (Section 7-2B)

3. Hemoglobin exhibits positive cooperativity in O_2 binding; that is, O_2 binding increases the affinity of the hemoglobin for additional O_2 molecules. This is accomplished through conformational changes in the globin subunits. The four globin chains of deoxyhemoglobin are in the T state, in which the heme is domed and the Fe is out of the plane of the heme toward His F8, to which it is liganded. O_2 binding to Fe causes the Fe to move into the plane of the heme and the heme to become more planar, thereby pulling His F8, and the F helix to which it is linked, toward it. The resulting change in tertiary structure is communicated to the other subunits primarily at the α_1–β_2 and α_2–β_1 interfaces. For example, when O_2 binds to a β subunit, its His FG4 (which is located in a loop at the end of the F helix), which contacts residue Thr C6 in the adjacent α chain, moves to contact Thr C3. This forces the α chain to assume the R conformation. This conformational shift also disrupts ion pairs that stabilize the T state. When at least one O_2 has bound to each $\alpha\beta$ dimer of hemoglobin, the entire

tetramer snaps into the R state. The affinity of hemoglobin for O_2 increases because all the subunits are now in the R state, which is the O_2-binding conformation. (Section 7-2C)

4. When hemoglobin binds O_2, the T→R conformational changes in the globin subunits disrupt networks of ion pairs that involve the C-terminal residues of each subunit. Several groups, including the N-terminal amino group of the α chains and the C-terminal His of the β chains, thereby become more acidic (deprotonated). For this reason, increasing the pH (decreasing $[H^+]$) promotes the T→R transition in hemoglobin (which favors O_2 binding). Decreasing the pH (increasing $[H^+]$) promotes the R→T transition (which favors the dissociation of O_2). This behavior (called the Bohr effect) is important for delivering O_2 to the tissues. Respiring tissues produce CO_2, which is converted to bicarbonate + H^+. This H^+ induces the hemoglobin to unload its bound O_2 to the tissues, where it is needed. Furthermore, when hemoglobin takes up H^+, more bicarbonate is formed, thereby drawing CO_2 from the tissues. The opposite reactions occur in the lungs: O_2 binding to hemoglobin causes the T→R transition, releasing the Bohr protons so that they can recombine with bicarbonate to drive off CO_2. (Section 7-2C)

5. BPG binds preferentially to deoxyhemoglobin, where it occupies the central cavity between the globin subunits. In the R (oxy) state, the central cavity is too narrow to accommodate BPG, and the N-terminal amino groups of the β subunits to which BPG binds have moved apart so that BPG cannot simultaneously bind both of them as it does in T-state hemoglobin. Hence BPG binding stabilizes the T (deoxy) state relative to the R state. BPG binding thereby induces the unloading of O_2 from hemoglobin. (Section 7-2C)

6. In the symmetry model of allosterism, all the subunits of the oligomer change conformation in a concerted manner in response to ligand binding. Only two conformational states for the oligomer are permitted, and ligand binding necessarily increases the affinity of the other subunits for the ligand. In the sequential model, ligand binding induces a conformational change in the subunit to which it binds, which affects neighboring subunits more than more distant subunits. As more ligand-binding sites in the oligomer are occupied, more conformational changes occur, until the entire oligomer has shifted conformation. In the sequential model, ligand binding to one subunit can either increase or decrease the affinity of the other subunits for the ligand. (Section 7-2E)

7. See Figure 7-23.

8. According to the sliding filament model of muscle contraction, each myosin molecule must repeatedly detach and reattach itself to new sites on the actin thin filament. This activity causes the thin and thick filaments to slide past one another, thereby causing the muscle to contract. The molecular events occur in a cycle: ATP binding to a myosin head causes it to release its bound actin. Subsequent ATP hydrolysis provides the free energy for myosin to assume a "cocked" conformation that allows it to bind weakly to an actin monomer further along the thin filament toward the Z disk. Myosin then releases P_i, which increase its affinity for actin. The ensuing conformational change causes the myosin head to return to its original position, thereby moving the thin filament toward the M disk. ADP release from myosin completes the reaction cycle. (Section 7-3C)

9. See Figure 7-34.

Questions

Myoglobin and Hemoglobin

1. Match the descriptions on the left with the terms on the right.

______ A component of cytochromes	A. methemoglobin
______ Binds O_2	B. myoglobin
______ Contains iron in the Fe(III) state	C. hemoglobin
______ Found in muscle only	D. hemoglobin S
______ Forms filaments in the deoxy state	E. heme

2. The heme moiety by itself can bind oxygen. What physiological function does the globin serve?

3. How do tissues with high metabolic activity facilitate oxygen delivery?

4. How would a lower p_{50} affect hemoglobin's oxygen acquisition in the lungs and oxygen delivery to the peripheral tissues?

5. Describe, on the molecular level, the role of myoglobin in O_2 transport in rapidly respiring muscle tissue.

6. What function(s) does carbamate formation in hemoglobin serve?

7. Which of the following modulators of O_2 binding to hemoglobin counteract each other? CO_2, H^+, BPG.

8. Explain why individuals with severe carbon monoxide poisoning are often given transfusions instead of oxygen-rich gas.

9. Match each of the structural elements of myoglobin or hemoglobin with its function below.

A. His F8	F. α_1–β_1 interface/α_2–β_2 interface
B. His E7	G. oxymyoglobin/deoxymyoglobin
C. E and F helices	H. oxyhemoglobin/deoxyhemoglobin
D. O_2–Fe(II)	I. α_1–β_2 interface/α_2–β_1 interface
E. Val E11	J. C-terminal salt bridges

______ Forms a coordination bond with Fe(II)
______ Involved in the binding of heme
______ Partially occludes the O_2-binding site
______ Conformations are nearly superimposable
______ Forms a hydrogen bond with O_2
______ Is associated with a rotational shift of the hemoglobin $\alpha_1\beta_1$ dimer with respect to the $\alpha_2\beta_2$ dimer
______ Disrupts the N- and C-terminal salt bridges in hemoglobin
______ Stabilizes the T state
______ Involved in the interactions of hemoglobin subunits

10. Oxygen binding to hemoglobin decreases the pK of the imidazole group of His 146β from 8.0 to 7.1. How does this contribute to the Bohr effect?

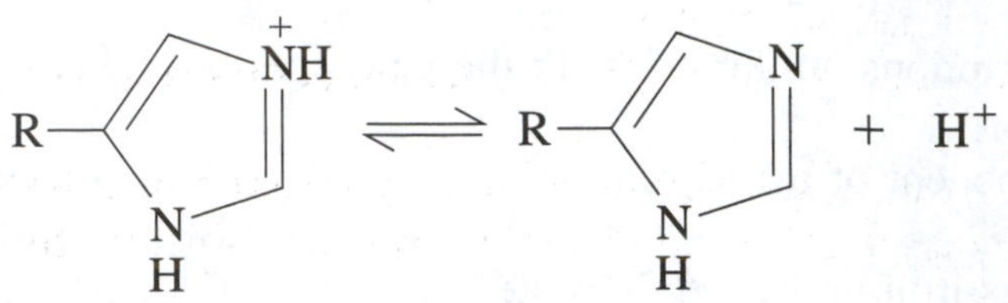

$$\text{R–(imidazole ring with } \overset{+}{\text{N}}\text{H and NH)} \rightleftharpoons \text{R–(imidazole ring with N and NH)} + \text{H}^+$$

11. Describe how BPG decreases the O_2-binding affinity of hemoglobin in terms of the $T \rightleftharpoons R$ equilibrium.

12. What physiological condition leads to hemoglobin S fiber formation in the capillaries?

13. Some sickle-cell individuals have significant levels of fetal hemoglobin in their erythrocytes. Why is this an advantage?

14. What kind of allosteric effect is inconsistent with the symmetry model?

15. How many conformational states are possible for a trimeric binding protein, whose three subunits each contain a ligand-binding site, when binding follows the (a) symmetry model or (b) the sequential model of allosterism?

Myosin and Actin

16. Describe how ATP hydrolysis is involved in muscle contraction.

17. How does calcium regulate muscle contraction?

Antibodies

18. In a typical protocol for preparing antibodies for laboratory use, an animal is injected with an antigen several times over a period of weeks or months. Why are multiple injections useful?

19. Indicate which class of immunoglobulin (IgA, IgD, IgE, IgG, or IgM) best corresponds to each characteristic listed below (more than one may be implicated by each description).
    (a) First to be secreted in response to an antigen
    (b) Implicated in allergic reactions
    (c) Occurs in the intestinal tract
    (d) Contains J chains
    (e) The most abundant antibody

20. The following questions refer to the production of monoclonal antibodies.
    (a) Why are myeloma cells used?
    (b) Do all cells growing in the selective medium produce antibodies to antigen X?
    (c) Why do only hybrid cells grow in the selective medium?

21. You wish to isolate a large amount of Fab fragments in order to examine their binding to protein X by X-ray crystallography. At first, you inject a large rabbit with protein X to obtain antibody. On further reflection, you decide to inject the antigen into a mouse in order to produce monoclonal antibodies.
    (a) How can you purify protein X–specific antibodies from rabbit serum or the hybridoma medium?
    (b) How do the rabbit and mouse antibody preparations differ?
    (c) Would the rabbit or mouse antibodies be more suitable for X-ray crystallography?

22. Why are the hypervariable sequences of immunoglobulins located in loops?

Answers to Questions

1.
   __E__ A component of cytochromes
   __B, C, D__ Binds O_2
   __A__ Contains iron in the Fe(III) state
   __B__ Found in muscle only
   __D__ Forms filaments in the deoxy state

2. The globin prevents the oxidation of its bound heme to the Fe(III) state, and in the case of hemoglobin, permits cooperative O_2 binding, which is responsible for the efficient transport of O_2 from the lungs to the tissues.

3. High metabolic activity generates CO_2, which reacts with water to form H_2CO_3, which in turn ionizes to yield $HCO_3^- + H^+$. The protons preferentially bind to T-state hemoglobin (the Bohr effect) and thereby cause O_2 to be released.

4. A lower p_{50}, or higher affinity of hemoglobin for O_2, would have almost no effect on O_2 uptake in the lungs, since hemoglobin is nearly saturated at arterial pO_2. However, the lower p_{50} would result in less O_2 released at the peripheral tissues, where the pO_2 is ~30 torr. The p_{50} of normal hemoglobin (26 torr) is an evolutionary compromise that allows hemoglobin to become saturated with O_2 in the lungs but to be deoxygenated in the oxygen-poor peripheral tissues.

5. The O_2 that is released in the capillaries diffuses to the mitochondria, where it is reduced to water. However, the solubility of O_2 in aqueous solution is too low to support its required rate of diffusion in rapidly respiring muscle. The presence of myoglobin, in effect, increases the solubility of O_2 in muscle tissue. Thus, the O_2 released by hemoglobin is passed from myoglobin molecule to myoglobin molecule in a kind of molecular bucket brigade until it is taken up by the mitochondria. In this way, myoglobin increases the rate that O_2 can diffuse through the tissues.

6. CO_2 reacts preferentially with amino groups in deoxyhemoglobin to form carbamates. This helps transport CO_2 and facilitates the release of O_2 by stabilizing the deoxy state.

7. None; CO_2, H^+, and BPG all bind preferentially to deoxyhemoglobin to reduce the affinity of hemoglobin for O_2.

8. CO binds to hemoglobin with so much higher affinity than O_2 that CO cannot be displaced by high concentrations of oxygen. For practical purposes, CO binding to hemoglobin is therefore irreversible (at least in the short term). Therefore, a fresh blood transfusion is required to counteract the effects of this poison.

9. __A__ Forms a coordination bond with Fe(II)
__A, C__ Involved in the binding of heme
__E__ Partially occludes the O_2-binding site
__G__ Conformations are nearly superimposable
__B__ Forms a hydrogen bond with O_2
__I__ Is associated with a rotational shift of the hemoglobin $\alpha_1\beta_1$ dimer with respect to the $\alpha_2\beta_2$ dimer.
__D__ Disrupts the N- and C-terminal salt bridges in hemoglobin
__J__ Stabilizes the T state
__F, I__ Involved in the interactions of hemoglobin subunits

10. The decrease in pK promotes deprotonation of the imidazole group at pH 7.4. The release of H^+ contributes to the Bohr effect.

11. BPG binds preferentially to the T state and stabilizes it through the formation of salt bridges between BPG and the β subunits. This makes the shift to the R state less energetically favorable, thereby decreasing hemoglobin's ability to bind O_2.

12. The low pO_2 of the capillaries, which leads to an increase in the concentration of deoxyhemoglobin, promotes polymerization of hemoglobin S.

13. Fetal hemoglobin (which contains γ globin chains rather than β chains) dilutes the hemoglobin S in erythrocytes, so that the deoxyhemoglobin S in the venous blood is less likely to achieve the critical concentration for fiber formation.

14. The symmetry model cannot account for negative cooperativity, in which the binding of a ligand to one subunit decreases the binding affinity of the other subunits.

15. (a) Only two conformational states are possible, either R or T, since all subunits change conformation simultaneously.
(b) Four conformations are possible, corresponding to a trimer in which 0, 1, 2, or 3 subunits have bound ligand.

16. Myosin moves along the actin filament via repeated cycles of conformational changes in the head region of the myosin molecule. This cycle of conformational changes is unidirectional because it is coupled to the irreversible hydrolysis of ATP. When the myosin head binds ATP, it releases the actin filament to which it is bound. ATP hydrolysis follows, resulting in a change in the conformation of the myosin head to the "cocked" or high-energy state. The myosin head then binds weakly to the actin filament at a new position. The strength of this binding interaction increases upon the release of P_i. The myosin head then undergoes a second major conformational change, producing the power stroke that causes the translocation of the actin filament relative to the myosin filament. Upon completion of the power stroke, the myosin head releases ADP but remains bound to the actin filament. It can

then bind another ATP molecule to begin the cycle anew. The thick filament's multiple myosin heads, each undergoing multiple ATP-driven reaction cycles, cause the thick filament to "walk" along the thin filament. As the thick and thin filaments slide past each other, the sarcomere decreases in length and the muscle thereby contracts.

17. The thin filaments of striated muscle contain actin in complex with tropomyosin and troponin. At resting Ca^{2+} concentrations, the muscle is relaxed because tropomyosin blocks the myosin binding sites on the actin filament. When the intracellular Ca^{2+} concentration increases in response to a nerve impulse, troponin C (a subunit of troponin) binds Ca^{2+}, causing a conformational change in troponin. This results in the movement of tropomyosin, which uncovers the myosin binding sites, and thus permits the myosin heads to interact with the thin filament.

18. A secondary immune response is greater than the first (see Figure 7-32), so more antibody can be recovered.

19. (a) IgM (b) IgE (c) IgA (d) IgA, IgM (e) IgG

20. (a) Antibody-producing lymphocytes have a limited proliferative capacity and are therefore unsuitable for growing in large numbers in culture to produce large amounts of antibody. Myeloma cells, like all cancer cells, have an unlimited proliferative capacity and impart their immortal phenotype to the hybrid cells.
    (b) All the lymphocytes harvested from the immunized animal can potentially fuse with the myeloma cells. The resulting hybridomas will therefore secrete a variety of different antibodies, only a small fraction of which are specific for protein X.
    (c) Unfused lymphocytes have no metabolic deficiency but grow for only a limited time, and then die out. In the selective medium, unfused myeloma cells cannot synthesize purines, which are necessary for DNA replication and cell division. Only the fused cells can both synthesize purines and proliferate without limit.

21. (a) Affinity chromatography using immobilized protein X would be the best procedure.
    (b) It is likely that protein X contains several antigenic features that are recognized by *B* cells and elicit antibody production. The anti-protein X antibodies isolated from the rabbit are therefore a mixture of different IgG molecules that recognize different antigenic features on protein X. Even antibodies that recognize the same portion of the antigen may have different amino acid sequences and different antigen-binding affinities. In contrast, each mouse monoclonal antibody preparation is homogeneous. Because monoclonal antibodies are all the products of identical *B* cells, they all have the same unique sequence and antigen-binding specificity.
    (c) One of the mouse monoclonal antibodies would be most suitable, since it would yield identical Fab fragments that would be more likely to form a regular crystal than the Fab fragments isolated from a heterogeneous population of rabbit antibodies.

22. The loops can accommodate a wide variety of amino acid sequences because they are on the surface of the protein. Such sequence variation in the β sheet core of the protein would likely disrupt its structure.

Chapter 8 Carbohydrates

This chapter is concerned with the structures and properties of carbohydrates. These molecules contain just three elements, namely carbon, hydrogen, and oxygen. Carbohydrates are not only important metabolic energy sources, as detailed in Chapters 14 and 15, but they also have key functions in molecular and cellular recognition events. You will first learn the structures and chemical characteristics of monosaccharides and some of their derivatives, and then of oligosaccharides and polysaccharides. This is followed by a presentation of the composition, structure, and function of molecules in which carbohydrates are covalently linked to polypeptides, including proteoglycans, bacterial cell wall components, and glycoproteins. It is important to realize that a large proportion of proteins contain covalently attached carbohydrates and that the structures of these carbohydrate chains can vary enormously.

Essential Concepts

Monosaccharides

1. Monosaccharides can be defined as aldehyde or ketone derivatives of straight-chain polyhydroxy alcohols containing a minimum of three carbon atoms. Sugars that have aldehyde groups are called aldoses, whereas those with ketone moieties are termed ketoses. Depending on the number of carbon atoms, monosaccharides are referred to as trioses, tetroses, pentoses, hexoses, etc.

2. D-Glucose has four chiral centers and is therefore one of 16 possible aldohexose stereoisomers. D sugars have the same absolute configuration at the asymmetric center most distant from the carbonyl group as does D-glyceraldehyde. Epimers are sugars in which the configuration around one carbon atom differs. Ketoses, which have a ketone function at C2, have one less chiral center than aldoses with the same number of carbons. Therefore, a ketohexose has only 8 possible stereoisomers.

3. Sugars can be represented in their cyclic hemiacetal and hemiketal forms as planar Haworth projections. Sugars that form six-membered rings are known as pyranoses, whereas those with five-membered rings are known as furanoses. A cyclic monosaccharide exists as either an α or a β anomer. Anomers freely interconvert in aqueous solution via the linear (open chain) form.

4. Five- and six-membered sugar rings are most abundant because of their stability. The tetrahedral bonding angles of carbon prevent the rings from being truly planar. The pyranose ring prefers the chair conformation and exists predominantly in the form that minimizes steric interactions among bulky ring substituents (i.e., bulky groups tend to occupy equatorial rather than axial positions).

5. Sugars can undergo reactions characteristic of aldehydes and ketones:
   (a) Oxidation of the aldehyde group of an aldose yields an aldonic acid. Thus, D-glucose oxidation results in D-gluconic acid.

(b) Oxidation of the primary hydroxyl group produces a uronic acid, such as D-glucuronic acid from D-glucose.

(c) Reduction of aldoses and ketoses yields polyhydroxy alcohols called alditols. Thus, D-glucose reduction gives glucitol (also known as sorbitol).

6. Monosaccharides in which an OH group is replaced by an H are called deoxy sugars. The most important of these is β-D-2-deoxyribose, a component of DNA.

7. In amino sugars, an OH group is replaced by an amino group that is usually acetylated. A common amino sugar is *N*-acetylglucosamine. *N*-Acetylneuraminic acid, an important constituent of glycoproteins (see below), is composed of an acetylated amino sugar, *N*-acetylmannosamine, covalently linked to pyruvic acid.

8. The anomeric carbon of a sugar can form a covalent bond with an alcohol to form an α or β glycoside. The bond is called a glycosidic bond. An *N*-glycosidic bond links an anomeric carbon and a nitrogen atom, as in the covalent bond between ribose and a purine or pyrimidine.

Polysaccharides

9. Polysaccharides are made up of monosaccharides covalently linked by glycosidic bonds. A homopolysaccharide contains one type of monosaccharide, whereas a heteropolysaccharide can contain diverse monosaccharides. Polysaccharides may be linear or branched, because glycosidic bonds can form between an anomeric carbon and any of the hydroxyl groups of another monosaccharide. Naturally occurring polysaccharides incorporate only a few types of monosaccharides and glycosidic linkages.

10. Disaccharides consist of two sugars linked by a glycosidic bond. One example is lactose, in which C1 of galactose is linked to C4 of glucose by an α-glycosidic bond. Lactose is a reducing sugar because the free anomeric carbon on the glucose residue can reduce a mild oxidizing agent. Another disaccharide is sucrose, common table sugar, in which the C1 anomeric carbons of glucose and fructose are joined by an α-glycosidic bond. Sucrose is therefore a nonreducing sugar.

11. Cellulose, the most abundant polysaccharide, is a large, linear polymer in which glucose units are linked by β(1→4) glycosidic bonds. Cellulose forms a highly hydrogen bonded structure of enormous strength that contributes to the rigidity of plant cell walls. Cows and other herbivores can utilize cellulose as a nutrient because they harbor microbes that produce cellulases which cleave the glycosidic bonds. Another widely distributed polysaccharide is chitin, which comprises the exoskeletons of many invertebrates. It is a homopolymer of *N*-acetylglucosamine residues linked in β(1→4) fashion.

12. Starch is the principal food reserve in plants. It has two structural forms: α-amylose, a linear polymer of α(1→4) linked glucose units, and amylopectin, an α(1→4) linked glucose polymer bearing periodic branches linked by α(1→6) bonds. Digestion of starch by animals begins with the action of salivary amylase, an enzyme that cleaves α(1→4) glycosidic bonds,

and is continued by pancreatic amylase, which produces $\alpha(1\rightarrow4)$ linked di- and trisaccharides, as well as oligosaccharides containing the $\alpha(1\rightarrow6)$ bonds. These latter polymers, known as dextrins, are further degraded by a debranching enzyme that can cleave $\alpha(1\rightarrow6)$ links. These oligosaccharides are eventually converted to glucose, which can be absorbed by the intestine.

13. Glycogen is a polysaccharide that is synthesized and stored by animals, primarily in skeletal muscle and liver. The structure of glycogen resembles that of amylopectin but is more highly branched. When needed for metabolic energy, glycogen is broken down through the combined action of glycogen phosphorylase, which cleaves $\alpha(1\rightarrow4)$ bonds, and glycogen debranching enzyme.

14. Glycosaminoglycans are major constituents of the extracellular matrix. Most of these rigid, linear polysaccharides are composed of alternating uronic acid and hexosamine residues. For example, hyaluronic acid is composed of D-glucuronate linked $\beta(1\rightarrow3)$ to *N*-acetylglucosamine which in turn is linked $\beta(1\rightarrow4)$ to the next glucuronate residue. Because of its polyanionic nature, hyaluronic acid forms viscoelastic solutions, a property that makes it an effective biological shock absorber and lubricant.

15. Other types of glycosaminoglycans, all of which are composed of sulfated disaccharide units, include chondroitin sulfates, dermatan sulfate, keratin sulfate, and heparin. The last named substance is found in mast cells and inhibits blood clotting.

Glycoproteins

16. A large proportion of all proteins have covalently bound carbohydrates and are therefore glycoproteins. The polypeptide chains of glycoproteins are encoded by nucleic acids, whereas the attached oligosaccharide chains are products of enzymatic reactions. This is the source of microheterogeneity, the variability in composition of the carbohydrate component in a population of glycoprotein molecules that all have the same polypeptide chain.

17. Proteoglycans are found mainly in the extracellular matrix and are combinations of proteins and glycosaminoglycans that associate by both covalent and noncovalent bonds. These molecules have a bottlebrush-like structure (e.g., Figure 8-13) in which up to 100 core proteins with attached glycosaminoglycans and both *N*-linked and *O*-linked oligosaccharides are linked to hyaluronate. This assembly of protein and carbohydrate is a huge, space-filling macromolecule. Proteoglycans are highly hydrated, so that, in combination with collagen, they account for the high resilience of cartilage.

18. Bacteria possess rigid cell walls that are responsible in part for their virulence. In gram-positive bacteria, the cell wall consists of polysaccharide and polypeptide chains that are covalently attached to form a baglike structure called peptidoglycan that completely envelops the cell. Gram-negative bacteria have a relatively thin peptidoglycan cell wall surrounded by a complex outer membrane.

19. The polysaccharide of some bacterial cell walls consists of alternating residues of β(1→4)-linked *N*-acetylmuramic acid and *N*-acetylglucosamine in which the *N*-acetylmuramic acid residues are linked via an amide bond to a tetrapeptide containing D-amino acids. A continuous meshlike framework is formed by cross-linking adjacent peptidoglycan chains through their tetrapeptide side chains (Figure 8-15). The enzyme lysozyme can degrade peptidoglycan by cleaving the glycosidic bond between *N*-acetylmuramic acid and *N*-acetylglucosamine. The antibiotic action of penicillin rests on its ability to inhibit the formation of cross-links in peptidoglycan.

20. Glycoproteins include nearly all membrane-bound and secreted eukaryotic proteins. The oligosaccharide chains are attached to the proteins by either *N*-glycosidic or *O*-glycosidic bonds. In an *N*-glycosidic bond, the amide group of an asparagine in the sequence Asn-X-Ser/Thr is linked to *N*-acetylglucosamine. *N*-Glycosylation occurs in stages:
    (a) An oligosaccharide rich in mannose and containing glucose and *N*-acetylglucosamine is attached cotranslationally, that is, while the polypeptide is being synthesized on ribosomes bound to the endoplasmic reticulum.
    (b) The oligosaccharide undergoes trimming, the enzymatic removal of some sugars, as the glycoprotein moves from the endoplasmic reticulum to the Golgi apparatus.
    (c) Further processing occurs in the Golgi apparatus, where monosaccharides such as *N*-acetylglucosamine, galactose, L-fucose, and *N*-acetylneuraminic acid are enzymatically added to the trimmed chain by glycosyltransferases. *N*-Linked glycoproteins exhibit great diversity in their oligosaccharide chains due to differences in the extent of processing.

21. *O*-Glycosidically linked oligosaccharide chains are covalently linked to a Ser or Thr side chain in a protein and vary considerably in structure. *O*-Linked oligosaccharides are added to completed polypeptides in the Golgi apparatus and are built up by stepwise addition of monosaccharides in reactions catalyzed by glycosyltransferases.

22. The number and structure of *N*- and *O*-linked oligosaccharides attached to a given polypeptide chain can vary, giving rise to glycoprotein variants called glycoforms.

23. Particular functions can sometimes be attributed to the presence of oligosaccharides. Oligosaccharides attached to proteins may modulate the conformational freedom of the polypeptide. The hydrophilic oligosaccharide chains may take up considerable volume and thereby tend to protect the protein from enzymatic attack or modify its activity.

24. The enormous number of oligosaccharide structures suggests that they contain biological information and are important in molecular recognition. All cells carry a coat of glycoconjugates (a mixture of glycoproteins and glycolipids). Proteins called lectins specifically recognize and bind to individual monosaccharides or small oligosaccharides in discrete glycosidic linkages at the cell surface. For example, the leukocyte selectins bind to cell-surface carbohydrates on endothelial cells, an interaction that helps direct leukocytes to a site of blood vessel injury.

25. Different oligosaccharide components of certain glycolipids distinguish the ABO blood group antigens.

26. Oligosaccharide chains mediate a variety of biological functions that depend on molecular recognition between proteins and carbohydrates. Among these are delivery of proteins to appropriate destinations within cells and the regulation of cell growth.

Guide to Study Exercises (text p. 218)

1. See Figure 8-3.

2. See Figure 8-3.

3. Cellulose, chitin, starch, and glycogen are similar in that all are polymers containing a single type of saccharide residue (either glucose or, in the case of chitin, a glucose derivative). Only glycogen and the amylopectin component of starch are branched; cellulose, chitin, and the α-amylose component of starch are linear polymers. The polymers differ in the glycosidic linkage between residues: A $\beta(1\rightarrow4)$ glycosidic bond links the glucose residues in cellulose and the *N*-acetylglucosamine residues in chitin, whereas an $\alpha(1\rightarrow4)$ linkage is the primary linkage in starch and glycogen, which also have $\alpha(1\rightarrow6)$ linkages at their branch points. Both cellulose and chitin are linear chains that assemble in stacked sheets. Starch and glycogen have an irregular helical structure such that they form globules in cells. Finally, the functions of the polymers differ: Cellulose provides structural support for plants, and chitin is a major component of the exoskeletons of arthropods and the cell walls of fungi and algae. In contrast, starch and glycogen are storage forms for the metabolic fuel glucose. Furthermore, the structural polymers, once formed, tend to be relatively permanent, whereas the storage polymers are readily synthesized and degraded. (Sections 8-2B and C)

4. Glycosaminoglycans are linear polymers of alternating uronic acid and hexosamine residues and are often sulfated. Their multiple negative charges cause them to assume an extended conformation and to be heavily hydrated. The large volume of bound water makes the glycosaminoglycans viscous and elastic. Furthermore, because of their long length, they exhibit shear-dependent viscosity, a property that suits their function as lubricants.

   Proteoglycans have similar properties due to their high content of glycosaminoglycans. These enormous molecules serve similar functions, for example, conferring resilience on cartilage. (Sections 8-2D and 8-3A)

5. *N*- and *O*-linked oligosaccharides differ in their structure and linkage to glycoproteins. *N*-Linked oligosaccharides consist of a $(\text{mannose})_3(\text{GlcNAc})_2$ core whose terminal GlcNAc is β-linked to the amide nitrogen of an Asn side chain. A variety of other sugars may be added to the core oligosaccharide to produce a branched structure. *O*-Linked oligosaccharides have a more variable size and structure but most commonly have GalNAc linked to the OH group of a Ser or Thr side chain. (Section 8-3C)

Questions

Monosaccharides

1. Indicate which of the following is an aldose, a ketose, a pentose, a hexose, a uronic acid, an alditol, a deoxy sugar, or a reducing sugar.

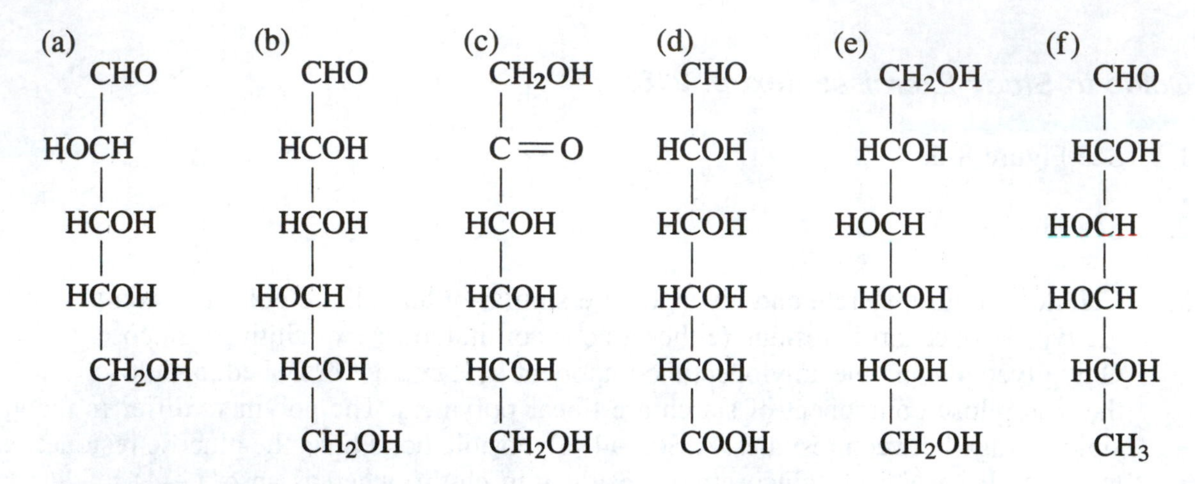

2. Draw the following monosaccharides as Haworth projections:
   (a) α anomer of D-ribose
   (b) β anomer of D-glucose
   (c) β anomer of D-fructose
   (d) methyl-β-D-galactose.

   Which of these compounds contains a glycosidic bond?

3. An equilibrium mixture of D-glucose contains approximately 63% β-D-glucopyranose and 36% α-D-glucopyranose. There are trace amounts of three other forms. What are they?

4. Although β-D-glucopyranose is the predominant form of glucose in solution, crystalline glucose consists almost exclusively of α-D-glucopyranose. What accounts for this difference?

5. Draw the most stable chair conformation of α-D-galactose.

Polysaccharides

6. Match each term at the top with its definition below.

   A. Alditol
   B. Epimer
   C. Glycan
   D. Anomer
   E. Glycoside

   ______ Differs in configuration at the anomeric carbon
   ______ Polyhydroxy alcohol

______ Product of condensation of anomeric carbon with an alcohol
______ Differs in configuration at one carbon atom
______ Polymer of monosaccharides

7. Which of the following di- and trisaccharides contains fructose, contains an α anomeric bond, or is a reducing sugar?

(a) sucrose

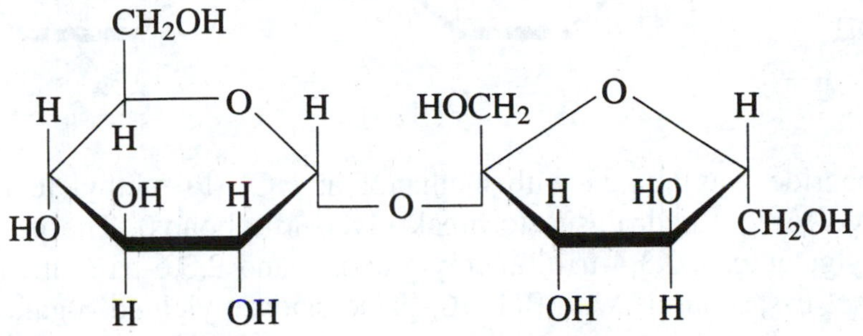

(b) cellobiose

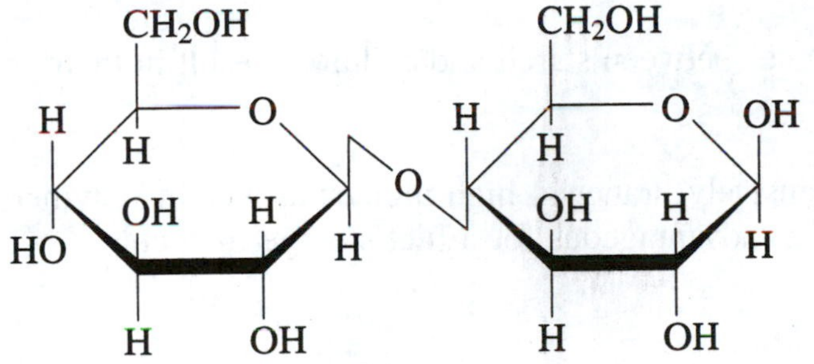

(c) lactose

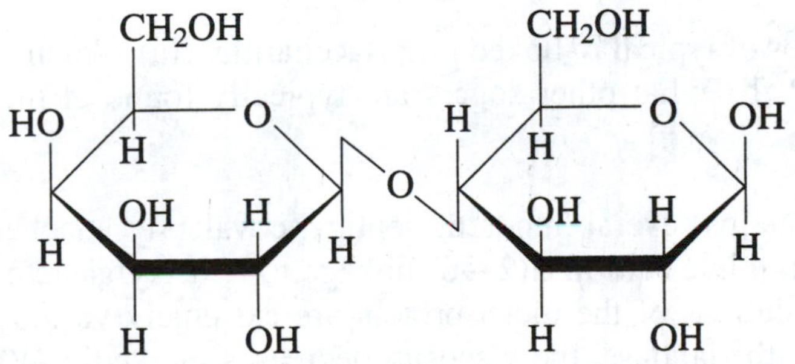

(d) melibiose

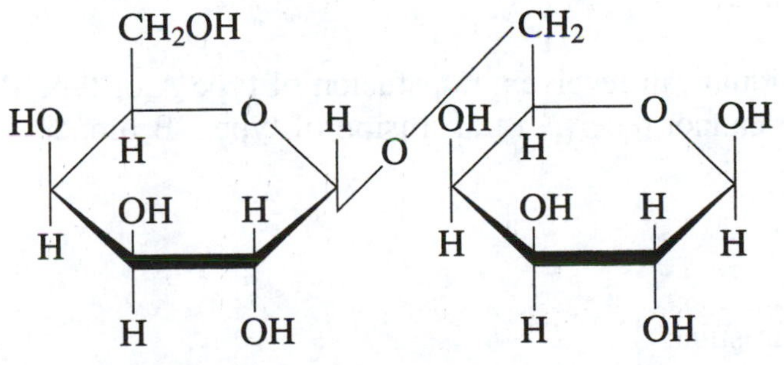

(e) raffinose

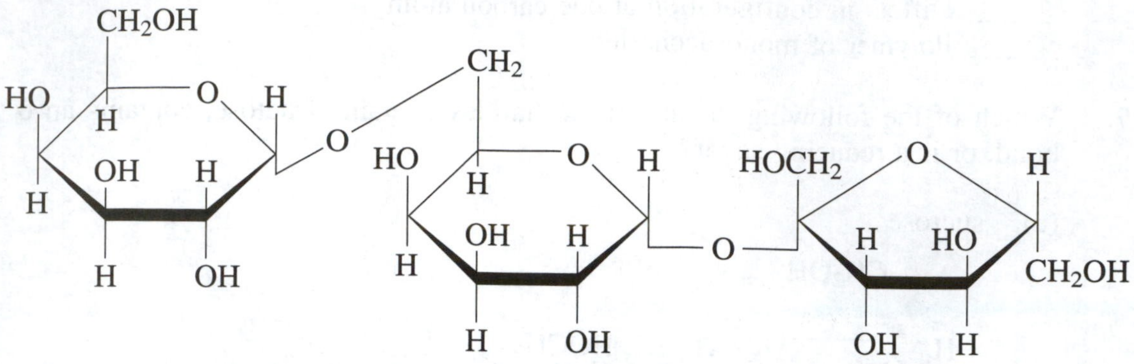

8. An unknown trisaccharide was treated with methanol in HCl (to methylate its free OH groups) then subjected to acid hydrolysis (to break glycosidic bonds). The products were 2,3,4,6-tetra-*O*-methylgalactose, 2,3,4-tri-*O*-methylglucose, and 2,3,6-tri-*O*-methylglucose. Treatment of the intact trisaccharide with β(1→6)-galactosidase yielded D-galactose and a disaccharide. Treatment of this disaccharide with α(1→4)-glucosidase yielded D-glucose. Draw the structure of this trisaccharide and give its systematic name.

9. How do the chemical differences between starch and cellulose result in their very different polymeric structures?

10. Glycogen and starch are extensively branched high-molecular-weight polymers. Give two reasons why such a structure is advantageous for a fuel-storage molecule.

Glycoproteins

11. How could lectins be used to purify polysaccharides and glycoproteins?

12. Figure 8-16 presents structures of typical *N*-linked oligosaccharides. (a) Which sugars make up the core oligosaccharide? (b) What other sugars are typically found at the termini of branched oligosaccharides?

13. A major protein of saliva contains several hundred identical covalently attached disaccharides containing *N*-acetylneuraminic acid in α(2→6) linkage to *N*-acetylgalactosamine that is linked to a Ser residue. Solutions of the intact protein are extremely viscous. However, when the protein is treated with sialidase, the viscosity decreases markedly. What features of the glycoprotein structure give rise to the high viscosity and why does sialidase treatment bring about the observed change?

14. Explain why a type AB individual can receive a transfusion of type A or type B blood, but a type A or type B individual cannot receive a transfusion of type AB blood.

Answers to Questions

1. (a) aldose, pentose, reducing sugar
   (b) aldose, hexose, reducing sugar

(c) ketose, hexose, reducing sugar
(d) aldose, hexose, uronic acid, reducing sugar
(e) hexose, alditol
(f) aldose, hexose, deoxy sugar, reducing sugar

2. (a)

HOH_2C O H
H H H OH
OH OH

(b)

CH_2OH
H O OH
H
OH H
HO H
H OH

(c)

HOH_2C O OH
H HO
H CH_2OH
OH H

(d)

CH_2OH
HO O OCH_3
H
OH H
H H
H OH

Only methyl-β-D-galactose (d) contains a glycosidic bond.

3. The three other forms are the open chain form of glucose, α-D-glucofuranose and β-D-glucofuranose.

4. The α anomer of glucose is less soluble than the β anomer and therefore comes out of solution more readily. As it crystallizes, the β-glucose remaining in solution interconverts with α-glucose in order to maintain the 36%α–63%β equilibrium ratio. Thus, the α anomer is continually generated and deposited in the crystal.

5.

OH
CH_2OH O
H H
H H
HO OH
H OH

The most stable conformation has its bulky CH_2OH group and two OH groups in equatorial positions, which minimizes their steric interference.

6. __D__ Differs in configuration at the anomeric carbon
__A__ Polyhydroxy alcohol
__E__ Product of condensation of anomeric carbon with an alcohol
__B__ Differs in configuration at one carbon atom
__C__ Polymer of monosaccharides

7. (a) Contains fructose, contains an α anomeric bond
(b) Is a reducing sugar
(c) Is a reducing sugar
(d) Contains an α anomeric bond, is a reducing sugar
(e) Contains fructose, contains an α anomeric bond

8.

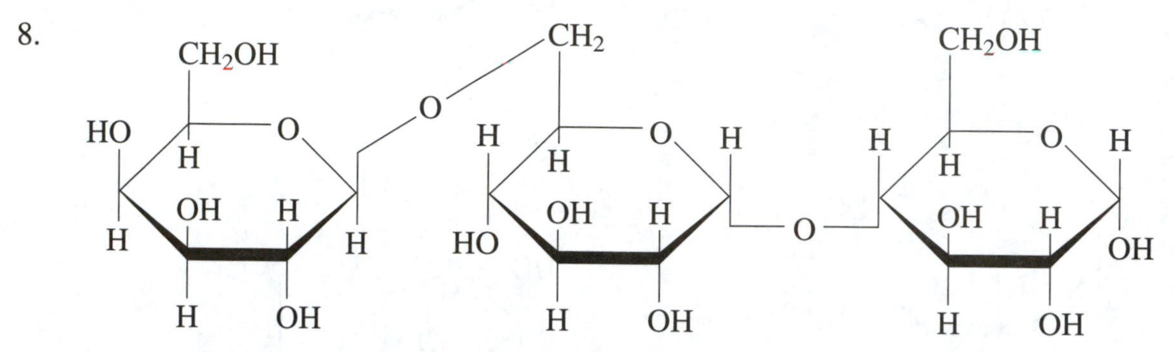

β-D-galactopyranosyl-(1→6)-α-D-glucopyranosyl-(1→4)-D-glucopyranose

9. The β(1→4) glycosidic bonds that join glucose groups in cellulose give rise to extended polymers that line up in parallel and are stabilized by extensive intrachain and interchain hydrogen bonds. This gives cellulose fibers unusual strength and renders them water insoluble. In contrast, the α(1→4) glycosidic bonds linking glucose residues in starch give rise to a totally different structure, namely a linear chain that assumes a relatively open helical conformation.

10. Two advantages are (1) the reduction in osmotic pressure in a cell when many residues are combined into a single polymeric molecule, and (2) the rapid mobilization of glucose when it is removed from the ends of many branches simultaneously.

11. Lectins are proteins that bind specific sugars with high affinity. Thus, a lectin attached to an immobile support can be used to selectively adsorb polysaccharides or glycoproteins via affinity chromatography. The adsorbed molecules can then be eluted by applying a solution containing an excess of the free sugar to which the lectin preferentially binds.

12. (a) The core oligosaccharide is $(mannose)_3(GlcNAc)_2$.
(b) Typical terminal sugars are galactose, *N*-acetylneuraminic acid (sialic acid), and fucose.

13. Each disaccharide unit bears a negative charge (from the ionized COOH group of *N*-acetylneuraminic acid). Because of the strong repulsion between charged groups, the protein assumes a rigid elongated shape, thereby accounting for the high viscosity of the solution. Removal of the *N*-acetylneuraminic acid residues by sialidase abolishes the large net negative charge and permits the protein to assume a more compact shape. As a result, the viscosity of the solution decreases.

14. Type AB individuals have both type A and type B carbohydrate structures on their cell surfaces, so they do not recognize either type A or type B blood cells as foreign. Consequently, they can receive either type A or type B blood. Type A individuals synthesize antibodies to type B carbohydrate antigens, so transfused blood cells bearing the type B carbohydrate antigen will agglutinate in the blood vessels. Similarly, type B individuals synthesize antibodies to the type A carbohydrate antigen and therefore cannot receive type A blood.

Chapter 9 Lipids

Lipids are vital components of all cells. Unlike nucleic acids, proteins, and carbohydrates, lipids do not have unifying structural features. They are defined operationally as a diverse group of nonpolar substances. They are therefore poorly or not at all water-soluble but dissolve readily in many organic solvents. Lipids are essential constituents of biological membranes. In addition, the acyl chains of lipids serve as energy sources. Finally, a number of cell signaling processes involve lipids. This chapter focuses on the structures and physical characteristics of lipids, as well as on the properties of the lipid bilayer. These subjects serve as an introduction to Chapter 10, in which biological membranes and protein–lipid interactions are considered.

Essential Concepts

Lipid Classification

1. Fatty acids are common components of other lipids, where they occur in ester or amide linkages. They are straight-chain carboxylic acids, usually having between 14 and 22 carbons. Eukaryotic fatty acids usually possess an even number of carbon atoms and may contain one or more double bonds. Although the systematic names of fatty acids reveal their structures, many fatty acids also have common names.

2. Saturated fatty acids, of which the most common are palmitic acid (C_{16}; hexadecanoic acid) and stearic acid (C_{18}; octadecanoic acid), contain no double bonds and are highly flexible molecules that tend to assume a fully extended conformation. In the pure compound, neighboring saturated fatty acyl chains pack together tightly. The resulting van der Waals interactions cause fatty acid melting points to increase as chain length increases.

3. The double bonds of unsaturated fatty acids nearly always have the *cis* configuration, which introduces a 30° bend in the acyl chain. This prevents unsaturated fatty acids from packing together as closely as saturated fatty acids. As a result, the melting point of an unsaturated fatty acid is always lower than the melting point for a saturated fatty acid with the same number of carbons.

4. Most fatty acid groups in eukaryotic lipids are unsaturated. The most common fatty acid with one double bond is oleic acid (C_{18}; 9-octadecenoic acid). Fatty acids with two or more double bonds are termed polyunsaturated.

5. Triacylglycerols, which contain three fatty acids esterified to the hydroxyl groups of D-glycerol, serve as energy reserves in plants and animals. A triacylglycerol generally contains more than one type of fatty acyl group. Mixtures of triacylglycerols may be fats (which are solid at room temperature) or oils (which are liquid at room temperature), depending on the properties of their component fatty acyl groups.

6. The highly reduced nature of triacylglycerols makes them an efficient metabolic energy store. Adipocytes are cells specialized for the biosynthesis, storage, and breakdown of triacylglycerols. Adipose tissue also provides thermal insulation.

7. The major lipid constituents of biological membranes are glycerophospholipids. These substances are composed of D-glycerol with fatty acids esterified to C1 and C2 and a phosphate group esterified to C3. The phosphate is almost always also esterified to a hydrophilic moiety. Thus, glycerophospholipids are amphipathic molecules with a polar head group and a nonpolar tail. Glycerophospholipids frequently contain saturated or monounsaturated acyl groups at C1 of glycerol and more highly unsaturated acyl moieties at C2.

8. Phospholipases are enzymes that hydrolyze glycerophospholipids. For example, phospholipase A_2, which occurs in bee and snake venoms, specifically cleaves the C2 ester linkage to yield a free fatty acid and a lysophospholipid. Several phospholipase-catalyzed hydrolysis products of glycerolipids play roles in cell signaling processes. These include 1,2-diacylglycerol and lysophosphatidic acid.

9. Plasmalogens are glycerophospholipids in which a fatty acyl group is linked to C1 via an α,β-unsaturated ether bond.

10. Sphingolipids all contain a long-chain nitrogen-containing alcohol, sphingosine. Ceramides have a fatty acyl group linked to the primary amino group of sphingosine and can be regarded as the basic building block of more complex sphingolipids. Among these are (a) sphingomyelin, in which ceramide is esterified to a phosphocholine or phosphoethanolamine head group; (b) cerebrosides, in which ceramide forms a glycosidic bond with either glucose or galactose; and (c) gangliosides, in which oligosaccharides containing one or more *N*-acetylneuraminic acid groups form the ceramide head group. Cerebrosides and gangliosides are therefore glycosphingolipids. On hydrolysis, sphingolipids, like glycerophospholipids, give rise to products that have signaling activity.

11. Steroids are derivatives of the cyclopentaneperhydrophenanthrene fused-ring system. Cholesterol, the most abundant animal steroid, is weakly amphiphilic because it has a hydroxyl group at C3 (it is therefore classified as a sterol). It is a major component of biological membranes in animals. In a cholesteryl ester, the hydroxyl group of cholesterol is esterified to a fatty acid.

12. Cholesterol is the precursor of steroid hormones, which include (a) glucocorticoids (e.g., cortisol), which modulate metabolic processes, inflammatory reactions, and stress responses; (b) mineralocorticoids (e.g., aldosterone), which regulate salt and water excretion; and (c) androgens and estrogens, which influence sexual development and reproductive functions.

13. Vitamins D_2 and D_3 are produced from steroid precursors through photolysis by ultraviolet light. These two vitamins are then converted to active forms by enzymatic hydroxylation. Active vitamin D promotes absorption of dietary Ca^{2+} and enhances the release of Ca^{2+} from bone into the blood.

14. Eicosanoids are derived from the highly unsaturated C_{20} lipid arachidonic acid after its release from membrane glycerophospholipids by phospholipase A_2. The eicosanoids include prostaglandins, thromboxanes, and leukotrienes, all of which are biologically active at extremely low concentrations. They are produced in a tissue-specific manner and play roles in inflammatory reactions, the regulation of the cardiovascular system, and reproduction.

15. The effectiveness of aspirin, ibuprofen, and acetaminophen as anti-inflammatory drugs is due to their ability to inhibit an enzyme required for prostaglandin synthesis.

Lipid Bilayers

16. The physical properties of lipids in aqueous solution cause them to aggregate. Water tends to exclude the hydrophobic portions of amphiphilic lipids, whereas the polar head groups remain in contact with the aqueous environment. Amphiphiles with a single nonpolar tail, such as soaps, many detergents, and lysophospholipids, form globular micelles. Amphiphiles with two hydrocarbon tails, such as glycerophospholipids, instead form disk-like micelles that are actually lipid bilayers. A bilayer of phospholipids in an aqueous milieu may form a vesicle with a solvent-filled interior, called a liposome. Liposomes are useful as models of biological membranes and as water-soluble drug delivery vehicles.

17. The movement of an amphiphilic lipid across a lipid bilayer, a process called transverse diffusion or a flip-flop, occurs infrequently, because the passage of a polar head group through the hydrophobic interior of the bilayer is thermodynamically unfavorable. In contrast, lipids readily diffuse laterally in the plane of the bilayer, indicating that the hydrocarbon core of the bilayer is highly fluid. The motion of the acyl chains is greatest in the center of the bilayer and decreases markedly near the polar head groups.

18. Bilayer fluidity is a function of temperature. At high temperatures, the lipids form a highly mobile liquid-crystal state, and at low temperatures, they form a gel-like solid as the hydrocarbon tails tightly associate. The transition temperature, the temperature at which the phase changes, depends on bilayer composition and increases with increasing acyl chain length and decreasing unsaturation of the component lipids. The presence of cholesterol also reduces a bilayer's transition temperature. Living organisms alter their membrane composition in order to maintain constant fluidity as the ambient temperature varies.

19. Cholesterol stabilizes the bilayer over a range of temperatures: At high temperatures, its rigid ring system interferes with fatty acyl chain mobility and thereby decreases membrane fluidity. At low temperatures, cholesterol prevents tight packing of adjacent hydrocarbon tails and thereby promotes membrane fluidity.

Guide to Study Exercises (text p. 237)

1. Lipids differ from the three other major classes of biological molecules primarily in their hydrophobicity. Because this property—rather than a common structure—unites lipids, they exhibit greater structural variety than do other classes of biological molecules. Lipids, unlike amino acids, carbohydrates, or nucleotides, do not form extended polymers.

2. The degree of unsaturation of a fatty acid is correlated with its melting point and overall fluidity: The greater the unsaturation, the lower the melting point and the more fluid the fatty acid. Consequently, unsaturated fatty acids and triacylglycerols that contain unsaturated fatty acid residues pack together less efficiently. (Section 9-1A)

3. The structures of triacylglycerols, glycerophospholipids, and sphingolipids are shown on pages 222, 223, and 226, respectively. The triacylglycerols are the simplest, with three fatty acyl groups esterified to a glycerol backbone. Glycerophospholipids consist of two fatty acyl groups esterified to a glycerol backbone whose C3 position is linked to a phosphate that is, in turn, linked to a polar group. Sphingolipids are similar to glycerophospholipids in overall shape since they contain two hydrocarbon tails and a phosphate derivative or carbohydrate as a head group.

   The physical properties of these three types of lipids depend on the length and saturation of their fatty acyl chains and on the identity of their head group. Triacylglycerols, which have no head group, are entirely nonpolar. Glycerolipids and sphingolipids that have phosphate derivatives as head groups are amphipathic molecules with nonpolar tails and polar or charged head groups. The sphingolipids known as gangliosides are attached to oligosaccharides whose large nonionic head groups dominate their structure. (Sections 9-1B, C, and D)

4. Steroids in the form of cholesterol are important components of animal cell membranes. In mammals, steroid hormones include the glucocorticoids (which regulate metabolism and stress responses), mineralocorticoids (which regulate water and salt balance), and the sex hormones (androgens and estrogens). Vitamin D, which is ultimately derived from the plant sterol ergosterol or the closely related 7-dehydrocholesterol, regulates Ca^{2+} metabolism. Eicosanoids, which are derivatives of arachidonic acid, regulate pain, fever, vasoconstriction, and other physiological processes. Both steroid hormones and eicosanoids are synthesized in a tissue-specific manner, but eicosanoids usually act close to their site of synthesis, whereas steroids travel throughout the body. (Sections 9-1E and F)

5. Lateral diffusion of lipids in a membrane is faster than transverse diffusion because the transbilayer movement of a membrane lipid requires that its hydrated head group pass through the hydrophobic interior of the bilayer, an energetically unfavorable process. (Section 9-2B)

Questions

Lipid Classification

1. Draw structures for the following fatty acids:
   (a) 18:0 octadecanoic acid
   (b) 20:4 5,8,11,14-eicosatetraenoic acid
   (c) 22:6 4,7,10,13,16,19-docosahexaenoic acid
   (d) 13-(2-cyclopentenyl)-tridecanoic acid (a cyclic fatty acid)

2. How do the following alterations in the structures of fatty acids affect their physical properties?
   (a) Increasing the chain lengths of saturated fatty acids
   (b) Increasing the number of double bonds in unsaturated fatty acids
   (c) Changing a *cis* double bond in a fatty acid to a *trans* double bond

3. Draw and name a typical triacylglycerol.

4. List the two major functions of triacylglycerols.

5. Why are glycerophospholipids considered amphiphilic molecules?

6. Draw the structures of the following lipids:
   (a) 1-palmitoyl-2-oleoyl-3-glycerophosphatidylethanolamine
   (b) 1-stearoyl-2-arachidonoyl-3-glycerophosphatidylinositol

7. What distinguishes a plasmalogen from other glycerophospholipids?

8. What chemical moieties do sphingomyelin and gangliosides have in common and which are unique to each?

9. The compound shown below has been advertised as a dietary supplement that purportedly prevents obesity, heart disease, and the ill effects of aging. Based on its structure, what physiological function is it most likely to actually perform?

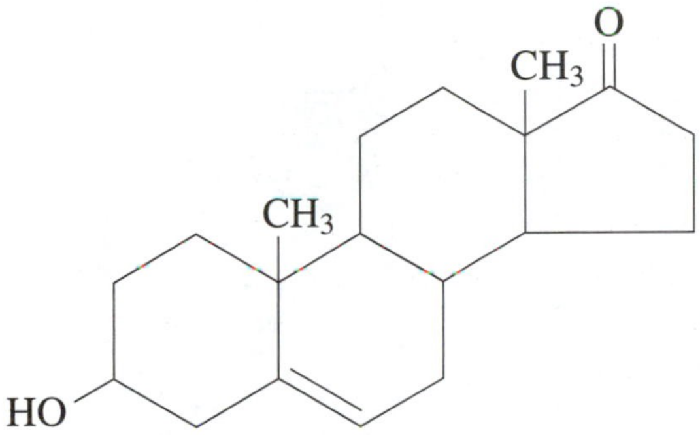

O
CH_3
CH_3
HO

10. Why are eicosanoids called local mediators rather than hormones?

11. Arachidonic acid can arise from phospholipid precursors by the action of:
    (a) Phospholipase A_1
    (b) Phospholipase A_2
    (c) Phospholipase C
    (d) Phospholipase D

12. In arachidonate metabolism, the production of ________, ________, and ________ is blocked by the drug ________, which inhibits the reaction catalyzed by ________.

Lipid Bilayers

13. Explain why single-tailed amphiphiles tend to form micelles whereas two-tailed amphiphiles tend to form bilayers.

14. Why is a polar solute unlikely to penetrate a lipid bilayer?

15. Describe the structural changes that occur when a pure phospholipid bilayer is warmed and passes through its transition temperature. Explain what would happen if the bilayer contained a significant amount of cholesterol.

Answers to Questions

1. (a)

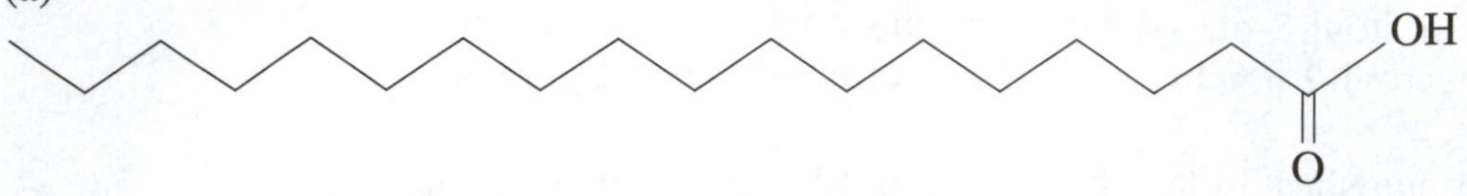

   (b)

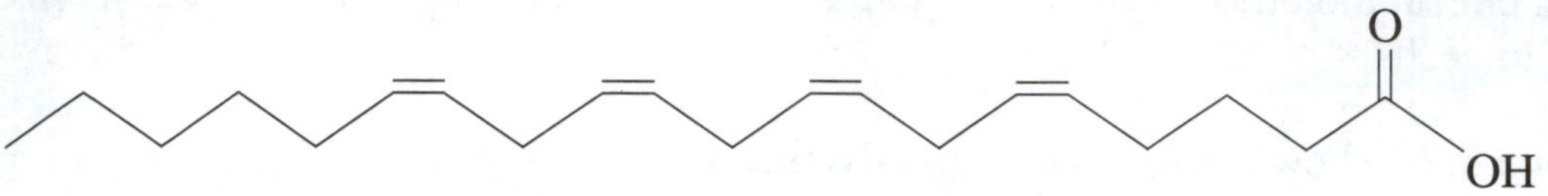

   (c)

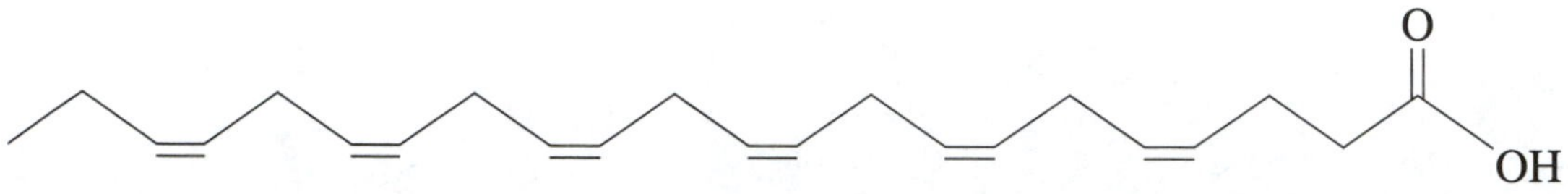

   (d)

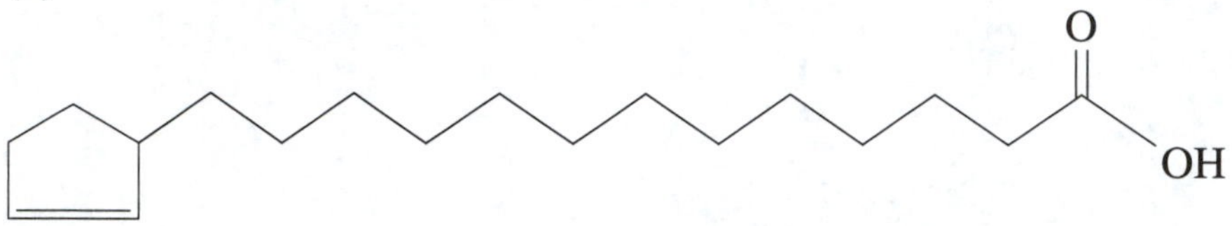

2. (a) The melting point increases with increasing hydrocarbon chain length. Fully saturated hydrocarbon chains have the least amount of steric hindrance and thus can pack together closely.
   (b) The melting point decreases with an increasing number of double bonds.
   (c) Fatty acid residues that have a *cis* double bond exhibit a 30° bend in their hydrocarbon chain that interferes with close packing. In contrast, *trans* double bonds distort the acyl chain conformation much less, allowing closer packing of molecules. Hence a fatty acid with *trans* double bonds has a higher melting point than the corresponding *cis* fatty acid of the same chain length and with the same number of double bonds.

3. For example, 1-palmitoyl-2-palmitoleoyl-3-oleoyl-glycerol

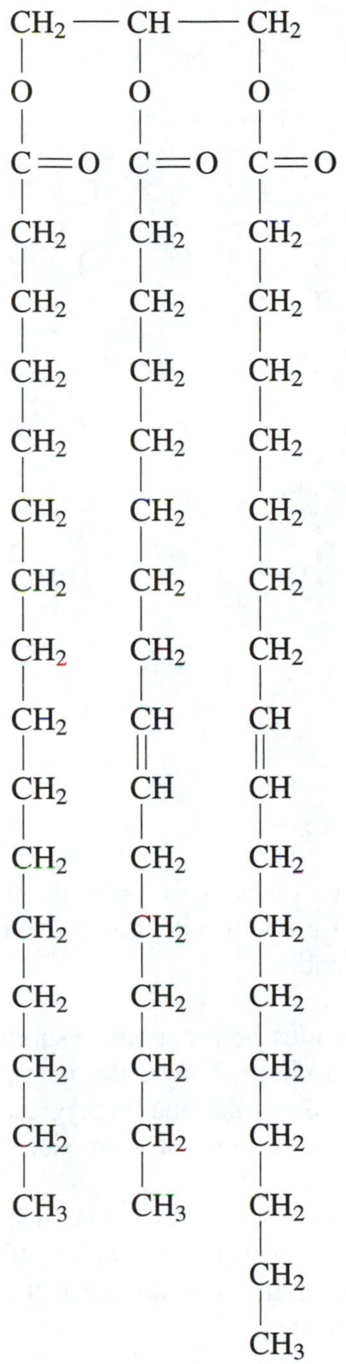

$$
\begin{array}{ccccc}
CH_2 & - & CH & - & CH_2 \\
| & & | & & | \\
O & & O & & O \\
| & & | & & | \\
C{=}O & & C{=}O & & C{=}O \\
| & & | & & | \\
CH_2 & & CH_2 & & CH_2 \\
| & & | & & | \\
CH_2 & & CH_2 & & CH_2 \\
| & & | & & | \\
CH_2 & & CH_2 & & CH_2 \\
| & & | & & | \\
CH_2 & & CH_2 & & CH_2 \\
| & & | & & | \\
CH_2 & & CH_2 & & CH_2 \\
| & & | & & | \\
CH_2 & & CH_2 & & CH_2 \\
| & & | & & | \\
CH_2 & & CH_2 & & CH_2 \\
| & & | & & | \\
CH_2 & & CH & & CH \\
| & & \| & & \| \\
CH_2 & & CH & & CH \\
| & & | & & | \\
CH_2 & & CH_2 & & CH_2 \\
| & & | & & | \\
CH_2 & & CH_2 & & CH_2 \\
| & & | & & | \\
CH_2 & & CH_2 & & CH_2 \\
| & & | & & | \\
CH_2 & & CH_2 & & CH_2 \\
| & & | & & | \\
CH_2 & & CH_2 & & CH_2 \\
| & & | & & | \\
CH_3 & & CH_3 & & CH_2 \\
 & & & & | \\
 & & & & CH_2 \\
 & & & & | \\
 & & & & CH_3
\end{array}
$$

4. Triacylglycerols serve as metabolic energy sources and thermal insulators.

5. Glycerophospholipids are amphiphilic because they are composed of a hydrophobic diacylglycerol "tail" attached to a hydrophilic phosphoryl derivative "head."

6. (a)

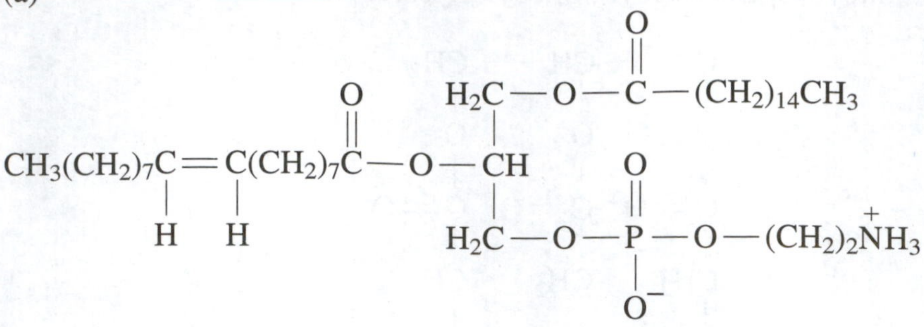

(b)

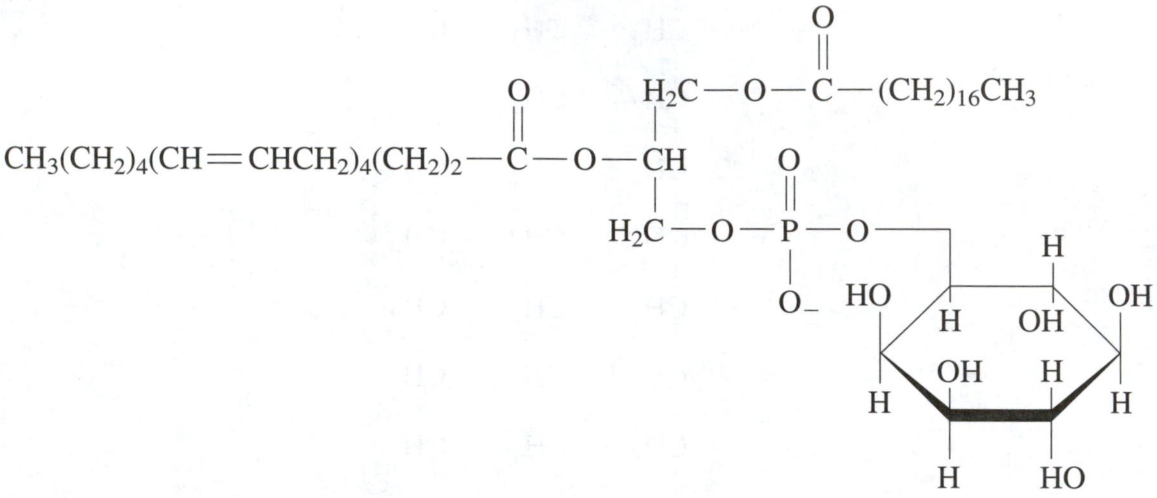

7. A plasmalogen differs from a diacylglycerophospholipid in that a hydrocarbon chain is attached to the C1 position of glycerol via an α,β-unsaturated ether bond in the *cis* configuration instead of an ester bond.

8. Sphingomyelin and gangliosides both contain a sphingosine moiety to which is attached a fatty acyl group in an amide linkage. They differ in that sphingomyelin has a phosphocholine (or less commonly a phosphoethanolamine) head group, whereas the ganglioside head group is an oligosaccharide that includes one or more sialic acid residues.

9. This compound is a sterol that closely resembles testosterone and estradiol (shown in Figure 9-11). It is most likely to exert its effects by influencing physiological processes that depend on the sex hormones. In fact, it is dehydroepiandrosterone (DHEA), a metabolic precursor of androgens and estrogens.

10. Although eicosanoids, like hormones, are synthesized by a variety of cells and exert their effects on other cells, they are considered local mediators rather than hormones because they act near their sites of synthesis rather than being carried throughout the body via the bloodstream, and because they decompose quickly.

11. (b)

12. In arachidonate metabolism, the production of prostacyclins, prostaglandins, and thromboxanes is blocked by the drug aspirin, which inhibits the reaction catalyzed by PGH_2 synthase.

13. The difference arises from geometrical considerations. Single-tailed amphiphiles, such as fatty acid anions, form micelles because of their tapered shape. Their hydrated head groups are wider than their tails, enabling them to pack efficiently into a spheroidal micelle. The cylindrical shape of two-tailed amphiphiles, such as glycerophospholipids, prevents their packing into a spheroidal micelle. Instead they pack together to form disk-like micelles, which are really extended bilayers.

14. In order for a polar solute to enter a lipid bilayer, its interactions with surrounding water molecules must first be disrupted (i.e., it must lose its hydration shell), and new interactions with the hydrophobic bilayer constituents would have to form. This would require an increase in free energy for the system; that is, the process is unfavorable.

15. When a pure phospholipid bilayer in an orderly gel-like state is warmed, it is converted (at its transition temperature) to a liquid crystal form that is more fluid. The presence of cholesterol in a phospholipid bilayer both decreases its fluidity and broadens the transition temperature range. These effects occur because the sterically rigid cholesterol molecules insert between the fatty acyl chains of the phospholipids and thus restrict their motion.

Chapter 10 Biological Membranes

The lipid bilayer introduced in Chapter 9 accounts for many of the fundamental properties of biological membranes. In particular, it provides a barrier to the movement of hydrophilic molecules and so not only enables cells to exclude components of the extracellular environment, but also permits compartmentation within the cell. An important principle of the organization and regulation of many cellular activities is that they take place in a spatially defined fashion. However, membranes are not merely inert permeability barriers but, under appropriate conditions, allow the passage of selected ions and molecules that contribute to the steady-state fluxes of energy and matter that are the essence of biological processes.

Biological membranes are composed of a great variety of proteins as well as lipids. These proteins are enzymes, facilitate transmembrane transport, and use information concerning the extracellular milieu to prompt the correct intracellular response. This chapter describes the features of membrane proteins that enable them to interact with lipids, and describes the overall arrangement of lipids and proteins in membranes. The structures and functions of lipoproteins are also covered. Lastly, the mechanisms involved in the transport of molecules across membranes are introduced.

Essential Concepts

Membrane Proteins

1. The types and relative amounts of membrane proteins and lipids vary among biological membranes. Membrane proteins, although extremely diverse, can be classified as integral, lipid-linked, or peripheral, depending on how they are associated with the membrane.

2. An integral protein is amphiphilic in that some regions of the molecule strongly associate with nonpolar membrane components by means of hydrophobic effects, whereas hydrophilic portions extend into the aqueous surroundings. Integral proteins can be solubilized and extracted from membranes only by using relatively harsh reagents, such as detergents and organic solvents, to disrupt the membrane structure.

3. Membrane proteins contribute to the asymmetry of a biological membrane because they either are present on one side of a membrane or, if they extend through it, are oriented in only one direction. The transmembrane domain of an integral protein may consist of one α helix (as in glycophorin A), a bundle of helices (as in bacteriorhodopsin), or a β barrel (as in porin). Both α helices and β barrels can form all their possible hydrogen bonds and are therefore stable in the interior of the lipid bilayer. In all cases, hydrophobic residues contact the hydrocarbon chains in the membrane core, whereas charged polar groups tend to predominate at the membrane surface.

4. Certain membrane proteins are covalently linked to a lipid moiety, which provides the protein with a hydrophobic anchor in the membrane. There are three main types of lipid-linked proteins:

(a) Prenylated proteins have an attached isoprenoid moiety, a polymer built of C_5 isoprene units. The most abundant of these are farnesyl (C_{15}) and geranylgeranyl (C_{20}) groups. As a rule, prenylation occurs at the C-terminus of a protein.

(b) Fatty acylated proteins have a covalently attached myristoyl or palmitoyl group. Myristoylation is a stable modification that occurs through an amide bond to the α amino group of an N-terminal glycine. Palmitoylation occurs via a thioester linkage to a cysteine and is reversible.

(c) Glycosylphosphatidylinositol (GPI)-linked proteins are located at the external surface of the cell and are anchored to the membrane by GPI, a glycolipid that is covalently attached to the C-terminus of the protein via an amide bond.

5. Peripheral membrane proteins associate with membrane surfaces via electrostatic and hydrogen bonding interactions. As a result, these proteins can be removed from membranes by relatively gentle methods, such as extraction with salt solutions or variations in pH, that do not greatly disturb native membrane structure. Once dissociated, they exhibit properties typical of water-soluble proteins.

Membrane Structure and Assembly

6. The widely accepted concept of biological membrane structure, for which there is much evidence, is the fluid mosaic model. The model envisions integral proteins that float in a sea of lipid and can diffuse laterally unless restrained by other cell constituents. The rate of membrane protein diffusion can be assessed by fluorescent photobleaching recovery, which measures the rate at which a fluorescent-labeled membrane component diffuses into an area of membrane that has been previously bleached by a laser beam.

7. The structure and function of the erythrocyte membrane have been extensively investigated, largely through the use of erythrocyte ghosts, which are cells from which hemoglobin and other soluble cell constituents have been removed by osmotic lysis. The erythrocyte's biconcave disk-like shape, which is essential for its O_2-carrying properties, is maintained by the membrane skeleton, a network of proteins located just beneath the membrane. The chief element in the skeleton is spectrin, a fibrous heterotetrameric protein that is cross-linked to other skeletal and membrane proteins, including actin and ankyrin. The proteins of the erythrocyte membrane skeleton occur in many other cell types.

8. The lipid and protein components of biological membranes are unevenly distributed. The cytoskeleton apparently restricts the movements of many membrane proteins and hence contributes to this heterogeneity. Localized membrane domains that have different lipid compositions may result from the association of certain lipids with membrane proteins.

9. Membrane constituents are asymmetrically distributed with respect to the outer and inner leaflets of the bilayer in natural membranes. For example, the carbohydrate components of plasma membrane glycoproteins and glycolipids are always located on the extracellular membrane surface. The inner and outer leaflets also contain lipids in different proportions.

10. Lipid asymmetry, which can be assessed through the use of specific phospholipases, results in part from the asymmetric synthesis of lipids on the cytoplasmic face of the plasma membrane (in prokaryotes) or the endoplasmic reticulum (in eukaryotes). Lipids are subsequently

redistributed by two mechanisms: (a) Enzymes called flipases facilitate the flip-flop of specific phospholipids from one leaflet to the other; and (b) ATP-dependent translocases establish a nonequilibrium distribution of lipids. In eukaryotes, lipids are transported as components of vesicles that bud off the endoplasmic reticulum and ultimately fuse with other cellular membranes.

11. Protein synthesis starts at the N-terminus and proceeds to the C-terminus of a polypeptide chain. Synthesis takes place either on free ribosomes (for soluble proteins) or on ribosomes bound to the endoplasmic reticulum (for integral membrane and secretory proteins). The signal hypothesis explains how polypeptides come to traverse the endoplasmic reticulum membrane:
    (a) Proteins synthesized on ER-bound ribosomes have an N-terminal amino acid sequence known as a signal peptide.
    (b) After an ~80-residue polypeptide chain has been synthesized, the signal peptide binds to the signal recognition particle (SRP). This event stops further polypeptide synthesis.
    (c) The SRP–ribosome complex, bearing the nascent polypeptide chain, binds to a docking protein (the SRP receptor) on the ER surface. Protein synthesis resumes, and the N-terminal sequence is translocated to the lumen of the ER via a transmembrane channel. At this point, the SRP dissociates from the complex.
    (d) Once in the ER lumen, the signal peptide is removed from the polypeptide by hydrolysis catalyzed by a signal peptidase.
    (e) The growing polypeptide undergoes folding and posttranslational modification, such as addition of core *N*-linked oligosaccharides.
    (f) When synthesis is complete, secretory proteins have passed entirely through the ER membrane into the lumen, whereas transmembrane proteins remain embedded in the membrane with their C-terminus on the cytoplasmic face.

12. Transmembrane and secretory proteins move from the endoplasmic reticulum to the Golgi apparatus as part of COPI- and COPII-coated vesicles (COP = *co*at *p*rotein), to be further processed, primarily by modification of their oligosaccharide chains. Clathrin-coated vesicles, which have a polyhedral structure, transport proteins from the Golgi apparatus to the plasma membrane and other destinations. After vesicle fusion, integral membrane proteins that faced the ER lumen face the extracellular space.

Lipoproteins and Receptor-Mediated Endocytosis

13. Many lipids are transported between tissues through the circulation as components of lipoproteins. Lipoproteins are globular structures that are composed of a hydrophobic interior containing triacylglycerols and cholesteryl esters encased in an amphiphilic outer layer of protein, phospholipid, and cholesterol. There are five classes of lipoproteins: (a) chylomicrons; (b) very low density lipoproteins (VLDL); (c) intermediate density lipoproteins (IDL); (d) low density lipoproteins (LDL); and (e) high density lipoproteins (HDL). The densities of lipoprotein classes increase as the quantity of lipid contained in the core decreases.

14. At least nine apolipoproteins comprise the protein components of human lipoproteins. Most of these have a high content of α helices, whose nonpolar and polar residues are on opposite sides of the α-helical cylinders. The nonpolar faces of the apolipoproteins interact with the phospholipid hydrophobic tails, and the polar faces interact with the phospholipid polar head groups.

15. Cholesterol can enter cells via receptor-mediated endocytosis of LDL. In this process, LDL binds to cell-surface receptors that recognize the apolipoprotein B-100 component of LDL. The LDL–receptor complex enters clathrin-coated pits, which pinch off from the plasma membrane to form clathrin-coated vesicles that are transported to endosomes. There, LDL dissociates from its receptor, which moves back to the plasma membrane. LDL is degraded in lysosomes, releasing cholesteryl esters that are then hydrolyzed to yield cholesterol.

16. Receptor-mediated endocytosis is a widely employed mechanism for the uptake of macromolecules.

Transport across Membranes

17. The hydrophobic interior of biological membranes renders them impermeable to ions and small polar compounds, such as amino acids, carbohydrates, and nucleotides, although water can traverse the bilayer with surprising ease. Polar substances cross membranes through the mediation of transport proteins.

18. The thermodynamics of diffusion across a membrane can be expressed in terms of a chemical equilibrium. The difference in concentration of substance A on two sides of a membrane generates a chemical potential difference, $\Delta\overline{G}_A$. When A is ionic, an electrical potential difference ($\Delta\Psi$) may also develop. Thus, the equation for the electrochemical potential difference contains terms for the concentration and the charge of substance A:

$$\Delta\overline{G}_A = RT \ln ([A]_{in}/[A]_{out}) + Z_A\mathcal{F}\Delta\Psi$$

where Z_A is the ionic charge of A and $\mathcal{F}$ is the Faraday constant ($96{,}485\ J \cdot V^{-1} \cdot mol^{-1}$). A negative value for $\Delta\overline{G}_A$ indicates spontaneous transport of substance A from the outside to the inside.

19. Transport may be classified as either mediated or nonmediated. Diffusion accounts for nonmediated transport. The chemical potential difference determines the direction of nonmediated transport, and the substance moves in the direction that will eliminate the concentration difference and at a rate proportional to the size of the gradient.

20. Mediated transport makes use of carrier molecules, called permeases, transporters, or translocases. In passive-mediated transport (also called facilitated diffusion), a molecule is transported from a high to a low concentration. In active transport, an energy-yielding process must be coupled to the movement of a substance from a lower to a higher concentration.

21. Ionophores are model carrier molecules that greatly increase the permeability of a membrane to certain ions. Some of these organic compounds (such as valinomycin) bind the ion on one side of the membrane and carry it to the other side; others (such as gramicidin A) form transmembrane channels through which the ion can pass. Porins are channel-forming membrane proteins that display some solute selectivity.

22. Integral membrane proteins that mediate passive transport cycle between conformational states in which binding sites for the molecule to be transported are alternately accessible on one side of the membrane and then on the other. For example, the glucose transporter binds

glucose at the external cell surface, then changes conformation so as to release glucose at the cytoplasmic surface. It then reverts to its previous conformation.

23. Passive transport proteins can operate in either direction, depending on the concentrations of the transported substance on both sides of the membrane. A uniport mechanism transfers a single atom or molecule across the membrane; symport involves two substances moving in the same direction; and antiport is the movement of two substances in opposite directions.

24. Active transport is coupled to the hydrolysis of ATP. A well-studied example is the plasma membrane (Na^+–K^+)–ATPase, which pumps 3 Na^+ out and 2 K^+ into the cell, thereby moving both ions against their concentration gradients, with every ATP hydrolyzed. The resulting electrochemical gradient of Na^+ and K^+ is the basis for the electrical excitability of nerve cells.

25. The mechanism of the (Na^+–K^+)–ATPase involves a series of reaction steps that normally operate in only one direction because ATP breakdown and ion movement are coupled vectorial processes. The Ca^{2+}–ATPase, which pumps cytosolic Ca^{2+} to the cell exterior against a large concentration gradient, functions in a similar manner to the (Na^+–K^+) pump.

26. In secondary active transport, the energy generated by an electrochemical gradient is utilized to drive another, endergonic transport process. One example is the uptake and concentration of glucose by the intestinal epithelial cells. This task is accomplished by using the Na^+ gradient produced by the (Na^+–K^+) pump. Another example is the bacterial lactose permease, which utilizes a proton gradient across the bacterial membrane to power the transport of lactose.

Key Equation

$$\Delta \overline{G}_A = RT \ln\left(\frac{[A]_{in}}{[A]_{out}}\right) + Z_A \mathcal{F} \Delta \Psi$$

Guide to Study Exercises (text p. 277)

1. Integral membrane proteins contain hydrophobic segments that span the width of the lipid bilayer and thereby anchor the protein in the hydrophobic core of a membrane. An integral membrane protein can be separated from the membrane only by disrupting the membrane.

   Peripheral membrane proteins, in contrast, are not anchored in the lipid bilayer but are associated with the outer surface of the membrane. A peripheral protein can therefore be separated from the membrane under relatively mild conditions. (Sections 10-1A and C)

2. Lipids can be covalently attached to proteins in three ways:
   (a) A prenyl group, such as a farnesyl or geranylgeranyl group, can be attached to a Cys residue at the protein's C-terminus.
   (b) A fatty acyl group, such as a myristoyl or palmitoyl group, can be attached to a protein via an amide linkage to an α-amino group or via thioester linkage to a Cys side chain.
   (c) Glycosylphosphatidylinositol can be linked to a protein's C-terminus. (Section 10-1B)

3. According to the fluid mosaic model, integral membrane proteins are free to diffuse laterally within a lipid bilayer due to its lateral fluidity. The movements of membrane proteins may be limited by their interactions with other membrane proteins or with cytoskeletal proteins underlying the membrane. (Sections 10-2A and B)

4. The cytoskeleton is an organized network of proteins associated with the cytosolic side of the plasma membrane. Some of its components are integral membrane proteins, others are peripheral proteins. The distributions of other membrane proteins are influenced by interactions between these proteins and elements of the cytoskeleton. The cytoskeleton may also limit the diffusion of membrane proteins to an extent determined by the degree to which they interact with cytoskeletal proteins. (Section 10-2B)

5. Lysosomal proteins, as are all proteins, are ribosomally synthesized, beginning with the N-terminus. An RNA–protein complex, the signal recognition particle (SRP), recognizes the N-terminal region of the nascent polypeptide, which contains a stretch of hydrophobic residues (the signal peptide). The SRP binds to the ribosome, thereby arresting polypeptide synthesis, and eventually becomes attached to the SRP receptor on the rough endoplasmic reticulum. This causes polypeptide synthesis to resume. The N-terminus of the polypeptide is extruded through a transmembrane protein channel into the lumen of the ER. A membrane-bound signal peptidase removes the signal peptide, and the remainder of the lysosomal protein is synthesized and enters the lumen of the ER, where it folds and undergoes posttranslational modification. It is then transported to the Golgi apparatus, where it undergoes further processing before being delivered to a lysosome. (Section 10-2D)

6. Synthesis of a cell-surface glycoprotein follows the same process described for a lysosomal protein in Study Exercise 5, up to the point of translocating through the ER membrane into the lumen. The protein destined for the cell surface is not completely extruded into the lumen but remains anchored in the membrane via an ~20-residue hydrophobic membrane anchor, with its N-terminal "extracellular" portion in the lumen and its C-terminal portion in the cytoplasm. A vesicle buds off from the ER and moves to the Golgi apparatus, carrying the membrane-bound protein with it. In the Golgi apparatus, the protein, which already bears a carbohydrate chain added in the ER, undergoes further glycosylation. A coated vesicle then transports the protein to the cell surface. After fusion of the vesicle and plasma membranes, the membrane-anchored protein becomes a cell-surface protein with its glycosylated domain exposed to the extracellular space. (Section 10-2D)

7. LDL is a protein–lipid complex that carries cholesterol as part of its solvent-exposed surface and cholesteryl esters in its hydrophobic interior. The lipoprotein travels in the bloodstream until an LDL receptor on a cell recognizes the apoB-100 protein component of the LDL. Receptor-bound LDL particles cluster into coated pits and are then engulfed by the cell through endocytosis. The resulting vesicle, which contains the LDL, fuses with an endosome, whose low pH induces the LDL to dissociate from its receptor. The LDL receptors are recycled to the cell surface while the apoB-100 is degraded and the lipids, including cholesterol, are released for use by the cell. (Section 10-3)

8. The mediated transport of a substance across a biological membrane occurs through the action of a carrier molecule, which is usually a protein. Mediated transport may be driven by

a concentration gradient such that the substance moves across the membrane from a region of high concentration to a region of low concentration. Alternatively, mediated transport may require the input of free energy from another exergonic process. Nonmediated transport of a substance across a membrane does not require a carrier molecule and occurs by simple diffusion according to the concentration gradient of the substance. (Section 10-4)

9. Ionophores, porins, and passive-mediated transport proteins all mediate the transmembrane movement of a substance that cannot pass through the membrane on its own due to its size and/or polarity. In all cases, the substance can cross the membrane in either direction, provided that it moves according to its chemical potential difference, and no other source of free energy is required. The three transporters differ in their structure and mechanism. Ionophores include small ion-binding molecules that can diffuse through a membrane, and small polypeptides that form a transmembrane channel through which an ion can diffuse. Porins are integral membrane proteins that form transmembrane channels that are always "open." However, the geometry of the channel and the amino acid residues lining it exert some solute selectivity. Transport proteins, unlike always-available ionophores and porins, do not have a single, always-open conformation. Instead, they alternate between two conformations. This allows the protein to bind a substance on one side of the membrane and change conformation so as to release it on the other side. (Section 10-4B)

10. Passive-mediated transport does not require an external source of free energy since it is driven only by the chemical potential difference of the transported substance. Active transport is an endergonic process that is coupled to the exergonic hydrolysis of ATP. The transport and ATPase functions are coupled so that both reactions go to completion in a vectorial manner. In secondary active transport, the free energy to drive transport comes from an electrochemical gradient that is established through the action of primary active transport proteins. In this case, the substance is transported against its concentration gradient as ions move to dissipate the electrochemical gradient. (Sections 10-4C and D)

Questions

Membrane Proteins

1. Summarize the physical properties of a typical integral protein, including the forces involved in its interaction with the membrane, its orientation, and its movement within the membrane.

2. Lipids linked covalently to proteins often ________________ the proteins in the membrane.

3. The three main types of lipid moieties that can be covalently attached to proteins are ________________, ________________, and ________________.

4. Mutated forms of the protein p21$^{c\text{-}ras}$ are associated with the development of a substantial proportion of tumors. However, a second mutation, involving the Cys residue in the C-terminal Cys-X-X-Y sequence of p21$^{c\text{-}ras}$ can inhibit its cancer-causing potential. What does this suggest about the molecular process leading to tumor formation?

Membrane Structure and Assembly

5. In verifying the fluid mosaic model of biological membranes, Michael Edidin observed that fusion of mouse and human cell membranes was temperature dependent, such that fusion did not occur below 15°C but proceeded readily at 37°C. Explain this difference in terms of membrane properties (see Figure 10-11).

6. Explain why an erythrocyte membrane that synthesizes too little spectrin is a sphere rather than a biconcave disk.

7. How did Kennedy and Rothman demonstrate that newly synthesized bacterial membrane phospholipids rapidly cross from the cytoplasmic side to the external side of the membrane (see Fig. 10-19)?

8. Explain why the flip-flop rate of phospholipids in biological membranes is far greater than in artificial lipid membranes.

9. For the signal hypothesis, match each of the following terms with its description below.

   ______ free ribosome
   ______ signal peptide
   ______ signal recognition particle (SRP)
   ______ core glycosylation
   ______ stop-transfer sequence
   ______ signal peptidase
   ______ membrane-bound ribosomes
   ______ SRP receptor

   A. A hydrophobic sequence in a transmembrane protein that arrests movement of the protein through the membrane.
   B. A posttranslational modification that occurs while peptide bond formation continues.
   C. Cleaves the signal peptide.
   D. An N-terminal hydrophobic sequence flanked by hydrophilic sequences.
   E. The site of membrane and secretory protein synthesis.
   F. Binds the SRP and reinitiates ribosomal polypeptide synthesis.
   G. The site of soluble protein synthesis.
   H. Binds to a ribosome bearing an extruded signal sequence and stops polypeptide synthesis.

10. A membrane glycoprotein-synthesizing system was devised that contained endoplasmic reticulum membranes and all the necessary ingredients for the *in vitro* synthesis of a protein. The complete synthesis of the protein molecule required 40 minutes. When the detergent Triton X-100 was added within a few minutes after the initiation of polypeptide synthesis, the resulting protein was devoid of glycosyl groups. (a) Explain the action of the detergent. (b) Would the protein have been glycosylated if the detergent had been added 35 minutes after initiation of protein synthesis?

11. The antibiotic monensin inhibits some steps of posttranslational protein modification in the Golgi apparatus. What effect would monensin have on protein targeting?

Lipoproteins and Receptor-Mediated Endocytosis

12. Why are cholesteryl esters located in the interior of a lipoprotein while cholesterol is located on the exterior?

13. If the LDL receptor underwent a mutation that increased its affinity for apoB-100, what would be the effect on serum cholesterol levels?

Transport across Membranes

14. What is the electrochemical potential difference when the intracellular $[Ca^{2+}] = 1\ \mu M$ and the extracellular $[Ca^{2+}] = 1$ mM? Assume $\Delta\Psi = -100$ mV (inside negative) and $T = 25°C$.

15. Rank the molecules below from lowest to highest according to their ability to diffuse across a lipid bilayer. Explain your rationale.

$$^{-}O{-}C({=}O){-}C({=}O){-}H$$
1

$$H_3C{-}C({=}O){-}NH_2$$
2

$$H_3CCH_2CH_2{-}C({=}O){-}NH_2$$
3

16. The following data (using arbitrary units) were obtained for the transmembrane movements of compounds A and B from outside to inside a cell:

A		*B*	
Extracellular concentration	***Flux into cell***	***Extracellular concentration***	***Flux into cell***
2	1.2	2	4.5
5	3.4	5	6.2
10	6.5	10	7.5

Which compound enters the cell by mediated transport? Explain.

17. Which graph below shows the expected relationship between temperature and flux (J, rate of flow) of an ion transported across a biological membrane via a carrier ionophore? Explain.

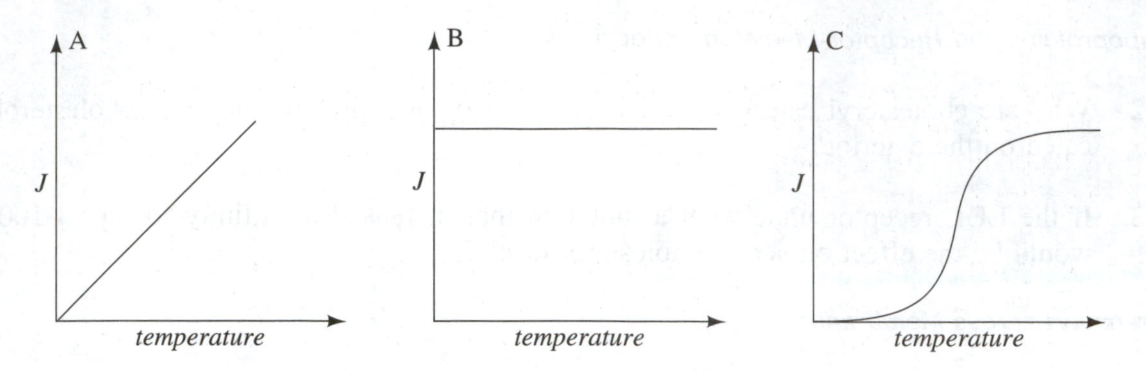

18. The model for glucose transport shown in Figure 10-35 shows two conformational states of the transporter when it is not bound to glucose. Is one of these conformational states likely to be preferred?

19. Does the activity of the $(Na^+–K^+)$–ATPase tend to make the cell interior electrically more negative or more positive with respect to the outside?

20. Determine whether each of the following transport systems is uniport, symport, or antiport. Which systems are active transport systems?
    (a) glucose transporter in erythrocytes
    (b) valinomycin
    (c) plasma membrane $(Na^+–K^+)$–ATPase
    (d) Na^+–glucose transporter of intestinal epithelium
    (e) *E. coli* lactose permease
    (f) $(H^+–K^+)$–ATPase of gastric mucosa

Answers to Questions

1. Integral proteins are strongly associated with membranes through hydrophobic effects and can typically be extracted only by agents such as detergents, organic solvents, or chaotropic substances, which disrupt membrane structure. Some integral proteins bind lipids so tightly that they are dissociated from them only under denaturing conditions. Integral proteins are amphiphilic molecules that are oriented in the membrane such that the regions buried in the interior of the membrane have a surface composed mainly of hydrophobic residues, whereas the portions exposed to the aqueous environment have mostly polar residues. These polar regions may occur on one side of the membrane or the other, or may asymmetrically extend from both sides (transmembrane proteins). Integral proteins can diffuse laterally in the plane of the membrane, but their flip-flop rate is minuscule.

2. Lipids linked covalently to proteins often anchor the proteins in the membrane.

3. The three main types of lipid moieties that can be covalently attached to proteins are isoprenoid groups, fatty acyl groups, and glycosylphosphatidylinositol (GPI) groups.

4. The cysteine residue in the C-terminal sequence is the site where p21$^{c\text{-}ras}$ is prenylated. Since this event permits the protein to insert into membranes, it appears that p21$^{c\text{-}ras}$ must be membrane-bound to promote tumor formation.

5. Fusion of the two membranes could not occur at low temperature because the fluidity of the membranes was too low to allow the free diffusion of lipids and proteins. At the higher temperature, the membranes were more fluid, making the lateral movement of lipids and proteins possible.

6. In normal erythrocytes, the cross-linking of spectrin and other membrane proteins creates a submembrane network of proteins, somewhat akin to a geodesic dome, that gives the cells a biconcave disk-like shape. A cell that contains too little spectrin to assemble a complete membrane skeleton is not constrained in shape and becomes a sphere, the simplest possible shape for a membrane enclosing the cell contents.

7. Growing bacteria were exposed briefly to radioactive phosphate so that newly synthesized phospholipids (mostly phosphatidylethanolamine in *E. coli*) would become labeled. Trinitrobenzenesulfonic acid (TNBS), a membrane-impermeable reagent that reacts with phosphatidylethanolamine (PE), was then added to the cells. At various times, samples of cells were tested for the presence of ^{32}P-labeled PE that also contained the TNB group. The appearance of doubly labeled PE after three minutes indicated that PE initially synthesized on the cytoplasmic side of the membrane (where it was labeled with ^{32}P) traversed the membrane and reacted with TNBS in the extracellular medium.

8. Biological membranes differ from artificial lipid membranes in having protein constituents (e.g., flipases and phospholipid translocases) that greatly accelerate the rate at which phospholipids move from one side of the bilayer to the other.

9. __G__ free ribosome
   __D__ signal peptide
   __H__ signal recognition particle (SRP)
   __B__ core glycosylation
   __A__ stop-transfer sequence
   __C__ signal peptidase
   __E__ membrane-bound ribosomes
   __F__ SRP receptor

10. (a) The membrane protein was synthesized on ribosomes bound to the endoplasmic reticulum and passed partially through the membrane into the ER lumen. Treatment with triton X-100 disrupted the membrane structure, which prevented the nascent protein from being glycosylated by the glycosyltransferases that are normally present in the ER lumen. Thus, posttranslational glycosylation requires an intact ER membrane.
    (b) Because glycosylation is initiated as the protein is being synthesized, the protein will almost certainly have been glycosylated by the ER enzymes before the membrane was disrupted by the detergent.

11. Disruption of posttranslational protein modification reactions would likely also disrupt protein-targeting mechanisms, many of which depend on modifications, such as glycosylation, that occur in the Golgi apparatus.

12. Cholesteryl esters are highly hydrophobic and therefore occupy the nonaqueous interior of the lipoprotein. Cholesterol, with an OH group, is weakly polar and therefore can interact with water molecules at the surface of the lipoprotein.

13. ApoB-100 is the LDL protein component that is recognized by the LDL receptor. Increased binding of LDL by its receptor would increase the cellular uptake of cholesterol-rich LDL. Consequently, the level of cholesterol in the serum would be lower than normal.

14. $$\begin{aligned}\Delta\overline{G} &= RT \ln ([Ca^{2+}]_{in}/[Ca^{2+}]_{out}) + Z\mathcal{F}\Delta\Psi \\ &= RT \ln (0.001/1) + (2)\mathcal{F}\Delta\Psi \\ &= (8.3145\ J \cdot K^{-1} \cdot mol^{-1})(298K)(-6.91) + (2)(96{,}485\ J \cdot V^{-1} \cdot mol^{-1})(-0.1V) \\ &= -17{,}121\ J \cdot mol^{-1} - 19297\ J \cdot mol^{-1} \\ &= -36{,}418\ J \cdot mol^{-1} = -36.4\ kJ \cdot mol^{-1}\end{aligned}$$

15. **1,2,3.** The charged molecule is least able to penetrate the hydrophobic core of the bilayer. Although compounds 2 and 3 bear the same polar groups, the larger hydrocarbon chain of compound 3 lets it diffuse more easily through the bilayer.

16. Compound B enters the cell by mediated transport since its rate of entry falls off as [B] increases. This is consistent with a carrier molecule that becomes saturated with B. Compound A enters by a nonmediated process, since its flux is directly related to its concentration at all values of [A].

17. C. A carrier ionophore must diffuse through the membrane. The gel-like state of the lipids at low temperatures slows ionophore movement. The steep portion of the curve corresponds to the transition of the membrane to a more fluid state that allows free ionophore diffusion.

18. Probably not, since this transporter allows equilibration of glucose concentration across the membrane in either direction.

19. The cell interior becomes more negative since 3 Na^+ exit the cell for every 2 K^+ that enter.

20. (a) uniport
    (b) uniport
    (c) antiport
    (d) symport
    (e) symport
    (f) antiport

    The active transport systems are c, d, e, and f.

Chapter 11 Enzymatic Catalysis

Enzymes are biological catalysts that increase the rates of biochemical reactions. In some cases, the enzyme-catalyzed reaction is nearly 10^{15} times faster than the uncatalyzed reaction. Enzymes are proteins (or in some cases, RNA) with very specific functions and are active under mild conditions. Enzymes function by lowering the free energy of the reaction's transition state. This chapter first discusses the various types of enzymes, their substrate specificity, the roles of coenzymes and cofactors, and the reaction coordinate. Next, the six modes of catalysis are described in detail. The chapter ends with full descriptions of the catalytic activity of lysozyme and serine proteases.

Essential Concepts

General Properties of Enzymes

1. Enzymes differ from ordinary catalysts in their higher reaction rates, their action under milder reaction conditions, their greater reaction specificities, and their capacity for regulation.

2. The IUBMB enzyme classification system divides enzymes into six groups (each with subgroups and sub-subgroups) based on the type of reaction catalyzed. A set of four numbers identifies each enzyme.

3. Enzymes are highly specific for their substrates and reaction products. Hence, the enzyme and its substrate(s) must have geometric, electronic, and stereospecific complementarity. Enzymes, for example, yeast alcohol dehydrogenase, can distinguish prochiral groups.

4. Many enzymes require cofactors for activity. Cofactors may be metal ions or organic molecules known as coenzymes. Many vitamins are coenzyme precursors. Coenzymes may be cosubstrates, which must be regenerated in a separate reaction, or prosthetic groups, which are permanently associated with the enzyme. An enzyme without its cofactor(s) is called an apoenzyme and is inactive, and the enzyme with its cofactor(s) is a holoenzyme and is active.

Activation Energy and the Reaction Coordinate

5. According to transition state theory, the reactants of a reaction pass through a short-lived high-energy state that is structurally intermediate to the reactants and products. This so-called transition state is the point of highest free energy in the reaction coordinate diagram. The free energy difference between the reactants and the transition state is the free energy of activation, $\Delta G^{\ddagger}$. The reaction rate decreases exponentially with the value of $\Delta G^{\ddagger}$; that is, the greater the value of $\Delta G^{\ddagger}$, the slower the reaction. A catalyst provides a reaction pathway whose $\Delta G^{\ddagger}$ is less than that of the uncatalyzed reaction and hence increases the rate that the reaction achieves equilibrium.

Catalytic Mechanisms

6. Enzymes use several types of catalytic mechanisms, including acid–base catalysis, covalent catalysis, metal ion catalysis, electrostatic catalysis, catalysis by proximity and orientation effects, and catalysis by preferential binding of the transition state.

7. Acid–base catalysis occurs when partial proton transfer from an acid and/or partial proton abstraction by a base lowers the free energy of a reaction's transition state. The catalytic rates of enzymes that use acid–base catalysis are pH-dependent. RNase A has two catalytic His residues, which act as general acid and general base catalysts.

8. In covalent catalysis, the reversible formation of a covalent bond permits the stabilization of the transition state of the reaction through electron delocalization. Nucleophilic attack on the substrate by the enzyme to form a Schiff base intermediate capable of stabilizing (lowering the free energy of) a developing negative charge is an example of covalent catalysis. Common nucleophiles, which are negatively charged or contain unshared electron pairs, include imidazole and sulfhydryl groups.

9. Metalloenzymes tightly bind catalytically essential transition metal ions. Metal-activated enzymes loosely bind alkaline and alkaline earth metal ions that play a structural role. Metal ions may orient substrates for reaction, mediate oxidation–reduction reactions, and electrostatically stabilize or shield negative charges. For example, the Zn^{2+} of carbonic anhydrase makes a bound water molecule more acidic, thereby increasing the concentration of the nucleophile OH^-.

10. Electrostatic catalysis occurs through the proper positioning of charged residues in the active site such that they stabilize the transition state.

11. Enzymes lower the activation energies of the reactions they catalyze by bringing their reactants into proximity, by properly orienting them for reaction, and most importantly, by freezing out the relative motions of the reactants and the enzyme's catalytic groups.

12. An enzyme's preferential binding of the transition state lowers $\Delta G^{\ddagger}$ and thereby increases the rate of the reaction. For this reason, an unreactive compound that mimics the transition state may be an effective enzyme inhibitor. Similarly, an antibody that binds with high affinity to a transition state analog of a reaction may also catalyze that reaction.

Lysozyme

13. Hen egg white lysozyme, an enzyme that cleaves the glycosidic linkage between NAG and NAM residues in bacterial cell walls, has a substrate-binding cleft that accommodates six sugar residues such that the cleavage occurs between the residues in subsites D and E.

14. In the Phillips mechanism for lysozyme, the NAM residue that binds in the D site is distorted toward the half-chair conformation. Glu 35 then transfers a proton to O1 to cleave the C1—O1 bond of the substrate (general acid catalysis). The resulting oxonium ion intermediate is stabilized by the ionized carboxyl group of Asp 52 (electrostatic catalysis).

The addition of water completes the catalytic cycle. Because the substrate is distorted toward the transition-state conformation on binding, transition state stabilization is an important catalytic mechanism.

Serine Proteases

15. Serine proteases are a widespread family of enzymes that have a common mechanism. The active-site Ser (identified through its inactivation by diisopropylphosphofluoridate), His (identified through affinity labeling with a chloromethylketone substrate analog), and Asp (identified by X-ray crystallography) form a hydrogen-bonded catalytic triad. Nonhomologous serine proteases have developed the same catalytic triad through convergent evolution.

16. The differing substrate specificities of trypsin, chymotrypsin, and elastase depend in part on the shapes and charge distribution of the substrate-binding pocket near the active site.

17. Catalysis by serine proteases is a multistep process in which nucleophilic attack on the scissile bond by Ser 195 (using the chymotrypsinogen numbering system) results in a tetrahedral intermediate that decomposes to an acyl–enzyme intermediate. The replacement of the amine product with water is necessary for the formation of a second tetrahedral intermediate, which yields the carboxyl product and regenerated enzyme.

18. Serine proteases use acid–base catalysis (involving the Ser-His-Asp triad), covalent catalysis (formation of the tetrahedral intermediates), and catalysis through binding of the transition state in the oxyanion hole. The tight binding of bovine pancreatic trypsin inhibitor to trypsin inhibits the enzyme by preventing full attainment of the tetrahedral intermediate as well as the entry of water into the active site.

Guide to Study Exercises (text p. 320)

1. Enzymes, which are either proteins or RNA, differ from other catalysts in that they increase reaction rates to a greater extent; they act under mild conditions (temperatures below 100°C, atmospheric pressure, near-neutral pH); they are much more specific for the substrates and reaction products; and their activity can be regulated. (Section 11-1)

2. An enzyme's substrate specificity depends on the arrangement of amino acid residues and the noncovalent interactions that occur at the substrate-binding site. Although an enzyme may change conformation on binding substrate, it must be complementary to the substrate with respect to geometry, stereochemistry, and charge distribution. (Section 11-1B)

3. Coenzymes are required for reactions that cannot readily be catalyzed by the functional groups that occur in the 20 standard amino acid residues. For example, oxidation–reduction reactions and many types of group-transfer reactions require cofactors such as metal ions and organic molecules (coenzymes). (Section 11-1C)

4. See Figure 11-5.

5. ΔG represents the difference in free energy between the reactants and products of a reaction and can be negative, positive, or zero. $\Delta G^{\ddagger}$, the free energy of activation, represents the difference in free energy between the reactants and the transition state and is always a positive quantity. ΔG indicates whether a reaction can proceed spontaneously (the reaction is favorable only when $\Delta G < 0$), whereas the rate of the reaction decreases exponentially with $\Delta G^{\ddagger}$ (the greater the value of $\Delta G^{\ddagger}$, the slower the rate). An enzyme may alter $\Delta G^{\ddagger}$ for a reaction, but it cannot alter its ΔG. (Section 11-2)

6. A nucleophile, which is an electron-rich group that is negatively charged or contains unshared electrons, can attack an electrophilic (electron poor) group to form a covalent bond. When the nucleophilic group of an enzyme attacks an electrophilic substrate, the resulting covalent bond allows the catalytic group (now an electrophile) to withdraw electrons, thereby facilitating the chemical transformation of the substrate. Because the covalent bond is unstable, the catalytic group is quickly eliminated, releasing the reaction product and the enzyme in its original nucleophilic form. (Section 11-3B)

7. Nonenzymatic catalysts commonly act via all modes of chemical catalysis except catalysis through stabilization of the transition state. This is because such a catalyst is unlikely to have the complex structure of an enzyme, whose binding site and arrangement of functional groups allow extensive interactions with the substrate throughout the catalytic process. (Section 11-3E)

8. Lysozyme accelerates the cleavage of its substrate through acid–base catalysis and electrostatic catalysis and by preferential binding of the transition state. Glu 35, a general acid catalyst, transfers its proton to O1 of the substrate's D ring to promote cleavage of the C1—O1 bond. Asp 52 acts as an electrostatic catalyst to stabilize the developing positive charge of the oxonium ion reaction intermediate. The D ring has already assumed a planar half-chair conformation, the conformation of the oxonium ion intermediate, in its initial binding to the enzyme, primarily through steric interference and through hydrogen bonding between O6 and the backbone NH of Val 109. Thus, lysozyme facilitates the reaction through stabilization of the transition state. (Section 11-4B)

9. The oxyanion hole of serine proteases refers to a portion of the active site that preferentially binds the tetrahedral reaction intermediate, thereby promoting catalysis through stabilization of the transition state. The substrate cannot occupy the oxyanion hole until it has been attacked by Ser 195. The resulting tetrahedral intermediate has a charged carbonyl oxygen that then moves into the hole to form three hydrogen bonds that are not present in the initial enzyme–substrate complex. (Section 11-5C)

10. Zymogens (inactive precursors) of digestive enzymes are maintained in their inactive state in the pancreas in three ways. (a) Trypsin, which converts trypsinogen, chymotrypsinogen, and other enzymes to their active forms, is not produced until the zymogens have been secreted into the duodenum, where enteropeptidase cleaves trypsinogen to form trypsin. (b) Trace amounts of trypsin that are generated within pancreatic cells are inhibited by specific inhibitor proteins such as BPTI. (c) Zymogen granules are resistant to proteolytic degradation and thereby encapsulate any inappropriately formed trypsin or other proteases that it has activated. (Section 11-5D)

Questions

General Properties of Enzymes

1. What is an enzyme's EC number?

2. Explain why enzymes are stereospecific.

3. Why is deuterium labeling useful for investigating the stereospecificity of an enzymatic reaction?

4. What is an apoenzyme and how does it differ from a holoenzyme? Which form is active?

5. What is the relationship between vitamins and coenzymes?

6. Proteins can be chemically modified by a variety of reagents that react with specific amino acid residues. How can such reagents be used to identify residues involved in an enzyme's activity? What are the shortcomings of this method?

Activation Energy and the Reaction Coordinate

7. What is the rate-determining step of an enzyme-catalyzed reaction?

8. Answer yes or no to the following questions and explain your answer.
   (a) Can the absolute value of ΔG for a reaction be larger than $\Delta G^{\ddagger}$?
   (b) Can $\Delta G^{\ddagger}$ for an enzyme-catalyzed reaction be greater than $\Delta G^{\ddagger}$ for the nonenzymatic reaction?
   (c) In a two-step reaction, such as the one diagrammed in Figure 11-4, must the intermediate (I) have less free energy than the reactant (A)?
   (d) In a multistep reaction, does the transition state with the highest free energy always correspond to the rate-determining step?

9. An increase in temperature increases the rate of a reaction. How does the temperature affect $\Delta G^{\ddagger}$?

10. $\Delta\Delta G^{\ddagger}$ for an enzymatic reaction at 25°C is 13 kJ/mole. (a) Calculate the rate enhancement. (b) What is $\Delta\Delta G^{\ddagger}$ when the rate enhancement is 10^5?

Catalytic Mechanisms

11. List the amino acid residues that are most likely to participate in general acid–base catalysis. Why isn't glycine among those listed?

12. If a protonated His residue acts as the proton donor in an acid-catalyzed enzymatic reaction, what happens to the enzyme's activity as the pH increases to a value that exceeds the pK_R of that residue?

13. The p*K* values of two essential catalytic residues in RNase A are 5.4 and 6.4. (a) Which corresponds to His 12 and which to His 119 (see Figure 11-8)? (b) Draw a titration curve for these two residues.

14. What is the difference between nucleophilic catalysis and general base catalysis?

15. A good covalent catalyst is highly nucleophilic and can form a good leaving group. What structural properties support these seemingly opposite characteristics?

16. Classify each of the following groups as an electrophile or nucleophile:
    (a) amine
    (b) carbonyl
    (c) cationic imine
    (d) hydroxyl
    (e) imidazole

17. For each of the following reactions, indicate the nucleophilic center and the electrophilic center. Draw curved arrows to indicate the movement of electrons and draw the reaction products.
    (a) carbonyl phosphate + ammonia $\rightleftharpoons$ carbamic acid + P_i

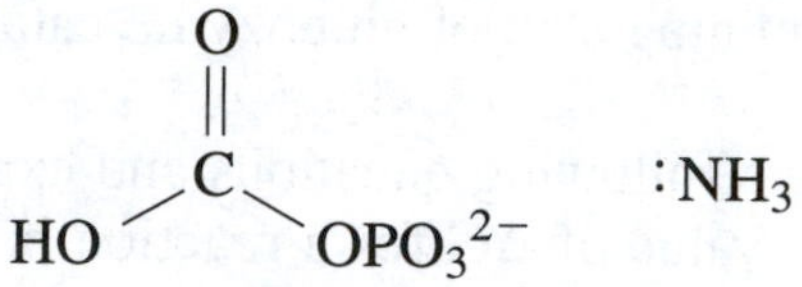

O
‖
HO–C–OPO_3^{2-} :NH_3

    (b) NADH + acetaldehyde $\rightleftharpoons$ ethanol + NAD^+

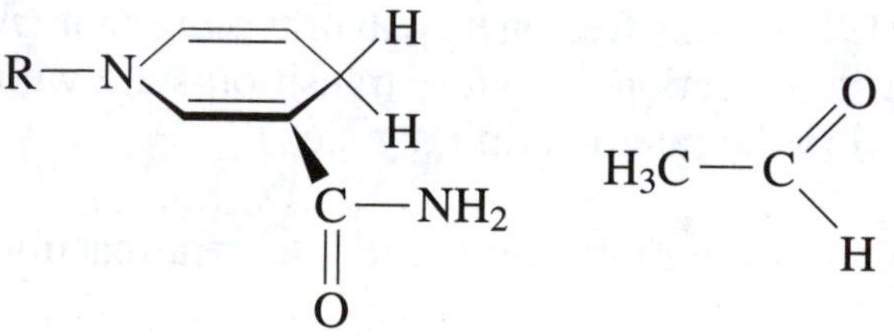

R–N (dihydropyridine ring, H, H, C(=O)–NH_2) H_3C–C(=O)–H

    (c) 2 amino acid $\rightleftharpoons$ dipeptide + H_2O

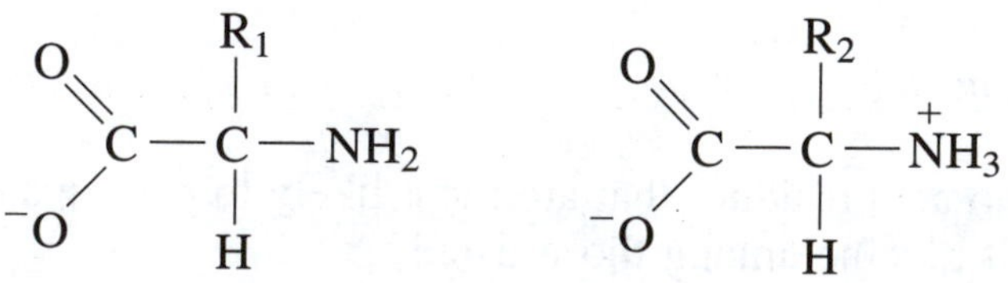

$^-$O–C(=O)–C(R_1)(H)–NH_2 $^-$O–C(=O)–C(R_2)(H)–$\overset{+}{N}H_3$

18. What is the role of Zn^{2+} in carbonic anhydrase?

Lysozyme

19. Why does lysozyme appear to bind only NAG residues at subsites C and E?

20. Would $(NAG)_6$ or $(NAM)_6$ be a better substrate for lysozyme and why?

21. Hydrogen bonding of substrates to enzymes often involves the polypeptide backbone rather than amino acid side chains. What backbone–substrate hydrogen bond helps distort NAM in the D subsite of lysozyme? Can this hydrogen bond form when *N*-acetylxylosamine is in the active site?

22. Draw the resonance forms of the half-chair conformation of the oxonium ion intermediate of the lysozyme reaction.

Serine Proteases

23. What two catalytic residues in chymotrypsin were identified by chemical modification? Could the same reagents used to identify these residues be used to label the catalytic residues of other serine proteases?

24. The different substrate specificities of chymotrypsin and trypsin have been attributed to the presence of different amino acid residues in the binding pocket. What problems can arise when site-directed mutagenesis is used to test predictions about the roles of such residues in substrate specificity?

25. The cleavage of the ester *p*-nitrophenylacetate (shown below) by chymotrypsin occurs in two stages. In the first stage, the product *p*-nitrophenolate is released in a burst, in amounts equivalent to the amount of active enzyme present. In the second stage, *p*-nitrophenolate is generated at a steady but much reduced rate. Explain this phenomenon in terms of the catalytic mechanism presented in Figure 11-26.

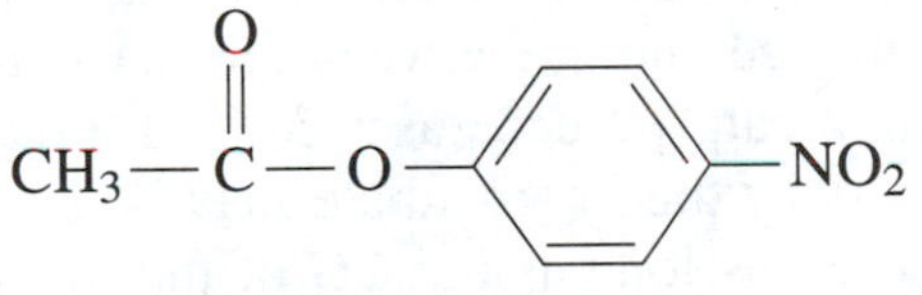

$$CH_3-\overset{\overset{\displaystyle O}{\|}}{C}-O-C_6H_4-NO_2$$

_p_-Nitrophenylacetate

26. What catalytic mechanism contributes the most to chymotrypsin's rate enhancement?

27. Lys 15 of bovine pancreatic trypsin inhibitor binds to the active site of trypsin but is not cleaved. Explain why the proteolytic reaction cannot proceed.

28. Since trypsin activation is autocatalytic, what is the role of enteropeptidase in activating trypsin?

Answers to Questions

1. The enzyme commission (EC) number is unique to each enzyme. An enzyme is assigned an EC number according to the type of reaction it catalyzes.

2. The protein (or RNA) enzyme is a chiral molecule whose binding clefts and catalytic residues are arranged in a specific three-dimensional asymmetric array. Hence, only substrates with the appropriate stereochemistry can bind to the enzyme, and the enzyme transforms the substrate to product according to the spatial arrangement of interacting functional groups.

3. Deuterium can be distinguished from hydrogen (usually by mass spectrometry). Substitution of a deuterium atom for a hydrogen atom usually does not significantly affect a reaction and provides a way to label a hydrogen-containing stereoisomer.

4. An apoenzyme is the protein portion of an enzyme that has lost its cofactor (a metal ion or coenzyme). A holoenzyme is an active enzyme containing both the protein and the cofactor.

5. Many coenzymes are synthesized from precursors that are vitamins (substances that an animal cannot synthesize and must obtain from its diet). However, not all coenzymes have vitamin precursors and not all vitamins are precursors of coenzymes.

6. Protein-modifying reagents can provide clues to the identities of catalytic residues. If chemical modification of a residue does not result in loss of activity, that residue can be ruled out as essential for catalysis. Loss of enzymatic activity on modification of a residue may indicate that the modified residue plays an essential role. However, chemical modification of residues at sites other than the active site may interfere with catalysis nonspecifically by disrupting protein structure.

7. The rate-determining step is the slowest of the steps in the reaction mechanism, the step with the greatest free energy of activation.

8. (a) Yes. ΔG is the difference in free energy between reactants and products, whereas $\Delta G^{\ddagger}$ is the difference in free energy between the reactants and the transition state.
   (b) No. By definition, a catalyst decreases $\Delta G^{\ddagger}$ of a reaction.
   (c) No. The free energy of the intermediate may be greater than that of the reactant. The reaction will proceed as long as the ΔG of the overall reaction A→P is negative.
   (d) No. The rate-determining step is the one whose $\Delta G^{\ddagger}$ is greatest. This does not always correspond to the step with the highest free energy, since $\Delta G^{\ddagger}$ depends on the difference in free energies between a reactant or an intermediate and the following transition state, not just on the free energy of this transition state.

9. $\Delta G^{\ddagger}$ is largely independent of temperature. An increase in temperature increases the rate of a reaction by increasing the number of reacting molecules that reach the transition state.

10. (a) The rate enhancement is

$$
\begin{aligned}
& e^{\Delta\Delta G^{\ddagger}/RT} \\
&= e^{(13\ \text{kJ}\cdot\text{mol}^{-1})/(8.3145\ \text{J}\cdot\text{K}^{-1}\cdot\text{mol}^{-1})(298\text{K})} \\
&= e^{5.25} \\
&= 190
\end{aligned}
$$

The enzyme-catalyzed reaction therefore proceeds 190 times faster than the uncatalyzed reaction.

(b) When the rate enhancement is 10^5,

$$\begin{aligned} 100{,}000 &= e^{\Delta\Delta G^{\ddagger}/RT} \\ \Delta\Delta G^{\ddagger} &= RT \ln 100{,}000 \\ &= (8.3145\ \text{J} \cdot \text{K}^{-1} \cdot \text{mol}^{-1})(298\text{K})(11.5) \\ &= 28.5\ \text{kJ} \cdot \text{mol}^{-1} \end{aligned}$$

11. Aspartate, cysteine, glutamate, histidine, lysine, and tyrosine are likely to participate in general acid–base catalysis. Glycine does not have an ionizable side chain.

12. As the pH approaches the pK_R, the His become progressively deprotonated and less effective as an acid catalyst. When pH = pK_R, 50% of the His residues are deprotonated and the enzyme is half-active. When the pH significantly exceeds pK_R, the His is entirely deprotonated and the enzyme is inactive.

13. (a) His 12 has a p*K* of 5.4, since it is active when unprotonated. It would lose activity when the pH decreases. His 119 has a p*K* of 6.4 since it is active when protonated. It would lose activity when the pH increases.

(b)

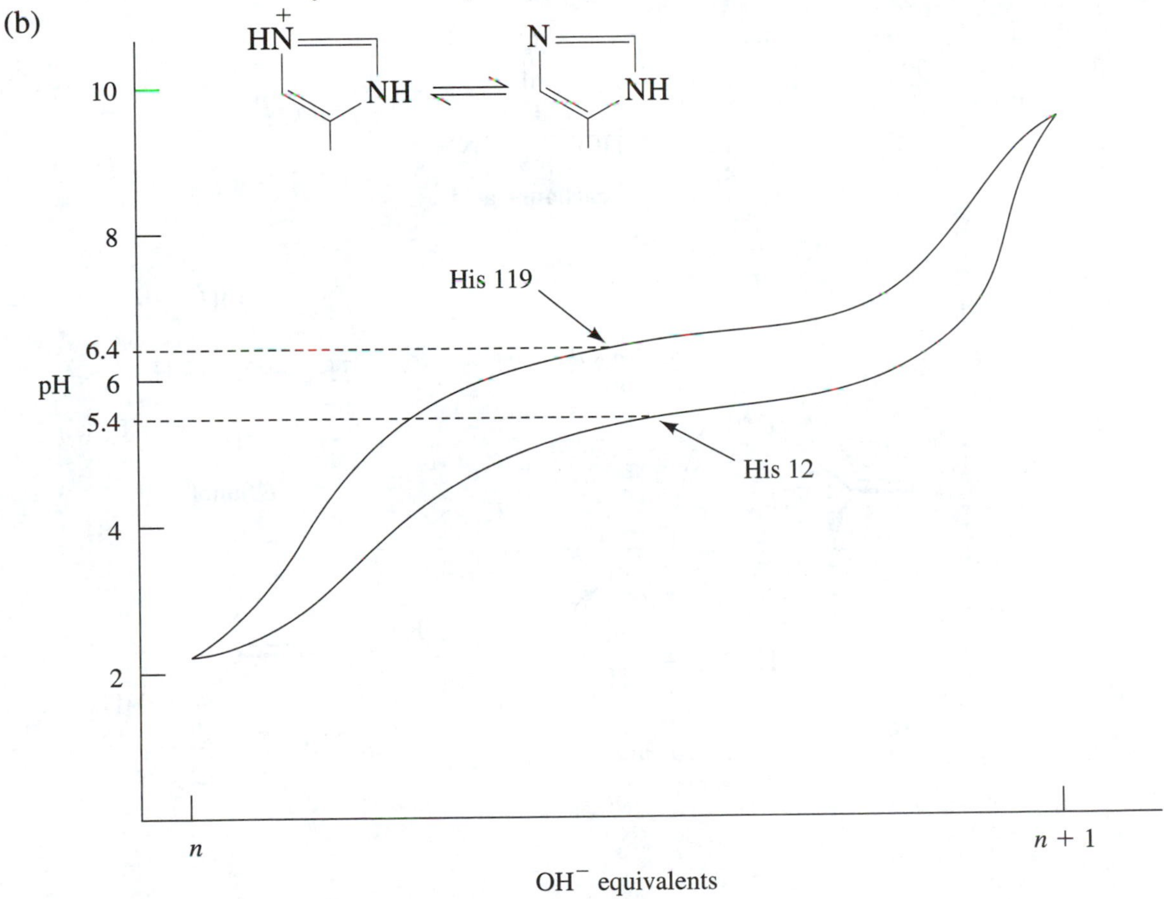

14. In general base catalysis, a proton is abstracted from the substrate, whereas in nucleophilic catalysis, a covalent bond forms.

15. Amino acid side chains that have highly delocalized electrons or high polarizabilities are good covalent catalysts. The hydroxyl group of serine, the carboxyl group of aspartate, and the thiol group of cysteine are highly polarizable groups, and the electrons are highly delocalized in the imidazole group of histidine.

16. (a) nucleophile
(b) electrophile
(c) electrophile
(d) nucleophile
(e) nucleophile

17. (a)

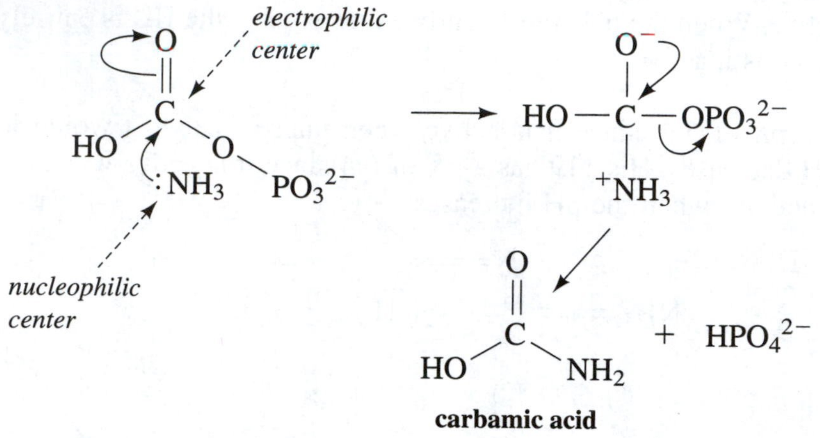

electrophilic center

nucleophilic center

$HO-C(=O)-O-PO_3^{2-}$ + $:NH_3$ → $HO-C(O^-)(NH_3^+)-OPO_3^{2-}$ → $HO-C(=O)-NH_2$ + HPO_4^{2-}

carbamic acid

(b)

nucleophilic center

electrophilic center

NADH + H_3C-CHO + H^+ → ethanol + NAD^+

ethanol

NAD^+

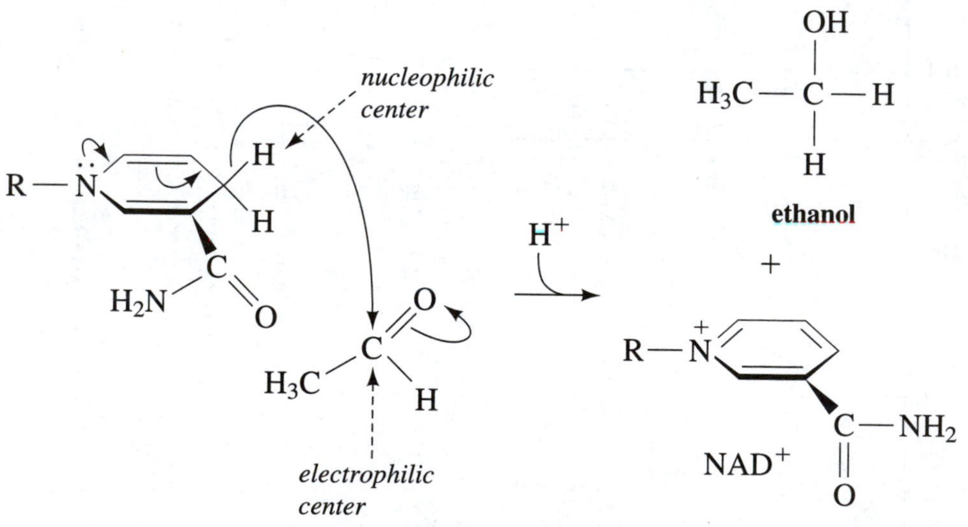

(c)

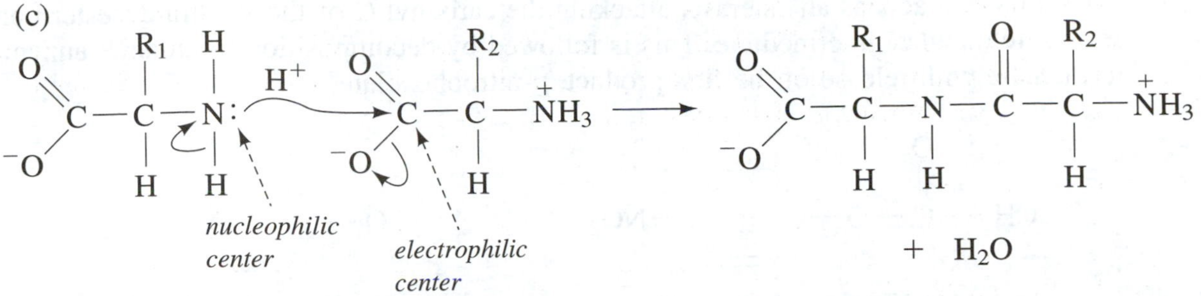

nucleophilic center

electrophilic center

$+ H_2O$

18. The Zn^{2+} in carbonic anhydrase polarizes a water molecule and thereby causes it to ionize. The resulting $Zn^{2+}—OH^-$ is the nucleophile that attacks carbon dioxide, converting it to HCO_3^-. The metal ion stabilizes the negative charge of the OH^-, which would otherwise not form at neutral pH.

19. The lactyl side chain of NAM residues sterically prevents NAM binding to subsites C and E. Hence only NAG residues bind to these subsites.

20. NAG_6 would be better because the C and E residues of NAM_6 would not bind to their subsites.

21. The backbone NH group of Val 109 forms a hydrogen bond with O6 of the D-site residue, helping stabilize its half-chair conformation. *N*-Acetylxylosamine lacks O6 and therefore cannot form this hydrogen bond.

22.

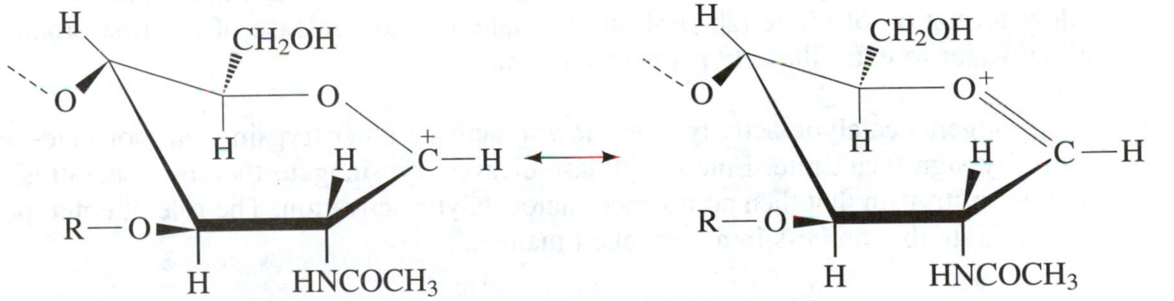

23. Ser 195 was identified through the use of diisopropylphosphofluoridate (DIPF), which reacts with the Ser—OH group at the active site of chymotrypsin and irreversibly inactivates the enzyme. His 57 was identified through affinity labeling using the substrate analog tosyl-L-phenylalanine chloromethylketone. DIPF also reacts with the active-site Ser in other serine proteases and therefore would label this residue. However, these other enzymes, not all of which share chymotrypsin's specificity for Phe-containing substrates, would require chloromethylketone analogs that incorporated residues corresponding to their substrate specificities, in order to react with their active-site His residues.

24. Site-directed mutations often change (or fail to change) enzymes in unexpected ways. For example, site-directed mutation of trypsin's Asp 189 to Ser did not change trypsin's specificity to that of chymotrypsin. Several other changes involving amino acids found on the surface loops surrounding the binding pocket were required for trypsin to emulate chymotrypsin.

25. Chymotrypsin acts as an esterase, attacking the carbonyl C of the substrate's ester bond to form a tetrahedral intermediate. This is followed by decomposition to an acyl–enzyme intermediate with release of the first product, *p*-nitrophenolate.

$$CH_3{-}\overset{\overset{\displaystyle O}{\|}}{C}{-}O{-}C_6H_4{-}NO_2 \quad + \quad HO{-}\text{Chymotrypsin} \longrightarrow {}^{-}O{-}C_6H_4{-}NO_2 \quad + \quad CH_3{-}\overset{\overset{\displaystyle O}{\|}}{C}{-}O{-}\text{Chymotrypsin}$$

p-Nitrophenylacetate + HO—Chymotrypsin ⟶ **p-Nitrophenolate** + acyl–chymotrypsin

Release of the second product, acetate, via hydrolysis of the acyl-enzyme intermediate, is much slower. Consequently, every active site attacks the substrate and releases the first product in a stoichiometric fashion, accounting for the initial burst of product formation. Because the second phase of the reaction is slow, the enzyme is only slowly regenerated and made available to catalyze additional rounds of the esterolytic reaction.

26. Stabilization of the transition state via the oxyanion hole is responsible for the largest portion of chymotrypsin's rate enhancement.

27. Bovine pancreatic trypsin inhibitor binds trypsin so tightly that the complex is too rigid to allow formation of the tetrahedral intermediate, to allow release of the first product, or to allow water to enter the active site of trypsin.

28. Trypsinogen's catalytic activity is too low to activate other trypsinogen molecules at a biologically significant rate. Enteropeptidase cleaves trypsinogen, thereby generating a small amount of trypsin that then commences autocatalytic activation. The role of enteropeptidase is to initiate this process in a controlled manner.

Chapter 12 Enzyme Kinetics, Inhibition, and Regulation

This chapter introduces chemical kinetics—the study of reaction rates—followed by the kinetics of enzymatic reactions. An enzyme-catalyzed reaction can be described by the Michaelis–Menten equation, which expresses the reaction velocity in terms of its Michaelis constant, K_M, and its maximum velocity, V_{max}. Detailed knowledge of the kinetics of a reaction can contribute to the understanding of its step-by-step reaction mechanism. The effects of different substrates, inhibitors, and other factors may also reveal an enzyme's physiological function. This knowledge can be exploited to develop drugs that are enzyme inhibitors. In this chapter, the three types of reversible enzyme inhibition and the equations that describe them are presented. Finally, the chapter describes the allosteric regulation of enzymes, using aspartate transcarbamoylase as an example.

Essential Concepts

Reaction Kinetics

1. A chemical reaction may proceed through several simple steps, called elementary reactions. The overall reaction pathway may therefore involve several short-lived intermediates.

2. The rate, or velocity (v), at which a reactant is consumed or a reaction product appears can be mathematically described. Thus, for the conversion of reactant A to product P,

$$v = \frac{d[\mathrm{P}]}{dt} = -\frac{d[\mathrm{A}]}{dt} = k[\mathrm{A}]$$

3. The rate of an elementary reaction varies with the concentration(s) of the reacting molecule(s). For example, for a single-reactant reaction (a unimolecular or first-order reaction), the rate is directly proportional to the concentration of the reactant. For a two-reactant reaction (a bimolecular or second-order reaction), the rate is directly proportional to the product of the concentrations of the reactants or to the square of the concentration of a reactant that reacts with itself.

4. The proportionality constant in the equation above, which is known as the rate constant, k, can be determined graphically. The rate equation for a first-order reaction is

$$\ln [\mathrm{A}] = \ln [\mathrm{A}]_o - kt$$

and that for a second-order reaction is

$$\frac{1}{[\mathrm{A}]} = \frac{1}{[\mathrm{A}]_o} + kt$$

where $[\mathrm{A}]_o$ is the initial concentration of the reactant and t is time. Consequently, if a plot of ln [A] versus t yields a straight line, the reaction is first-order, and if a plot of 1/[A] versus t yields a straight line, the reaction is second-order. The slope of the line reveals the value of the corresponding rate constant, k.

5. The kinetics of enzyme-catalyzed reactions are more complicated because the enzyme and substrate (reactant) combine to form a complex that then decomposes to product and free enzyme. For reactions involving a single substrate, S, the reaction velocity is typically measured under conditions where $[S] \gg [E]$. At very high substrate concentrations, the velocity is independent of [S] and the enzyme is said to be saturated with substrate.

6. The Michaelis–Menten equation, which describes an enzymatic reaction, is based on the assumption that the enzyme–substrate complex maintains a steady state; that is, its concentration does not change. This assumption is valid over most of the course of a typical enzymatic reaction.

7. The Michaelis–Menten equation is

$$v_o = \frac{V_{max}\,[S]}{K_M + [S]}$$

where v_o is the initial velocity of the reaction (before more than ~10% of the substrate has been consumed), V_{max} is the maximum rate of the reaction, and K_M is the Michaelis constant. This equation describes a rectangular hyperbola (the shape of the curve generated by a plot of v_o versus [S]) whose asymptote is V_{max}.

8. The Michaelis constant, K_M, is unique to each enzyme–substrate pair. Its value is the substrate concentration at which the reaction velocity is half-maximal. It is therefore a measure of the affinity of the enzyme for its substrate.

9. The catalytic constant (k_{cat}), or turnover number, of an enzyme can be derived from V_{max}:

$$k_{cat} = \frac{V_{max}}{[E]_T}$$

where $[E]_T$ is the total enzyme concentration. The overall catalytic efficiency of an enzyme can be expressed as k_{cat}/K_M, which is an apparent second-order rate constant for the enzymatic reaction.

10. The kinetic parameters for an enzymatic reaction can be determined by taking the reciprocal of the Michaelis–Menten equation:

$$\frac{1}{v_o} = \left(\frac{K_M}{V_{max}}\right)\frac{1}{[S]} + \frac{1}{V_{max}}$$

A plot of $1/v_o$ versus $1/[S]$, a so-called Lineweaver–Burk or double-reciprocal plot, yields a straight line whose slope and intercepts yield the values of K_M and V_{max}.

11. Steady state kinetics cannot unambiguously establish a reaction mechanism because there are an infinite number of mechanisms that are consistent with a given set of kinetic data. However, mechanisms that are not consistent with the kinetic data can be ruled out.

12. Many enzymes have multiple (usually two) substrates and products. For example, transferase reactions are bisubstrate reactions. In a Sequential reaction, all the substrates bind to the enzyme before products are formed. In Sequential reactions, a particular order of substrate addition may be obligatory (an Ordered mechanism) or not (a Random mechanism).

13. In a Ping Pong reaction, one or more products of a transferase reaction are released before all substrates bind.

Enzyme Inhibition

14. Many drugs and pharmaceutical compounds alter the activities of specific enzymes. Detailed information about the mechanism of an enzyme can aid in the design of drugs with the desired properties. Studies of the kinetics of enzymes in the presence of specific inhibitors help reveal how the inhibitor acts and provide a way to quantitatively compare the effects of different inhibitors. An inhibitor may reversibly interact with an enzyme to interfere with its substrate binding, its catalytic activity, or both.

15. Three modes of reversible enzyme inhibition can be distinguished by their effects on the kinetic behavior of enzymes: competitive, uncompetitive, and mixed (noncompetitive) inhibition. Double-reciprocal plots of data collected in the presence of different concentrations of an inhibitor reveal the value of K_I, the dissociation constant of the inhibitor from the enzyme.

16. A competitive inhibitor competes with a normal substrate for binding to the enzyme. It therefore reduces the apparent affinity of the enzyme for its substrate (increases K_M). A large excess of substrate can overcome the effect of the inhibitor, so V_{max} is not affected.

17. An uncompetitive inhibitor binds only to the enzyme–substrate complex and apparently distorts the active site. It decreases the apparent K_M and V_{max}.

18. A mixed inhibitor binds to both free and substrate-bound enzymes and may interfere with both substrate binding and catalysis. As a result, the apparent V_{max} decreases, and the apparent K_M may increase or decrease. When only V_{max} is affected, the inhibition is said to be noncompetitive.

Regulation of Enzyme Activity

19. Enzyme activity can be controlled either by altering the amount of enzyme available for reaction or by modifying its catalytic activity through allosteric effects (or by covalent modification). For example, in feedback regulation, the end product of a metabolic pathway inhibits the first committed step in the pathway. This ensures that the pathway is active when product concentrations are low but is inactive when product concentrations exceed the levels needed by the cell.

20. Aspartate transcarbamoylase (ATCase) provides an example of allosteric regulation that includes feedback inhibition by CTP (the ultimate product of the pyrimidine synthesis pathway, which begins with ATCase) and activation by ATP (which ensures that the concentrations of pyrimidine nucleotides keep pace with those of purines). The binding of either effector molecule (ATP or CTP) to the regulatory subunits of ATCase induces changes in the enzyme's quaternary structure that alter the activity of the catalytic subunits. ATP stabilizes the R (high activity) state, whereas CTP stabilizes the T (low activity) state of this allosteric enzyme.

Key Equations

$$\ln [A] = \ln [A]_o - kt$$

$$\frac{1}{[A]} = \frac{1}{[A]_o} + kt$$

$$v_o = \frac{V_{max}[S]}{K_M + [S]}$$

$$k_{cat} = \frac{V_{max}}{[E]_T}$$

$$\frac{1}{v_o} = \left(\frac{K_M}{V_{max}}\right)\frac{1}{[S]} + \frac{1}{V_{max}}$$

Guide to Study Exercises (text p. 348)

1. First-order reaction: $\ln [A] = \ln [A]_o - kt$ or $[A] = [A]_o e^{-kt}$
   Second-order reaction: $1/[A] = 1/[A]_o + kt$
   (Section 12-1A)

2. Instantaneous velocity is the rate of a reaction at a particular point in time. It varies over the course of the reaction. The initial velocity of a reaction, v_o, is defined as the rate when $t = 0$, but really refers to the rate at a point after the enzymatic reaction has achieved a steady state (usually less than 1 sec). In practice, v_o is taken to be the reaction velocity before more than ~10% of the substrate has been converted to product. The maximal velocity is a property of enzyme-catalyzed reactions and refers to the velocity when $[S] \gg K_M$, that is, at substrate concentrations at which the enzyme is saturated with substrate. (Sections 12-1A and B)

3. See pp. 327–328.

4. K_M, the substrate concentration at which the rate of an enzyme-catalyzed reaction is half-maximal, is a measure of an enzyme's affinity for a substrate. The smaller the K_M, the higher the affinity.

The value of k_{cat}/K_M is indicative of an enzyme's catalytic efficiency. It is an apparent second-order rate constant for the reaction E + S → P. Hence, it takes into account the enzyme's affinity for the substrate (K_M) as well as the rate that the enzyme converts ES to P (k_{cat}). (Section 12-1B)

5. The Lineweaver–Burk plot is $1/v_o = (K_M/V_{max})(1/[S]) + 1/V_{max}$. Therefore, if $1/v_o$ is plotted as a function of 1/[S], the slope of the line is K_M/V_{max}, the extrapolated intercept on the 1/[S] axis is $-1/K_M$, and the intercept on the $1/v_o$ axis is $1/V_{max}$. (Section 12-1C)

6. See p. 334.

7. A competitive inhibitor increases K_M but does not affect V_{max}. An uncompetitive inhibitor decreases both K_M and V_{max}. A mixed inhibitor decreases V_{max} and may increase or decrease K_M. (Section 12-2)

8. An inhibitor binds reversibly to an enzyme, whereas an inactivator reacts irreversibly with an enzyme to inactivate it. (Section 12-2)

9. Enzyme activity can be regulated by (a) altering the amount of enzyme present through changes in its rate of synthesis or degradation, and (b) influencing its substrate-binding or catalytic properties through allosteric effectors or by covalent modification. (Section 12-3)

10. In ATCase, three dimers of regulatory subunits link the subunits of two catalytic trimers. Allosteric effects are made possible by intersubunit contacts that communicate changes in one subunit to the other 11 subunits. The activator ATP preferentially binds to and stabilizes the R (high activity) conformation of ATCase, whereas the inhibitor CTP preferentially binds to and stabilizes the T (low activity) conformation. Hence, ATP or CTP binding to a regulatory subunit alters the substrate affinity of all the catalytic subunits. The quaternary structural change of the T → R transition is primarily a counter-rotation of the regulatory dimers of ~15° that is accompanied by the separation of the catalytic trimers by ~11 Å. In the R state, the two domains of each catalytic subunit have swung together to assume a catalytically active conformation. (Section 12-3)

Questions

Reaction Kinetics

1. For each of the following reactions, write a rate equation and determine the reaction order.
   (a) A → P
   (b) A + B → P + Q
   (c) 2A → P

2. List two different ways to measure the progress of a chemical reaction.

3. A first-order reaction has a $t_{1/2}$ of 20 minutes.
   (a) What is the rate constant k?
   (b) What time is required to form 20% of the product?

(c) What time is required to form 80% of the product?
(d) How much starting material remains after 15 min?
(e) Compare the rate constant for this reaction to that of the decay of ^{32}P, which has a half-life of 14 days.

4. The energy of binding a transition state complex ($X^{\ddagger}$) can be determined by writing an equilibrium expression for the formation of the complex.
   (a) For the reaction $A \rightleftharpoons X^{\ddagger}$, what is the equilibrium expression?
   (b) What is the expression for the free energy of binding to the transition state ($\Delta G^{\ddagger}$)?
   (c) What factors are needed to equate the rate constant k to the free energy of activation?

5. For the following reaction:

$$E + S \underset{k_{-1}}{\overset{k_1}{\rightleftharpoons}} ES \overset{k_2}{\rightleftharpoons} P + E$$

   (a) What is meant by the term "enzyme–substrate complex"?
   (b) Write a rate equation for the production of ES.
   (c) What is the rate of product formation from ES?
   (d) If all the enzyme is bound to substrate, what is the effect of adding more substrate on the forward rate of the reaction?

6. What is meant by (a) the steady state assumption, (b) K_M, (c) k_{cat}, (d) turnover number, (e) catalytic efficiency, and (f) diffusion-controlled limit?

7. The following data were obtained for the reaction $A \rightleftharpoons B$, catalyzed by the enzyme Aase. The reaction volume was 1 mL and the stock concentration of A was 5.0 mM. Seven separate reactions were examined, each containing a different amount of A. The reactions were initiated by adding 2.0 μL of a 10 μM solution of Aase. After 5 minutes, the amount of B was measured.

Reaction	*Volume of A added (μL)*	*Amount of B present at 5 minutes (nmoles)*
1	8	26
2	10	29
3	15	39
4	20	43
5	40	56
6	60	62
7	100	71

   (a) Calculate the initial velocity of each reaction (in units of $\mu M \cdot min^{-1}$).
   (b) Determine the K_M and V_{max} of Aase from a Lineweaver–Burk plot.
   (c) Calculate k_{cat}.

8. Can you use kinetic data to prove that a particular model for an enzymatic reaction mechanism is correct? Explain.

9. Why is it possible for Sequential bisubstrate reactions to be Ordered or Random, whereas a Ping Pong reaction always has an invariant order of substrate addition and product release?

Enzyme Inhibition

10. There are three general mechanisms for the reversible inhibition of enzymes that follow the Michaelis–Menten model. How does the mode of inhibitor–enzyme binding differ among the three mechanisms?

11. The catalytic behavior of an enzyme may depend on ionizable amino acids. Therefore, a change in pH may influence an enzyme's catalytic behavior. How can you tell whether pH affects substrate binding or catalytic activity?

12. The movement of glucose across the erythrocyte membrane is "catalyzed" by a transport protein (Section 10-4B and Box 10-2).
    (a) What is the kinetic behavior of this process?
    (b) Can glucose transport be subject to competitive, uncompetitive, or mixed inhibition? Explain.

13. In hen egg white lysozyme (Section 11-4), the substitution of Ala for Asn at position 37 or for Trp at position 62 may alter the enzyme's kinetics. What changes would you predict and why?

Regulation of Enzyme Activity

14. Draw velocity versus [Asp] curves for the reaction catalyzed by the ATCase catalytic trimer and by the intact enzyme. Explain why the curves differ.

15. How do carbamoyl phosphate, aspartate, ATP, and CTP affect the $T \rightleftharpoons R$ equilibrium of ATCase?

Answers to Questions

1. (a) $v = -d[\mathrm{A}]/dt = k[\mathrm{A}]$ This is a first-order reaction.
   (b) $v = -d[\mathrm{A}]/dt = -d[\mathrm{B}]/dt = k[\mathrm{A}][\mathrm{B}]$ This is a second-order reaction (A and B must collide to form product).
   (c) $v = -d[\mathrm{A}]/dt = k[\mathrm{A}]^2$ This is also a second-order reaction (A must collide with another molecule of A for the reaction to proceed).

2. The progress of a reaction can be followed by measuring the rates of the appearance of the product(s) or the disappearance of the reactant(s). In practice, any physical property, such as light absorbance, pH, or an NMR signal, can be followed, provided that it changes in proportion to the concentration(s) of the reactant(s) or product(s).

3. (a) For a first-order reaction, $t_{1/2} = 0.693/k$ (Equation 12-9). Therefore, $k = 0.693/20$ min $= 0.035$ min^{-1}.
   (b) Since $\ln([A]/[A]_o) = -kt$ (from Equation 12-6), $t = \ln([A]/[A]_o)/-k$. When 20% of A has been converted to product, $[A]/[A]_o = 0.8$ and $t = (\ln 0.8)/(-0.035\ \text{min}^{-1}) = (-0.22)/(-0.035\ \text{min}^{-1}) = 6.4$ min.
   (c) When 80% of A has been converted to product, $[A]/[A]_o = 0.2$ and $t = (\ln 0.2)/(-0.035\ \text{min}^{-1}) = (-1.61)/(-0.035\ \text{min}^{-1}) = 46$ min.
   (d) When $t = 15$ min, $\ln([A]/[A]_o) = (-0.035\ \text{min}^{-1})(15\ \text{min}) = -0.525$. Since $e^{-0.525} = 0.59$, 59% of A remains at 15 minutes.
   (e) For the decomposition of ^{32}P, $k = (0.693/14\ \text{days})\ (1\ \text{day}/1440\ \text{min}) = 3.4 \times 10^{-5}$ min^{-1}. This is ~1000 times slower than the reaction described above.

4. (a) $K^{\ddagger} = [X^{\ddagger}]/[A]$.
   (b) $\Delta G^{\ddagger} = -RT \ln K^{\ddagger}$.
   (c) As described in Box 12-2, $d[P]/dt = k[A] = k'[X^{\ddagger}]$ where k' is the rate constant for the decomposition of $X^{\ddagger}$ to from products. k' can be expressed in terms of the Boltzman constant (k_B) and Planck's constant (h): $k' = k_BT/h$. Thus, since $[X^{\ddagger}] = K^{\ddagger}[A]$ and $K^{\ddagger} = e^{-\Delta G^{\ddagger}/RT}$,

$$\frac{d[P]}{dt} = \frac{k_BT}{h} e^{-\Delta G^{\ddagger}/RT} [A]$$

and

$$k = \left(\frac{k_BT}{h}\right) e^{-\Delta G^{\ddagger}/RT}$$

5. (a) The enzyme–substrate (ES) complex is the species formed by the interaction between an enzyme and its substrate.
   (b) The rate equation for the net formation of ES is the rate of formation of ES minus the rate of ES degradation: $d[ES]/dt = k_1[E][S] - k_{-1}[ES] - k_2[ES]$.
   (c) The rate of product formation, $d[P]/dt = v = k_2[ES]$.
   (d) Increasing substrate concentration when all the enzyme is in the ES complex does not increase the rate of the reaction since then $v = V_{max} = k_2[E]_T$ where $[E]_T$ is the total enzyme concentration; that is, the enzyme is working at its maximal rate; it can work no faster.

6. (a) The steady state assumption assumes that during the course of an enzyme-catalyzed reaction, the concentration of the ES complex does not change.
   (b) K_M is the Michaelis constant: $K_M = (k_{-1} + k_2)/k_1$. K_M is the substrate concentration at which the reaction velocity is half-maximal.
   (c) The k_{cat} is the maximum velocity divided by the total enzyme concentration: $k_{cat} = V_{max}/[E]_T$. It is the number of reaction processes (turnovers) that each active site catalyzes per unit time. For the simple kinetic scheme used to derive the Michaelis–Menten equation, $k_{cat} = k_2$.
   (d) The turnover number is the same as k_{cat}.

(e) Catalytic efficiency, calculated as k_{cat}/K_M, is the apparent second-order rate constant for the reaction of E + S and indicates how often the enzyme catalyzes a reaction upon encountering its substrate.

(f) If an enzyme catalyzes a reaction every time it collides with its substrate, it has reached catalytic perfection and the rate is controlled by how often the molecules collide, that is, by their rate of diffusion. At this point, the rate is said to have reached its diffusion-controlled limit.

7. (a) To calculate v_o for Reaction 1, for example,

$$v_o = (26 \text{ nmol}/5 \text{ min})/(1.0 \text{ mL}) \times (10^3 \text{ mL}/1 \text{ L}) \times (0.001 \text{ µmol/nmol})$$
$$= 5.2 \text{ µM} \cdot \text{min}^{-1}$$

Reaction	*v_o (µM · min^{-1})*
1	5.2
2	5.8
3	7.8
4	8.6
5	11.2
6	12.4
7	14.2

(b) First, calculate [S] for each reaction. In Reaction 1, for example, [A] = (0.008 mL)(5 mM)/(1 mL) × (1000 µM/1 mM) = 40 µM. Next, convert the data to values of 1/[S] and 1/*v*.

Reaction	*[S] (µM)*	*1/[S] (µM^{-1})*	*v (µM · min^{-1})*	*1/v (min · µM^{-1})*
1	40	0.025	5.2	0.192
2	50	0.02	5.8	0.172
3	75	0.0133	7.8	0.128
4	100	0.010	8.6	0.116
5	200	0.005	11.2	0.089
6	300	0.0033	12.4	0.081
7	500	0.002	14.2	0.070

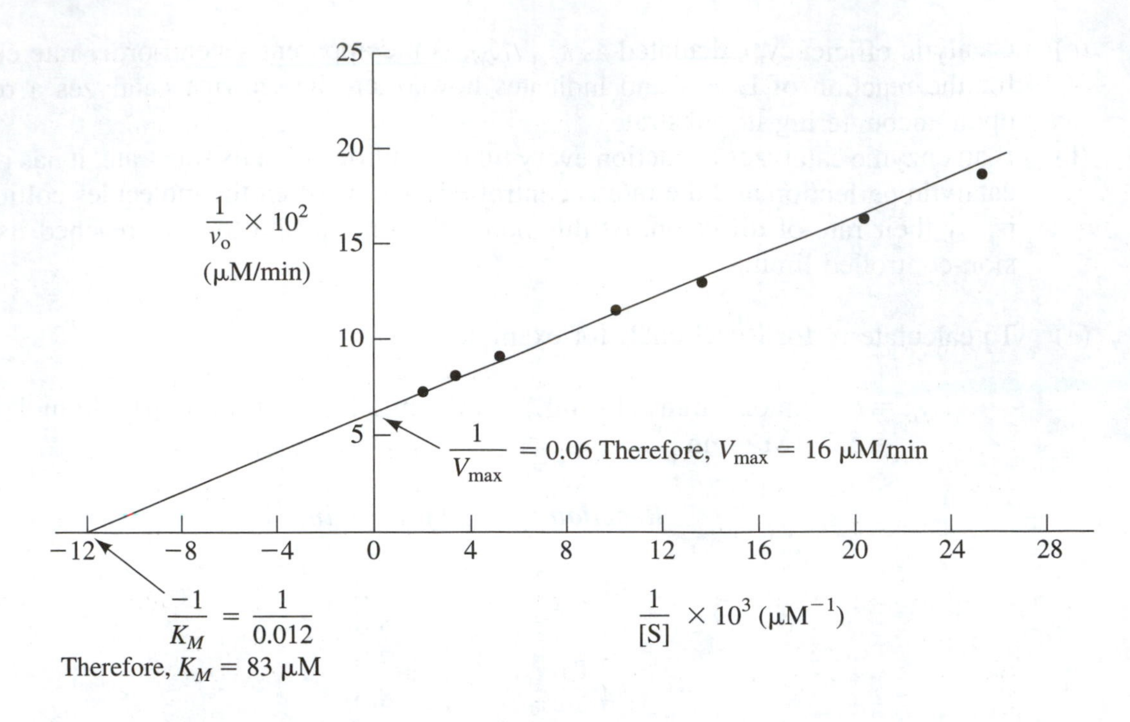

(c) First calculate $[E]_T$ = (0.002 mL)(10 μM Aase)/(1 mL) = 0.02 μM. Using the value of V_{max} determined above and Equation 12-27, $k_{cat} = V_{max}/[E]_T$ = (16 μM · min^{-1})/(0.02 μM) = 800 min^{-1}.

8. Kinetics can support but cannot prove a particular reaction mechanism. A single kinetic model may explain several mechanisms, so additional experiments must be performed to establish a particular mechanism. However, kinetic data can rule out mechanisms that are inconsistent with the observed behavior.

9. In a Sequential reaction, both substrates must bind before any product is released. Hence it is possible for the two substrates to bind in an Ordered or Random fashion. In a Ping Pong reaction, a group is transferred from the first substrate to the enzyme to form the first product; then a second substrate binds and is converted to a second product. Consequently, the second product cannot bind first or yield the second product until the first substrate has bound and been converted to the first product.

10. The three general mechanisms of inhibition are competitive, uncompetitive, and mixed inhibition. In competitive inhibition, the inhibitor competes with the substrate for binding to the active site, increasing the apparent K_M. The inhibitor binds only to the free enzyme and not to the ES complex. In uncompetitive inhibition, the inhibitor binds only to the ES complex to inhibit the enzyme. Although the inhibitor does not directly affect substrate binding, it decreases the apparent K_M. In mixed inhibition, the inhibitor can bind to both the free enzyme and the ES complex. Binding of a mixed inhibitor may alter the binding of the substrate and therefore may alter the apparent K_M. If the inhibitor does not alter the apparent K_M, inhibition is said to be noncompetitive.

11. Aside from extremely low or high pH's, which can denature an enzyme, changes in pH may affect the protonation/deprotonation of residues involved in substrate binding and/or catal-

ysis. Constructing a Lineweaver–Burk plot at several different pH values should yield a set of lines that indicate whether K_M (substrate binding) and/or V_{max} (catalytic activity) is affected (just as Lineweaver–Burk plots can reveal the different types of enzyme inhibition).

12. (a) The process obeys Michaelis–Menten (saturation) kinetics:

$$\text{Glucose}_{out} + \text{T}_{out} \underset{k_{-1}}{\overset{k_1}{\rightleftharpoons}} \text{glucose}_{out} \cdot \text{T}_{out} \xrightarrow{k_2} \text{glucose}_{in} \cdot \text{T}_{in} \rightarrow \text{glucose}_{in} + \text{T}_{in}$$

(b) Yes. All three inhibitory modes are possible. An inhibitor could compete with glucose for binding to the transport protein (competitive inhibition); it could bind to the transporter–glucose complex and interfere with the conformational change that exposes glucose to the other side of the membrane (uncompetitive inhibition); or it could interfere with the transporter's function with glucose bound or not bound (mixed inhibition).

13. These substitutions would weaken substrate binding to lysozyme by removing hydrogen bonds that hold the substrate in place. The K_M would increase, but V_{max} would remain nearly the same, since the catalytic residues Asp 52 and Glu 35 are not changed.

14. The catalytic trimer does not display cooperativity in the absence of the regulatory subunits. Hence the rate profile is hyperbolic, much like the binding of O_2 to myoglobin. Intact ATCase exhibits a sigmoidal curve characteristic of cooperative substrate binding, which is possible when the regulatory subunits mediate conformational changes between catalytic subunits.

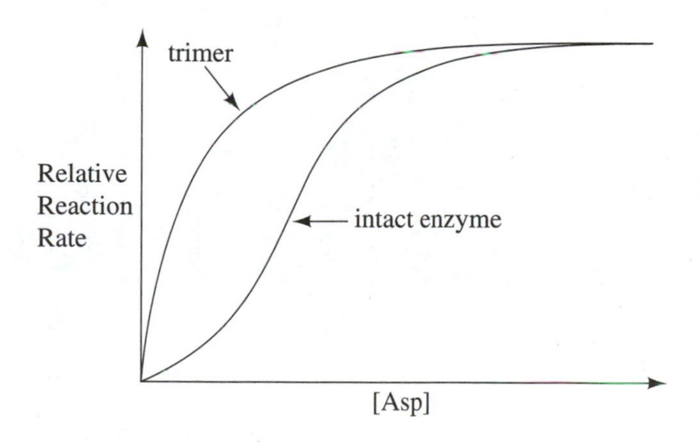

15. ATP has higher affinity for and stabilizes the R (more active) state, and CTP has higher affinity for and stabilizes the T (less active) state.

Chapter 13 Introduction to Metabolism

This chapter provides a brief overview of the biological strategies and thermodynamics of metabolism. Metabolism is the overall biochemical processes that living systems use to acquire and use free energy. Organisms break down macromolecules to a common set of smaller molecules, or metabolites, which then serve as precursors for new biosynthesis. This chapter introduces some basic thermodynamic features of metabolic pathways and the mechanisms that have evolved to allow an organism to control the flow of a few common metabolites through different metabolic pathways.

Organisms harness the free energy from the degradation of macromolecules by trapping it in certain nucleotides (ATP, NAD^+, and FAD) and certain thioesters, which then make free energy available to energy-requiring pathways. The chapter describes these energy transmitters in terms of their thermodynamic features and their chemical properties. Some phosphorylated compounds have significant negative free energies of hydrolysis, which are described as their phosphoryl group-transfer potentials, their tendency to transfer their phosphoryl group to another compound. ATP, with its intermediate phosphoryl group-transfer potential, is the principal energy currency of life. Thioesters are also "high-energy" compounds. One of these, coenzyme A, shuttles acyl groups in metabolic processes.

Oxidation–reduction reactions are the most important process through which living organisms acquire and use free energy. The chapter reviews the principles of redox reactions, including the Nernst equation, and describes the mathematical relationship between free energy and reduction potentials. The chapter concludes with a brief look at the methods used to map the labyrinth of biochemical pathways in living cells and to understand the regulation of biochemical pathways.

Essential Concepts

1. Metabolism, the network of all biochemical reactions in cells, can be divided into two parts:
   (a) Catabolism (degradation), in which free energy is released as organic molecules are broken down into smaller constituents.
   (b) Anabolism (biosynthesis), in which biological molecules are synthesized from smaller, simpler molecules.

2. In general, catabolic reactions release free energy, which can then be used to drive endergonic synthetic reactions. This coupling requires free energy transmitters including nucleotides (ATP, NADH, NADPH, and $FADH_2$) and thioesters (e.g., coenzyme A).

Overview of Metabolism

3. Organisms use different strategies for capturing free energy from their environment:
   (a) Autotrophs synthesize all their macromolecules from simple molecules obtained from their environment. The chemolithotrophs oxidize inorganic compounds, and the photoautotrophs use light to drive synthetic reactions.
   (b) Heterotrophs obtain free energy from the oxidation of organic compounds (usually produced by autotrophs).

4. Organisms can be further classified by their requirements for oxygen. Obligate aerobes require oxygen for the oxidation of nutrients. Obligate anaerobes are poisoned by oxygen and must use another electron acceptor for the oxidation of nutrients. Facultative anaerobes can oxidize nutrients both in the absence and presence of oxygen.

5. Metabolic pathways are compartmentalized in the cytosol of prokaryotic and eukaryotic cells. Eukaryotic cells further compartmentalize metabolic pathways in membrane-bound organelles. In multicellular organisms, many metabolic pathways are compartmentalized in different tissues.

6. In any given metabolic pathway, most reactions are near equilibrium, such that the law of mass action largely dictates the flow rate (flux) of metabolites.

7. In each metabolic pathway, there is at least one reaction that is far from equilibrium, in which the reactants accumulate above their equilibrium values and $\Delta G \ll 0$. Such reactions are referred to as rate-determining steps since they control flux in the pathway. The flux changes only with a change in the enzyme's ability to increase the reaction rate.

8. Metabolic pathways have three key characteristics:
   (a) They are irreversible.
   (b) They have an exergonic step that serves as the first committed step and ensures irreversibility.
   (c) Catabolic and anabolic pathways involving the interconversion of two metabolites differ in key exergonic reactions.

9. Control of flux in the rate-determining step requires control of the enzyme catalyzing it by one or more of the following mechanisms:
   (a) Allosteric control by feedback regulation from an end product of the pathway.
   (b) Covalent modification of the enzyme, which may increase or decrease its ability to accelerate a reaction.
   (c) Substrate cycles in which interconversion of two substrates utilizes different rate-determining enzymes.
   (d) Genetic control, which regulates the steady-state levels of the enzyme.

   Mechanisms (a)–(c) respond quickly to changes in physiological states (seconds to minutes), whereas mechanism (d) involves slower, long-term adaptive changes to new physiological states (minutes to days).

"High-Energy" Compounds

10. The free energy derived from the degradation of organic compounds is transiently captured in "high-energy" compounds whose subsequent breakdown provides the free energy to drive otherwise endergonic reactions.

11. ATP is the primary energy currency of cells. Its energy resides in the thermodynamic instability of its two phosphoanhydride bonds. The free energy of hydrolysis of ATP, in which the phosphate group is transferred to water, is called its phosphoryl group-transfer poten-

tial. ATP has an intermediate phosphoryl group-transfer potential, making it a conduit for the transfer of free energy from higher-energy compounds to lower-energy compounds.

12. The high-energy character of the phosphoanhydride bonds results from
    (a) Increased resonance stabilization of the hydrolysis products.
    (b) The destabilizing effect of electrostatic repulsions between the charged phosphates at neutral pH.
    (c) Increased solvation energy of the hydrolysis products.

13. Many endergonic reactions in cells are coupled to the hydrolysis of ATP or pyrophosphate (PP_i) so that the net reaction is exergonic. In some biosynthetic reactions, the transfer of a nucleotidyl group "activates" the substrate for further reaction (e.g., the polymerization reactions of polysaccharides and the formation of aminoacyl–tRNA for protein synthesis).

14. ATP can be replenished by transfer of a phosphoryl group to ADP from a compound with a higher phosphoryl group-transfer potential. Such a transfer is called substrate-level phosphorylation. The concentrations of ATP and other nucleotides are maintained in part by the activity of kinases.

15. The transfer of acyl groups requires their "activation" by formation of a thioester bond to a sulfur-containing compound such as coenzyme A. The hydrolysis of thioesters is about as exergonic as the hydrolysis of ATP. Hence, thioester cleavage drives the otherwise endergonic transfer of the acyl group.

Oxidation–Reduction Reactions

16. Oxidation–reduction reactions are the principal source of free energy for life. The oxidation of organic compounds is coupled to the reduction of the nucleotide cofactors NAD^+ (and $NADP^+$) and FAD.

17. A measure of the potential electrical energy (electromotive force or reduction potential) in an electrochemical cell is described by the Nernst equation:

$$\Delta\mathcal{E} = \Delta\mathcal{E}^{\circ\prime} - \frac{RT}{n\mathcal{F}} \ln\left(\frac{[A_{red}][B_{ox}^{n+}]}{[A_{ox}^{n+}][B_{red}]}\right)$$

Here, $\Delta\mathcal{E}$ is the reduction potential, $\Delta\mathcal{E}^{\circ\prime}$ is the reduction potential when all the components are in their biochemical standard states, $\mathcal{F}$ is the faraday (96,485 $J \cdot V^{-1} \cdot mol^{-1}$), n is the number of moles of electrons transferred per mole of reactants reduced, and R is the gas constant.

18. $\Delta\mathcal{E}$ is related to the free energy change in a redox reaction by the following relationship:

$$\Delta G = -n\mathcal{F}\Delta\mathcal{E}$$

Electrons flow spontaneously from a compound with the lower reduction potential to a compound with the higher reduction potential.

Experimental Approaches to the Study of Metabolism

19. Metabolites can be traced by labeling them with isotopes of certain atoms (e.g., C, S, P, and H) that can be detected by their radioactivity or through nuclear magnetic resonance spectroscopy. Radioactive tracers are especially useful for establishing precursor–product relationships and for examining the rates of biochemical transformation in living cells and tissues.

20. Other methods for analyzing a metabolic pathway include the use of metabolic inhibitors, which inhibit specific enzymes, and genetic mutations in enzymes involved in the pathway. In recent years, genetic engineering has become a powerful new tool for studying metabolism, as the gene for a specific enzyme can be added, deleted, or specifically altered.

Key Equations

$$\Delta G = -n\mathcal{F}\Delta\mathcal{E}$$

$$\Delta\mathcal{E} = \Delta\mathcal{E}^{\circ\prime} - \frac{RT}{n\mathcal{F}} \ln\left(\frac{[A_{red}][B_{ox}^{n+}]}{[A_{ox}^{n+}][B_{red}]}\right)$$

$$\Delta\mathcal{E}^{\circ} = \mathcal{E}^{\circ}_{(e^- \text{ acceptor})} - \mathcal{E}^{\circ}_{(e^- \text{ donor})}$$

Guide to Study Exercises (text p. 380)

1. Autotrophs can synthesize all the molecules they require from simple molecules that are available from their environment. They obtain the free energy to do so by oxidizing inorganic compounds or by absorbing light energy.

   Heterotrophs cannot synthesize all their molecules from environmentally available precursors, nor can they harness the energy of inorganic compounds or the sun. Instead, they rely on autotrophs to supply them with organic compounds that they break down to obtain free energy and precursors for synthesizing other compounds. (Section 13-1A)

2. Reactions that operate near equilibrium are freely reversible. Since their ΔG values are close to zero, flux in either direction is possible, with the direction of flux determined by the relative concentrations of reactants and products (at equilibrium there is no net flow). Such a reaction does not offer an opportunity for metabolic control, but because it operates in either direction, it can participate in two opposing pathways.

   Reactions that function far from equilibrium have $\Delta G \ll 0$ and are therefore irreversible. These steps provide control points for metabolic pathways. Because the enzyme that catalyzes such a reaction is insensitive to changes in substrate concentrations, it can control the flux of substrate through the pathway in response to other factors, such as changes in the amount of enzyme present, the presence of allosteric effectors, and covalent modification. (Sections 13-1C and D)

3. ATP is a “high-energy” compound because its breakdown is highly exergonic. The free energy of hydrolysis of one of its phosphoanhydride bonds (which approximates the free energy of cleaving one of these bonds as part of another reaction) is $\sim 30\ kJ \cdot mol^{-1}$ under standard biochemical conditions. Cleavage of these bonds is thermodynamically favored because the reaction products are resonance stabilized, experience less electrostatic repulsion, and are better solvated. (Section 13-2A)

4. An exergonic process can drive an endergonic process only if the two processes are linked by a common intermediate. This can happen when a product of an exergonic reaction is a reactant for an endergonic reaction, or when a product of an endergonic reaction is a reactant for an exergonic reaction. Thus, the exergonic reaction can either “push” or “pull” the endergonic reaction. The coupled reactions will proceed if the net change in free energy is negative; that is, if the magnitude of ΔG for the exergonic reaction is greater than that of the endergonic reaction. (Section 13-2B)

5. The coenzymes NAD^+ and FAD are reduced with electrons obtained from the exergonic oxidation of metabolic fuels, so that their reduced forms (NADH and $FADH_2$) are a form of stored free energy. When the reduced coenzymes give up their electrons in order to return to their oxidized forms, that free energy is harvested for the endergonic synthesis of ATP from ADP + P_i. This is accomplished through the exergonic movement of the electrons through a series of electron carriers that establish a transmembrane proton concentration gradient whose dissipation drives ADP phosphorylation (oxidative phosphorylation). (Section 13-3A)

6. The Nernst equation (Equation 13-7) expresses the electromotive force (or reduction potential; $\Delta\mathcal{E}$, in units of volts) of an oxidation–reduction reaction as the difference between the standard reduction potential ($\Delta\mathcal{E}^\circ$, or $\Delta\mathcal{E}^{\circ\prime}$ when all components are in their biochemical standard state) and a variable term that takes into account the concentrations of the reactants (the logarithmic term), the temperature (T, in Kelvin), and the number of electrons transferred (n). These quantities are scaled by the gas constant ($R = 8.3145\ J \cdot K^{-1} \cdot mol^{-1}$) and the faraday ($\mathcal{F} = 96{,}485\ J \cdot V^{-1} \cdot mol^{-1}$). (Section 13-3B)

7. $\Delta\mathcal{E}$ is related to ΔG as $\Delta G = -n\mathcal{F}\Delta\mathcal{E}$ so that the free energy change of an oxidation–reduction reaction depends on the number of electrons transferred (n, multiplied by a constant, the faraday, $\mathcal{F}$) and the reduction potential difference between the electron acceptor and the electron donor ($\Delta\mathcal{E}$). ΔG is negative, and the reaction is spontaneous, only when $\Delta\mathcal{E}$ is positive, that is, when electrons move from a substance with a low reduction potential to a substance with a high reduction potential. (Sections 13-3B and C)

8. The transformation of a precursor compound to a product can be followed if an atom in the precursor that also appears in the product is replaced by an uncommon isotope. This allows the precursor, intermediates, and product to be detected through NMR or radioactivity. A small amount of labeled compound is added to the experimental system (cell extract, tissue culture, or whole organism) and the compounds containing the label are followed over a period of time. In this way, it is possible to trace the decrease in precursor concentration and the increase in concentration of the product. (Section 13-4A)

Questions

Overview of Metabolism

1. In searching for life on Mars in the 1970s, the Viking spacecraft tested the Martian soil for rapid oxidation–reduction reactions. Explain why such reactions might indicate the presence of life.

2. Use the following terms to fill in the diagram below: ADP, P_i, $NADP^+$, ATP, NADPH, carbohydrates, proteins, lipids, acetyl-CoA, catabolism, anabolism.

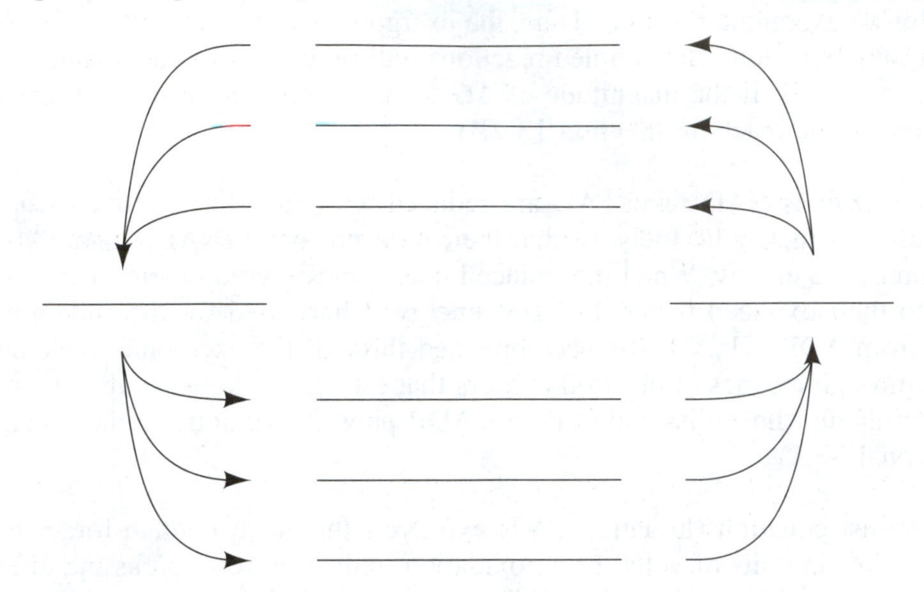

3. Phosphofructokinase (PFK) catalyzes the reaction

$$\text{ATP} + \text{fructose-6-phosphate (F6P)} \rightleftharpoons \text{fructose-1,6-bisphosphate (FBP)} + \text{ADP}$$

Explain why the reaction rate is relatively insensitive to changes in the concentrations of F6P or FBP. What does this tell you about PFK?

"High-Energy" Compounds

4. In the cell, divalent cations such as Mg^{2+} bind to the anionic phosphate groups of ATP. How would the removal of the divalent cations affect the ΔG for the hydrolysis of ATP?

5. What processes maintain the cellular concentration of ATP?

6. Why don't "high-energy" compounds such as phosphoenolpyruvate and phosphocreatine (Figure 13-7) break down quickly under physiological conditions?

7. Explain how phosphocreatine acts as an ATP "buffer."

8. Many metabolic reactions are actually coupled reactions. A common coupled reaction is substrate phosphorylation by ATP. For example, the oxidation of glucose begins with its phosphorylation to glucose-6-phosphate (G6P).
   (a) For the reaction

$$\text{Glucose} + P_i \rightleftharpoons \text{glucose-6-phosphate} + H_2O \qquad \Delta G = 14\ \text{kJ} \cdot \text{mol}^{-1}$$

   Calculate the ratio of [G6P]/[glucose][P_i] at equilibrium at 25°C.

   (b) In muscle cells at 37°C, the steady-state ratio of [ATP]/[ADP] is 12. Assuming that glucose and G6P achieve equilibrium values in muscle, what is the ratio of [G6P] to [glucose]?

9. Aldolase catalyzes the reaction

$$\text{Fructose-1,6-bisphosphate (FBP)} \rightleftharpoons \text{glyceraldehyde-3-phosphate (GAP)} + \text{dihydroxyacetone phosphate (DHAP)}$$

$\Delta G^{\circ\prime}$ for this reaction is 22.8 kJ · mol^{-1}. In the cell at 37°C, the ΔG for this reaction is -5.9 kJ · mol^{-1}. What is the ratio [GAP][DHAP]/[FBP]?

Oxidation–Reduction Reactions

10. Using the curved-arrow convention, show the transfer of electrons in the reduction of pyruvate to lactate in the presence of NADH + H^+.

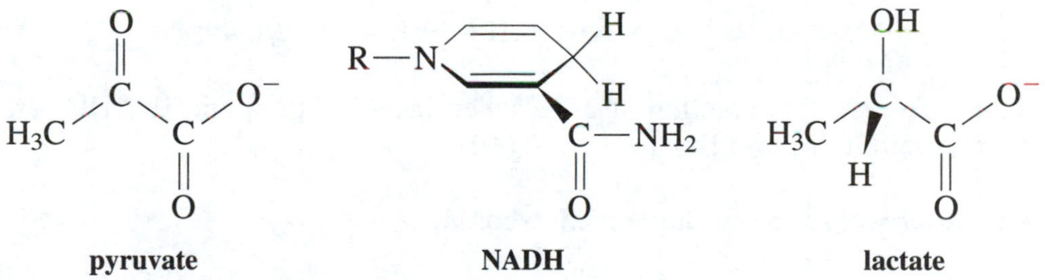

11. In the diagram on the next page, indicate the direction of flow of electrons and the voltage (on the meter) for the following half-reactions under standard conditions:

$$Zn^{2+} + 2e^- \rightarrow Zn \qquad \mathcal{E}^\circ = -0.763\ \text{V}$$
$$Ni^{2+} + 2e^- \rightarrow Ni \qquad \mathcal{E}^\circ = -0.250\ \text{V}$$

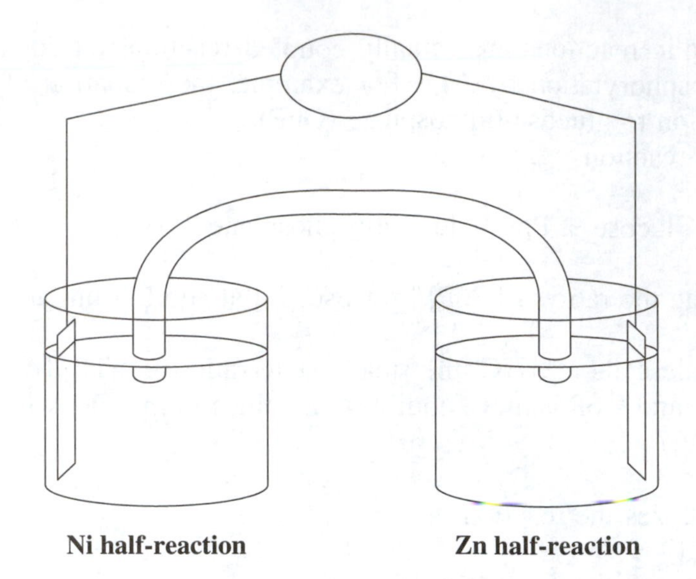

12. At pH 0,

$$\frac{1}{2}\, O_2 + 2\, H^+ + 2\, e^- \rightleftharpoons H_2O \qquad \mathscr{E}° = 1.23 \text{ V}$$

Is oxygen reduction more favored at pH 0 or at pH 7? Explain in electrochemical terms as well as in terms of chemical equilibria.

13. Consider the reaction in which acetoacetate is reduced by NADH to β-hydroxybutyrate:

$$\text{Acetoacetate} + \text{NADH} + H^+ \rightleftharpoons \beta\text{-hydroxybutyrate} + \text{NAD}^+$$

Calculate ΔG for this reaction at 25°C when [acetoacetate] and [NADH] are 0.01 M, and [β-hydroxybutyrate] and [NAD^+] are 0.001M.

Experimental Approaches to the Study of Metabolism

14. Radioactive isotope tracers and metabolic inhibitors have been essential to the elucidation of metabolic pathways. Can both kinds of agents be used to determine the order of metabolites in a metabolic pathway?

15. Many biosynthetic pathways have been elucidated by the analysis of genetic mutations in organisms such as *Neurospora crassa* and *Escherichia coli*. How would you elucidate the steps of a hypothetical biosynthetic pathway in *N. crassa* in which compound A leads to compound Z?

16. You have isolated four mutants in amino acid metabolism of the mold *N. crassa*. Mutant 1 requires two compounds for growth, X and Z. Mutant 2 only requires X. Mutants 3 and 4 require only Z. Mutant 3 accumulates a compound, W, that supports the growth of Mutant 4 but not that of Mutants 1 or 2. Mutant 4 accumulates a compound, Y, that alone supports the growth of Mutant 1.

(a) Diagram the biosynthetic pathway connecting compounds W, X, Y, and Z, indicating the step at which each mutant is blocked.
(b) According to the diagram in (a), what is the first committed step in the synthesis of Z?

Answers to Questions

1. The fundamental processes by which living organisms acquire and use free energy depend on oxidation–reduction reactions. For example, the oxidation of a metabolic fuel generates a reduced cofactor, such as NADH, whose subsequent reoxidation generates ATP. Rapid (i.e., enzyme-catalyzed) oxidation–reduction reactions therefore would be consistent with (but would not prove) the presence of life on Mars.

2. 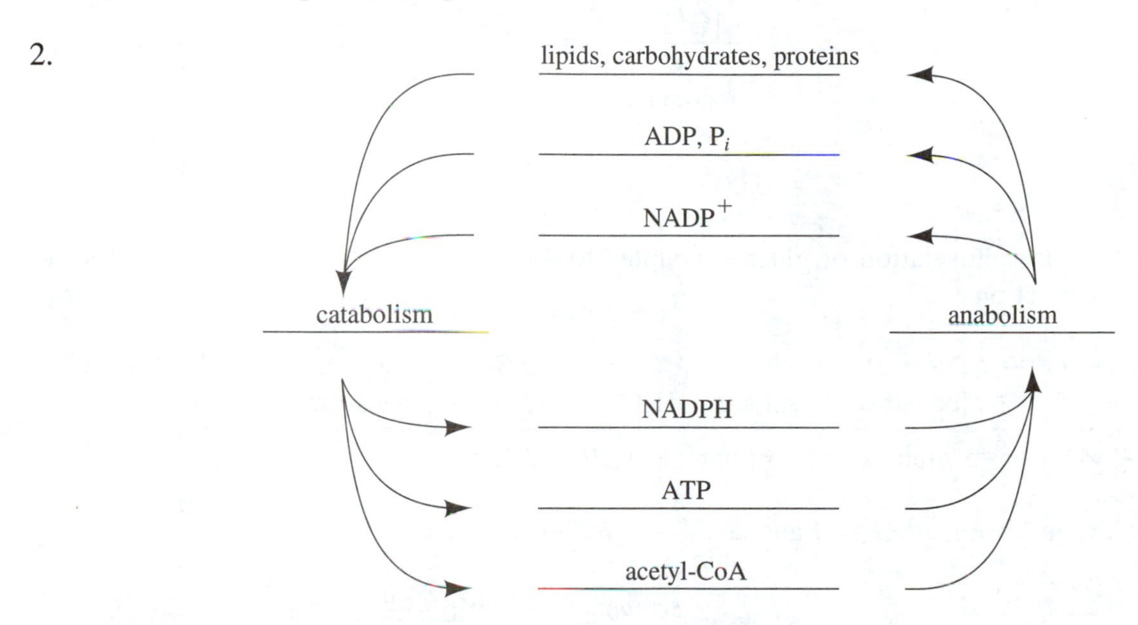

3. PFK operates far from equilibrium, so it is not sensitive to changes in substrate concentrations. It most likely catalyzes a rate-determining step of a metabolic pathway. (In fact, it is part of the glycolytic pathway, and its activity is under allosteric control.)

4. ΔG would become more negative because, without the shielding effect of the divalent cations, the phosphate groups in ATP would experience more repulsion.

5. Processes that maintain the cellular concentration of ATP are oxidative phosphorylation (the synthesis of ATP from ADP + P_i as driven by the free energy of dissipation of a transmembrane proton concentration gradient); substrate-level phosphorylation (direct transfer of a phosphoryl group from a "high-energy" compound such as phosphoenolpyruvate to ADP); and reactions catalyzed by kinases such as nucleoside diphosphate kinase and adenylate kinase.

6. These "high-energy" compounds have large negative values for ΔG of hydrolysis. Therefore, their breakdown is thermodynamically spontaneous (exergonic). However, the kinetics of their breakdown depend on the availability and activity of enzymes to catalyze such reactions. In the absence of the appropriate enzymatic activity, the compounds are quite kinetically stable.

7. The reaction catalyzed by creatine kinase

$$\text{ATP} + \text{creatine} \rightleftharpoons \text{phosphocreatine} + \text{ADP}$$

is at equilibrium and is freely reversible in the cell. Hence, the concentration of phosphocreatine is directly sensitive to changes in [ATP]. When ATP is plentiful, phosphocreatine is formed by the forward reaction. When [ATP] drops, the reaction proceeds in reverse, so that phosphocreatine can transfer its phosphoryl group to ADP to produce more ATP. The extent of this reaction depends on the decrease in [ATP].

8. (a) At equilibrium, $\Delta G^{\circ\prime} = -RT \ln K_{eq}$,

$$\begin{aligned}\text{Hence } K_{eq} &= \frac{[\text{G6P}]}{[\text{glucose}][\text{P}_i]} = e^{-\Delta G^{\circ\prime}/RT} \\ &= e^{-(14{,}000\ \text{J}\cdot\text{mol}^{-1})/(8.3145\ \text{J}\cdot\text{K}^{-1}\cdot\text{mol}^{-1})(298\text{K})} \\ &= e^{-5.65} \\ &= 0.0035\end{aligned}$$

(b) The phosphorylation of glucose coupled to ATP hydrolysis is the sum of the following reactions:

$\text{ATP} + \text{H}_2\text{O} \rightleftharpoons \text{ADP} + \text{P}_i$	$\Delta G^{\circ\prime} = -30.5\ \text{kJ}\cdot\text{mol}^{-1}$ (Table 13-2)
$\text{glucose} + \text{P}_i \rightleftharpoons \text{glucose-6-phosphate} + \text{H}_2\text{O}$	$\Delta G^{\circ\prime} = 14\ \text{kJ}\cdot\text{mol}^{-1}$
$\text{ATP} + \text{glucose} \rightleftharpoons \text{glucose-6-phosphate} + \text{ADP}$	$\Delta G^{\circ\prime} = -16.5\ \text{kJ}\cdot\text{mol}^{-1}$

At equilibrium, $\Delta G = 0$ and $\Delta G^{\circ\prime} = -RT \ln K_{eq}$. Thus,

$$K_{eq} = e^{-\Delta G^{\circ\prime}/RT} = \frac{[\text{ADP}][\text{G6P}]}{[\text{ATP}][\text{glucose}]}$$

Since [ATP]/[ADP] = 12,

$$\begin{aligned}\frac{[\text{G6P}]}{[\text{glucose}]} &= 12\ e^{-\Delta G^{\circ\prime}/RT} \\ &= 12\ e^{-(-16{,}500\ \text{J}\cdot\text{mol}^{-1})/(8.3145\ \text{J}\cdot\text{K}^{-1}\cdot\text{mol}^{-1})(310\text{K})} \\ &= 12\ e^{6.40} \\ &= 7233\end{aligned}$$

9. Use Equation 13-1:

$$\Delta G = \Delta G^{\circ\prime} + RT \ln\left(\frac{[\text{GAP}][\text{DHAP}]}{[\text{FBP}]}\right)$$

$$\frac{\Delta G - \Delta G^{\circ\prime}}{RT} = \ln\left(\frac{[\text{GAP}][\text{DHAP}]}{[\text{FBP}]}\right)$$

$$\frac{(-5900\ \text{J}\cdot\text{mol}^{-1}) - (22{,}800\ \text{J}\cdot\text{mol}^{-1})}{(8.3145\ \text{J}\cdot\text{K}^{-1}\cdot\text{mol}^{-1})(310\text{K})} = \ln\left(\frac{[\text{GAP}][\text{DHAP}]}{[\text{FBP}]}\right)$$

$$-11.13 = \ln\left(\frac{[\text{GAP}][\text{DHAP}]}{[\text{FBP}]}\right)$$

$$e^{-11.13} = \frac{[\text{GAP}][\text{DHAP}]}{[\text{FBP}]}$$

$$1.46 \times 10^{-5} = \frac{[\text{GAP}][\text{DHAP}]}{[\text{FBP}]}$$

10.

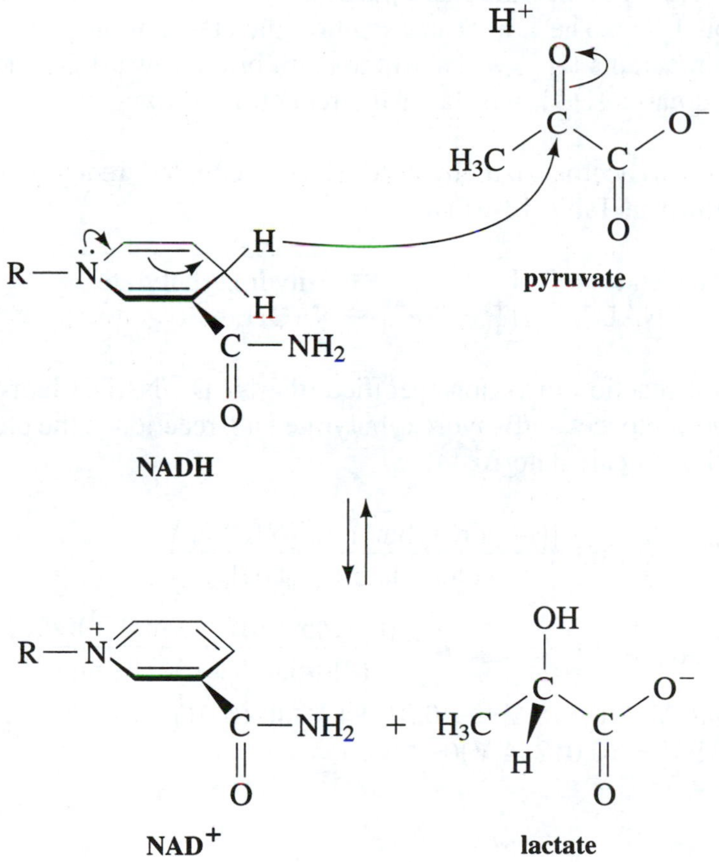

H^+

NADH + pyruvate ⇌ NAD^+ + lactate

NADH **pyruvate**

NAD^+ **lactate**

11. Electrons flow from substances of lower reduction potential to substances of higher reduction potential. $\Delta\mathscr{E}° = -0.250\text{ V} - (-0.763\text{ V}) = 0.513\text{ V}$.

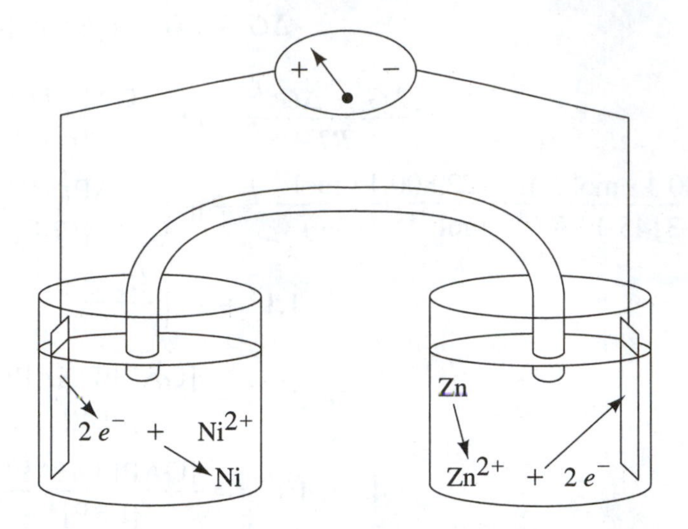

12. At pH 7.0, $\mathscr{E}°' = 0.815$ V (Table 13-3). Therefore, reduction is more favorable at pH 0 (where $\mathscr{E}° = 1.23$ V) since the more positive the reduction potential, the more negative the ΔG (Equation 13-6). The law of mass action dictates that an increase in the concentration of one of the reactants (H^+) will shift the equilibrium toward product. Thus, decreasing the pH, which increases [H^+], will favor the reduction of oxygen.

13. The reduction of acetoacetate by NADH is a coupled redox reaction, where the half-reactions (shown in Table 13-3) are

$$\text{Acetoacetate} + 2\,H^+ + 2\,e^- \rightleftharpoons \beta\text{-hydroxybutyrate} \qquad \mathscr{E}°' = -0.346\text{ V}$$
$$NAD^+ + H^+ + 2\,e^- \rightleftharpoons \text{NADH} \qquad \mathscr{E}°' = -0.315\text{ V}$$

For the overall reaction direction specified, the NAD^+/NADH half-reaction is the electron donor, and the acetoacetate/β-hydroxybutyrate half-reaction is the electron acceptor. Use the Nernst equation to calculate $\Delta\mathscr{E}$:

$$\Delta\mathscr{E} = \Delta\mathscr{E}°' - \frac{RT}{n\mathscr{F}} \ln\left(\frac{[\beta\text{-hydroxybutyrate}][NAD^+]}{[\text{acetoacetate}][\text{NADH}]}\right)$$
$$= (\mathscr{E}°'_{e^-\text{ acceptor}} - \mathscr{E}°'_{e^-\text{ donor}}) - \frac{(8.3145\text{ J}\cdot\text{K}^{-1}\cdot\text{mol}^{-1})(298\text{K})}{(2)(96{,}485\text{ J}\cdot\text{V}^{-1}\cdot\text{mol}^{-1})} \ln\left[\frac{(0.001)(0.001)}{(0.01)(0.01)}\right]$$
$$= (-0.346\text{ V} + 0.315\text{ V}) - (0.01284\text{ V}) \ln(0.01)$$
$$= -0.031\text{ V} - (0.01284\text{ V})(-4.605)$$
$$= 0.028\text{ V}$$

Next, use Equation 13-6 to calculate ΔG:

$$\begin{aligned}\Delta G &= -n\mathcal{F}\Delta\mathcal{E} \\ &= -(2)(96{,}485 \text{ J} \cdot \text{V}^{-1} \cdot \text{mol}^{-1})(0.028 \text{ V}) \\ &= -5403 \text{ J} \cdot \text{mol}^{-1} \\ &= -5.4 \text{ kJ} \cdot \text{mol}^{-1}\end{aligned}$$

14. Radioactive tracers can be used to determine the order of metabolic transformations in a pathway. An inhibitor that blocks a step of the pathway causes earlier intermediates to accumulate but does not necessarily reveal their order, and cannot reveal information about the order of metabolites following the blocked step.

15. To study the metabolic steps between compounds A and Z, first generate mutants by irradiating the cells or treating them with a chemical mutagen. Next, screen cells for their inability to grow in the absence of compound Z, the end product of the pathway. This would yield cells with defects in the A→Z pathway. To identify mutations in enzymes that catalyze individual steps of the pathway, screen the mutant cells for their inability to grow in the absence of each of the compounds suspected to be intermediates in the transformation of A to Z.

16. (a) Since Mutant 1 requires two compounds for growth, the pathway for their synthesis is probably branched, such that both are derived from a common precursor lacking in Mutant 1. Mutant 2 is blocked in the pathway leading to X, while Mutants 3 and 4 are blocked in the pathway leading to Z. The step blocked in Mutant 4 precedes that blocked in Mutant 3, since Mutant 3 accumulates compound W, which supports the growth of Mutant 4. Since compound Y, which accumulates in Mutant 4, supports the growth of Mutant 1, it must be the common precursor of X and Z. Therefore, the most likely pathway is

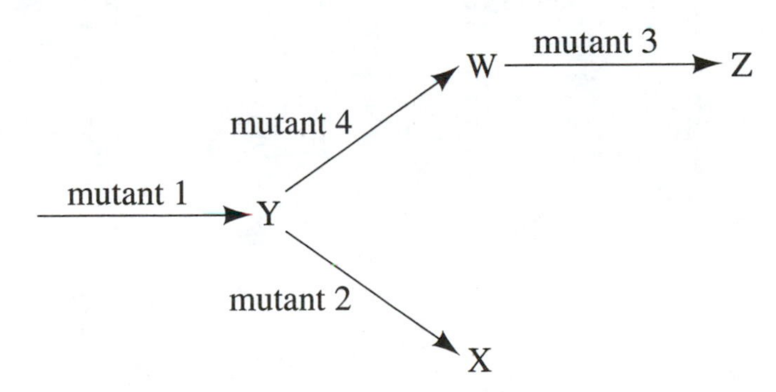

(b) The first committed step in the synthesis of Z is the step that converts Y to W, the step missing in Mutant 4.

Chapter 14 Glucose Catabolism

Glycolysis, or the biochemical conversion of glucose to pyruvate, is one of the best understood metabolic pathways. The mechanisms and structures of the ten glycolytic enzymes are known in some detail, and they serve as models for the study of other enzymes with similar reaction mechanisms. The regulation of metabolic pathways, a topic introduced in Chapter 13, is illustrated here. This chapter also describes the fate of pyruvate (the end product of glycolysis), the entry of other sugars into the glycolytic pathway, and the pentose phosphate pathway that also catabolizes glucose.

Essential Concepts

Overview of Glycolysis

1. Glucose is a major source of metabolic energy in many cells. The energy released during its conversion (oxidation) to pyruvate is conserved in the form of ATP and the reduced coenzyme NADH. Ten enzymes catalyze the glycolytic pathway, which occurs in both prokaryotes and eukaryotes and is almost universal.

2. Stage I of glycolysis is a preparatory state in which glucose is "activated" by phosphorylation by ATP and broken down into two C_3 sugars. Stage II produces "high-energy" intermediates that phosphorylate ADP to form ATP, for a net gain of 2 ATP. One glucose molecule yields two pyruvate molecules and requires the oxidizing power of two NAD^+. The overall equation for glycolysis is

$$\text{Glucose} + 2\ NAD^+ + 2\ ADP + 2\ P_i \rightarrow 2\ \text{pyruvate} + 2\ NADH + 2\ ATP + 2\ H_2O + 4\ H^+$$

The Reactions of Glycolysis

3. Hexokinase, the first enzyme in the pathway, transfers a phosphoryl group from ATP to the C6-OH group of glucose to produce glucose-6-phosphate (G6P). Hexokinase undergoes a large conformational change on binding glucose, which excludes water from the active site and promotes the specific transfer of the phosphoryl group from ATP to glucose.

4. Phosphoglucose isomerase catalyzes the conversion of glucose-6-phosphate to fructose-6-phosphate (F6P). This reaction proceeds via an enediolate intermediate.

5. Phosphofructokinase (PFK) converts fructose-6-phosphate to fructose-1,6-bisphosphate (FBP) by another phosphoryl-group transfer from ATP. This reaction, which is irreversible, is the first committed step of the pathway. The phosphofructokinase reaction is the rate-determining step of glycolysis and the principal regulatory point.

6. Aldolase cleaves fructose-1,6-bisphosphate to two C_3 compounds: glyceraldehyde-3-phosphate (GAP) and dihydroxyacetone phosphate (DHAP). Class I aldolases (in animals and plants) operate via Schiff base and enamine intermediates. Class II aldolases use Zn^{2+} or Fe^{2+} to stabilize the enolate intermediate.

7. Triose phosphate isomerase catalyzes the interconversion of glyceraldehyde-3-phosphate and dihydroxyacetone phosphate through an enediolate intermediate. This α/β barrel enzyme has achieved catalytic perfection. Its activity allows the products of the first stage of glycolysis to proceed through the second stage as glyceraldehyde-3-phosphate.

8. Glyceraldehyde-3-phosphate dehydrogenase (GAPDH) catalyzes the formation of the first glycolytic intermediate that has sufficient free energy to synthesize ATP from ADP. Glyceraldehyde-3-phosphate is converted to 1,3-bisphosphoglycerate (1,3-BPG) by the reduction of NAD^+ and the addition of inorganic phosphate. The sulfhydryl group of an enzyme Cys residue attacks GAP, which is then reduced by NAD^+ to form an acyl thioester intermediate that is attacked by P_i.

9. 1,3-Bisphosphoglycerate, a "high-energy" mixed anhydride of a phosphate and a carboxylic acid, is the substrate for phosphoglycerate kinase, which transfers the phosphoryl group at C1 to ADP to generate 3-phosphoglycerate (3PG) and the first ATP product of glycolysis.

10. 3-Phosphoglycerate is converted to 2-phosphoglycerate (2PG) by phosphoglycerate mutase. The active form of this enzyme contains a phosphohistidine residue whose phosphoryl group is transferred to 3PG to produce 2,3-bisphosphoglycerate (2,3-BPG), which then transfers the phosphoryl group at the 3 position back to the histidine, yielding 2PG.

11. Enolase dehydrates (removes water from) 2-phosphoglycerate to form phosphoenolpyruvate (PEP). This reaction produces the second "high-energy" intermediate of glycolysis.

12. The free energy of phosphoenolpyruvate is released in the reaction catalyzed by pyruvate kinase. The transfer of the phosphoryl group to ADP produces the second ATP product of glycolysis and an enol product whose tautomerization to the keto form yields pyruvate. Most of the free energy of the pyruvate kinase reaction is supplied by this tautomerization step. Keep in mind that because the initial substrate of glycolysis is a C_6 compound that is converted to two C_3 compounds, the first stage of glycolysis consumes 2 ATP but the second stage generates 4 ATP (2 for each GAP), for a net yield of 2 ATP.

Fermentation: The Anaerobic Fate of Pyruvate

13. The NADH produced by glycolysis must be converted back to NAD^+ in order for the glyceraldehyde-3-phosphate dehydrogenase reaction to proceed. Consequently, pyruvate is not the end product of glucose metabolism but can undergo one of three processes to regenerate NAD^+: homolactic fermentation, alcoholic fermentation, or oxidative metabolism. In oxidative metabolism, pyruvate is oxidized to CO_2 via the citric acid cycle.

14. In muscle cells, when O_2 is in short supply, lactate dehydrogenase reduces pyruvate to lactate, with the concomitant oxidation of NADH to NAD^+. The lactate that builds up in muscles upon strenuous exertion can either be reconverted to pyruvate later or carried by the blood to the liver, where it can be converted back to glucose by a process called gluconeogenesis.

15. Under anaerobic conditions, yeast carry out alcoholic fermentation. In the first reaction of this process, pyruvate is decarboxylated to acetaldehyde and carbon dioxide. Pyruvate de-

carboxylase catalyzes this reaction with the aid of the coenzyme thiamine pyrophosphate, which stabilizes the reaction's carbanion intermediate. In the second reaction, acetaldehyde is reduced with NADH to form ethanol and NAD^+, as catalyzed by alcohol dehydrogenase. These two reactions have been known for thousands of years: The released CO_2 raises bread, and the ethanol is used to make alcoholic beverages.

16. The anaerobic catabolism of glucose can be 100 times faster than the catabolism of glucose in the presence of oxygen. However, fermentation produces 2 ATP per glucose, whereas oxidative metabolism (via the citric acid cycle and oxidative phosphorylation) generates 38 ATP for each glucose molecule that is converted to CO_2 and H_2O.

Control of Glycolysis

17. The reaction catalyzed by phosphofructokinase, which has a large negative ΔG, is the first committed step in glycolysis and the primary control point for the pathway. PFK is allosterically inhibited by ATP, which binds to an inhibitory site and stabilizes PFK's T (less active) state. This is an example of feedback inhibition, since ATP is a product of the pathway. AMP, ADP, and fructose-2,6-bisphosphate (F2,6P) relieve the inhibition of PFK by ATP by preferentially binding to the R (more active) state. PFK thereby senses the energy state of the cell and adjusts the flux through glycolysis accordingly.

18. Additional control of glycolytic flux is provided by a substrate cycle. PFK catalyzes the reaction F6P + ATP $\rightarrow$ FBP + ADP, whereas fructose-1,6-bisphosphatase (FBPase) catalyzes the opposing reaction FBP + H_2O $\rightarrow$ F6P + P_i. The sum of these reactions is the hydrolysis of ATP. Both enzymes exist in the same cell and their relative activity is under hormonal and neuronal control. Substrate cycling, which produces heat, can provide a form of nonshivering thermogenesis.

Metabolism of Hexoses Other than Glucose

19. Three other sugars—fructose, galactose, and mannose—are major sources of cellular energy. In muscle, fructose can be directly phosphorylated by hexokinase to F6P. However, liver glucokinase cannot directly phosphorylate fructose. In the liver, fructokinase phosphorylates C1 to generate fructose-1-phosphate. Fructose-1-phosphate aldolase generates dihydroxyacetone phosphate and glyceraldehyde by an aldol cleavage. Glyceraldehyde kinase phosphorylates C3 of glyceraldehyde to produce the glycolytic intermediate GAP. Three other enzymes (alcohol dehydrogenase, glycerol kinase, and glycerol phosphate dehydrogenase) convert glyceraldehyde to dihydroxyacetone phosphate, which is then converted to GAP by triose phosphate isomerase. Individuals who have defective fructose-1-phosphate aldolase have fructose intolerance and quickly develop a strong distaste for anything sweet.

20. Galactose, which differs from glucose in the configuration of the OH group at C4, requires four reactions to enter glycolysis. First, galactose is converted to galactose-1-phosphate by galactokinase. Next, galactose-1-phosphate uridylyl transferase transfers the UMP group of UDP–glucose to produce UDP–galactose and glucose-1-phosphate. An epimerase converts UDP–galactose to UDP–glucose. Finally, phosphoglucomutase converts glucose-1-phosphate to glucose-6-phosphate.

21. Mannose, the C2 epimer of glucose, is recognized by hexokinase. The resulting mannose-6-phosphate is then converted to the glycolytic intermediate fructose-6-phosphate by mannose isomerase.

The Pentose Phosphate Pathway

22. Besides ATP, cells require the reducing power of NADPH for the biosynthesis of macromolecules (anabolism). NADPH is used in biosynthesis, whereas NADH is used in oxidative metabolism (catabolism). Cells keep the $[NAD^+]/[NADH]$ ratio near 1000 (which favors metabolite oxidation) and the $[NADP^+]/[NADPH]$ ratio near 0.01 (which favors reductive biosynthesis). The oxidation of glucose by the pentose phosphate pathway generates NADPH.

23. The first stage of the pentose phosphate pathway consists of three steps, its oxidative reactions:
    (a) Glucose-6-phosphate is oxidized to 6-phosphoglucono-δ-lactone by glucose-6-phosphate dehydrogenase, producing the first NADPH.
    (b) 6-Phosphogluconolactonase hydrolyzes the lactone (cyclic ester) to yield 6-phosphogluconate.
    (c) 6-Phosphogluconate dehydrogenase then catalyzes the oxidative decarboxylation of 6-phosphogluconate by $NADP^+$ to yield ribulose-5-phosphate, CO_2, and the second NADPH.

24. Stage two of the pentose phosphate pathway is catalyzed by two enzymes that act on ribulose-5-phosphate (Ru5P): Ribulose-5-phosphate epimerase converts Ru5P to xylulose-5-phosphate (Xu5P), and ribulose-5-phosphate isomerase converts Ru5P to ribose-5-phosphate (R5P). The R5P can be used to produce nucleosides for RNA and DNA synthesis.

25. In the third stage of the pentose phosphate pathway, 3 five-carbon sugars are converted to 2 fructose-6-phosphate and 1 glyceraldehyde-3-phosphate. First, transketolase transfers a two-carbon unit from Xu5P to R5P, yielding the seven-carbon sugar sedoheptulose-7-phosphate (S7P) and glyceraldehyde-3-phosphate (GAP). Next, transaldolase transfers a three-carbon unit from S7P to GAP to form fructose-6-phosphate (F6P) and the four-carbon sugar erythrose-4-phosphate (E4P). Finally, another transketolase reaction converts Xu5P and E4P to F6P and GAP.

26. The reversible nature of the second and third stages of the pentose phosphate pathway permits the cell to meet its need for R5P (a nucleic acid precursor) and NADPH. For example, if the need for NADPH is greater than that for R5P, the excess R5P is converted to F6P and GAP for consumption via glycolysis. Conversely, if the demand for R5P outstrips the need for NADPH, the glycolytic intermediates F6P and GAP can be diverted to the pentose phosphate pathway to synthesize R5P.

27. The flux of glucose-6-phosphate through the pathway is controlled by the rate of the glucose-6-phosphate dehydrogenase (G6PD) reaction. This enzyme is regulated by the availability of its substrate $NADP^+$ so that the pathway flux increases in response to increasing levels of $NADP^+$ (which indicates increased cellular demand for NADPH).

28. A deficiency of G6PD is the most common human enzyme deficiency. The resulting shortage of NADPH increases the sensitivity of red blood cells to oxidative stress, since NADPH is required to maintain the supply of reduced glutathione, which removes organic hydroperoxides that occasionally form. Certain compounds such as the antimalarial drug primaquine stimulate peroxide formation, which induces hemolytic anemia in G6PD-deficient individuals. However, mutations in G6PD confer resistance to malaria in heterozygous females.

Guide to Study Exercises (text p. 424)

1. See Figure 14-1.

2. The fate of pyruvate depends on the energy needs of the cell and the availability of oxygen. Under aerobic conditions, pyruvate can be completely oxidized to CO_2 by the citric acid cycle, which ultimately generates large amounts of ATP. Under anaerobic conditions, pyruvate is metabolized to a lesser extent and no additional free energy is recovered in the form of ATP. In yeast, alcoholic fermentation converts pyruvate to CO_2 and ethanol by the actions of pyruvate decarboxylase and alcohol dehydrogenase. In muscle, pyruvate may be reversibly converted to lactate by the action of lactate dehydrogenase. (Section 14-3)

3. Phosphofructokinase (PFK) is an allosteric enzyme. The inhibitor ATP (which is also a substrate) binds to an inhibitory site, thereby stabilizing the T state of PFK, which has a lower affinity for its other substrate, fructose-6-phosphate. ADP, AMP, and fructose-2,6-bisphosphate reverse the inhibitory effects of ATP. The R (more active) state of PFK is stabilized by F6P, whose phosphoryl group forms an ion pair with an Arg side chain in another subunit. The R→T conformational change, which replaces this ion pair with unfavorable electrostatic interactions between F6P and the enzyme, is prevented when the activator AMP or ADP is bound to the R state enzyme. (Section 14-4A)

4. A substrate cycle permits greater variations in flux than regulation of a single enzyme. The coordinated regulation of two opposing reactions in a substrate cycle creates a situation that resembles a near-equilibrium reaction, whose flux can vary widely. The simultaneous operation of the two reactions results in the apparently wasteful consumption of ATP. This loss of free energy is offset by the ability to dramatically alter the flux through the pathway on short notice. An additional advantage of a substrate cycle is that two—rather than one—enzymes are involved, which allows a greater variety of hormonal or neuronal control mechanisms to fine-tune the activity of the pathway. (Section 14-4B)

5. Fructose enters the glycolytic pathway in a straightforward manner in muscle, by being converted to fructose-6-phosphate by the action of hexokinase. In liver, which has glucokinase rather than hexokinase, fructose is first converted to fructose-1-phosphate, which then undergoes aldol cleavage to dihydroxyacetone phosphate (which is converted to the glycolytic intermediate glyceraldehyde-3-phosphate) and glyceraldehyde. The glyceraldehyde may be phosphorylated to GAP for entry into glycolysis or it may be transformed into dihydroxyacetone phosphate by a three-reaction pathway that converts it first to glycerol and then glycerol-3-phosphate. This latter compound is required for lipid synthesis.

   Galactose requires four enzymatic reactions to enter glycolysis. First, galactose is phosphorylated at C1. The resulting galactose-1-phosphate is transferred to the UMP group of

UDP–glucose, thereby releasing glucose-1-phosphate. The UDP–galactose undergoes epimerization to regenerate UDP–glucose. The glucose-1-phosphate released from UMP is converted by hexokinase to the glycolytic intermediate glucose-6-phosphate.

Mannose is recognized by hexokinase, which produces mannose-6-phosphate. An isomerization reaction yields fructose-6-phosphate, which can continue through glycolysis. (Section 14-5)

6. The pentose phosphate pathway consists of three sets of reactions (see Figure 14-29). In the first stage, glucose-6-phosphate is oxidized and decarboxylated in three reactions to produce 2 NADPH, CO_2, and ribulose-5-phosphate. In the second stage, ribulose-5-phosphate is converted to ribose-5-phosphate and xylulose-5-phosphate. If ribose-5-phosphate is not siphoned off for nucleotide biosynthesis, these five-carbon sugars are produced in a ratio of two xylulose-5-phosphate to one ribose-5-phosphate so that they can be quantitatively converted to glycolytic intermediates. In the third stage, the five-carbon sugars undergo rearrangements catalyzed by transaldolase and transketolase so that 3 five-carbon sugars are converted to 2 fructose-6-phosphate and 1 glyceraldehyde-3-phosphate. (Section 14-6)

7. When ribose-5-phosphate is not needed, the pentose phosphate pathway yields 2 NADPH for each glucose-6-phosphate, and the carbon skeleton of the G6P substrate is entirely recycled back to glycolytic intermediates by the activities of ribulose-5-phosphate isomerase, ribulose-5-phosphate epimerase, transaldolase, and transketolase.

   When R5P is needed for nucleotide biosynthesis, the ribulose-5-phosphate produced by the oxidative reactions is converted to R5P rather than xylulose-5-phosphate. In addition, the glycolytic intermediates fructose-6-phosphate and glyceraldehyde-3-phosphate can be converted to R5P by the reversible reactions catalyzed by transaldolase and transketolase. (Section 14-6D)

Questions

Overview of Glycolysis

1. Describe the two-stage "chemical strategy" of glycolysis and write a balanced equation for each phase.

2. The two "high-energy" compounds produced in glycolysis are ____________________ and ____________________.

The Reactions of Glycolysis

3. Examine the following five glycolytic intermediates:
   (a) Name each intermediate.
   (b) Write the order in which they appear in glycolysis.
   (c) Which intermediate is a reactant in substrate-level phosphorylation?
   (d) List the glycolytic enzyme for which each intermediate is a substrate.
   (e) Which phosphorylated intermediates of glycolysis are not shown here?

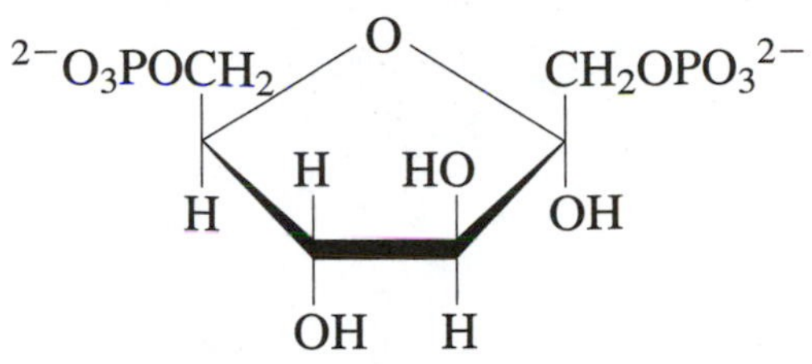

$^{2-}O_3POCH_2$ O $CH_2OPO_3^{2-}$ H H HO OH OH H

A

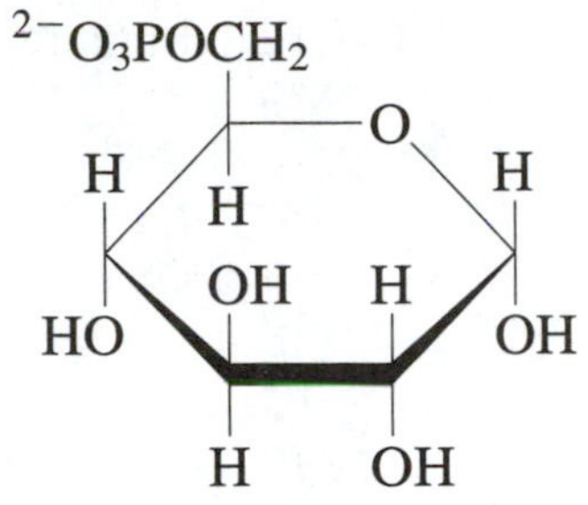

$^{2-}O_3POCH_2$ O H H H OH H HO OH H OH

B

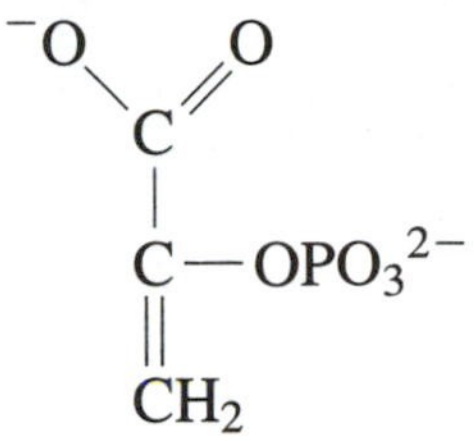

^-O O C $C-OPO_3^{2-}$ CH_2

C

$H_2C-OPO_3^{2-}$ HO—C—H C ^-O O

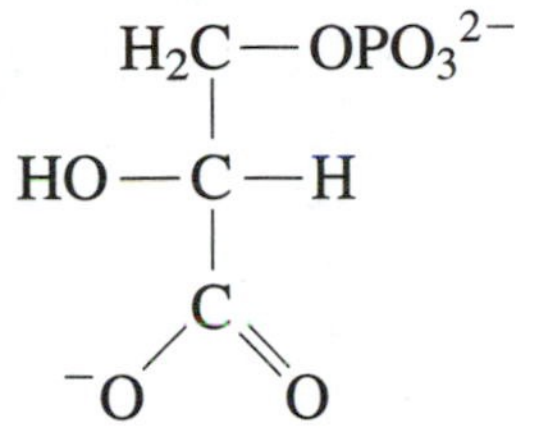

D

$H_2C-OPO_3^{2-}$ HO—C—H C H O

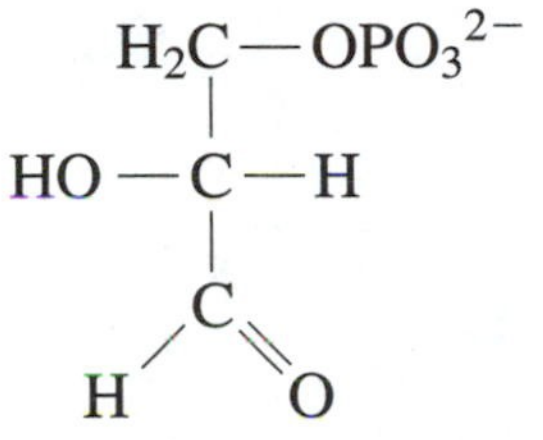

E

4. List the four kinases of glycolysis.

5. There are _____ isomerization reactions in glycolysis. The enzymes that catalyze them are ____.

6. How do Class II aldolases differ from Class I aldolases?

7. For the reaction catalyzed by triose phosphate isomerase, what is the equilibrium ratio of reactants and products under standard biochemical conditions? How does this ratio differ from the ratio observed in the cell at 37°C?

8. How does the GAPDH-catalyzed exchange of ^{32}P between P_i and 1,3-bisphosphoglycerate corroborate the existence of an acyl–enzyme intermediate in the GAPDH reaction?

9. 2,3-BPG is an intermediate in the reaction catalyzed by phosphoglycerate mutase. Why does the cell require trace amounts of 2,3-BPG?

10. If the cytosolic $[NAD^+]/[NADH]$ ratio is 100 and the $[ATP]/[ADP][P_i]$ ratio is 10, what is the actual (not equilibrium) ratio of [GAP]/[3PG] at 37°C in the cell? Assume that $[H^+] = 1$, its value in the biochemical standard state.

11. Fluoride ions specifically inhibit enolase in the presence of P_i in cell extracts.
    (a) Explain why both 2PG and 3PG accumulate in the presence of F^- and P_i.
    (b) Explain why 1,3-BPG does not accumulate.

12. Match each ^{14}C-labeled glucose with the ^{14}C-labeled pyruvate produced by glycolysis.

A

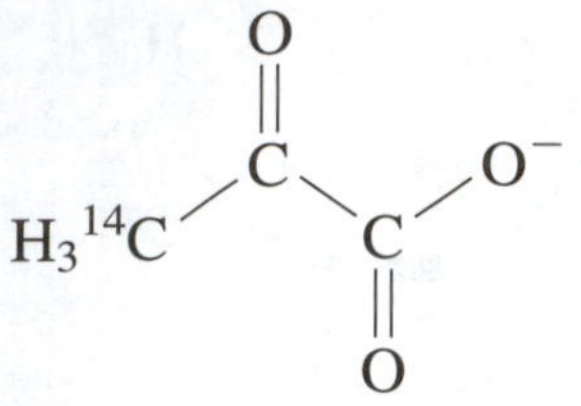

O

$H_3{}^{14}C$ C O$^-$

C

O

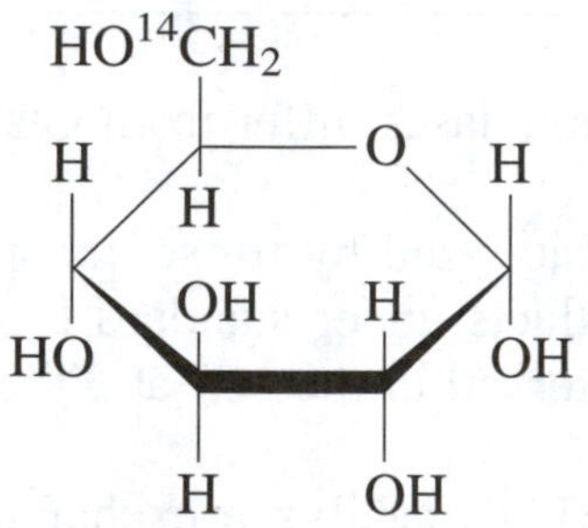

B

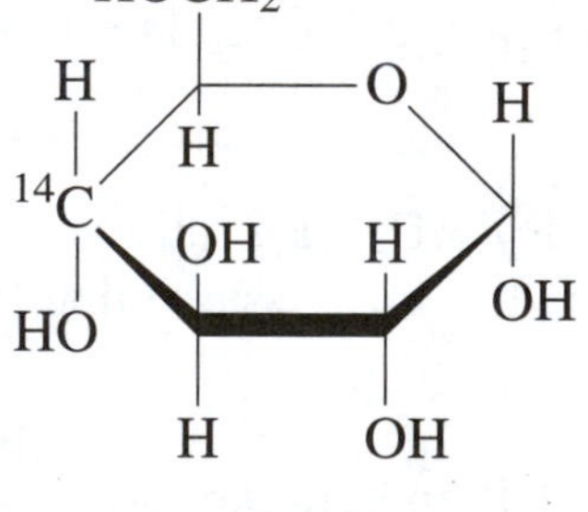

O

^{14}C

H_3C C O$^-$

O

C

O

C

H_3C ^{14}C O$^-$

O

(a) $HOCH_2$

H O H

H ^{14}C

OH H

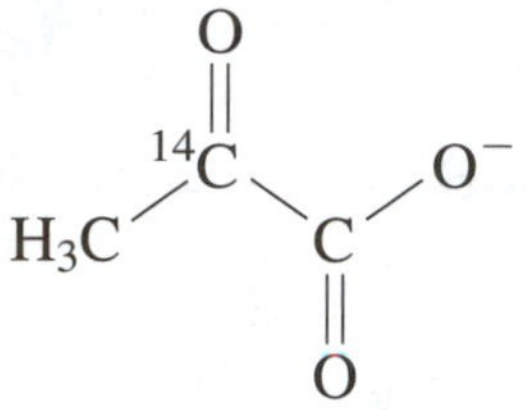

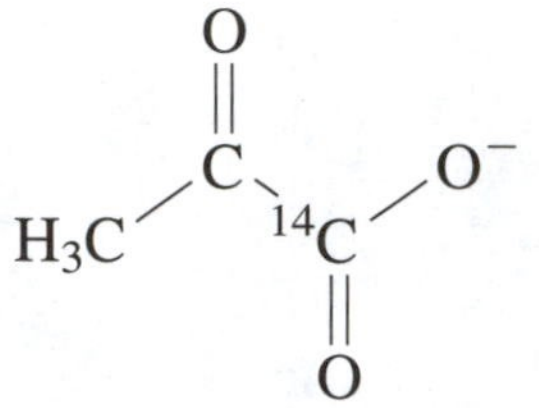

HO OH

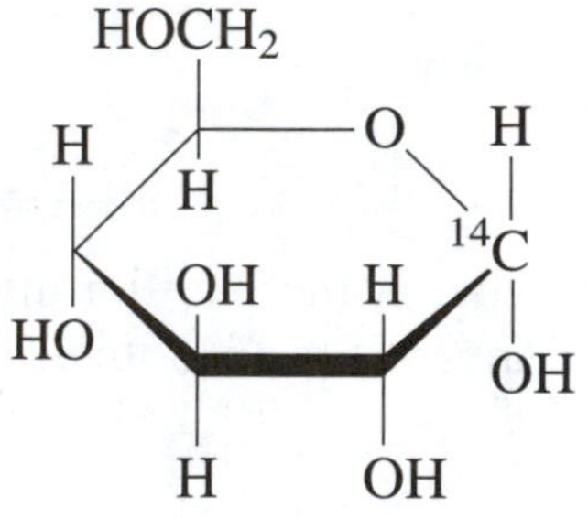

H OH

(c) $HO^{14}CH_2$

H O H

H

OH H

HO OH

H OH

(b) $HOCH_2$

H O H

H

^{14}C

OH H

HO OH

H OH

(d) $HOCH_2$

H O H

H

OH H

HO ^{14}C OH

H OH

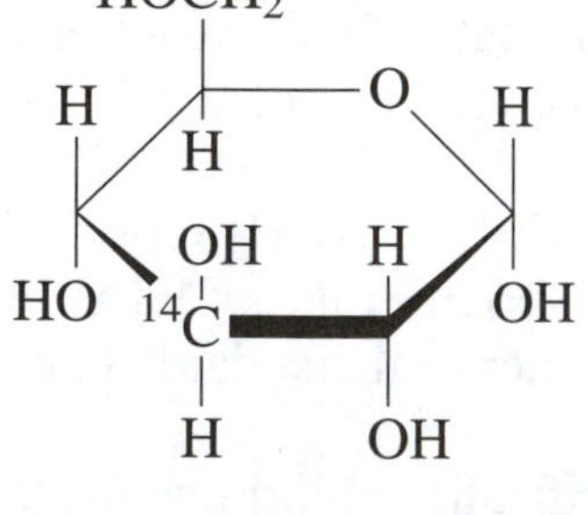

Fermentation: The Anaerobic Fate of Pyruvate

13. How does a muscle cell maintain the $[NAD^+]/[NADH]$ ratio during the catabolic breakdown of glucose?

14. Which of the carbon(s) of glucose must be labeled with ^{14}C for the end products of alcoholic fermentation to be unlabeled ("cold") ethanol and $^{14}CO_2$?

15. Compare the rates of ATP production in fermentation versus oxidative phosphorylation. Which process is utilized in rapid bursts of muscular activity?

Control of Glycolysis

16. In many metabolic pathways the first reaction is the rate-determining step of the pathway.
    (a) What is the rate-determining step of glycolysis?
    (b) What rationale can you offer for this "unusual" regulation?

17. What is meant by a futile cycle?

Metabolism of Hexoses Other than Glucose

18. The catabolism in the liver of which sugar—mannose, fructose, or galactose—bears the closest similarity to the first two reactions of glycolysis?

19. Defend or refute the following statement: The association of NAD^+ with UDP–galactose-4-epimerase suggests the generation of an additional NADH in the oxidation of galactose to pyruvate.

20. Why is galactosemia especially dangerous to nursing infants?

21. Some organisms use glycerol as a carbon energy source, and it is also an intermediate in fructose metabolism.
    (a) Write equations for the reactions required to oxidize glycerol to pyruvate.
    (b) Compare the ATP yield of two molecules of glycerol versus one molecule of glucose in glycolysis.
    (c) Does the anaerobic fermentation of glycerol maintain the redox balance of the cell? Explain.

The Pentose Phosphate Pathway

22. The diagram below shows the interconversions of the nonoxidative reactions of the pentose phosphate pathway.

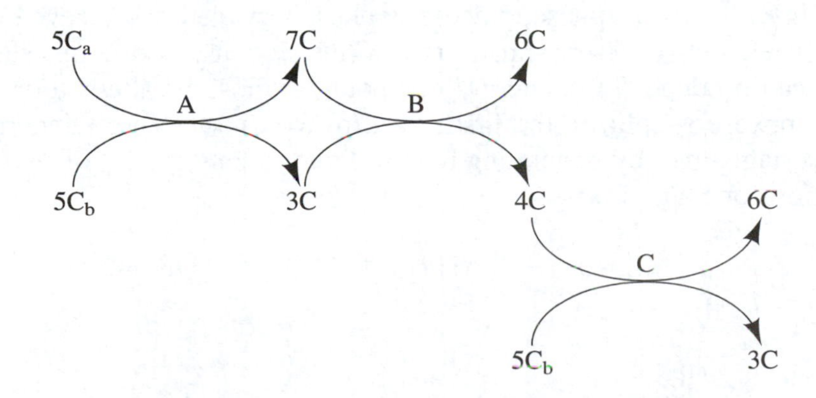

(a) Which sugar phosphates correspond to the 5C compounds?
(b) Which sugar phosphate corresponds to the 6C compound?
(c) Which sugar phosphate corresponds to the 3C compound?
(d) Which reaction(s) involves transketolase? Transaldolase?

23. The nonoxidative reactions of the pentose phosphate pathway convert pentose phosphates into hexose phosphates.
    (a) For every 3 glucose phosphates that enter the pentose phosphate pathway, how many fructose-6-phosphates are recovered?
    (b) For every 3 glucose-6-phosphates that enter the pentose phosphate pathway, how many glyceraldehyde-3-phosphates are recovered?

24. Ribulose-5-phosphate is converted to xylulose-5-phosphate and ribose-5-phosphate by an epimerase and an isomerase, respectively. What distinguishes these isomerizations?

25. You obtain a mutant transketolase from yeast that binds R5P and E4P poorly. What unique side product of the reaction catalyzed by this enzyme might you find in these cells?

26. Circle the appropriate choice in parentheses: Transaldolase transfers a (1, 2, 3)-carbon unit from a(n) (ketose / aldose) to a(n) (ketose / aldose) to form a(n) (ketose / aldose).

27. Circle the appropriate choice in parentheses: Transketolase transfers a (1, 2, 3)-carbon unit from a(n) (ketose / aldose) to a(n) (ketose / aldose) to form a(n) (ketose / aldose).

28. Which step commits glucose to oxidation via the pentose phosphate pathway? How is this enzyme regulated?

29. The conversion of G6P to R5P via the reactions of glycolysis and the pentose phosphate pathway, *without* the production of NADPH, can be summarized as

$$5\ \text{G6P} + \text{ATP} \rightleftharpoons 6\ \text{R5P} + \text{ADP} + \text{H}^+$$

What reaction requires ATP? How do you account for the stoichiometry?

Answers to Questions

1. In the energy investment phase, glucose is phosphorylated and cleaved to yield two molecules of glyceraldehyde-3-phosphate. Two ATPs are consumed in this stage so that the second stage can produce "high-energy" compounds whose breakdown drives ATP synthesis. The initial hexose is split, in the first stage, to two trioses. These undergo the reactions of the second stage, thereby generating four ATPs for a net "profit" of two ATPs per glucose. The equation for Stage I is

$$\text{Glucose} + 2\ \text{ATP} \rightarrow 2\ \text{GAP} + 2\ \text{ADP} + 2\ \text{H}^+$$

The equation for Stage II is

$$2\ \text{GAP} + 2\ \text{NAD}^+ + 4\ \text{ADP} + 2\ \text{HPO}_4^{2-} \rightarrow 2\ \text{pyruvate} + 2\ \text{NADH} + 4\ \text{ATP} + 2\ \text{H}_2\text{O} + 2\ \text{H}^+$$

2. ATP and NADH

3. (a) **A** Fructose-1,6-bisphosphate
   B Glucose-6-phosphate
   C Phosphoenolpyruvate
   D 3-Phosphoglycerate
   E Glyceraldehyde-3-phosphate
   (b) **B, A, E, D, C**
   (c) **C**
   (d) **A** Aldolase
   B Phosphoglucose isomerase
   C Pyruvate kinase
   D Phosphoglycerate mutase
   E Glyceraldehyde-3-phosphate dehydrogenase
   (e) Fructose-6-phosphate, dihydroxyacetone phosphate, 1,3-bisphosphoglycerate, and 2-phosphoglycerate.

4. Hexokinase, phosphofructokinase, phosphoglycerate kinase, and pyruvate kinase.

5. There are three isomerization reactions in glycolysis. The enzymes that catalyze them are phosphoglucose isomerase, triose phosphate isomerase, and phosphoglycerate mutase.

6. Class II aldolases stabilize the enolate intermediate by using a metal ion to polarize the carbonyl oxygen, rather than forming a covalently bound Schiff base intermediate, as in Class I aldolases.

7. Triose phosphate isomerase catalyzes the interconversion of GAP and DHAP. Its $\Delta G^{\circ\prime} = 7.9$ kJ $\cdot$ mol^{-1} and $\Delta G = 4.4$ kJ $\cdot$ mol^{-1} (Table 14-1). Since $\Delta G = -RT \ln K$, $\ln K = -\Delta G/RT$ and $K = e^{-\Delta G/RT}$. Under standard conditions,

$$K = \frac{[\text{GAP}]}{[\text{DHAP}]} = e^{-(7900\ \text{J}\cdot\text{mol}^{-1})/(8.3145\ \text{J}\cdot\text{K}^{-1}\cdot\text{mol}^{-1})(310\text{K})}$$
$$= 0.047$$

In the cell, where the reaction is not at equilibrium,

$$\Delta G = \Delta G^{\circ\prime} + RT \ln \frac{[\text{GAP}]}{[\text{DHAP}]}$$

$$\frac{[\text{GAP}]}{[\text{DHAP}]} = e^{(\Delta G - \Delta G^{\circ\prime})/RT}$$

$$\frac{[\text{GAP}]}{[\text{DHAP}]} = e^{(4400-7900\ \text{J}\cdot\text{mol}^{-1})/(8.3145\ \text{J}\cdot\text{K}^{-1}\cdot\text{mol}^{-1})(310\text{K})}$$
$$= 0.26$$

Therefore, the cellular ratio [GAP]/[DHAP] is nearly 6 times larger than under standard conditions, so that in the cell, formation of GAP is favored.

8. This isotope exchange reaction is consistent with the existence of an acyl–enzyme intermediate through the following series of events: For exchange to occur, 1,3-bisphosphoglycerate (1,3-BPG) must react with GAPDH to form an acyl–enzyme intermediate and eliminate P_i. (Fig. 14-9, Reaction 5 in reverse). In this reaction, the ^{32}P-labeled P_i reacts with the acyl-enzyme intermediate to yield ^{32}P-labeled 1,3-BPG.

9. 2,3-BPG is occasionally released by bisphosphoglycerate mutase. The resulting enzyme is inactive because it requires a phospho-His residue to phosphorylate 2PG. The presence of trace amounts of 2,3-BPG permits this substance to rebind to the enzyme and thereby activate it by forming the phospho-His residue.

10. The combined glyceraldehyde-3-phosphate dehydrogenase/phosphoglycerate kinase reaction is

$$\text{GAP} + \text{NAD}^+ + \text{ADP} + \text{P}_i \rightarrow \text{3PG} + \text{NADH} + \text{ATP} + \text{H}^+$$

$\Delta G^{\circ\prime} = -16.7 \text{ kJ} \cdot \text{mol}^{-1}$ and $\Delta G = -1.1 \text{ kJ} \cdot \text{mol}^{-1}$ (Table 14-1).

Since $\Delta G = \Delta G^{\circ\prime} + RT \ln Q$, where

$$Q = \frac{[\text{3PG}][\text{NADH}][\text{ATP}][\text{H}^+]}{[\text{GAP}][\text{NAD}^+][\text{ADP}][\text{P}_i]}$$

$$\begin{aligned} Q &= e^{(\Delta G - \Delta G^{\circ\prime})/RT} \\ &= e^{(-1100 + 16{,}700 \text{ J}\cdot\text{mol}^{-1})/(8.3145 \text{ J}\cdot\text{K}^{-1}\cdot\text{mol}^{-1})(310\text{K})} \\ &= 425 \end{aligned}$$

Since $$Q = \frac{[\text{3PG}][\text{NADH}][\text{ATP}][\text{H}^+]}{[\text{GAP}][\text{NAD}^+][\text{ADP}][\text{P}_i]} = \frac{[\text{3PG}]}{[\text{GAP}]} \times \frac{1}{100} \times 10 \times 1 = 425$$

$$\frac{[\text{GAP}]}{[\text{3PG}]} = \frac{1}{4250} = 2.35 \times 10^{-4}$$

11. (a) Inhibition of enolase causes its substrate, 2PG, to accumulate. Because the preceding reaction (catalyzed by phosphoglycerate mutase) is near equilibrium, 2PG equilibrates with 3PG so that [3PG] increases as [2PG] increases.
    (b) The formation of 3PG from 1,3-BPG is endergonic, so this reaction (catalyzed by phosphoglycerate kinase) does not proceed in reverse. Therefore, an increase in [3PG] does not cause [1,3-BPG] to increase.

12. (a) A
    (b) C
    (c) A
    (d) C

13. During the catabolic breakdown of glucose, NAD^+ is reduced to NADH, thereby lowering the $[NAD^+]/[NADH]$ ratio. Under anaerobic conditions, pyruvate can be converted to lactate with the concomitant oxidation of NADH to NAD^+. Under aerobic conditions, oxidative phosphorylation regenerates the NAD^+. Both of these processes will increase the $[NAD^+]/[NADH]$ ratio.

14. C3 or C4 of glucose must be labeled with ^{14}C.

15. The rate of ATP production during anaerobic fermentation can be as much as 100 times greater than during oxidative phosphorylation. Therefore, anaerobic fermentation provides the bulk of ATP during rapid bursts of muscle activity, even though the yield of ATP per glucose is much lower than in oxidative phosphorylation.

16. (a) The phosphofructokinase reaction is the rate-determining step.
    (b) The first reaction of glycolysis, catalyzed by hexokinase, is not a suitable control point for the overall pathway because significant amounts of sugar enter glycolysis as glucose-6-phosphate, the product of this reaction. For example, galactose is converted to glucose-6-phosphate to enter glycolysis. Not all this glucose-6-phosphate proceeds through glycolysis: Some is shunted to the pentose phosphate pathway. Furthermore, mannose and, in muscle, fructose enter glycolysis as fructose-6-phosphate, which freely interconverts with glucose-6-phosphate. Therefore, the production of fructose-1,6-bisphosphate by PFK is the first committed step of glycolysis and the most effective point for regulation.

17. Substrate cycles were originally referred to as futile cycles because of their apparent futile waste of free energy.

18. Mannose.

19. No additional NADH is produced. The NAD^+ is probably reduced and then reoxidized in the sequential oxidation and reduction of C4 during epimerization of the hexose.

20. The primary sugar in human milk is lactose, a disaccharide of glucose and galactose.

21. (a) Glycerol can be converted to pyruvate through reactions catalyzed by glycerol kinase, glycerol phosphate dehydrogenase, triose phosphate isomerase, and the enzymes of Stage II of glycolysis (see Fig. 14-26):

$$\begin{aligned}
\text{Glycerol} + \text{ATP} &\rightarrow \text{glycerol-3-phosphate} + \text{ADP} \\
\text{Glycerol-3-phosphate} + \text{NAD}^+ &\rightarrow \text{DHAP} + \text{NADH} + \text{H}^+ \\
\text{DHAP} &\rightarrow \text{GAP} \\
\text{GAP} + \text{NAD}^+ + 2\ \text{ADP} + \text{P}_i &\rightarrow \text{pyruvate} + \text{NADH} + 2\ \text{ATP} + \text{H}_2\text{O} + 2\ \text{H}^+ \\
\hline
\text{Glycerol} + 2\ \text{NAD}^+ + \text{ADP} + \text{P}_i &\rightarrow \text{pyruvate} + 2\ \text{NADH} + \text{ATP} + \text{H}_2\text{O} + 3\ \text{H}^+
\end{aligned}$$

    (b) The net yield of ATP is the same for the oxidation of 1 glucose or 2 glycerol to pyruvate.
    (c) For every pyruvate formed from glycerol, 2 NAD^+ are reduced to NADH. Alcoholic fermentation regenerates only 1 NAD^+ per pyruvate, so this pathway does not maintain the redox balance of the cell.

22. (a) $5C_a$ is ribose-5-phosphate, and $5C_b$ is xylulose-5-phosphate.
    (b) 6C is fructose-6-phosphate.
    (c) 3C is glyceraldehyde-3-phosphate.
    (d) A and C are transketolase reactions, and B is a transaldolase reaction.

23. (a) Two
    (b) One directly, although four more can be produced via glycolysis from the two F6P molecules also produced.

24. The conversion of Ru5P to Xu5P is a racemization that inverts the configuration at the C3 chiral center. The isomerization of Ru5P to R5P converts a ketose to an aldose by shifting the position of a double bond.

25.

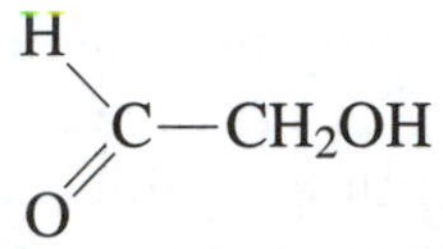

H
C—CH_2OH
O

2-hydroxy-ethanol (hydroxy-acetaldehyde)

As in the pyruvate decarboxylase reaction, an aldehyde might be expected to be released from the TPP cofactor of transketolase in the absence of the second substrate.

26. (3); (ketose); (aldose); (ketose and an aldose)

27. (2); (ketose); (aldose); (ketose and an aldose)

28. The committed step is the exergonic reaction mediated by G6PD, which is regulated by the availability of $NADP^+$.

29. (1) The 5 G6P are converted to 5 F6P by phosphoglucose isomerase.
    (2) One F6P is converted to FBP, which requires ATP for the reaction catalyzed by phosphofructokinase.
    (3) The FBP is converted to 2 GAP by aldolase and triose phosphate isomerase.
    (4) The transaldolase and transketolase reactions of the pentose phosphate pathway operate in reverse to convert 2 F6P + 1 GAP to 2 Xu5P and 1 R5P. Since the starting materials are 4 F6P and 2 GAP, the result is 4 Xu5P and 2 R5P.
    (5) The 4 Xu5P can be isomerized to 4 R5P, for a total yield of 6 R5P from the original 5 G6P.

Chapter 15 Glycogen Metabolism and Gluconeogenesis

This chapter discusses glycogen breakdown and synthesis, gluconeogenesis, and oligosaccharide synthesis, with an emphasis on the mechanisms that regulate these metabolic pathways. Six major enzymes are involved in the breakdown and synthesis of glycogen: glycogen phosphorylase, glycogen debranching enzyme, and phosphoglucomutase for glycogen breakdown, and UDP–glucose pyrophosphorylase, glycogen synthase, and glycogen branching enzyme for glycogen synthesis. The chapter explains how the synthesis of glycogen from glucose-1-phosphate requires the free energy of nucleotide hydrolysis in a reaction that is the opposite of the exergonic breakdown of glycogen. The mechanisms of glycogen phosphorylase and glycogen synthase are examined, including the role of the oxonium ion transition state in each case. Box 15-1 explores the strategies for optimizing the branched structure of glycogen. The chapter then discusses the regulation of glycogen phosphorylase and glycogen synthase by allosteric effectors and covalent modification. Here you are introduced to the notion of an enzyme cascade where extracellular signals (hormones) initiate the activation of successive kinases, resulting in the reciprocal activation of glycogen phosphorylase and inactivation of glycogen synthase. In this context, you are introduced to the second messengers cyclic AMP and Ca^{2+}. The role of phosphoprotein phosphatase-1 in modulating the ratio of phosphorylated to dephosphorylated enzymes is also explored as an additional layer of complexity in the regulation of the enzymes involved in glycogen breakdown and synthesis. Box 15-2 shows how inherited metabolic diseases contribute to our understanding of glycogen metabolism.

The chapter then moves on to gluconeogenesis, the process by which pyruvate and related metabolites can be converted to glucose. Gluconeogenesis uses many of the same enzymes as glycolysis but requires other enzymes to bypass the exergonic steps of glycolysis. The chapter discusses the interesting role of the allosteric effector fructose-2,6-bisphosphate in the regulation of gluconeogenesis and glycolysis in liver and heart muscle. The last section of this chapter discusses the roles of nucleotide sugars and dolichol pyrophosphate in oligosaccharide synthesis.

Essential Concepts

1. The mobilization of glucose from glycogen stores in the liver provides a constant supply of glucose to the central nervous system and red blood cells, which use glucose as their sole energy source. Under fasting conditions, amino acids (mainly from muscle protein degradation) serve as precursors for new glucose synthesis (gluconeogenesis).

2. Glucose-6-phosphate (G6P) is a key branch point in glucose metabolism in the liver, as it can be polymerized to glycogen, degraded to pyruvate, converted to ribose-5-phosphate, or hydrolyzed to glucose.

Glycogen Breakdown

3. Glycogen breakdown (glycogenolysis) utilizes three enzymes:
   (a) Glycogen phosphorylase, which catalyzes the phosphorolysis at the nonreducing ends of glycogen to yield glucose-1-phosphate (G1P).
   (b) Glycogen debranching enzyme, which transfers a trisaccharide and hydrolyzes the $\alpha(1\rightarrow6)$ linkage at branch points.
   (c) Phosphoglucomutase, which converts G1P to G6P.

4. Phosphorylase utilizes the cofactor pyridoxal-5′-phosphate (PLP) in the general acid–base catalytic mechanism in glycogen phosphorolysis. Inorganic phosphate attacks the terminal glucose residue, which passes through an oxonium ion intermediate before being released as G1P. The activity of the dimeric enzyme is regulated by allosteric effectors and by phosphorylation/dephosphorylation of the protein at Ser 14.

5. Glycogen debranching enzyme acts as an α(1→4) transglycosylase by transferring a trisaccharide from a limit branch (a four- or five-glucose-residue segment that phosphorylase cannot further degrade) to the nonreducing end of another branch. The remaining glucosyl residue, which is attached to glycogen by an α(1→6) linkage, is hydrolyzed at a separate active site on the enzyme to release free glucose.

6. Phosphoglucomutase converts G1P to G6P by way of a G1,6P intermediate. A phosphate group from a Ser residue is transferred to C6 of G1P, followed by transfer of the C1 phosphate back to the Ser residue, in a manner similar to the phosphoglycerate mutase reaction.

7. The liver expresses glucose-6-phosphatase, an enzyme that hydrolyzes G6P. The resulting free glucose equilibrates with glucose in the blood so that the breakdown of liver glycogen leads to an elevation of blood glucose levels.

Glycogen Synthesis

8. Glycogen synthesis requires three enzymes to covert G1P to glycogen. First, UDP–glucose pyrophosphorylase catalyzes the transfer of UMP from UTP to the phosphate group of G1P to form UDP–glucose and PP_i. PP_i is eventually hydrolyzed to P_i by inorganic pyrophosphatase, which provides the exergonic push for this reaction.

9. Glycogen synthase catalyzes a transfer reaction in which the glucosyl residue of UDP–glucose is added to the nonreducing end of glycogen through an α(1→4) bond. Glycogen synthase can only extend a pre-existing α(1→4)-linked chain. The glycogen molecule originates through the action of the protein glycogenin, which assembles a seven-residue glycogen "primer" for glycogen synthase to act on.

10. Glycogen branching enzyme transfers a seven-residue segment from the end of an α(1→4)-linked glucan chain to the C6-hydroxyl group of a glucosyl residue on the same chain or another chain, thereby forming an α(1→6)-linked branch.

Control of Glycogen Metabolism

11. The opposing processes of glycogen breakdown and synthesis are coordinately controlled by allosteric regulation and covalent modification of key enzymes. The major allosteric regulators are ATP, AMP, and G6P. Covalent modification occurs with the transfer of P_i from ATP to certain enzymes by the action of specific kinases.

12. Glycogen phosphorylase exists in two forms: phosphorylase *a* (the phosphorylated, more active enzyme) and phosphorylase *b* (the dephosphorylated, less active enzyme). Each form also has two conformations: the T (relatively inactive) state and the R (relatively active) state. AMP promotes the T→R conformational change and thereby activates phosphorylase

b, whereas ATP and G6P inhibit this conformational change. Phosphorylase *a* is less sensitive to allosteric effectors and is mainly in the R state. However, high concentrations of glucose promote the R→T transition.

13. Covalent modification regulates glycogen phosphorylase and glycogen synthase in a reciprocal fashion. Glycogen phosphorylase tends to be more active when it is phosphorylated, so the ratio of phosphorylase *a* to phosphorylase *b* largely determines the rate of glycogen phosphorolysis. This ratio is set by the activity of phosphorylase kinase, which is regulated by cAMP-dependent protein kinase (cAPK), and by protein phosphatase-1 (PP-1). In contrast, glycogen synthase is activated by dephosphorylation so that cAPK promotes glycogen breakdown while inhibiting glycogen synthesis, and PP-1 inhibits glycogen breakdown while promoting glycogen synthesis.

14. The hormones glucagon and epinephrine activate adenylate cyclase, a transmembrane protein that converts ATP to cyclic AMP (cAMP). Elevated levels of cAMP bind to the regulatory subunits of cAPK, which causes its two catalytic subunits to dissociate from its regulatory dimer. This dissociation activates the catalytic subunits, which phosphorylate phosphorylase kinase, thereby activating it.

15. Phosphorylase kinase is also activated by Ca^{2+}, which binds to calmodulin, a ubiquitous Ca^{2+}-binding protein that interacts with numerous proteins and is a subunit of phosphorylase kinase. Elevated calcium levels can arise hormonally via epinephrine or from the neuronal impulses that trigger muscle contraction.

16. Protein phosphatase-1 (PP-1) can dephosphorylate glycogen phosphorylase, phosphorylase kinase, and glycogen synthase. In muscle, insulin-stimulated protein kinase activates PP-1 by phosphorylating its G subunit, which binds to glycogen in muscle . The cAPK-mediated phosphorylation of another site on the G subunit causes PP-1 to be released in the cytoplasm, where it cannot dephosphorylate glycogen-bound enzymes. In addition, phosphoprotein phosphatase inhibitor 1 (inhibitor-1) inhibits PP-1. Inhibitor-1 activity is stimulated by phosphorylation by cAPK, which helps preserve the phosphorylated (active) forms of phosphorylase kinase and phosphorylase *a*.

17. Glycogen synthase is inactivated by phosphorylation by the same enzyme system that phosphorylates glycogen phosphorylase. Hence, activation of phosphorylase kinase, which activates phosphorylase *a*, inactivates glycogen synthase. This regulatory mechanism provides a rapid and large-scale control of flux in the substrate cycle that links glycogen and G1P. Glycogen synthase activity is also controlled by other kinases.

18. Hormones, including glucagon, insulin, epinephrine, and norepinephrine, ultimately control glycogen metabolism. These hormones bind to transmembrane protein receptors and initiate a series of reactions that lead to the production of molecules called second messengers (e.g., cAMP and Ca^{2+}), which modulate the activities of numerous intracellular proteins. Glucagon binding to its receptor in the liver results in an elevation of cAMP, which favors glycogen breakdown. Epinephrine and norepinephrine bind to α- and β-adrenergic receptors in the liver and to β-adrenergic receptors in muscle. The binding of these hormones to β-adrenergic receptors increases [cAMP], whereas their binding to α-adrenergic receptors increases cytosolic [Ca^{2+}]. Insulin binding to its receptor in tissues other than the liver decreases [cAMP] and promotes glycogen synthesis.

Gluconeogenesis

19. The principal noncarbohydrate precursors of glucose are lactate, pyruvate, and amino acids. In animals, these compounds (except for leucine and lysine) are converted, at least in part, into oxaloacetate, which is required for gluconeogenesis.

20. The conversion of pyruvate (or lactate) to glucose follows a pathway that is the reverse of glycolysis except where it bypasses the exergonic steps catalyzed by pyruvate kinase, phosphofructokinase, and hexokinase.

21. To bypass the pyruvate kinase reaction, pyruvate is first carboxylated by pyruvate carboxylase in a reaction that is driven by ATP hydrolysis. The enzyme's biotin prosthetic group is converted to a carboxybiotinyl group in order to transfer CO_2 to pyruvate. The product of this reaction is oxaloacetate, which is subsequently decarboxylated to phosphoenolpyruvate (PEP). The HCO_3^- added to pyruvate therefore leaves as CO_2. This second reaction, catalyzed by PEP carboxykinase (PEPCK) is driven by the hydrolysis of GTP.

22. Oxaloacetate is produced in the mitochondria, while the reactions that convert PEP to glucose occur in the cytosol. In species with mitochondrial PEPCK, the PEP formed in the mitochondria is exported to the cytosol via a specific transporter. In species with cytosolic PEPCK, oxaloacetate is first converted to malate or aspartate, each of which has a transporter that allows mitochondrial-cytosolic exchange, and, in the cytosol, is reconverted to oxaloacetate.

23. PEP is converted to fructose-1,6-bisphosphate (FBP) by the enzymes of glycolysis operating in reverse. FBP is then hydrolyzed to fructose-6-phosphate and P_i by the action of fructose bisphosphatase (FBPase-1). Similarly, G6P is hydrolyzed by glucose-6-phosphatase, yielding glucose and P_i. These reactions therefore bypass the exergonic hexokinase and phosphofructokinase reactions.

24. Gluconeogenesis and glycolysis are reciprocally regulated in the liver. The principal allosteric regulator is the metabolite fructose-2,6-bisphosphate (F2,6P), which is a potent activator of PFK-1 and inhibitor of FBPase-1. F2,6P is formed and degraded by a bifunctional protein referred to as PFK-2/FBPase-2. cAPK phosphorylates this protein, thereby activating FBPase-2 and inactivating PFK-2, which results in a net decrease in F2,6P (and favors gluconeogenesis). In heart muscle, the situation is reversed, so that phosphorylation activates PFK-2, which facilitates the muscle's ability to extract energy from glucose via glycolysis.

25. Glucose metabolism is also regulated by other mechanisms:
    (a) Acetyl-CoA activates pyruvate carboxylase.
    (b) Alanine inhibits pyruvate kinase. The amino group of alanine is transferred to an α-keto acid by transamination to yield pyruvate and a new amino acid. The resulting pyruvate then serves as a substrate for gluconeogenesis.
    (c) Long-term regulation of gluconeogenesis occurs via changes in gene expression. Prolonged low concentrations of insulin or high concentrations of cAMP stimulate the transcription of the genes for PEPCK, FBPase, and glucose-6-phosphatase, and repress the transcription of the genes for glucokinase, PFK, and the PFK-2/FBPase-2 bifunctional enzyme.

Other Carbohydrate Biosynthetic Pathways

26. The formation of glycosidic bonds in oligo- and polysaccharides is facilitated by nucleotide sugars, as in the polymerization of glucose by glycogen synthase. The principal nucleotides used are ADP and CDP. Nucleotide sugars are glycosyl donors in the synthesis of *O*-linked oligosaccharides and in the processing of *N*-linked oligosaccharides.

27. *N*-linked oligosaccharides are initially built on dolichol, an isoprenoid lipid carrier in the endoplasmic reticulum (ER). This process begins in the cytosol but finishes in the lumen of the ER, as the dolichol-oligosaccharide flips back and forth across the ER membrane. After the oligosaccharide reaches 14 residues, it is transferred to a protein, leaving dolichol pyrophosphate.

28. Lactose synthesis in mammals involves the mammary gland protein α-lactalbumin, which changes the substrate specificity of a galactosyltransferase so that it synthesizes lactate rather than *N*-acetyllactosamine.

Guide to Study Exercises (text p. 464)

1. Glucose-6-phosphate is obtained by the phosphorylation of glucose by hexokinase, by the phosphorolysis of glycogen (catalyzed by glycogen phosphorylase) followed by isomerization (catalyzed by phosphoglucomutase), and by gluconeogenesis. G6P can be converted to glucose in the liver by the action of glucose-6-phosphatase; it can enter the catabolic pathways of glycolysis and the pentose phosphate pathway; and it can be used to synthesize glycogen after being converted to G1P by phosphoglucomutase.

2. The structure of glycogen, a branched polymer whose chains each have two branches of 8–14 residues, satisfies several (often conflicting) criteria. First, glycogen must be a polymer in order to accommodate a large number of glucose residues in a small volume. The large number of branches in the outermost layer provides a large number of points of attack for glycogen phosphorylase to mobilize glucose. However, the branching cannot be so dense that the enzyme cannot access the glycosidic bonds. Furthermore, the branches must be long enough to support the rapid release of many glucose residues before debranching is required. (Box 15-2)

3. Glycogen degradation is catalyzed by three enzymes. Glycogen phosphorylase catalyzes the phosphorolysis of the $\alpha(1\rightarrow4)$ glycosidic bonds, thereby removing glucose residues (as glucose-1-phosphate) to within four or five residues of a branch point. Glycogen debranching enzyme (acting as a transglycosylase) transfers a trisaccharide from a shortened branch to the end of another branch. The same enzyme, now acting as an $\alpha(1\rightarrow6)$ glucosidase, hydrolyzes the residue remaining at the branch point and releases it as glucose. Phosphoglucomutase reversibly converts G1P to glucose-6-phosphate, which can enter glycolysis or the pentose phosphate pathway.

   Glycogen synthesis requires three other enzymes. G1P is "activated" by the attachment of a UTP-derived UMP group to produce UDP–glucose, in a reaction catalyzed by UDP–glucose pyrophosphorylase. The reaction is driven by the subsequent hydrolysis of the PP_i released from UTP. Next, glycogen synthase catalyzes formation of $\alpha(1\rightarrow4)$ glycosidic bonds to extend a chain in a pre-existing glycogen molecule. Branches are created when branching enzyme (a transglycosylase) removes a seven-residue segment from the end of a chain and reattaches it in an $\alpha(1\rightarrow6)$ linkage at a point farther up the same chain or another chain. (Sections 15-1 and 2)

4. Opposing metabolic pathways must differ in at least one step because each process must be exergonic. This means that an exergonic step of the degradative pathway cannot simply operate in reverse for the biosynthetic pathway (for which this would be an endergonic step) but must be bypassed by another, exergonic reaction. For example, in glycogen degradation, the exergonic step is catalyzed by glycogen phosphorylase. Therefore, in glycogen synthesis, this step is bypassed by linking the glucose to a nucleotide (UDP–glucose), which can then serve as a substrate for glycogen synthase. These two steps are exergonic. (Section 15-2).

5. In a simple allosteric system, the activity of each enzyme in a metabolic pathway varies in response to the presence of allosteric activators and inhibitors. These compounds bind to the enzyme in a stoichiometric fashion, so their concentrations must remain constant for their effects on the enzyme to persist. A phosphorylation/dephosphorylation system allows more sensitive regulation of a pathway for two reasons. First, it can operate in addition to allosteric effects, which permits a greater range of regulatory responses. Second, phosphorylation and dephosphorylation, which result from a hormonal or neuronal signal, can be amplified or dampened in a catalytic fashion. Thus, the state of phosphorylation of an enzyme does not require a constant stoichiometric concentration of another substance. (Section 15-3)

6. In muscle, phosphoprotein phosphatase-1 (whose action deactivates glycogen phosphorylase and activates glycogen synthase) is active only when it associates with glycogen through its G subunit. Insulin (which signals glucose availability) prevents this association and therefore promotes glycogen synthesis. Epinephrine (which opposes the action of insulin) leads to the dissociation of PP-1 from glycogen and therefore promotes glycogen breakdown. PP-1 is also inhibited by phosphoprotein phosphatase inhibitor 1, which is activated by cAMP (which also activates glycogen phosphorylase). Thus, the increased inhibition of phosphoprotein phosphatase-1 leads to increased glycogen breakdown.

   In liver, PP-1 is inhibited by binding to glycogen phosphorylase *a*. When phosphorylase *a* is in its active R form, its Ser 14 phosphoryl group is inaccessible to PP-1. When phosphorylase *a* converts to the less active T form, the Ser 14 phosphoryl group becomes accessible to PP-1, which then catalyzes its dephosphorylation, thereby converting phosphorylase *a* to phosphorylase *b*. Because liver cells contain ten times more glycogen phosphorylase than PP-1, ~90% of the glycogen phosphorylase must be converted to the *b* form before PP-1 is released and can then dephosphorylate and activate glycogen synthase. (Section 15-3B)

7. Gluconeogenesis from pyruvate uses seven of the ten glycolytic enzymes, operating in reverse (see Figure 15-22). Four other enzymes are required to bypass the three exergonic reactions of glycolysis. These are pyruvate carboxylase and PEPCK, which together convert pyruvate to PEP at the expense of 2 ATP equivalents; fructose bisphosphatase, which converts fructose-1,6-bisphosphate to fructose-6-phosphate; and glucose-6-phosphatase, which converts glucose-6-phosphate to glucose. (Section 15-4)

8. Oxaloacetate, the product of the pyruvate carboxylase reaction (it is also an intermediate of the citric acid cycle), is produced in the mitochondrion. Depending on the species, oxaloacetate may be converted to PEP in the mitochondrion or in the cytosol, and the remaining gluconeogenic reactions occur in the cytosol. In species with cytosolic PEPCK, mitochondrial oxaloacetate must be converted to malate or aspartate, both of which have specific transporters. In the cytosol, the malate or aspartate is converted back to oxaloacetate. When the malate shuttle is used, NADH is oxidized in the mitochondrion (to reduce oxaloacetate to malate), and NAD^+ is reduced in the cytosol. The net result is the transfer of a reducing equivalent from the mitochondrion to the cytosol. Thus, the malate shuttle supplies the cytosolic NADH required for gluconeogenesis (in the glyceraldehyde-3-phosphate dehydrogenase reaction). If gluconeogenesis begins with lactate rather than pyruvate, its oxidation to pyruvate generates the required NADH, so that the aspartate shuttle (which does not involve any redox reactions) can be used instead of the malate shuttle to transport oxaloacetate from the mitochondrion to the cytosol. (Section 15-4A)

9. Fructose-2,6-bisphosphate (F2,6P) allosterically activates phosphofructokinase (PFK, a glycolytic enzyme) and inhibits FBPase (a gluconeogenic enzyme). Therefore, the balance between glycolysis and gluconeogenesis depends on the concentration of F2,6P, which is synthesized and degraded by a bifunctional enzyme containing PFK-2 and FBPase-2 activities. The balance between F2,6P synthesis and degradation in turn depends on the phosphorylation state of this enzyme, which is ultimately under hormonal control. For example, glucagon binding to liver cell receptors leads to an increase in cAMP, which activates cAPK to phosphorylate the bifunctional enzyme. This activates FBPase-2 and inactivates PFK-2 so that [F2,6P] decreases. As a result, PFK activity decreases and FBPase activity increases, which favors gluconeogenesis rather than glycolysis. (Section 15-4C)

Questions

Glycogen Breakdown

1. List the three enzymes required for the breakdown of glycogen.

2. What feature of the structure of glycogen phosphorylase is consistent with the observation that the enzyme cannot cleave a glycosidic bond within five residues of a branch point?

3. What role does pyridoxal phosphate (PLP) play in the mechanism of glycogen phosphorylase?

4. What would be the product of the nonenzymatic phosphorolysis of glycogen?

5. In the diagram of glycogen shown below, circle the substrates for glycogen debranching enzyme.

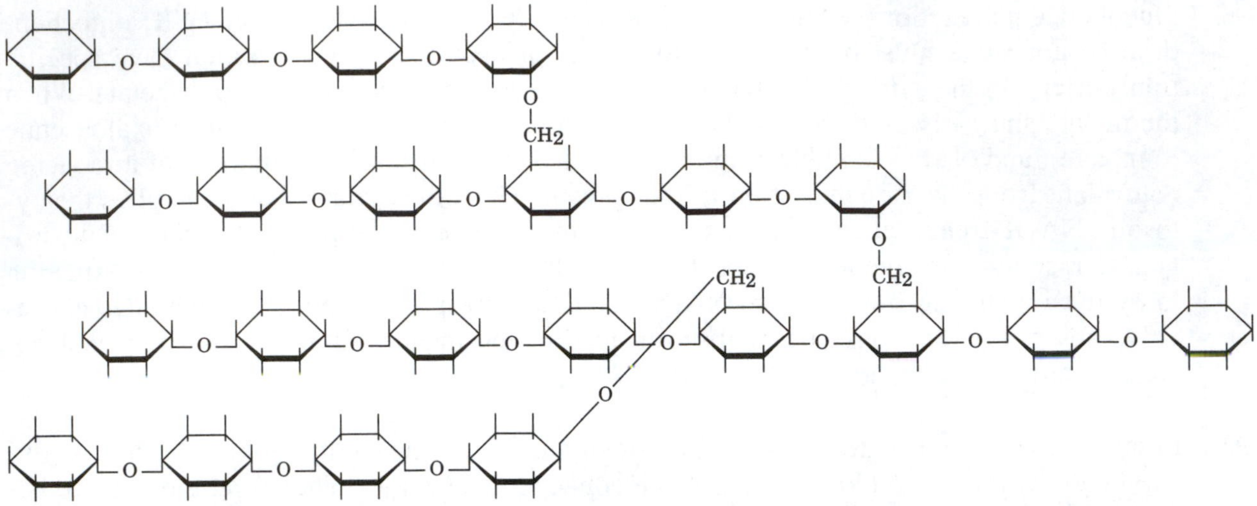

6. The rate of debranching is much slower than that of phosphorolysis. Explain how highly branched glycogen molecules release glucose-1-phosphate at a greater rate than relatively unbranched ones.

7. Write an equation for the equilibrium constant for the phosphorylase reaction. $\Delta G^{\circ\prime}$ for this reaction is $+3.1\ \text{kJ} \cdot \text{mol}^{-1}$, yet glycogen breakdown in the liver and muscle is thermodynamically favored at 37°C. What is the minimum ratio of $[P_i]/[G1P]$ required to make the phosphorylase reaction exergonic? Assume that the concentration of glycogen does not change significantly.

Glycogen Synthesis

8. How is the thermodynamic barrier to glycogen synthesis overcome by cells?

9. In the diagram below, each circle represents a glycosyl unit. By the action of branching enzyme, show the most branched structure this molecule can assume. Indicate the reducing and nonreducing ends of the molecule and the position where a glycogenin molecule would be found. There are 50 glycosyl residues.

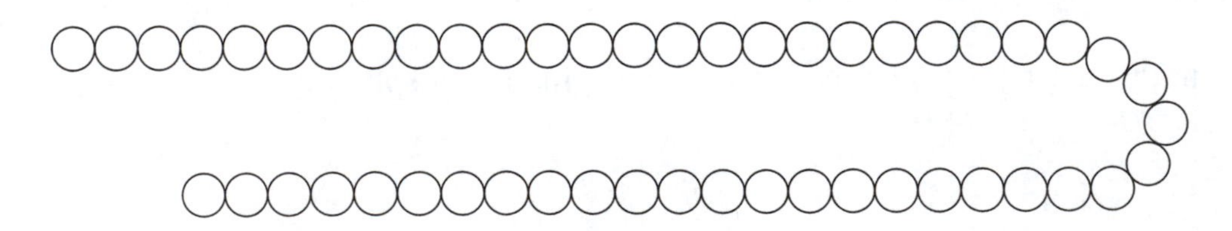

10. Below is a diagram showing the synthesis and breakdown of glycogen $(\text{glucose})_n$. Fill in the blanks and identify the enzyme that catalyzes each numbered reaction.

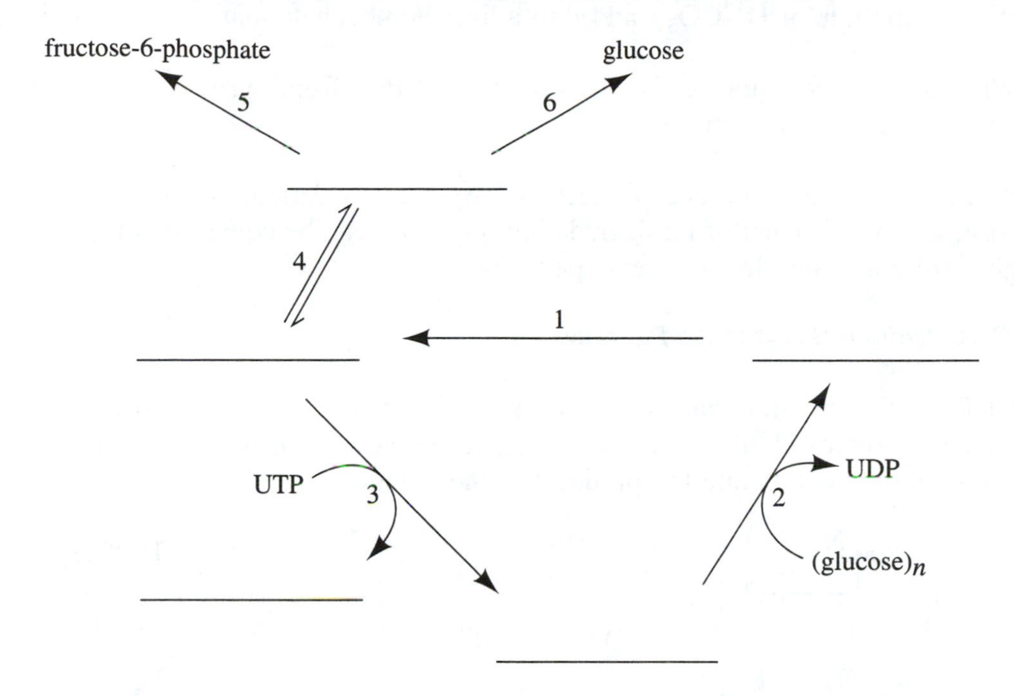

Control of Glycogen Metabolism

11. What structural features distinguish the T conformation of glycogen phosphorylase from the R conformation of the enzyme?

12. The addition of a mild detergent to a liver extract elevates cAPK activity and renders it insensitive to cAMP. Explain.

13. Glycogen phosphorylase is activated in vigorously active muscle without significant changes in intracellular [cAMP]. Explain.

14. Compare the structures of calmodulin (Figure 15-16) and troponin C (Figure 7-30).
    (a) How do their structures differ?
    (b) How are their functions similar?

15. The presence of epinephrine results in the stimulation of PFK-2 in heart muscle. How does epinephrine affect glycolysis in this organ?

16. Which target enzyme in glycogen metabolism requires both α- and β-adrenergic receptors to be activated for full enzyme activity?

17. Which of the hereditary glycogen storage diseases results in a pronounced decrease of stored glycogen?

Gluconeogenesis

18. What is the fate of $H^{14}CO_3^-$ added to a liver homogenate that is active in gluconeogenesis?

19. Write a balanced equation for the formation of PEP from pyruvate and compare it with the reverse reaction of glycolysis.

20. While acetyl-CoA, the end product of fatty acid oxidation, cannot be converted into glucose, another product of fat degradation, glycerol, can be converted to glucose. Where does glycerol enter the gluconeogenic pathway?

Other Carbohydrate Biosynthetic Pathways

21. UDP–galactose can donate its sugar residue to glucose to form the disaccharide lactose. For the structures of UDP–galactose and glucose below, indicate the reactive electrophilic and nucleophilic centers and the products of the reaction.

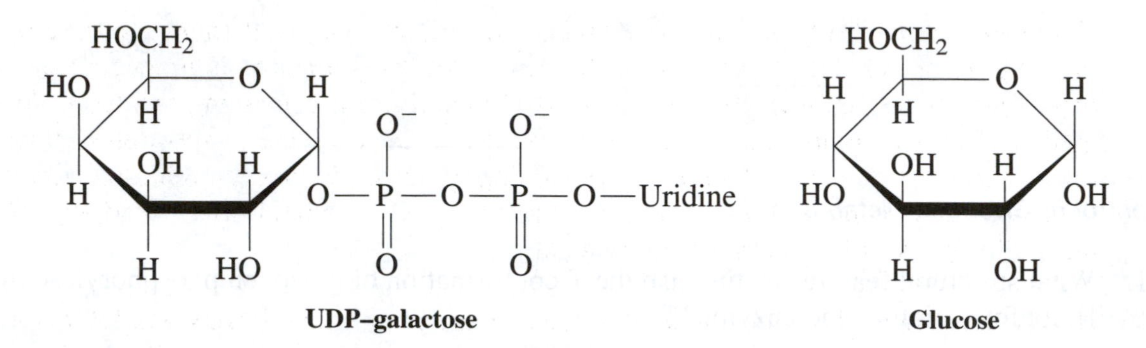

Answers

1. Glycogen phosphorylase, glycogen debranching enzyme, and phosphoglucomutase.

2. A narrow crevice on the surface of glycogen phosphorylase, which connects the glycogen storage site and the active site, can accommodate up to five residues of an unbranched glycogen chain but cannot accommodate a branched chain.

3. The phosphoryl group of PLP participates in general acid–base catalysis by donating a proton to the anionic P_i that reacts with glycogen to release G1P.

4. The reaction product would be a mixture of α and β anomers of G1P because the reaction intermediate is an oxonium ion whose C1 can react with phosphate approaching either face of the sugar residue.

5.

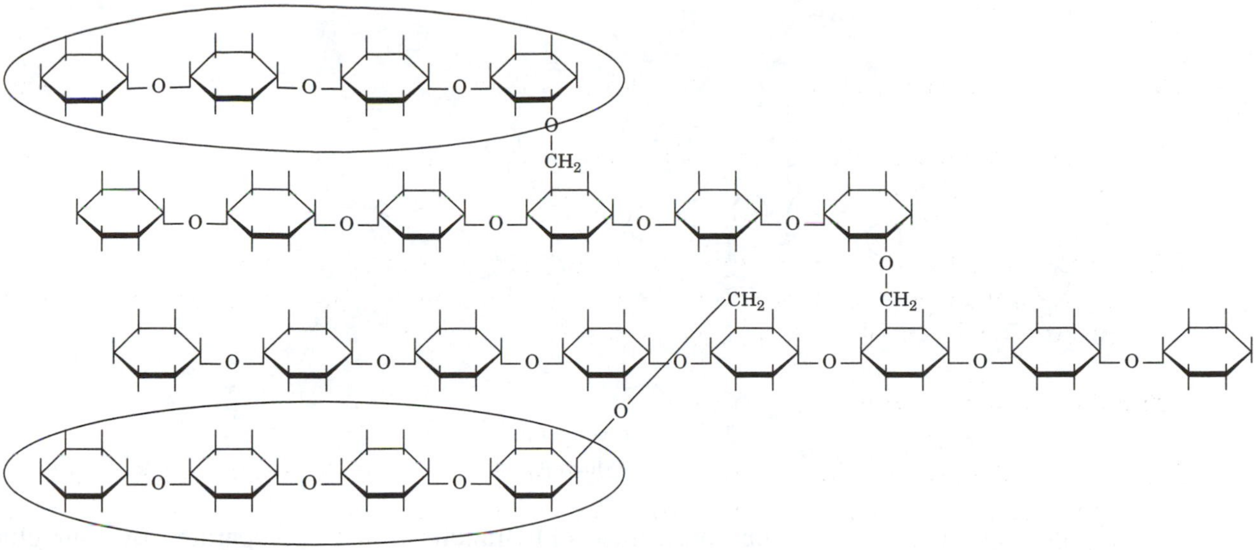

Glycogen debranching enzyme has two catalytic functions: (1) It transfers three $\alpha(1\rightarrow4)$-linked glucose residues from a "limit branch" of glycogen to the nonreducing end of another branch, and (2) it hydrolyzes the $\alpha(1\rightarrow6)$ bond of the remaining residue to form free glucose. In the diagram above, the top and bottom branches serve as substrates in which the trisaccharide is transferred and the remaining glucose residue is released. Once these branches have been eliminated, continued phosphorolysis would produce additional "limit branch" substrates for debranching enzyme.

6. Highly branched glycogen molecules have more nonreducing ends available for phosphorylase to produce G1P, so the rate of G1P release remains high until a limit branch is encountered.

7.

$$K_{eq} = \frac{[\text{glycogen}_{n-1}][\text{G1P}]}{[\text{glycogen}_n][\text{P}_i]} \approx \frac{[\text{G1P}]_{eq}}{[\text{P}_i]_{eq}}$$

$$\Delta G = \Delta G^{\circ\prime} + RT \ln \frac{[\text{G1P}]}{[\text{P}_i]}$$

In order for ΔG to be less than zero, $-RT \ln ([\text{G1P}]/[\text{P}_i])$ must be at least as great as $\Delta G^{\circ\prime}$.

$$\begin{aligned}
-RT \ln ([\text{G1P}]/[\text{P}_i]) &\geq \Delta G^{\circ\prime} \\
-\ln ([\text{G1P}]/[\text{P}_i]) &\geq \Delta G^{\circ\prime}/RT \\
\ln ([\text{G1P}]/[\text{P}_i]) &\leq -\Delta G^{\circ\prime}/RT \\
[\text{G1P}]/[\text{P}_i] &\leq e^{-\Delta G^{\circ\prime}/RT} \\
[\text{G1P}]/[\text{P}_i] &\leq e^{-(3100\ \text{J}\cdot\text{mol}^{-1})/(8.3145\ \text{J}\cdot\text{K}^{-1}\cdot\text{mol}^{-1})(310\text{K})} \\
[\text{G1P}]/[\text{P}_i] &\leq 0.30
\end{aligned}$$

Therefore, $[\text{G1P}]/[\text{P}_i]$ can be no greater than 0.30, so $[\text{P}_i]/[\text{G1P}]$ must be at least $1/0.30 = 3.33$.

8. A "high-energy" compound is required to bypass the exergonic glycogen phosphorylase reaction. The formation of UDP–glucose from G1P and UTP yields PP_i, whose hydrolysis provides the thermodynamic "pull" to add another glucose residue to glycogen.

9. nonreducing ends

glycogenin;
reducing end

Branching enzyme has three constraints: (1) Branches must be separated by four glucosyl residues; (2) seven residues are transferred at a time to make a branch; and (3) the donating chain must be at least 11 residues long. The first branch is made by moving a seven-residue segment from the reducing end. The most highly branched structure results when branches are added to branches for a final product with six $\alpha(1\rightarrow6)$ branch points, as shown above.

10.

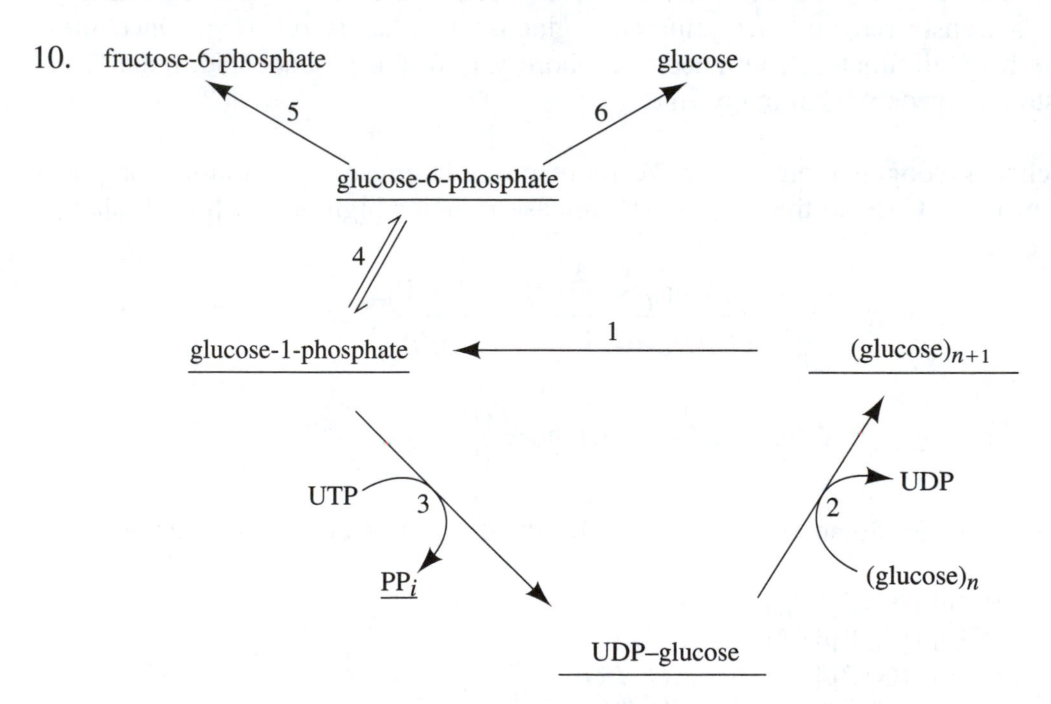

1 Glycogen phosphorylase
2 Glycogen synthase
3 UDP–glucose pyrophosphorylase
4 Phosphoglucomutase
5 Phosphoglucose isomerase
6 Glucose-6-phosphatase

11. In the T form, the active site is less accessible to its substrates compared to the R form due to the presence of a loop that covers the T state active site so as to prevent access of substrate to it. In the R state, the tower helices have tilted and pulled apart relative to their positions in the T state, thereby inducing an ~10° counter-rotation of the two subunits. This also displaces and disorders the loop covering the active site, thus making the active site accessible to substrate. In addition, the side chain of Arg 569, which is located in the active site, rotates in such a way that it increases the R-state enzyme's affinity for its P_i substrate.

12. The mild detergent probably causes the dissociation of the subunits of cAPK (a process which normally requires cAMP) such that the freed catalytic subunits are catalytically active.

13. Ca^{2+} released during muscle contraction activates phosphorylase kinase via binding of Ca^{2+} to calmodulin (the δ subunit of phosphorylase kinase). Phosphorylase kinase is therefore active even without cAPK-catalyzed phosphorylation of its α and β subunits.

14. (a) In troponin C, two globular domains are connected by a nine-turn helix. The two Ca^{2+}-binding domains differ such that one domain binds two Ca^{2+} at low $[Ca^{2+}]$ and the other domain binds two Ca^{2+} only at higher $[Ca^{2+}]$. In calmodulin, a seven-turn helix connects two highly similar Ca^{2+}-binding domains.
    (b) In both proteins, Ca^{2+} binding induces conformational changes that are necessary to alter the activities of other proteins with which troponin C and calmodulin interact.

15. Epinephrine-dependent stimulation of PFK-2 leads to an increase in [F2,6P], which activates PFK-1. The result is increased flux through glycolysis.

16. The full activity of phosphorylase kinase requires both the presence of Ca^{2+} (whose concentration increases in response to hormone binding to α-adrenergic receptors) and phosphorylation by cAPK (which is activated by the binding of the cAMP second messenger whose synthesis is stimulated by hormone binding to β-adrenergic receptors).

17. Only Type 0 (glycogen synthase deficiency) causes a decrease in stored glycogen.

18. The $H^{14}CO_3^-$ is added to pyruvate to yield oxaloacetate via the pyruvate carboxylase reaction. The ^{14}C is then released as CO_2 by the PEP carboxykinase reaction, which converts oxaloacetate to PEP + CO_2.

19. *gluconeogenesis:*

$$\text{pyruvate} + HCO_3^- + \text{ATP} \rightleftharpoons \text{oxaloacetate} + \text{ADP} + P_i$$
$$\text{oxaloacetate} + \text{GTP} \rightleftharpoons \text{PEP} + \text{GDP} + CO_2$$
$$\overline{\text{pyruvate} + HCO_3^- + \text{ATP} + \text{GTP} \rightleftharpoons \text{PEP} + \text{ADP} + \text{GDP} + P_i + CO_2}$$

glycolysis:

$$H^+ + \text{PEP} + \text{ADP} \rightleftharpoons \text{ATP} + \text{pyruvate}$$

The formation of PEP from pyruvate requires the investment of two ATP equivalents ("high-energy" phosphoanhydride bonds), whereas PEP's reaction to form pyruvate yields only one ATP.

20. Glycerol enters the gluconeogenic pathway as dihydroxyacetone phosphate (DHAP; see Figure 14-26 on p. 413). Glycerol is a substrate for glycerol kinase, which catalyzes the reaction

$$\text{Glycerol} + \text{ATP} \rightleftharpoons \text{glycerol-3-phosphate} + \text{ADP}$$

Glycerol phosphate dehydrogenase then catalyzes the reaction

$$\text{Glycerol-3-phosphate} + \text{NAD}^+ \rightleftharpoons \text{DHAP} + \text{NADH} + \text{H}^+$$

21.

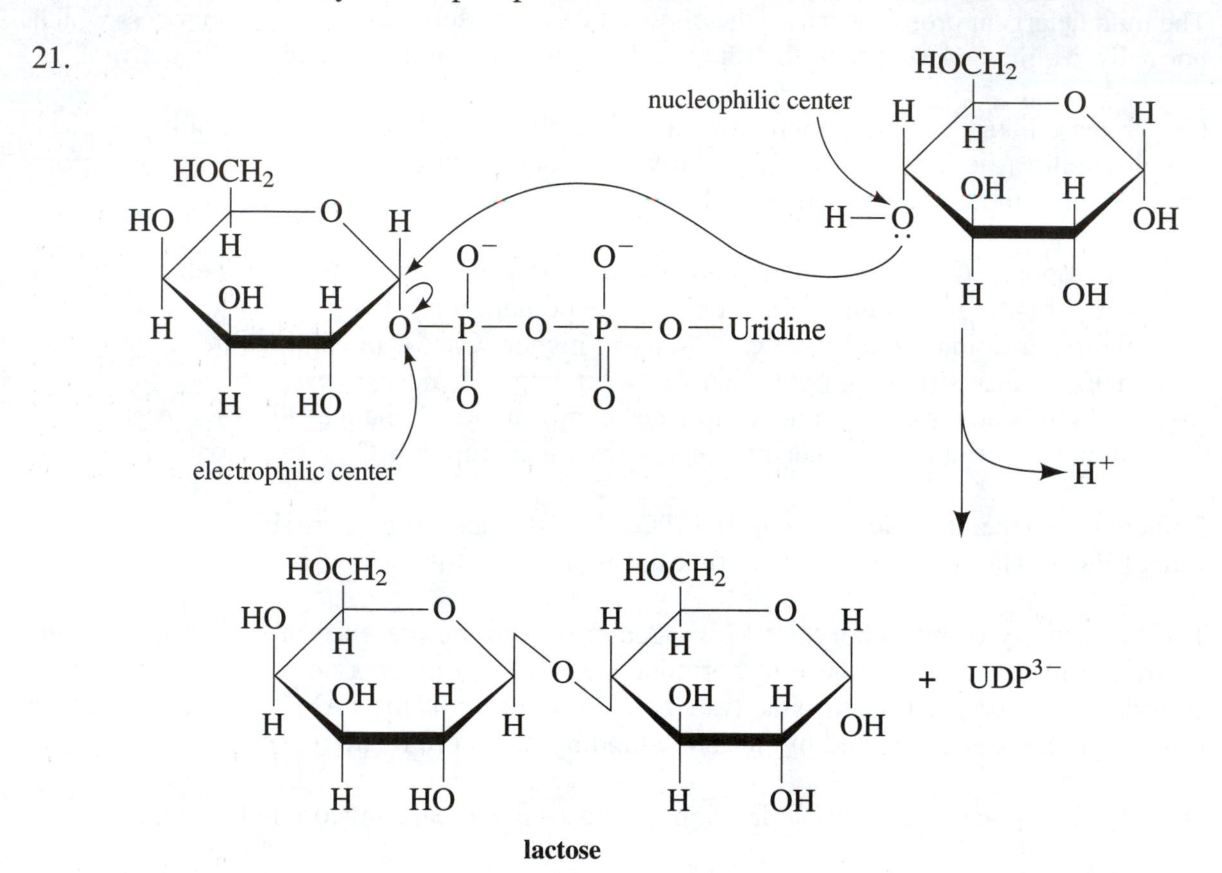

Chapter 16 Citric Acid Cycle

This chapter discusses the citric acid cycle as both a catabolic and an anabolic process. The citric acid cycle was elucidated by several investigators, but the key insights were provided by Hans Krebs, so the cycle is often referred to as the Krebs cycle. It is also called the tricarboxylic acid (TCA) cycle, which refers to its first intermediate, citrate. The chapter first provides an overview of the key features of the citric acid cycle. It then takes a closer look at each step of the cycle, beginning with a discussion of the pyruvate dehydrogenase complex that converts pyruvate to acetyl-CoA, the fuel molecule that enters the cycle. In this chapter, you will encounter several coenzymes that are critical to the functioning of the citric acid cycle, all derived from water-soluble vitamins. A discussion of the regulation of the citric acid cycle then ensues, in which you will encounter familiar regulatory mechanisms as well as new ones. The chapter then turns to the anabolic features of the citric acid cycle. The citric acid cycle is one of the key metabolic hubs in the cell: Its intermediates are generated by a variety of degradative reactions, and they provide precursors for several biosynthetic pathways. As citric acid cycle intermediates are diverted to anabolic pathways, they are replaced through anaplerotic reactions. The final section of the chapter introduces the glyoxylate cycle found in plants. This pathway is the only known pathway for the net conversion of acetyl-CoA to glucose. The chapter also discusses arsenic poisoning in the citric acid cycle (Box 16-1) and the stereospecificity of the cycle's enzymes (Box 16-2). Box 16-3 introduces the metabolon hypothesis, which centers around the idea of substrate channeling and the organization of large multienzyme complexes that carry out several different catalytic activities.

Essential Concepts

Overview of the Citric Acid Cycle

1. The citric acid cycle is a central pathway for recovering energy from the three major metabolic fuels: carbohydrates, fatty acids, and amino acids. These fuels are broken down to yield acetyl-CoA, which enters the citric acid cycle by condensing with the C_4 compound oxaloacetate. The citric acid cycle is a series of reactions in which 2 CO_2 are released for every acetyl-CoA that enters the cycle, so that oxaloacetate is always reformed. Hence, the cyclical series of reactions acts catalytically to process acetyl-CoA continuously.

2. The oxidation of the acetyl carbon skeleton in the citric acid cycle is coupled to the reduction of NAD^+ and FAD. Oxidation of the resulting NADH and $FADH_2$ by the electron-transport chain supplies free energy for ATP synthesis and regenerates NAD^+ and FAD for the oxidation of additional acetyl-CoA. In aerobic respiration, O_2 serves as the terminal acceptor of the acetyl group's electrons. In anaerobic respiration, this function is carried out by molecules such as NO_3^-, SO_4^{2-}, and Fe^{3+}.

3. One complete round of the cycle yields 2 CO_2, 3 NADH, 1 $FADH_2$, and 1 GTP (which is the equivalent of 1 ATP). Hence, the net reaction of the citric acid cycle is

$$3\ NAD^+ + FAD + GDP + P_i + \text{acetyl-CoA} + 2\ H_2O \rightarrow 3\ NADH + FADH_2 + GTP + CoA + 2\ CO_2 + 3\ H^+$$

The carbon atoms of the acetyl group entering the cycle do not exit the cycle as CO_2 in the first round but do so in subsequent rounds.

Synthesis of Acetyl-Coenzyme A

4. In eukaryotes, all the enzymes of the citric acid cycle (and the pyruvate dehydrogenase complex) occur in the inner compartment or matrix of the mitochondrion. The pyruvate produced by glycolysis in the cytosol is transported into the mitochondrion via a pyruvate$-H^+$ symport protein.

5. The pyruvate dehydrogenase complex is a large multienzyme complex containing multiple copies of three enzymes (E_1, E_2, and E_3) organized in a polyhedral array. For example, the *E. coli* pyruvate dehydrogenase complex consists of 24 copies of E_1 and 24 copies of E_2 arranged about the corners of concentric cubes, and 12 copies of E_3.

6. The pyruvate dehydrogenase complex catalyzes five reactions in the oxidative decarboxylation of pyruvate. Five different coenzymes are involved.
   (a) Pyruvate dehydrogenase (E_1) decarboxylates pyruvate in a reaction identical to that catalyzed by pyruvate decarboxylase in alcoholic fermentation (Figure 14-10). However, in the pyruvate dehydrogenase reaction, the hydroxyethyl group remains linked to the thiamine pyrophosphate (TPP) prosthetic group rather than being released as acetaldehyde.
   (b) Dihydrolipoyl transacetylase (E_2) transfers the hydroxyethyl group from TPP to a lipoic acid residue that is tethered to an enzyme Lys side chain via an amide linkage (lipoamide) to form acetyl-dihydrolipoamide.
   (c) E_2 then transfers the acetyl group to CoA to form acetyl-CoA and dihydrolipoamide.
   (d) Dihydrolipoyl dehydrogenase (E_3) oxidizes the dihydrolipoamide group of E_2 via the reduction of E_3's reactive Cys—Cys disulfide bond.
   (e) E_3 is then reoxidized by NAD^+, in a reaction involving the transient reduction and reoxidation of the enzyme's FAD group, to regenerate the reactive disulfide bond, thereby preparing the pyruvate dehydrogenase complex for another round of hydroxyethyl transfer and oxidation.

Enzymes of the Citric Acid Cycle

7. Citrate synthase catalyzes the condensation of acetyl-CoA and oxaloacetate to form citrate in a highly exergonic reaction. The enzyme undergoes a large conformational change as part of an Ordered Sequential reaction mechanism.

8. Aconitase catalyzes the isomerization of citrate to isocitrate via stereospecific dehydration and rehydration.

9. Isocitrate dehydrogenase catalyzes the oxidative decarboxylation of isocitrate to produce α-ketoglutarate and the first CO_2 and NADH of the citric acid cycle.

10. α-Ketoglutarate dehydrogenase (a multienzyme complex similar to the pyruvate dehydrogenase complex) catalyzes the oxidative decarboxylation of α-ketoglutarate by a mechanism identical to that of the pyruvate dehydrogenase complex. The reaction products are succinyl-CoA and the second CO_2 and second NADH of the cycle. Note that the two carbons released as CO_2 in this round of the citric acid cycle are not the carbons that entered the cycle as acetyl-CoA.

11. Succinyl-CoA synthetase catalyzes a coupled reaction in which thioester hydrolysis is coupled to the phosphorylation of GDP via succinyl-phosphate and phospho-His intermediates. This enzyme therefore catalyzes substrate-level phosphorylation. The GTP produced in this step is easily interconverted with ATP by nucleoside diphosphate kinase.

12. The remaining reactions of the citric acid cycle regenerate oxaloacetate.
    (a) Succinate dehydrogenase catalyzes the stereospecific dehydrogenation of succinate to fumarate. The enzyme's FAD prosthetic group is reduced in this redox reaction and is reoxidized when it gives up its electrons to the respiratory electron-transport chain.
    (b) Fumarase catalyzes the hydration of fumarate to malate.
    (c) Malate dehydrogenase catalyzes the oxidation of malate, which regenerates oxaloacetate and produces the third NADH of the cycle. This highly endergonic reaction is driven by the exergonic reaction that follows, namely the condensation of oxaloacetate with acetyl-CoA, which begins the next round of the citric acid cycle.

Regulation of the Citric Acid Cycle

13. The citric acid cycle generates ATP through substrate-level phosphorylation (GTP production) and by the subsequent reoxidation of its 3 NADH and $FADH_2$ products by the electron-transport chain. Each NADH generates ~3 ATP and $FADH_2$ generates ~2 ATP, so each round of the citric acid cycle yields ~12 ATP.

14. The pyruvate dehydrogenase complex is regulated by product inhibition by NADH and acetyl-CoA, and, in eukaryotes, by covalent modification via the phosphorylation/dephosphorylation of E_1. Pyruvate dehydrogenase kinase represses E_1 activity by phosphorylating it at a specific Ser residue, and pyruvate dehydrogenase phosphatase stimulates E_1 activity by dephosphorylating it.

15. The citric acid cycle is regulated principally at the steps catalyzed by citrate synthase, isocitrate dehydrogenase, and α-ketoglutarate dehydrogenase. Regulatory mechanisms include:
    (a) **Substrate availability.** The most critical regulators are acetyl-CoA, oxaloacetate, and the ratio of [NADH] to [NAD^+].
    (b) **Product inhibition.** Citrate competes with oxaloacetate in the citrate synthase reaction, and succinyl-CoA and NADH inhibit α-ketoglutarate dehydrogenase.
    (c) **Competitive feedback inhibition.** Succinyl-CoA competes with acetyl-CoA in the citrate synthase reaction.
    (d) **Allosteric regulation.** ADP activates isocitrate dehydrogenase, while ATP allosterically inhibits it. Ca^{2+} activates pyruvate dehydrogenase, isocitrate dehydrogenase, and α-ketoglutarate dehydrogenase.

Reactions Related to the Citric Acid Cycle

16. The citric acid cycle is amphibolic (both anabolic and catabolic). As an anabolic cycle, the citric acid cycle provides intermediates for gluconeogenesis (oxaloacetate, which must be first converted to malate for export from the mitochondrion), amino acid biosynthesis (oxaloacetate, α-ketoglutarate), porphyrin synthesis (succinyl-CoA), and lipid biosynthesis (citrate). The citric acid cycle also functions catabolically to complete the degradation of carbohydrates and fatty acids (which yield acetyl-CoA) and amino acids that are converted to fumarate, succinyl-CoA, α-ketoglutarate, and oxaloacetate.

17. Anaplerotic reactions replenish citric acid cycle intermediates that have been siphoned off into anabolic reactions. The most important of these reactions is catalyzed by pyruvate carboxylase:

$$\text{Pyruvate} + CO_2 + \text{ATP} + H_2O \rightarrow \text{oxaloacetate} + \text{ADP} + P_i$$

18. Acetyl-CoA cannot serve as a precursor for gluconeogenesis in animals, but it can do so in plants via the glyoxylate pathway. This pathway is a variation of the citric acid cycle that takes place in two organelles, the mitochondrion and the glyoxysome (a specialized peroxisome found in plants). The glyoxysome contains isocitrate lyase (which cleaves isocitrate to succinate and glyoxylate) and malate synthase (which condenses glyoxylate with acetyl-CoA to form malate). The net reaction for the glyoxylate pathway is

$$2\ \text{Acetyl-CoA} + 2\ NAD^+ + \text{FAD} + 3\ H_2O \rightarrow \text{oxaloacetate} + 2\ \text{CoA} + 2\ \text{NADH} + FADH_2 + 4\ H^+$$

Note that in the glyoxylate pathway, no carbons are lost as CO_2.

Guide to Study Exercises (text p. 491)

1. The five reactions of the pyruvate dehydrogenase complex are:
   (a) Pyruvate dehydrogenase (E_1) decarboxylates pyruvate, releasing CO_2 and leaving a hydroxyethyl group bound to its TPP prosthetic group.
   (b) Dihydrolipoyl transacetylase (E_2) transfers the hydroxyethyl group from TPP to its lipoamide prosthetic group, producing acetyl-dihydrolipoamide.
   (c) E_2 then transfers the acetyl group to CoA, yielding acetyl-CoA and a dihydrolipoamide group.
   (d) Dihydrolipoyl dehydrogenase (E_3) oxidizes the dihydrolipoamide group of E_2 by a disulfide interchange with two E_3 Cys residues.
   (e) The two reduced Cys residues are oxidized by NAD^+ through the intermediacy of FAD, producing NADH and regenerating E_3. (Section 16-2B)

2. See Figure 16-2.

3. $$\text{Pyruvate} + 4\ NAD^+ + \text{FAD} + \text{GDP} + P_i + 2\ H_2O \rightarrow 4\ \text{NADH} + FADH_2 + \text{GTP} + 3\ CO_2 + 3\ H^+$$

$$\text{Acetyl-CoA} + 3\ \text{NAD}^+ + \text{FAD} + \text{GDP} + \text{P}_i + 2\ \text{H}_2\text{O} \rightarrow 3\ \text{NADH} + \text{FADH}_2 + \text{GTP} + \text{CoA} + 2\ \text{CO}_2 + 3\ \text{H}^+$$

$$\text{Glucose} + 10\ \text{NAD}^+ + 2\ \text{FAD} + 2\ \text{ADP} + 2\ \text{GDP} + 4\ \text{P}_i + 2\ \text{H}_2\text{O} \rightarrow 10\ \text{NADH} + 2\ \text{FADH}_2 + 2\ \text{ATP} + 2\ \text{GTP} + 10\ \text{H}^+ + 6\ \text{CO}_2$$

(Section 16-1)

4. Flux through the citric acid cycle is regulated at four points. Entry of acetyl units into the cycle is controlled by product inhibition and covalent modification of the pyruvate dehydrogenase complex. The flux of metabolites through the cycle itself is regulated at the steps catalyzed by citrate synthase, isocitrate dehydrogenase, and α-ketoglutarate dehydrogenase. The activity of these enzymes is controlled by substrate availability and feedback inhibition by products of that reaction or a reaction farther along the cycle. (Section 16-4)

5. Ca^{2+} activates the citric acid cycle by activating isocitrate dehydrogenase and α-ketoglutarate dehydrogenase. It also activates the pyruvate dehydrogenase complex by inhibiting pyruvate dehydrogenase kinase and activating pyruvate dehydrogenase phosphatase. NADH inhibits citrate synthase and inhibits isocitrate dehydrogenase and α-ketoglutarate dehydrogenase by product inhibition. It competes with NAD^+ for binding to pyruvate dehydrogenase, thereby slowing this reaction. In combination with acetyl-CoA, NADH drives the E_2 and E_3 reactions of the pyruvate dehydrogenase complex backward, preventing the entry of additional acetyl-CoA into the citric acid cycle. NADH and acetyl-CoA also inactivate pyruvate dehydrogenase by activating the kinase that phosphorylates it. Acetyl-CoA itself regulates the citrate synthase reaction according to its availability and competes with CoA for binding to the pyruvate dehydrogenase complex. (Section 16-4)

6. The intermediates of a cyclic pathway that are removed to serve as precursors for other pathways must be replenished by anaplerotic reactions that synthesize these intermediates from other compounds. (Sections 16-5A and B)

7. See Figure 16-16. The glyoxylate pathway, which operates in plants, requires the enzymes of the citric acid cycle as well as the glyoxysome enzymes isocitrate lyase and malate synthase. The conversion of isocitrate to succinate and the C_2 compound glyoxylate, which is catalyzed by isocitrate lyase, bypasses the two CO_2-generating steps of the citric acid cycle. Malate synthase then combines glyoxylate with acetyl-CoA to form malate. This allows the net synthesis of oxaloacetate, a gluconeogenic precursor, from acetyl-CoA. (Section 16-5C)

Questions

Overview of the Citric Acid Cycle

1. The citric acid cycle can be divided into two phases with respect to the oxidation of acetyl-CoA. Describe each phase and write its balanced equation.

2. Early experiments showed that malonate, which inhibits succinate dehydrogenase, blocks cellular respiration. This led to the idea that succinate participates in oxidative metabolism as an intermediate and not as just another metabolic fuel. Using isotopically-labeled reagents, what observation(s) would demonstrate the validity of this interpretation?

Synthesis of Acetyl-Coenzyme A

3. Name the three enzymes that form the pyruvate dehydrogenase complex.

4. For each reaction listed below, indicate the appropriate enzyme(s) in the pyruvate dehydrogenase complex and the relevant cofactor(s), if applicable.

	Reaction	*Enzyme*	*Cofactor*
(a)	Oxidative formation of an enzymatic disulfide bond	______	______
(b)	Transfer of hydroxyethyl group bound to TPP	______	______
(c)	Liberation of CO_2	______	______
(d)	Oxidation of dihydrolipoamide	______	______
(e)	Formation of acetyl-CoA	______	______

5. What are the roles of FAD and NAD^+ in the pyruvate dehydrogenase catalytic mechanism?

6. You are given two preparations of purified pyruvate dehydrogenase complex enzymes with all the required cofactors. You add pyruvate to each preparation and then measure the rate of production of acetyl-CoA and acetaldehyde under aerobic conditions.

		Acetyl-CoA (molecules/s)	*Free acetaldehyde (molecules/s)*
Preparation	**A**	10^5	10^{-6}
Preparation	**B**	10^{-2}	10^3

How might the pyruvate dehydrogenase complex enzymes differ in each preparation?

7. Which of the following labeled glucose molecules would yield $^{14}CO_2$ following glycolysis and the pyruvate dehydrogenase reaction?
(a) 1-[^{14}C]-glucose
(b) 3-[^{14}C]-glucose
(c) 4-[^{14}C]-glucose
(d) 6-[^{14}C]-glucose

Enzymes of the Citric Acid Cycle

8. What is the energetic function of the thioester bond of CoA in the citrate synthase reaction?

9. For the half-reaction

$$FAD + 2\,e^- + 2\,H^+ \rightleftharpoons FADH_2$$

$\mathcal{E}^{\circ\prime} \approx 0.00$ V for FAD bound to succinate dehydrogenase.

(a) Calculate $\Delta G^{\circ\prime}$ for the oxidation of succinate to fumarate by enzyme-bound FAD.
(b) How does this compare to the $\Delta G^{\circ\prime}$ for the oxidation of succinate to fumarate by free NAD^+?
(c) Based on these results, explain why nature chose FAD rather than NAD^+ as the oxidizing agent in the succinate dehydrogenase reaction.

10. What is the fate of C4 (the carboxyl that is β to the carbonyl) of oxaloacetate?

Regulation of the Citric Acid Cycle

11. How would the rapid accumulation of succinyl-CoA affect the rate of glucose oxidation?

Reactions Related to the Citric Acid Cycle

12. List four anabolic pathways that utilize citric acid cycle intermediates as starting material.

13. How does acetyl-CoA affect the activity of pyruvate carboxylase? Why is this advantageous for the cell?

14. Write a balanced equation for the synthesis of glucose from acetyl-CoA via the glyoxylate cycle.

15. Which reactions of the glyoxylate cycle deviate from the citric acid cycle?

16. Which pathway intermediates pass between the plant mitochondrion and the glyoxysome during the functioning of the glyoxylate cycle?

Answers to Questions

1. The first phase involves the net oxidation of acetyl-CoA to two molecules of CO_2 and the regeneration of CoA (Reactions 1–5 of the cycle).

$$\text{Acetyl-CoA} + \text{oxaloacetate} + \text{GDP} + 2\,\text{NAD}^+ + \text{H}_2\text{O} + \text{P}_i \rightarrow \text{succinate} + \text{GTP} + \text{CoASH} + 2\,\text{NADH} + 2\,\text{CO}_2 + 2\,\text{H}^+$$

The second phase involves the regeneration of oxaloacetate from succinate (Reactions 6–8).

$$\text{Succinate} + \text{FAD} + \text{NAD}^+ + \text{H}_2\text{O} \rightarrow \text{oxaloacetate} + \text{NADH} + \text{FADH}_2 + \text{H}^+$$

2. To demonstrate that succinate is an intermediate in glucose oxidation, show that, in the presence of malonate, cell extracts containing newly added ^{14}C-labeled acetyl-CoA (labeled at either of the acetyl C atoms) and any of the other citric acid cycle intermediates yield ^{14}C-labeled succinate.

3. Pyruvate dehydrogenase (E_1), dihydrolipoyl transacetylase (E_2), and dihydrolipoyl dehydrogenase (E_3).

4.

	Reaction	Enzyme	Cofactor
(a)	Oxidative formation of an enzymatic disulfide bond	E_3	FAD, NAD^+
(b)	Transfer of hydroxyethyl group bound to TPP	E_2	lipoic acid
(c)	Liberation of CO_2	E_1	TPP
(d)	Oxidation of dihydrolipoamide	E_2, E_3	
(e)	Formation of acetyl-CoA	E_2	CoA, lipoic acid

5. FAD functions to oxidize the E_3 sulfhydryl groups that were reduced as they oxidized the sulfhydryl groups of dihydrolipoamide. NAD^+ then oxidizes the $FADH_2$. Both oxidants serve to regenerate the reactive functional groups of E_3.

6. The data suggest that in preparation **B,** E_1 (pyruvate dehydrogenase) is somehow unable to transfer its bound hydroxyethyl carbanion to the lipoamide cofactor of E_2 (dihydrolipoyl transacetylase). Instead, free acetaldehyde is released from the TPP group of E_1 (as occurs in the pyruvate decarboxylase reaction; Section 14-3B).

7. (b) and (c) would yield $^{14}CO_2$ after the pyruvate dehydrogenase reaction since the C3 and C4 atoms of glucose both become C1 of glyceraldehyde-3-phosphate and then pyruvate. Pyruvate dehydrogenase liberates C1 of pyruvate as CO_2.

8. The large, negative free energy of hydrolysis of the thioester bond provides the thermodynamic force for the otherwise endergonic condensation of the acetyl group and oxaloacetate under conditions of low oxaloacetate concentration. The oxaloacetate concentration is low because of the endergonic nature of the preceding malate dehydrogenase reaction.

9. (a) For the reaction succinate + FAD $\rightarrow$ fumarate + $FADH_2$, succinate is the electron donor and FAD is the electron acceptor. $\mathscr{E}°'$ for the succinate $\rightarrow$ fumarate half-reaction is 0.031 V (Table 13-3). Therefore,

$$\begin{aligned}
\Delta\mathscr{E}°' &= \mathscr{E}°'_{FAD} - \mathscr{E}°'_{succinate} = 0.00\text{ V} - 0.031\text{ V} = -0.031\text{ V} \\
\Delta G°' &= -n\mathscr{F}\Delta\mathscr{E}°' \\
&= -(2)(96{,}485\text{ J}\cdot\text{V}^{-1}\cdot\text{mol}^{-1})(-0.031\text{ V}) \\
&= 5982\text{ J}\cdot\text{mol}^{-1} = 6.0\text{ kJ}\cdot\text{mol}^{-1}
\end{aligned}$$

(b) When NAD^+ is the electron acceptor ($\mathscr{E}°' = -\ 0.315$ V),

$$\begin{aligned}
\mathscr{E}°' &= -0.315\text{ V} - 0.031\text{ V} = -0.346\text{ V} \\
\Delta G°' &= -(2)(96{,}485\text{ J}\cdot\text{V}^{-1}\cdot\text{mol}^{-1})(-0.346\text{ V}) \\
&= 66768\text{ J}\cdot\text{mol}^{-1} = 67\text{ kJ}\cdot\text{mol}^{-1}
\end{aligned}$$

Therefore, under standard biochemical conditions, the oxidation of succinate by FAD is slightly disfavored ($\Delta G°' > 0$). However, the oxidation of succinate by NAD^+ is strongly disfavored ($\Delta G°' \gg 0$).

(c) Succinate lacks the reducing power (has a reduction potential that is too high) to reduce NAD^+ to NADH but has sufficient reducing power to reduce FAD to $FADH_2$.

10. It is lost as CO_2 in the isocitrate dehydrogenase reaction in the first turn of the cycle (see Figure 16-2).

11. The rate of glucose oxidation would decrease because succinyl-CoA inhibits its own synthesis by the α-ketoglutarate dehydrogenase reaction, and it inhibits the citrate synthase reaction via feedback inhibition. The rapid accumulation of succinyl-CoA would also deplete the mitochondrial pool of CoA, thereby slowing the production of acetyl-CoA from glucose-derived pyruvate.

12. Glucose synthesis (gluconeogenesis) utilizes oxaloacetate; lipid biosynthesis utilizes acetyl-CoA derived from citrate; amino acid biosynthesis utilizes α-ketoglutarate and oxaloacetate, and porphyrin biosynthesis utilizes succinyl-CoA (Figure 16-15).

13. Acetyl-CoA activates pyruvate carboxylase. Increases in [acetyl-CoA] are indicative of an inability of the citric acid cycle to oxidize acetyl-CoA as fast as it is being produced (from glycolysis and fatty acid oxidation). Hence, activating pyruvate carboxylase, which adds more oxaloacetate to the pool of citric acid cycle intermediates will catalytically accelerate acetyl-CoA oxidation.

14. Two rounds of the glyoxylate cycle are necessary to yield two oxaloacetate, which give rise to one glucose by gluconeogenesis.

    glyoxylate cycle:

$$\text{4 Acetyl-CoA} + 4\ \text{NAD}^+ + 2\ \text{FAD} + 6\ \text{H}_2\text{O} \rightarrow \text{2 oxaloacetate} + 4\ \text{CoA} + 4\ \text{NADH} + 2\ \text{FADH}_2 + 8\ \text{H}^+$$

    gluconeogenesis:

$$\text{2 Oxaloacetate} + 2\ \text{GTP} + 2\ \text{ATP} + 2\ \text{NADH} + 4\ \text{H}_2\text{O} + 6\ \text{H}^+ \rightarrow \text{glucose} + 2\ \text{GDP} + 2\ \text{ADP} + 2\ \text{NAD}^+ + 4\ \text{P}_i + 2\ \text{CO}_2$$

    net:

$$\text{4 Acetyl-CoA} + 2\ \text{GTP} + 2\ \text{ATP} + 2\ \text{NAD}^+ + 2\ \text{FAD} + 10\ \text{H}_2\text{O} \rightarrow \text{glucose} + 4\ \text{CoA} + 2\ \text{CO}_2 + 2\ \text{GDP} + 2\ \text{ADP} + 2\ \text{NADH} + 2\ \text{FADH}_2 + 2\ \text{H}^+ + 4\ \text{P}_i$$

15. *Isocitrate lyase*: isocitrate → succinate + glyoxylate
    Malate synthase: acetyl-CoA + glyoxylate → malate + CoA

16. Aspartate and α-ketoglutarate move from the mitochondrion to the glyoxysome, and succinate and glutamate move from the glyoxysome to the mitochondrion.

Chapter 17 Electron Transport and Oxidative Phosphorylation

This chapter introduces you to the remarkable process by which cells harness the free energy of oxidation and use it to synthesize ATP. (You have already seen how ATP can be synthesized by the phosphorylation of ADP by a metabolite with a higher phosphoryl group-transfer potential.) The reduced cofactors generated in metabolic reactions, NADH and $FADH_2$, are reoxidized in the mitochondrion by a set of reactions in which electrons flow through a series of redox carriers, finally reducing oxygen to water. During electron transport, an electrochemical potential is developed across the inner mitochondrial membrane by the vectorial transfer of protons. This proton gradient is stable because the inner mitochondrial membrane is impermeable to ions. The free energy of the electrochemical proton gradient is utilized by an ATP synthase to catalyze the endergonic reaction ADP + P_i → ATP. The coupling of the electrochemical gradient to ATP synthesis is described by the chemiosmotic hypothesis, which is supported by considerable evidence.

Essential Concepts

1. The complete oxidation of glucose carbons by glycolysis and the citric acid cycle can be written as

$$C_6H_{12}O_6 + 6\ H_2O \rightarrow 6\ CO_2 + 24\ H^+ + 24\ e^-$$

The reducing equivalents (electrons) are captured in the form of reduced coenzymes (NADH and $FADH_2$), which eventually transfer the electrons to molecular oxygen:

$$6\ O_2 + 24\ H^+ + 24\ e^- \rightarrow 12\ H_2O$$

This process regenerates NAD^+ and FAD and generates a proton concentration gradient across the inner mitochondrial membrane, whose dissipation provides the free energy for ATP synthesis. This process is known as oxidative phosphorylation.

The Mitochondrion

2. The mitochondrion is surrounded by a relatively porous outer membrane. An inner membrane is folded to form cristae and encloses the gel-like matrix, which contains the enzymes of the citric acid cycle and fatty acid oxidation. The matrix also contains genetic machinery (DNA, RNA, and ribosomes), reflecting the bacterial origin of this organelle. The proteins involved in electron transport and oxidative phosphorylation are located in the inner mitochondrial membrane. The inner and outer membranes also contain proteins that mediate the transport of ions and metabolites.

3. NADH produced in the cytosol as a result of glycolysis must enter the mitochondrion to be aerobically oxidized. There are two shuttles for NADH.
   (a) The malate–aspartate shuttle allows NADH to be indirectly transported into the mitochondrion by reducing oxaloacetate to malate in the cytosol and transporting it into

the mitochondrion, where it is reoxidized to produce oxaloacetate and NADH. The oxaloacetate is converted by transamination to aspartate and transported out again.

(b) The glycerophosphate shuttle first reduces cytosolic dihydroxyacetone phosphate to 3-phosphoglycerate and NAD^+. The 3-phosphoglycerate is oxidized by an inner mitochondrial membrane enzyme, flavoprotein dehydrogenase, which introduces electrons directly into the electron-transport pathway.

4. Most of the ATP generated in the mitochondria is used in the cytosol. The ADP–ATP translocator exports ATP out of the matrix while importing ADP. ATP has one more negative charge than ADP, so transport is electrogenic. Transport is driven by the electrochemical potential of the proton concentration gradient (positive outside). The proton gradient also favors the transport of P_i into the matrix by a P_i–H^+ symport system.

5. The mitochondrial $[Ca^{2+}]$ is controlled by two transporters. The influx transporter responds to cytosolic $[Ca^{2+}]$, and transport is driven by the membrane potential (negative inside). Ca^{2+} exits the mitochondrion in exchange for Na^+ by an antiport mechanism that operates at its maximum velocity. Thus, the mitochondrion acts as a Ca^{2+} buffer. Large influxes of Ca^{2+} activate citric acid cycle enzymes.

Electron Transport

6. In an electron transfer reaction, electrons flow from a substance with a lower reduction potential to a substance with a higher reduction potential. The standard reduction potential, $\mathscr{E}^{\circ\prime}$, is a measure of a substance's affinity for electrons. For a redox reaction, $\Delta\mathscr{E}^{\circ\prime} = \mathscr{E}^{\circ\prime}_{(e^- \text{ acceptor})} - \mathscr{E}^{\circ\prime}_{(e^- \text{ donor})}$. When $\Delta\mathscr{E}^{\circ\prime}$ is positive, the reaction is spontaneous, since $\Delta G^{\circ\prime} = -n\mathscr{F}\Delta\mathscr{E}^{\circ\prime}$, where n is the number of electrons transported and $\mathscr{F}$ is the faraday (96,485 $\text{J} \cdot \text{V}^{-1} \cdot \text{mol}^{-1}$). The transfer of electrons from NADH to O_2 ($\Delta\mathscr{E}^{\circ\prime} = 1.13$ V and $\Delta G^{\circ\prime} = -218\ \text{kJ} \cdot \text{mol}^{-1}$) provides enough free energy to synthesize three ATP molecules.

7. Four large protein complexes in the inner mitochondrial membrane are involved in transferring electrons from reduced coenzymes to O_2. Complexes I and II transfer electrons to the lipid-soluble electron carrier ubiquinone (coenzyme Q or CoQ), which transfers electrons to Complex III. From there, electrons pass to cytochrome *c*, a peripheral membrane protein with a heme prosthetic group, which transfers electrons to Complex IV. The reactions of Complexes I–IV are as follows:

(I) $\text{NADH} + \text{CoQ}\ (ox) \rightarrow \text{NAD}^+ + \text{CoQ}\ (red)$

$$\Delta\mathscr{E}^{\circ\prime} = 0.360 \text{ V and } \Delta G^{\circ\prime} = -69.5 \text{ kJ} \cdot \text{mol}^{-1}$$

(II) $\text{FADH}_2 + \text{CoQ}\ (ox) \rightarrow \text{FAD} + \text{CoQ}\ (red)$

$$\Delta\mathscr{E}^{\circ\prime} = 0.085 \text{ V and } \Delta G^{\circ\prime} = -16.4 \text{ kJ} \cdot \text{mol}^{-1}$$

(III) $\text{CoQ}\ (red) + \text{cytochrome } c\ (ox) \rightarrow \text{CoQ}\ (ox) + \text{cytochrome } c\ (red)$

$$\Delta\mathscr{E}^{\circ\prime} = 0.190 \text{ V and } \Delta G^{\circ\prime} = -36.7 \text{ kJ} \cdot \text{mol}^{-1}$$

(IV) Cytochrome *c* (*red*) + $\frac{1}{2}$ O_2 → cytochrome *c* (*ox*) + H_2O

$$\Delta\mathscr{E}°' = 0.580 \text{ V and } \Delta G°' = -112 \text{ kJ} \cdot \text{mol}^{-1}$$

8. Complex I is an enormous protein complex containing flavin mononucleotide (FMN, which is FAD minus its AMP group) and multiple iron–sulfur clusters (which are one-electron carriers). The two electrons donated by NADH are transferred through these redox-active prosthetic groups and then to CoQ. As electrons are transferred, four protons are pumped from the matrix to the intermembrane space, most likely by protein conformational changes similar to those in bacteriorhodopsin.

9. Complex II, which contains the citric acid cycle enzyme succinate dehydrogenase, transfers electrons from succinate to FAD and then to CoQ. No protons are translocated by Complex II, which serves mainly to feed electrons into the electron transport chain.

10. Complex III (cytochrome bc_1 or cytochrome *c* reductase) contains two *b*-type cytochromes, cytochrome c_1, and an iron–sulfur protein, which contains a [2Fe–2S] cluster. Electron flow from CoQ through Complex III follows a bifurcated cyclic pathway known as the Q cycle. In the first round of the Q cycle, fully reduced ubiquinone (ubiquinol; QH_2) donates one electron to the iron–sulfur protein, which then transfers it to cytochrome c_1 and then to cytochrome *c*. This one-electron donation yields the ubisemiquinone anion ($Q^{\cdot-}$), which donates its remaining electron to the low potential cytochrome *b* (b_L), and then to the high potential cytochrome *b* (b_H). The resulting ubiquinone diffuses to the other side of the membrane, where it accepts the electron from b_L to reform $Q^{\cdot-}$. A second round of electron transfers completes the cycle: Another QH_2 donates its electrons, one to the iron–sulfur protein and one to cytochrome b_L. The net result is that two electrons are transferred, one at a time, to two cytochrome *c* molecules, and four protons are transferred from the matrix to the intermembrane space, two from each QH_2 that participates in the Q cycle.

11. Cytochrome *c* shuttles electrons between Complexes III and IV. Cytochrome *c* is a small water-soluble protein whose heme group is largely buried in a crevice surrounded by a ring of Lys residues. Both cytochrome c_1 and cytochrome *c* oxidase (Complex IV) have a corresponding patch of negatively charged amino acid residues to facilitate cytochrome *c* binding and electron transfer.

12. Complex IV (cytochrome *c* oxidase) has four redox centers [cytochrome *a*, cytochrome a_3, Cu_A (which contains two Cu ions), and Cu_B], and it carries out the following reaction:

$$4 \text{ Cytochrome } c\ (Fe^{2+}) + 4\ H^+ + O_2 \rightarrow 4 \text{ cytochrome } c\ (Fe^{3+}) + 2\ H_2O$$

O_2 reduction takes place at the cytochrome a_3–Cu_B binuclear complex, which mediates four one-electron transfer reactions. Four protons are consumed in the production of H_2O, and four additional proteins are pumped from the matrix to the intermembrane space (two for each pair of electrons that enter the electron-transport chain).

Oxidative Phosphorylation

13. ATP synthase (Complex V) phosphorylates ADP by a mechanism driven by the free energy of electron transport, which is conserved in the formation of an electrochemical proton gradient across the inner mitochondrial membrane. The two processes are coupled as described by the chemiosmotic hypothesis. Four observations support this hypothesis:
    (a) Mitochondrial ATP formation requires an intact inner membrane.
    (b) The inner membrane is impermeable to ions, so an electrochemical gradient across the membrane can be sustained.
    (c) Electron transport pumps protons out of the mitochondrion to create a measurable electrochemical gradient.
    (d) Agents that increase the permeability of the inner mitochondrial membrane to protons inhibit ATP synthesis but not electron transport.

14. The protonmotive force results from the difference in concentration of protons (pH) in the matrix and the intermembrane space and from the difference in charge (membrane potential, $\Delta\Psi$) across the membrane. Thus, $\Delta G = 2.3\,RT\,[\text{pH}(in) - \text{pH}(out)] + Z\mathcal{F}\Delta\Psi$, where Z is the charge of the proton. $\Delta\Psi$ is positive when a proton is transported from negative to positive, or against its potential. Thus, pumping protons out of the matrix (against the gradient) is an endergonic process, whereas transporting them back in (with the gradient) is an exergonic process. About three protons are needed to supply sufficient energy to synthesize one ATP from ADP + P_i.

15. ATP synthase, also called F_1F_0-ATPase, has two functional units. The F_0 component comprises the transmembrane proton channel. Dicyclohexylcarbodiimide and oligomycin binding to F_0 inhibits proton translocation and thereby inhibits ATP synthesis. The F_1 component is a water-soluble protein of subunit composition $\alpha_3\beta_3\gamma\delta\varepsilon$ that associates with the membrane via F_0 to form a lollipop-like structure.

16. The 3 α, 3 β, and γ subunits of F_1 form a pseudo-symmetrical structure of alternating α and β subunits in a ring with the elongated γ subunit in its central hole. The three pairs of $\alpha\beta$ subunits exhibit pseudo-threefold rotational symmetry.

17. The binding change mechanism describes ATP synthesis in terms of three processes:
    (a) Translocation of protons carried out by F_0.
    (b) Formation of the phosphoanhydride bond of ATP, catalyzed by F_1.
    (c) Interaction of F_0 and F_1 to couple the dissipation of the proton gradient to formation of the phosphoanhydride bond.

18. According to the binding change mechanism, ATP is synthesized as each $\alpha\beta$ protomer shifts through three conformations in sequence. The three possible conformations are called open (O), loose (L), and tight (T). ADP and P_i bind to a protomer with the L conformation and are converted to ATP when the conformation shifts to the T state. The free energy of the proton concentration gradient converts the T state to the O state (this is the rate-limiting step), thereby releasing ATP. This three-step mechanism is consistent with the pseudo-threefold axis of rotation. The γ subunit rotates with respect to the $\alpha_3\beta_3$ assembly, and the geometric relationship of individual $\alpha\beta$ protomers to the γ subunit dictates their conformational state. Three protons, each of which promotes one conformational shift, are required to synthesize one ATP.

19. The ratio of the amount of ATP produced to the amount of substrate oxidized (measured as oxygen consumed) is called the P/O ratio. [The P/O ratio refers to atomic oxygen, O, rather than molecular oxygen, O_2, because each substrate (NADH or $FADH_2$) transfers two electrons, not four.] Depending on where a substrate's electrons enter the electron-transport chain, the P/O ratio is ~3, ~2, or ~1. For example, the two electrons transferred from NADH through Complexes I, III, and IV pump 10 protons, which yields ~3 ATP, whereas the two electrons transferred from $FADH_2$ through Complexes II, III, and IV pump 6 protons, which yields ~2 ATP. The complete oxidation of glucose therefore yields 38 ATP. The P/O ratio is not necessarily a whole number, because protons are contributed to the gradient by more than one process and some protons leak back into the matrix.

20. Electron transport and oxidative phosphorylation are normally strongly coupled processes; that is, neither process occurs in the absence of the other. This is because, if the rate of electron transport were to outpace the rate of ATP synthesis, the proton gradient would build up to the level that it would resist additional proton pumping by Complexes I, III, and IV and hence the rate of electron transport would be slowed. However, when uncoupling agents, which dissipate the proton gradient, are added to respiring mitochondria, electron transport proceeds unchecked while ATP synthesis stops. The free energy of electron transport is then redirected from ATP synthesis to generate heat. 2,4-Dinitrophenol is an uncoupling agent because it carries protons through the membrane from the intermembrane space to the matrix, thereby providing a route for dissipation of the gradient that bypasses F_0.

Control of ATP Production

21. Electron transfer from NADH to cytochrome *c* is nearly at equilibrium. In contrast, the cytochrome *c* oxidase reaction is irreversible and hence its rate depends on the concentration of its substrate, reduced cytochrome *c*. Increased NADH concentrations and decreased ATP concentrations lead to the production of more reduced cytochrome *c* and hence to increased electron transfer rates. Thus, the overall rate of oxidative phosphorylation depends on the ratios [NADH]/[NAD^+] and [ATP]/[ADP][P_i], which in turn may depend on the activities of the respective mitochondrial transporters.

22. The coordinated control of oxidative metabolism centers on several key enzymes: hexokinase (HK), phosphofructokinase (PFK), pyruvate kinase (PK), pyruvate dehydrogenase (PDH), citrate synthase (CS), isocitrate dehydrogenase (IDH) and α-ketoglutarate dehydrogenase (KDH). High levels of ATP inhibit PFK and PK while high [NADH]/[NAD^+] ratios inhibit PDH, IDH, and KDH. Citrate inhibits both PFK and CS. The need for ATP, represented by high concentrations of either AMP or ADP, activates PFK, PK, PDH, and IDH, while Ca^{2+} stimulates PDH, IDH, and KDH.

Physiological Implications of Aerobic Metabolism

23. Anaerobic glycolysis produces 2 ATP per glucose consumed, whereas oxidative metabolism generates 38 ATP per glucose, a 19-fold increase. However, there are several drawbacks of oxygen-based metabolism. Many organisms depend on oxidative metabolism and would perish without a steady supply of oxygen. Reactive oxygen species generated by incomplete oxygen reduction are potentially dangerous.

24. Oxygen is used by the enzyme cytochrome P450 to detoxify many harmful compounds such as polycyclic aromatic hydrocarbons, polychlorinated biphenyls, phenobarbital, and steroids. Cytochrome P450 catalyzes the addition of one atom of diatomic oxygen into the substrate as a hydroxyl group, while the other atom is converted to water. Hydroxylation generally renders the substrate more water-soluble and hence more easily excreted.

25. Reactive oxygen species include the superoxide radical $O_2^{\cdot -}$ (produced by the reaction $O_2 + e^- \rightarrow O_2^{\cdot -}$), which is a precursor of even more powerful oxidizing species such as $HO_2\cdot$ and $\cdot OH$. These free radicals readily extract electrons from other substances, creating a chain reaction. Neurodegenerative conditions such as Parkinson's, Alzheimer's, and Huntington's diseases are associated with mitochondrial oxidative damage. Free radical reactions arising from normal oxidative metabolism appear to be partially responsible for the aging process.

26. Antioxidants limit oxidative damage by destroying free radicals. Superoxide dismutase (SOD) catalyzes the production of oxygen and hydrogen peroxide from superoxide:

$$2\ O_2^{\cdot -} + 2\ H^+ \rightarrow H_2O_2 + O_2$$

This enzyme electrostatically guides its substrate to the active site to catalyze a reaction near the diffusion-controlled limit. Mutations in Cu,Zn SOD are associated with amyotrophic lateral sclerosis (Lou Gehrig's disease).

27. Catalase and glutathione peroxidase degrade hydroperoxides. Some types of glutathione peroxidase require selenium, so Se also appears to be an antioxidant.

Key Equation

$$\Delta G = 2.3\ RT\ [\text{pH}(in) - \text{pH}(out)] + Z\mathcal{F}\Delta\Psi$$

Guide to Study Exercises (text p. 527)

1. In the oxidation of glucose to 2 pyruvate by glycolysis, four electrons are transferred to 2 NAD^+ to produce 2 NADH at the step catalyzed by glyceraldehyde-3-phosphate dehydrogenase. The conversion of 2 pyruvate to 2 acetyl-CoA by pyruvate dehydrogenase transfers four more electrons to 2 NAD^+. During two rounds of the citric acid cycle (which completes the oxidation of the carbon atoms originally from glucose), 12 electrons are transferred to 6 NAD^+ (in the reactions catalyzed by isocitrate dehydrogenase, α-ketoglutarate dehydrogenase, and malate dehydrogenase), and four electrons are transferred to 2 FAD in the reactions catalyzed by succinate dehydrogenase. Thus, the 24 electrons from glucose yield 10 NADH and 2 $FADH_2$. These reduced cofactors are reoxidized by the electron-transport chain. NADH gives up its electrons to Complex I, and the $FADH_2$ is part of Complex II, whose electrons are transferred to the electron carrier CoQ (which also receives electrons from Complex I). Electrons then travel from CoQ to Complex III, to cytochrome *c*, and then to Complex IV, which carries out the four-electron reduction of O_2 to H_2O. (Chapter introduction and Section 17-2B)

2. The two electrons carried by NADH are transferred together to FMN, which is a two-electron acceptor group in Complex I. $FMNH_2$ then transfers one electron at a time to the first of a series of iron–sulfur clusters. Each electron is then passed through the iron–sulfur clusters to the mobile electron carrier coenzyme Q, which is a two-electron carrier. Reduced CoQ then transfers its two electrons to Complex III. The flow of electrons follows a bifurcated cyclic pathway (the Q cycle) in which the first electron reduces the Rieske iron–sulfur protein and then cytochrome c_1, which in turn reduces cytochrome *c*. Simultaneously, the second electron passes from $CoQ^{\overline{\cdot}}$ to cytochrome b_L and then to cytochrome b_H and back to $CoQ^{\overline{\cdot}}$. A second round of the cycle involving a second reduced CoQ results in the reduction of a second cytochrome c_1 and the reduction of the $CoQ^{\overline{\cdot}}$ back to reduced CoQ. In this way, the electrons of the two-electron carrier NAD^+ are transferred to two molecules of the one-electron carrier cytochrome *c*. (Sections 17-2C and E)

3. The translocation of protons from the matrix to the intermembrane space by Complexes I and IV depends on protein conformational changes that occur in conjunction with the reduction and reoxidation of its redox centers. In contrast, the protons translocated by Complex III are ferried from the matrix to the intermembrane space via the Q cycle by binding to a redox cofactor—CoQ—rather than the protein. (Sections 17-2C, E, and F)

4. According to the chemiosmotic theory, the free energy of electron transport is conserved in the formation of a transmembrane proton concentration gradient that is established when protons are pumped from the mitochondrial matrix to the intermembrane space by the action of the electron-transport complexes. The free energy of the gradient is harnessed to drive the phosphorylation of ADP to produce ATP. (Section 17-3A)

5. In the binding change mechanism, each of the three αβ protomers of F_1F_0-ATPase assumes one of three conformations. The free energy of proton translocation through F_0 causes each protomer to shift its conformation so that its transit through the three conformations in succession accomplishes ATP synthesis. Thus, a protomer in the L conformation binds the substrates ADP and P_i loosely. It then changes to the T conformation, which binds the substrates tightly and is catalytically active. The resulting ATP is released when the protomer shifts to the O conformation. (Section 17-3B)

6. ATP is synthesized when reduced molecules donate their electrons to the electron-transport chain, which ultimately reduces O_2 to H_2O. However, the amount of ADP phosphorylated is indirectly related to the amount of O_2 reduced because phosphorylation is driven by the free energy of a transmembrane proton concentration gradient that is established by the action of more than one electron-transport protein and may be dissipated by more than one mechanism. (Section 17-3C)

7. Oxidative phosphorylation is linked to electron transport by the arrangement of protein complexes in the inner mitochondrial membrane such that electron transport through Complexes I, III, and IV generates a transmembrane proton concentration gradient whose dissipation through the F_0 channel drives ADP phosphorylation by F_1F_0-ATPase. This coupling depends on the impermeability of the membrane, which allows the electron-transport complexes to increase the concentration of protons on the cytoplasmic side of the membrane, and which prevents the protons from re-entering the matrix except through F_1F_0-ATPase. Electron transport and oxidative phosphorylation can be uncoupled by an agent that dissipates the proton

concentration gradient. The result is that electron transport proceeds without the buildup of the proton gradient, and hence no ATP is synthesized. (Section 17-3D)

8. The primary advantage of O_2-based metabolism is that under aerobic conditions, the complete catabolism of 1 glucose yields 38 ATP, whereas the oxidation of glucose to lactate under anaerobic conditions yields only 2 ATP. One disadvantage of aerobic metabolism is that O_2 must be constantly available to serve as the terminal acceptor of electrons from substrate oxidation. Another disadvantage is that reactive oxygen species produced by the incomplete reduction of O_2 can damage cellular lipids, proteins, and DNA. (Section 17-5)

Questions

The Mitochondrion

1. Draw a cross-section of a mitochondrion and label the following structural features:

Outer membrane (OM)	Inner membrane (IM)
Matrix (M)	Intermembrane space (IMSP)
ATP synthase complex (ASC)	Direction of proton flux
ATP and P_i transporters (T)	Cristae (CR)

2. Match the following enzyme or other molecule with its location:

______ Pyruvate dehydrogenase	A. Cytosol
______ 3-Phosphoglycerate dehydrogenase	B. Mitochondrial outer membrane
______ Flavoprotein dehydrogenase	C. Mitochondrial inner membrane
______ Malate dehydrogenase	D. Mitochondrial intermembrane space
______ Cytochrome *c*	E. Mitochondrial matrix
______ Cytochrome c_1	
______ Fatty acid oxidation enzymes	
______ Mitochondrial DNA	
______ ADP–ATP translocator	
______ Mitochondrial porin	

3. About half the volume of the mitochondrial matrix is water, and the rest is protein. If a single protein of molecular mass 40,000 were as concentrated, what would be its molar concentration? Assume the protein's density is $1.37\ g \cdot mL^{-1}$.

4. Oxidative phosphorylation requires the transfer of electrons donated by NADH. (a) Is NADH imported directly into the mitochondria? Explain. (b) Describe two import mechanisms that transfer cytosolic electrons from NADH into the mitochondrion. (c) Why is it important to maintain a relatively constant level of cytosolic NAD^+?

5. What controls the rate of Ca^{2+} influx into the mitochondrial matrix? How does muscle activity change the respiration rate?

Electron Transport

6. Which reactions of the citric acid cycle donate electron pairs to the mitochondrial electron-transport chain?

7. The half-cell reduction potential is provided by the Nernst equation (Equation 13-8):

$$\mathscr{E}_A = \mathscr{E}^{\circ}{}_A - \frac{RT}{n\mathscr{F}} \ln\left(\frac{[A_{red}]}{[A_{ox}^{n+}]}\right)$$

(a) On the graph below, plot the reduction potentials for the $FADH_2$/FAD half-cell ($\mathscr{E}^{\circ\prime} = -0.219$ V) when the [$FADH_2$]/[FAD] ratios are 100, 10, 5, 2, 1, 0.5, 0.2, 0.1, and 0.01 at 25°C versus the percent reduction.

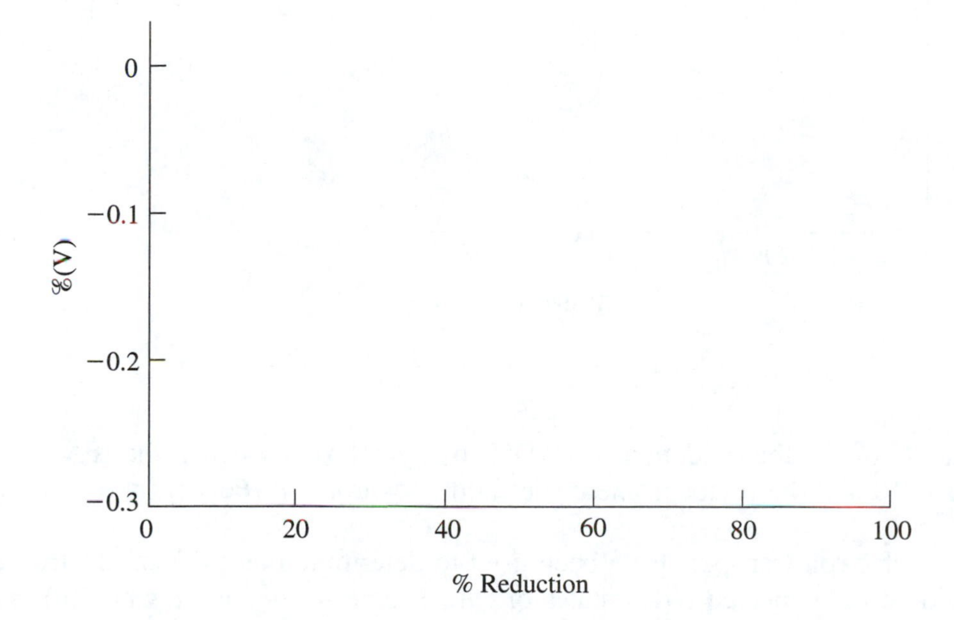

(b) Using the same [Reduced]/[Oxidized] ratios, plot the reduction potentials of cytochrome *c* ($\mathcal{E}°' = 0.235$ V).

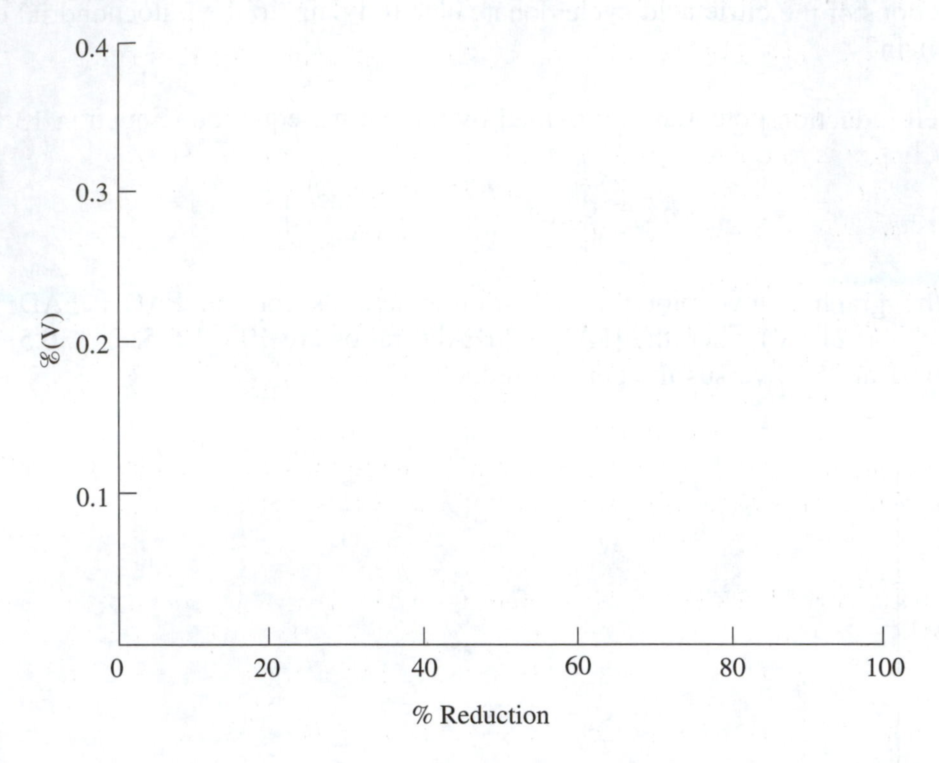

(c) What is $\Delta\mathcal{E}$ for the oxidation of $FADH_2$ by cytochrome *c* when the $[FADH_2]/[FAD]$ ratio is 10 and the [cytochrome *c* (Fe^{2+})]/[cytochrome *c* (Fe^{3+})] ratio is 0.1?

8. Inhibitors of electron transport have been used to determine the order of electron carriers. What would be the expected redox states of cytochromes *a*, b_L, and *c* when (a) myxothiazol, (b) antimycin A, or (c) rotenone is added to succinate-driven respiring mitochondria?

9. What is the most abundant type of redox carrier in Complex I?

10. Which mitochondrial electron carriers are potentially proton carriers? Which are more abundant—proton carriers or electron carriers? What does this suggest about the mechanism of transmembrane proton transport?

11. What amino acid residues would you expect to find at the cytochrome *c*–binding sites of Complex III and Complex IV?

12. Which redox group(s) in Complex IV accepts electrons from cytochrome *c*? Which redox group(s) binds oxygen during its four-electron reduction?

Oxidative Phosphorylation

13. What key observations support the chemiosmotic hypothesis?

14. Can a pH gradient exist without $\Delta\Psi$? Can $\Delta\Psi$ exist without a ΔpH?

15. (a) What is a P/O ratio?
 (b) What happens to oxygen consumption when electron donors and inorganic phosphate are present in a suspension of mitochondria in the absence of ADP?
 (c) What happens when ADP is added?

16. (a) Valinomycin, an antibiotic ionophore, allows the free passage of only K^+ ions across a membrane. If K^+ and valinomycin are added to respiring, fully coupled ATP-synthesizing mitochondria, what happens to the pH gradient and the $\Delta\Psi$?
 (b) Nigericin, another ionophore, exchanges one K^+ for one H^+. How does this affect ATP synthesis and electron transport in mitochondria?
 (c) Gramicidin allows the free passage of many small molecules and ions across the membrane. What happens to ATP production and electron transport in the presence of gramicidin?

17. The free energy-requiring step in the synthesis of ATP is not the formation of ATP from ADP and P_i ($\Delta G \approx 0$), but the release of tightly bound ATP. Explain why this is not inconsistent with the $+30.5 \text{ kJ} \cdot \text{mol}^{-1}$ free energy of formation of ATP in solution.

18. What happens to the electron transport rate when DCCD is added to actively respiring mitochondria?

19. What does an H^+/P ratio measure? Why would it be impractical to determine an H^+/P ratio?

Control of ATP Production

20. The conversion of glucose to 2 lactate has $\Delta G°' = -196 \text{ kJ} \cdot \text{mol}^{-1}$. The complete oxidation of glucose to 6 CO_2 has $\Delta G°' = -2823 \text{ kJ} \cdot \text{mol}^{-1}$. Compare the efficiencies of ATP synthesis by each of these processes under standard conditions.

21. On an average day, an adult dissipates about 7000 kJ of free energy. Assuming that this occurs under standard conditions,
 (a) How many moles of ATP must be hydrolyzed to provide this quantity of free energy?
 (b) What is the mass of this quantity of ATP?
 (c) If the amount of ATP in an adult is about 0.1 mole, how many times per day, on average, is a molecule of ADP recycled? The molecular mass of ATP is 507.

22. What is the irreversible step in electron transport and how is its rate controlled?

Physiological Implications of Aerobic Metabolism

23. Cytochrome P450 catalyzes a reaction in which two electrons supplied by NADPH reduce the heme Fe atom so that it can then reduce O_2 preparatory to the hydroxylation of a substrate molecule. Substrate binding to the enzyme displaces a water molecule that forms a ligand to the heme iron atom. This changes the reduction potential of the Fe from -0.300 V to -0.170 V. Why is this change necessary for efficient catalysis?

24. Rats that are fed a "cafeteria" diet (in which food is always available) tend to die sooner than rats whose dietary intake is limited. Propose an explanation for this observation.

Answers to Questions

1.

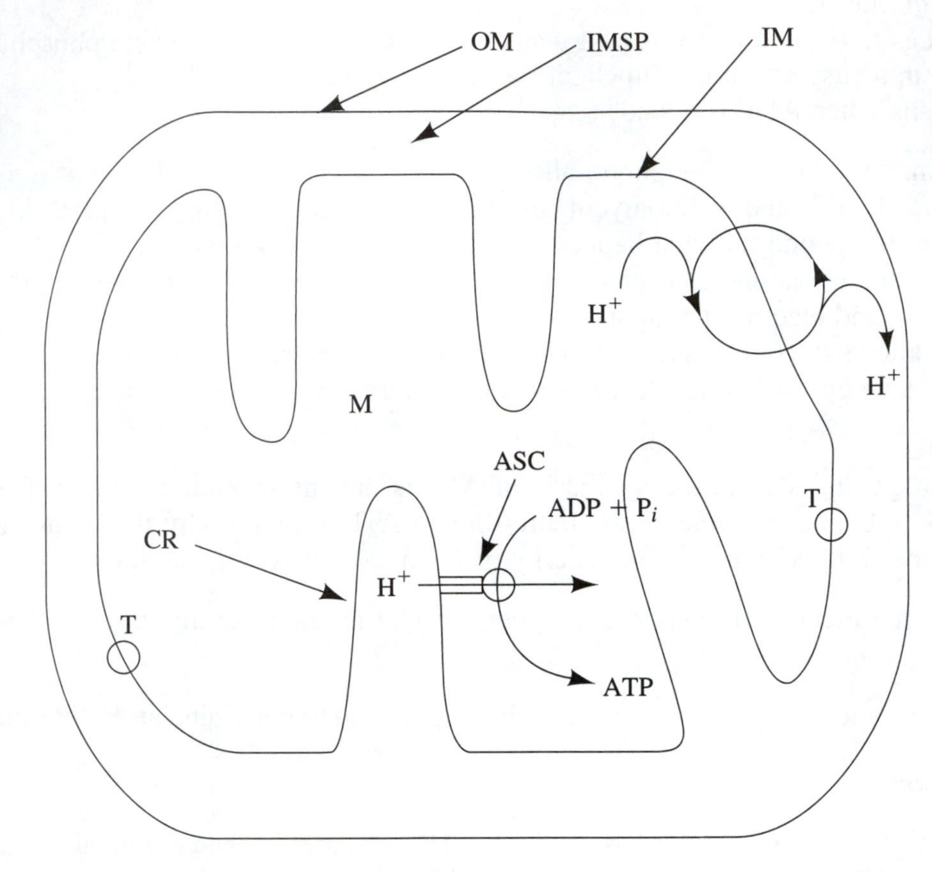

2.
E Pyruvate dehydrogenase
A 3-Phosphoglycerate dehydrogenase
C Flavoprotein dehydrogenase
E, A Malate dehydrogenase
D Cytochrome *c*
C Cytochrome c_1
E Fatty acid oxidation enzymes
E Mitochondrial DNA
C ADP–ATP translocator
B Mitochondrial porin

3. The protein's molar concentration would be

$$\frac{1}{40{,}000\ \text{g} \cdot \text{mol}^{-1}} \times \frac{1.37\ \text{g}}{\text{mL}} \times \frac{1000\ \text{mL}}{\text{L}} \times 50\%$$
$$= 0.0171\ \text{mol} \cdot \text{L}^{-1}$$
$$= 17.1\ \text{mM}$$

4. (a) The negatively charged phosphate groups of NADH (Fig. 3-4) prevent its diffusion across the inner mitochondrial membrane, and there are no NADH transport proteins to facilitate its transport.
   (b) The malate–aspartate shuttle allows the indirect import of NADH reducing equivalents (Fig. 15-27). The reduction of oxaloacetate to malate by NADH followed by the facilitated transport of malate across the inner mitochondrial membrane yields NADH in the mitochondrial matrix when the malate is reoxidized to oxaloacetate. Oxaloacetate then returns to the cytosol by being converted to aspartate, for which there is a transporter. The glycerophosphate shuttle utilizes NADH to convert dihydroxyacetone phosphate to 3-phosphoglycerate by a flavoprotein dehydrogenase that donates electrons to the electron-transport chain in a manner similar to succinate dehydrogenase (Fig. 17-4).
   (c) Cytosolic NAD^+ is required for the glyceraldehyde-3-phosphate dehydrogenase reaction of glycolysis. Limited $[NAD^+]$ would shut down glycolysis.

5. The Ca^{2+} influx into the mitochondrial matrix is regulated by cytosolic $[Ca^{2+}]$, which is less than the K_M for Ca^{2+} transport, so that Ca^{2+} influx is approximately first-order. During muscle activity, cytoplasmic $[Ca^{2+}]$ increases, which then increases the mitochondrial $[Ca^{2+}]$. Ca^{2+} stimulates the enzymes of the citric acid cycle, leading to an increase in production of reduced coenzymes, whose reoxidation demands an increase in the respiration rate.

6. Isocitrate dehydrogenase, α-ketoglutarate dehydrogenase, and malate dehydrogenase produce NADH, which transfers its electrons to Complex I. Succinate dehydrogenase, whose FAD group is reduced in the oxidation of succinate to fumarate, is a component of Complex II.

7. (a) The percent reduction is

$$\frac{[\text{Reduced}]}{[\text{Reduced}] + [\text{Oxidized}]} \times 100 \quad \text{or} \quad \frac{[\text{FADH}_2]}{[\text{FADH}_2] + [\text{FAD}]} \times 100$$

For the $FADH_2$/FAD half-cell, the Nernst equation is

$$\mathscr{E} = (-0.219\text{ V}) - \frac{(8.3145\text{ J}\cdot\text{K}^{-1}\cdot\text{mol}^{-1})(298\text{K})}{(2)(96{,}485\text{ J}\cdot\text{V}^{-1}\cdot\text{mol}^{-1})}\ln\left(\frac{[\text{FADH}_2]}{[\text{FAD}]}\right)$$

$$\mathscr{E} = -0.219\text{ V} - (0.0128\text{ V})\ln\left(\frac{[\text{FADH}_2]}{[\text{FAD}]}\right)$$

[FADH$_2$]/[FAD]	*% Reduction*	*$\mathscr{E}$ (V)*
100	99.0	−0.278
10	90.9	−0.248
5	83.3	−0.240
2	66.7	−0.228
1	50.0	−0.219
0.5	33.3	−0.210
0.2	16.7	−0.198
0.1	9.1	−0.190
0.01	0.99	−0.160

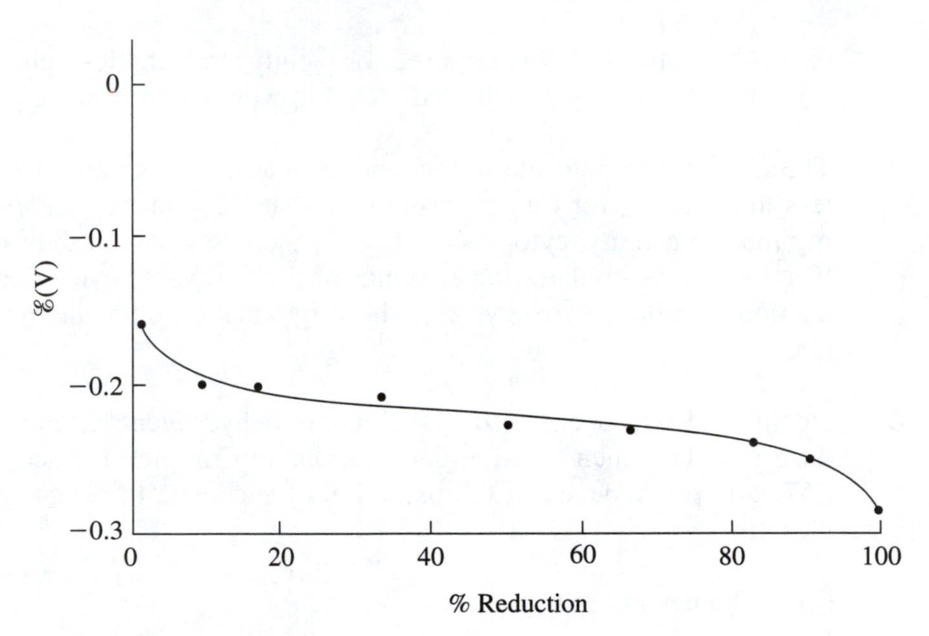

(b) For cytochrome c,

$$\mathscr{E} = (0.235 \text{ V}) - \frac{(8.3145 \text{ J} \cdot \text{K}^{-1} \cdot \text{mol}^{-1})(298\text{K})}{(1)(96{,}485 \text{ J} \cdot \text{V}^{-1} \cdot \text{mol}^{-1})} \ln\left(\frac{[\text{cyto } c \text{ (Fe}^{2+})]}{[\text{cyto } c \text{ (Fe}^{3+})]}\right)$$

$$\mathscr{E} = 0.235 \text{ V} - (0.0257 \text{ V}) \ln\left(\frac{[\text{cyto } c \text{ (Fe}^{2+})]}{[\text{cyto } c \text{ (Fe}^{3+})]}\right)$$

[cyto c *(Fe^{2+})]/[cyto* c *(Fe^{3+})]*	*% Reduction*	*$\mathscr{E}$ (V)*
100	99.0	0.117
10	90.9	0.176
5	83.3	0.194
2	66.7	0.217
1	50.0	0.235
0.5	33.3	0.253
0.2	16.7	0.276
0.1	9.1	0.294
0.01	0.99	0.353

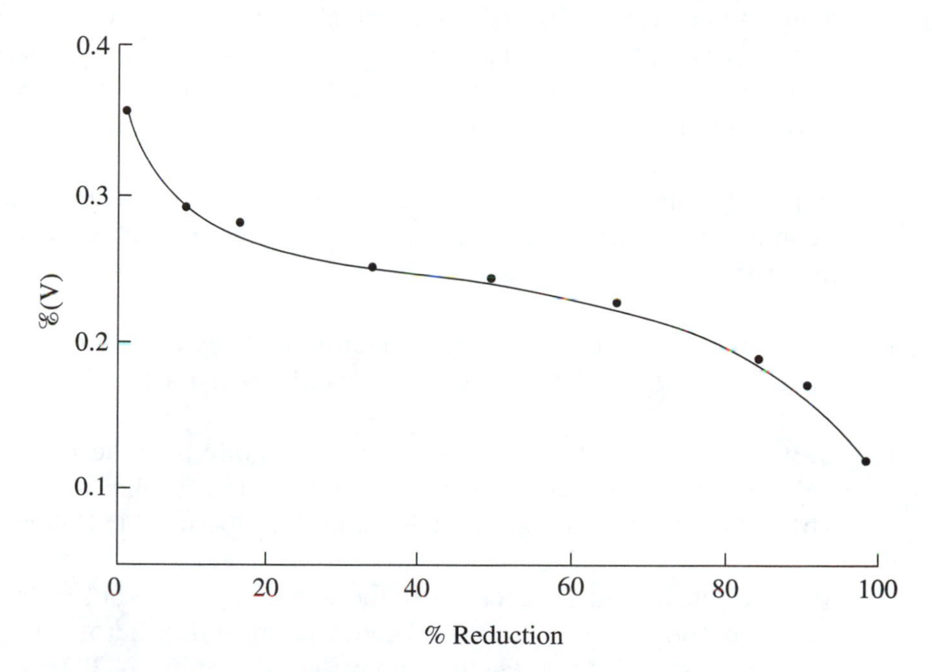

(c) When $[FADH_2]/[FAD] = 10$, $\mathscr{E}_{FAD} = -0.248$ V
When [cyto *c* (Fe^{2+})]/[cyto *c* (Fe^{3+})] = 0.1, $\mathscr{E}_{\text{cyto } c} = 0.294$ V
According to Equation 13-10,

$$\begin{aligned}\Delta\mathscr{E} &= \mathscr{E}_{(e-\text{ acceptor})} - \mathscr{E}_{(e-\text{ donor})}\\ &= \mathscr{E}_{\text{cyto } c} - \mathscr{E}_{\text{FAD}}\\ &= 0.294 \text{ V} - (-0.248 \text{ V})\\ &= 0.542 \text{ V}\end{aligned}$$

8. (a) Myxothiazol inhibits the electron flow from $CoQH_2$ to Complex III. Hence cytochrome *a* (a component of Complex IV) cannot obtain electrons and would be largely in its oxidized state. Similarly, cytochrome b_L (part of Complex II) and cytochrome *c* (which links Complexes III and IV) would be largely in their oxidized states.

(b) Antimycin A blocks electron transport in Complex III from heme b_H to CoQ or $CoQ^{\cdot-}$. Thus, cytochrome b_L would be reduced since it could not pass its electrons on to cytochrome b_H. However, cytochromes *a* and *c* would be oxidized since they are both downstream of the block.

(c) Rotenone blocks electron transport in Complex I. However, since electrons are being introduced into the electron-transport chain at Complex II, electron transport can proceed all the way to O_2. Therefore, all three cytochromes would be predominantly in their reduced forms.

9. The most abundant type of electron carrier in Complex I is iron–sulfur clusters.

10. Only two electron carriers in the mitochondrial electron-transport chain can also carry protons, FMN and ubiquinone (CoQ). The paucity of proton carriers compared to electron carriers suggests that transmembrane proton transport involves some sort of protein-mediated proton pumping rather than only direct transport by the proton carriers.

11. Since cytochrome *c* contains positively charged Lys residues around its heme crevice, its redox partners must have negatively charged residues such as Glu and Asp at their cytochrome *c*–binding sites.

12. The Cu_A center accepts the first electron from cytochrome *c*. Oxygen binds to the partially reduced Fe(II)–Cu(I) form of the cytochrome a_3-Cu_B binuclear complex.

13. Key observations that support the chemiosmotic hypothesis are:
    (a) Oxidative phosphorylation requires an intact inner membrane;
    (b) the inner mitochondrial membrane is impermeable to ions such as H^+, OH^-, K^+, and Cl^-;
    (c) electron transport results in the transport of H^+ out of the mitochondrial matrix; and
    (d) compounds that increase proton permeability across the membrane uncouple phosphorylation from electron transport and inhibit ATP synthesis.

14. A pH gradient can exist without $\Delta\Psi$. If a counterion such as Cl^- moved in the same direction as the H^+, or if K^+ moved in the opposite direction, a $[H^+]$ gradient could form without altering the net charge on either side of the membrane. Similarly, the transmembrane movement of ions other than H^+ could generate $\Delta\Psi$ without generating a proton concentration gradient.

15. (a) The P/O ratio is a measure of how many ATPs are synthesized for every O atom reduced.
    (b) Oxygen consumption stops when the production of ATP from ADP cannot be carried out. This is because electron transport and proton translocation cannot take place independently. Hence, when the proton gradient has built up to the point that electron transport has insufficient free energy to translocate additional protons across the inner mitochondrial membrane, electron transport must stop.
    (c) When ADP is added, ATP can be synthesized. O_2 consumption then resumes because the dissipation of the proton concentration gradient by ATP synthase reduces the free energy of proton translocation across the inner mitochondrial membrane to the point that electron transport can continue.

16. (a) The pH gradient would remain unchanged. However, the $\Delta\Psi$ would collapse because K^+ would equilibrate across the membrane in response to the $\Delta\Psi$.
    (b) Nigericin would dissipate the proton gradient by exchanging protons for K^+ and hence would arrest ATP synthesis. The rate of electron transport would increase because there would be no buildup of a proton gradient to hold electron transport in check. However, $\Delta\Psi$ would remain intact because there would be no net change in charge across the membrane.
    (c) Gramicidin would cause the collapse of both the $\Delta\Psi$ and the pH gradient, so ATP synthesis would stop but electron transport would accelerate.

17. In solution, the ATP-forming reaction is ADP + P_i → ATP + H_2O. Recall that an enzymatic reaction may progress through several steps, but it cannot alter the ΔG for a reaction. For the ATP synthase–catalyzed reaction, the binding of substrates and release of products are just two of the steps in the overall process. Letting E represent the enzyme, the reaction can be written as

$$\text{ADP} + \text{P}_i + \text{E} \rightarrow \text{ADP} \cdot \text{P}_i \cdot \text{E} \rightarrow \text{ATP} \cdot \text{H}_2\text{O} \cdot \text{E} \rightarrow \text{ATP} + \text{H}_2\text{O} + \text{E}$$

where each step has a different ΔG value. However, the overall reaction remains

$$\text{ADP} + \text{P}_i + \text{E} \rightarrow \text{ATP} + \text{H}_2\text{O} + \text{E}$$

and hence the overall free energy change is still 30.5 kJ · mol^{-1}.

18. DCCD reacts with a Glu residue in a subunit of F_0 that forms the proton channel through the membrane. DCCD binding inhibits ATP formation. The proton gradient therefore builds up to the point that it arrests further electron transport.

19. The H^+/P ratio is a measure of the number of protons transported across the inner membrane for each ATP molecule synthesized. This could be difficult to determine since the measurement of pH would not take into account $\Delta\Psi$, which also contributes to the proton-motive force. In addition, the measurement would be highly sensitive to cytosolic pH changes that were not involved in mitochondrial electron transport.

20. $\Delta G^{\circ\prime} = +30.5$ kJ · mol^{-1} for the synthesis of ATP from ADP + P_i. The conversion of glucose to lactate by glycolysis is accompanied by the synthesis of 2 ATP. Thus, the efficiency of ATP production under standard conditions is [(2 × 30.5 kJ · mol^{-1})/(196 kJ · mol^{-1})] × 100 = 31%. When glucose is completely oxidized, the yield is 38 ATP. The efficiency of this process is [(38 × 30.5 kJ · mol^{-1})/(2823 kJ · mol^{-1})] × 100 = 41%. Therefore, not only does oxidative phosphorylation yield more ATP than glycolysis, but at least under standard conditions, it does so with greater efficiency.

21. (a) The energy is supplied by 7000 kJ/30.5 kJ · mol^{-1} = 230 moles of ATP.
    (b) 230 moles × 507 g · mol^{-1} = 11700 g = 117 kg
    (c) Since 230 moles are needed each day, 0.1 mol of ADP must be recycled 230/0.1 = 2300 times.

22. The irreversible step in electron transport is the formation of water from oxygen. The rate of cytochrome *c* oxidase is controlled by the ratio of reduced to oxidized cytochrome *c*, which is in turn controlled by the $[NADH]/[NAD^+]$ and $[ATP]/[ADP][P_i]$ ratios.

23. The change in redox potential of the Fe atom promotes its ability to accept the electrons from NADPH ($\mathscr{E}°' = -0.320$). Recall that electrons flow spontaneously from a substance with a more negative redox potential to a substance with a more positive redox potential.

24. Rats that consume large quantities of metabolic fuels have a higher rate of oxidative metabolism and hence generate more oxygen radicals than rats that consume less food and have lower rates of oxidative metabolism. The cumulative oxidative damage would be greater in the cafeteria-fed rats, which therefore die sooner.

Chapter 18 Photosynthesis

In the preceding chapters, you learned that free energy is derived from reduced foodstuffs such as glucose and that the energy produced from catabolic pathways is used to generate both ATP and NAD(P)H. Photosynthesis, an ancient and important process, allows energy to be harvested directly from the most abundant and renewable source, the sun. Photosynthesis is a light-driven process in which carbon dioxide is "fixed" to produce carbohydrates. This occurs in two phases: (1) The light reactions (requiring light) produce ATP and NADPH, and (2) the dark reactions (not requiring light) use ATP and NADPH to synthesize carbohydrates. This chapter describes how different pigments (e.g., chlorophylls in plants and bacteria) efficiently capture light energy and redistribute it to specific reaction centers. Purple photosynthetic bacteria contain one photosystem that recycles its electrons, whereas higher plants have two photosystems that use water as a source of electrons to reduce NADPH. The oxidation of water in higher plants generates O_2 as a by-product of photosynthesis. As in mitochondria, the topology of chloroplasts is central to the biochemistry of photosynthesis, starting with the light-driven reactions in the thylakoid membrane and finishing with the dark reactions in the stroma. The dark reactions occur via the Calvin cycle, a set of reactions that synthesize glyceraldehyde-3-phosphate from 3 CO_2. The chapter also discusses the control of the Calvin cycle along with a variant called the C_4 pathway.

Essential Concepts

1. Photosynthesis is divided into two processes:
   (a) In the light reactions, organisms capture light energy to synthesize ATP and generate reducing equivalents in the form of NADPH.
   (b) In the dark reactions, carbon dioxide is converted to carbohydrates using the ATP and NADPH generated in the light reactions. Although the dark reactions are not light-driven, they only occur when it is light and hence are better described as light-independent.

Chloroplasts

2. Plants differ from bacteria by providing a separate organelle for the photosynthetic machinery, the chloroplast. A chloroplast is enveloped by a highly permeable outer membrane and a nearly impermeable inner membrane. The inner membrane encloses the stroma, which contains the soluble enzymes of carbohydrate synthesis, and the thylakoid membrane, which is organized in stacks of pancake-like disks (grana) that enclose the thylakoid compartment and that are linked by unstacked stromal lamellae. The proteins that capture light energy and mediate electron-transport processes are embedded in the thylakoid membrane.

3. Various pigment molecules absorb light of different wavelengths. The principal photosynthetic pigment is chlorophyll, a cyclic tetrapyrrole that ligands a central Mg^{2+} ion. Photosynthetic organisms also contain other pigments, such as carotenoids, phycoerythrin, and phycocyanin, which together with chlorophyll absorb most of the visible light in the solar spectrum.

4. Multiple pigment molecules are arranged in light-harvesting complexes (LHCs), which are proteins that act as antennae to gather light energy and redirect it to photosynthetic reaction centers, where the light energy is converted to chemical energy in the form of ATP and NADPH. The accessory pigments in the LHCs boost light absorption at wavelengths at which chlorophyll does not absorb strongly.

The Light Reactions

5. Photons propagate as discrete energy packets called quanta, whose energy, E, is given by Planck's law: $E = h\nu = hc/\lambda$, where h is Planck's constant (6.626×10^{-34} J · s), c is the speed of light (2.998×10^{8} m · s^{-1}), λ is the wavelength of light, and ν is the frequency of the radiation.

6. When a molecule absorbs a photon, one of its electrons is promoted to a higher energy orbital. The excited electron can return to the ground state in several ways:
   (a) In internal conversion, electronic energy is converted to kinetic energy (heat).
   (b) Fluorescence results when the molecule emits a photon at a lower wavelength.
   (c) The excitation energy can be transferred to another molecule by exciton transfer (resonance energy transfer). This occurs in the transfer of light energy from LHCs to the photosynthetic reaction center.
   (d) The molecule may undergo photooxidation by transfer of an electron to another molecule. The excited chlorophyll at the reaction center transfers electrons in this manner.

7. The bacterial photosynthetic reaction center of *Rps. viridis* consists of a series of prosthetic groups arranged with nearly twofold symmetry: 2 closely associated bacteriochlorophyll *a* (BChl *a*) molecules known as the special pair, 2 bacteriopheophytin *a* (BPheo *a*; BChl *a* that lacks an Mg^{2+} ion), 2 additional BChl *a* molecules, a menaquinone, a ubiquinone, and an Fe(II) ion.

8. In purple photosynthetic bacteria, the special pair undergoes photooxidation virtually every time it absorbs a photon. The transferred electron is first passed to the BPheo *a* on the "right" side of the photosynthetic reaction center and then to the menaquinone and then the ubiquinone to yield a semiquinone radical anion (Q_B^{-}). A second photon absorption then transfers a second electron to yield Q_B^{2-}, which picks up two protons from the cytoplasm to form QH_2 and then exchanges with the membrane-bound pool of ubiquinone.

9. The electrons ejected from the special pair return to the photosynthetic reaction center via an electron-transport chain consisting of a cytochrome bc_1 complex and cytochrome c_2. Electron flow follows a Q cycle in cytochrome bc_1, which translocates four protons to the periplasmic space for every two electrons transferred. The free energy of the resulting transmembrane proton gradient drives ATP synthesis.

10. In plants and cyanobacteria, photooxidation occurs at two reaction centers, and electron transport is noncyclical. The path of electrons from water to NADPH is described by the Z-scheme, in which photosystem II (PSII) passes its electrons to the cytochrome b_6f complex via the mobile electron carrier ubiquinol (QH_2), and cytochrome b_6f then transfers these electrons to photosystem I (PSI) via the mobile Cu-containing protein plastocyanin. Since

PSII and PSI are thereby "connected in series," the energy of each electron is boosted by two photon absorptions.

11. PSII includes the Mn-containing oxygen-evolving center (OEC), which cycles through five electronic states (S_0-S_4) in the conversion of H_2O to O_2 and is driven by four consecutive excitations of the PSII reaction center (called P680). The four electrons released from H_2O follow a path similar to that of the bacterial photosynthetic reaction center, eventually reaching the membrane plastoquinone pool.

12. Electron transport through the cytochrome b_6f complex (which resembles the mitochondrial Complex III) generates a transmembrane proton gradient via a Q cycle. Eight protons are translocated to the thylakoid lumen for the four electrons released from each H_2O. Plastocyanin, a peripheral membrane protein, has a Cu redox center that ferries one electron at a time from cytochrome b_6f to PSI.

13. PSI contains multiple pigments and redox groups, including chlorophylls, carotenoids, [4Fe–4S] clusters, and phylloquinone. Photooxidation of the PSI special pair (called P700) allows the electron received from plastocyanin to pass through a series of electron carriers in one of two routes:
    (a) In the noncyclic pathway, electrons flow through PSI to the [2Fe–2S]-containing one-electron carrier ferredoxin (Fd), which is a soluble stromal protein. Two Fd's deliver their electrons to ferredoxin–$NADP^+$ reductase, which thereupon carries out the two-electron reduction of $NADP^+$ to NADPH.
    (b) In the cyclic pathway, electrons return from PSI through cytochrome b_6f to the plastoquinone pool and thereby participate in the Q cycle. This pathway augments the proton gradient across the thylakoid membrane and hence contributes additional free energy for the synthesis of ATP but does not yield NADPH.

14. The free energy of the proton gradient is tapped by chloroplast CF_1CF_0 ATP synthase, which closely resembles the mitochondrial F_1F_0-ATPase. Approximately 12 protons enter the thylakoid lumen for each O_2 generated in noncyclic electron transport (4 H^+ from the OEC reaction and 8 H^+ from the Q cycle). The synthesis of ATP requires the transport of ~3 protons from the thylakoid lumen to the stroma.

The Dark Reactions

15. CO_2 is incorporated into carbohydrates by carboxylation of a 5-carbon sugar, ribulose-5-phosphate (R5P). The resulting 6-carbon compound is split into two molecules of 3-phosphoglycerate (3PG), which is then converted to glyceraldehyde-3-phosphate (GAP). Some of the GAP is diverted to carbohydrate synthesis, and the rest is converted back to Ru5P. This set of 13 reactions, called the Calvin cycle, has two stages:
    (a) In the production phase, 3 Ru5P react with 3 CO_2 to yield 6 GAP (for a net yield of 1 GAP from 3 CO_2), at a cost of 9 ATP and 6 NADPH.
    (b) In the recovery phase, the carbons of 5 GAP are shuffled via aldolase- and transketolase-catalyzed reactions to reform 3 Ru5P, without consuming ATP or NADPH.

    The GAP product of the cycle is then converted to glucose-1-phosphate, a precursor of sucrose and starch.

16. Ribulose bisphosphate carboxylase, which accounts for up to 50% of leaf proteins, catalyzes the carboxylation of ribulose-1,5-bisphosphate (RuBP). Enzymatic abstraction of a proton from RuBP generates an enediolate that attacks CO_2. H_2O then attacks the resulting β-keto acid to yield two 3PG.

17. The activity of RuBP carboxylase is controlled in several ways so that the dark reactions proceed only when the light reactions are able to provide the ATP and NADPH necessary to drive them:
    (a) RuBP carboxylase is most active at pH 8.0, which occurs when protons are pumped out of the stroma during the light reactions.
    (b) The Mg^{2+} that enters the stroma to compensate for the efflux of H^+ stimulates RuBP carboxylase.
    (c) Plants synthesize 2-carboxyarabinitol-1-phosphate, an inhibitor of RuBP carboxylase, only in the dark.

18. The Calvin cycle enzymes fructose bisphosphatase and sedoheptulose bisphosphatase are also activated by increases in pH, $[Mg^{2+}]$, and [NADPH]. The redox state of ferredoxin is sensed by a thiol-exchange cascade involving ferredoxin–thioredoxin reductase, thioredoxin, and disulfide groups on the bisphosphatases, so that the Calvin cycle enzymes are stimulated when Fd is reduced (i.e., when the light reactions are occurring).

19. RuBP carboxylase can also react with oxygen, which competes with CO_2 at the carboxylase active site. This process, called photorespiration, converts RuBP into 3PG and 2-phosphoglycolate, a two-carbon compound. A series of reactions in the chloroplast and peroxisome convert two 2-phosphoglycolate to two glycine, which are converted in the mitochondria to serine + CO_2. The serine is converted back to the Calvin cycle intermediate 3PG by reactions that require NADH and ATP. Thus, photorespiration consumes O_2 and produces CO_2, at the expense of ATP and NADH, thereby reversing the results of photosynthesis. All known RuBP carboxylases have this activity, which may protect chloroplasts from photoinactivation when high light intensity has greatly reduced the local CO_2 concentration.

20. The rate of photorespiration becomes significant on hot bright days when photosynthesis has depleted the level of CO_2 at the chloroplast and raised the concentration of O_2. However, C_4 plants such as corn and sugar cane prevent photorespiration by concentrating CO_2 at the chloroplast. They do so by using phosphoenolpyruvate (PEP) carboxylase to make oxaloacetate. Oxaloacetate is converted to malate in the mesophyll cells (which lack RuBP carboxylase) and transported to the bundle-sheath cells, where the Calvin cycle operates. There, malic enzyme cleaves malate into pyruvate and CO_2. The CO_2 is thereby delivered to RuBP carboxylase at a high enough concentration to essentially eliminate photorespiration. The pyruvate is transported back to the mesophyll cells and converted to PEP at the expense of two "high-energy" bonds. In the tropics, C_4 plants grow faster than so-called C_3 plants. In more temperate climates, where the rate of photorespiration is reduced, C_3 plants have an advantage because they require less energy to fix CO_2.

21. Many desert succulent plants conserve water by opening their stomata only at night to acquire CO_2. The CO_2 is converted to PEP at night and is released as CO_2 in the day, to be fixed by RuBP carboxylase. This process is called Crassulacean acid metabolism (CAM) because it was discovered in the family Crassulaceae.

Key Equation

$$E = h\nu = \frac{hc}{\lambda}$$

Guide to Study Exercises (text p. 560)

1. The light reactions and dark reactions are the two stages of photosynthesis. In the light reactions, specialized pigment molecules capture light energy and are thereby photooxidized. The transfer of their electrons through a series of electron carriers results in the reduction of $NADP^+$ to NADPH and the generation of a transmembrane proton concentration gradient whose free energy is harnessed by CF_1CF_0-ATPase to phosphorylate ADP. In the dark reactions, the NADPH and ATP produced in the light reactions are used to reduce CO_2 and incorporate it into carbohydrates. The light reactions occur only in the presence of light energy. In contrast, the dark reactions do not strictly require light, but they are regulated so that they proceed when the light reactions are active and hence producing the NADPH and ATP required for the dark reactions. (Chapter introduction and Section 18-3B)

2. Light-harvesting complexes (LHCs) are complexes of pigment molecules that function as light-absorbing antennae that gather the energy of photons and pass it to a photosynthetic reaction center. LHCs are transmembrane proteins that typically contain a variety of pigment molecules whose number and arrangement have been optimized for efficient light absorption and energy transfer. Without the LHCs, the photosynthetic reaction centers would intercept too few photons to support life. (Section 18-1B)

3. A molecule that has absorbed light energy becomes electronically excited. This energy can be dissipated by the following processes:
   (a) Internal conversion, which is the conversion of electronic energy to kinetic energy (heat).
   (b) Fluorescence, which is the emission of a photon whose energy is lower (of longer wavelength) than that of the photon absorbed by the molecule.
   (c) Exciton transfer (resonance energy transfer), which is the transfer of absorbed energy to an unexcited molecule with similar electronic properties.
   (d) Photooxidation, which occurs when an electron held by the excited molecule is transferred to another molecule. (Section 18-2A)

4. The purple bacterial reaction center is a transmembrane protein containing 4 bacteriochlorophyll (BChl), 2 bacteriopheophytin (BPheo), an Fe(II) ion, and 2 ubiquinone (or a ubiquinone and a menaquinone). When the so-called special pair of BChl absorbs a photon, it becomes energized (has a more negative reduction potential) and transfers an electron to a BPheo group on its "right" side, possibly via the intervening accessory BChl. The electron then migrates to the adjacent quinone and then to the other quinone. Another such electron transfer, resulting from the absorption of a second photon, is required to completely reduce the terminal quinone acceptor to a quinol. The electrons lost by the special pair are returned to it via a series of electron transfers involving membrane-bound ubiquinol, a cytochrome bc_1 complex, and cytochrome c_2. Electron flow through the cytochrome bc_1 complex follows the Q cycle pathway, which transfers the two electrons from ubiquinol to two

molecules of cytochrome c_2, each of which returns an electron to the special pair. The overall pathway of electron transfer is cyclic, with no net oxidation or reduction of its components. (Section 18-2B)

5. The Z-scheme describes the noncyclic path of electrons through the photosynthetic machinery of green plants and cyanobacteria. The zig-zag appearance of the Z-scheme on a reduction potential diagram reflects the shifts in reduction potential that occur when its two photosynthetic reaction centers absorb light energy. Overall, electrons flow from groups with lower reduction potentials to groups with higher reduction potentials, so that the net result of the pathway is the reduction of $NADP^+$ by electrons derived from H_2O. The Z-scheme links three membrane-bound protein complexes with mobile electron carriers. The oxygen-evolving complex (OEC) of photosystem II (PSII) catalyzes the conversion of 2 H_2O to $O_2 + 4\ H^+$ in a catalytic cycle driven by the excitation of the PSII reaction center. The four electrons released from H_2O are transferred by the excited reaction center through the redox groups of PSII to a membrane-bound pool of plastoquinone. From there, electrons pass to the cytochrome b_6f complex, where they follow a cyclic pathway (the Q cycle) that results in the reduction of the mobile electron carrier plastocyanin. This peripheral membrane protein shuttles one electron at a time to the reaction center of photosystem I (PSI). The absorption of a photon decreases the reduction potential of the PSI reaction center so that it can transfer the electron through a series of redox groups. The electron eventually reaches the soluble one-electron carrier ferredoxin. Two reduced ferredoxin proteins carry their electrons to the stromal enzyme ferredoxin–$NADP^+$ reductase, which carries out the two-electron reduction of $NADP^+$. (Section 18-2C)

6. Noncyclic electron transport in PSI results in the net transfer of electrons to ferredoxin and then to $NADP^+$. In cyclic electron flow, electrons ejected from PSI return to the cytochrome b_6f complex, where they participate in the Q cycle. This results in the translocation of two protons from the stroma to the thylakoid lumen for each electron. Therefore, cyclic electron flow augments the transmembrane proton concentration gradient and results in the synthesis of more ATP. The balance between cyclic and noncyclic electron transfer determines the relative amounts of NADPH and ATP produced by the light reactions. (Section 18-2C)

7. In both photophosphorylation and oxidative phosphorylation, the free energy of a transmembrane proton concentration gradient drives the synthesis of ATP from ADP + P_i as catalyzed by a membrane protein consisting of a pseudo-threefold symmetric ATP synthase in complex with a transmembrane proton channel. In both cases, ATP synthesis requires an intact membrane and can be uncoupled from electron transport by agents that dissipate the proton gradient.

   Photophosphorylation differs from oxidative phosphorylation in that the proton gradient involves the chloroplast stroma (which is analogous to the mitochondrial matrix) and the thylakoid space, which is not in contact with the cytoplasm (as is the mitochondrial intermembrane space). Another difference is that the thylakoid membrane is permeable to Mg^{2+} and Cl^- so that the membrane potential ($\Delta\Psi$) does not contribute significantly to the electrochemical gradient in the chloroplast. Finally, the stoichiometry of ATP molecules synthesized per electron transferred is relatively constant in mitochondria, but it varies in chloroplasts depending on the proportion of cyclic to noncyclic electron flow. (Section 18-2D)

8. In the first stage of the Calvin cycle, three molecules of ribulose-5-phosphate (Ru5P) react with three CO_2 to yield six molecules of glyceraldehyde-3-phosphate (GAP) at the expense of 9 ATP and 6 NADPH. This process is equivalent to the conversion of 3 CO_2 to 1 GAP.

   In the second stage of the Calvin cycle, the carbon skeletons of 5 GAP are rearranged to yield 3 Ru5P. The reactions in this stage, most of which are reversible, do not require ATP or NADPH. (Section 18-3A)

9. Photosynthesis is regulated so that the dark reactions occur only when the free energy they require (in the form of ATP and NADPH) is available from the light reactions. This is accomplished through several mechanisms:
   (a) Ribulose bisphosphate (RuBP) carboxylase activity increases in response to light by the increase in stromal pH and the influx of Mg^{2+} ions that occur when protons are pumped across the thylakoid membrane. The increase in pH and [Mg^{2+}] also activates the Calvin cycle enzymes FBPase and SBPase.
   (b) Plants synthesize 2-carboxyarabinitol-1-phosphate (CA1P), which inhibits RuBP carboxylase, only in the dark.
   (c) The productivity of the light reactions, represented by the redox state of ferredoxin, is communicated to thioredoxin. This small protein activates FBPase and SBPase by a disulfide interchange reaction. (Section 18-3B)

10. Photorespiration, a pathway that consumes NADH and ATP, may occur when there is insufficient CO_2. So-called C_4 plants minimize photorespiration by a mechanism that increases the availability of CO_2 at the chloroplast. These plants take up CO_2 in mesophyll cells (which lack RuBP carboxylase) by condensing it with phosphoenolpyruvate to yield the C_4 compound oxaloacetate. The oxaloacetate is reduced to malate, which enters the bundle-sheath cells (which contain RuBP carboxylase) and is decarboxylated to produce pyruvate and CO_2 at a higher concentration than that directly available from the atmosphere.

Questions

1. Define the terms light reactions and dark reactions. Do the dark reactions occur in the dark? Explain.

Chloroplasts

2. Draw a cross-section of a chloroplast and indicate the locations of the following proteins and other structural features:

Outer membrane	Thylakoid lumen
Stromal lamella	Photosystem I
Inner membrane	Photosystem II
Intermembrane space	Cytochrome b_6f
Grana	CF_1CF_0-ATP synthase
Stroma	Direction of proton pumping

The Light Reactions

3. Calculate the energy of a mole of photons with wavelengths of (a) 400 nm, (b) 500 nm, (c) 600 nm, and (d) 700 nm.

4. Purple photosynthetic bacteria have different pigments than higher plants. Why is this an advantage for these bacteria?

5. What distinguishes the chlorophyll in a reaction center from the antennae chlorophyll?

6. The initial electron transfers in the bacterial photosynthetic reaction center are extremely rapid, but the lifetime of the terminal semiquinone is relatively long. (a) Why is it essential for the electrons to quickly leave the vicinity of the special pair? (b) Why does the terminal semiquinone persist?

7. Compare the electron flow in purple photosynthetic bacteria to that in higher plant chloroplasts. What is the origin of the electrons and their eventual fates? How many protons are translocated?

8. What is the standard reduction potential for the oxidation of H_2O (see Table 13-3)? Can this value be obtained from purple bacterial photosynthesis? Compare this to two-center photosynthesis.

9. The number of O_2 molecules released per photon absorbed by a suspension of algae can be measured at different wavelengths. When algae are illuminated by 700 nm light, very little O_2 is produced. However, when they are also illuminated by 500 nm light, the O_2 production is well in excess of that produced with only the 500 nm light. Explain.

10. What is the fate of water-derived electrons in chloroplasts treated with DCMU? What simple screening method could be used to identify plants that had been exposed to DCMU?

11. What do the *S* states represent in the oxygen-evolving center (OEC)? What *S* state predominates in dark-adapted chloroplasts?

12. What chloroplast component generates the majority of the proton gradient used for ATP production? Does it have a mitochondrial counterpart?

13. Why does the Cu atom of plastocyanin have an unusually high standard reduction potential?

14. What are the similarities and differences between photosystem I and photosystem II?

15. Estimate the minimum reduction potential for P680. Estimate the maximum reduction potential for ferredoxin. Explain your answers.

16. Describe the distribution of LHCs between grana (stacked membranes) and stromal lamellae (unstacked membranes) in chloroplasts under a bright sun and in shady light.

The Dark Reactions

17. What is the first stable radioactive sugar intermediate seen when $^{14}CO_2$ is added to algae such as *Chlorella*? When the supply of $^{14}CO_2$ is cut off, what compound accumulates? What do these results suggest about the pathway of CO_2 incorporation into carbohydrates?

18. Using Figure 18-20, write out the 13 reactions of the Calvin cycle. What is the main difference between Stage I and Stage II reactions? Write a net equation for each stage.

19. Glycolysis and the Calvin cycle are opposing pathways. Which reactions form a potential futile cycle? How are these reactions controlled in plant chloroplasts?

20. What is photorespiration? What is the ultimate result of this process?

21. The concentration of atmospheric CO_2 has been increasing for many decades. If this trend continues, how might it affect the relative abundance of C_3 and C_4 plants?

Answers to Questions

1. The light reactions depend on light for activity and include the reactions of the photosystems and the electron transport chain. The dark reactions are those of the Calvin cycle, which convert CO_2 to GAP.

   The dark reactions do not require light for their mechanisms. However, the light reactions regulate the dark reactions to ensure that the cell maintains adequate levels of ATP and NADPH. Thus the dark reactions do not actually occur in the dark.

2. 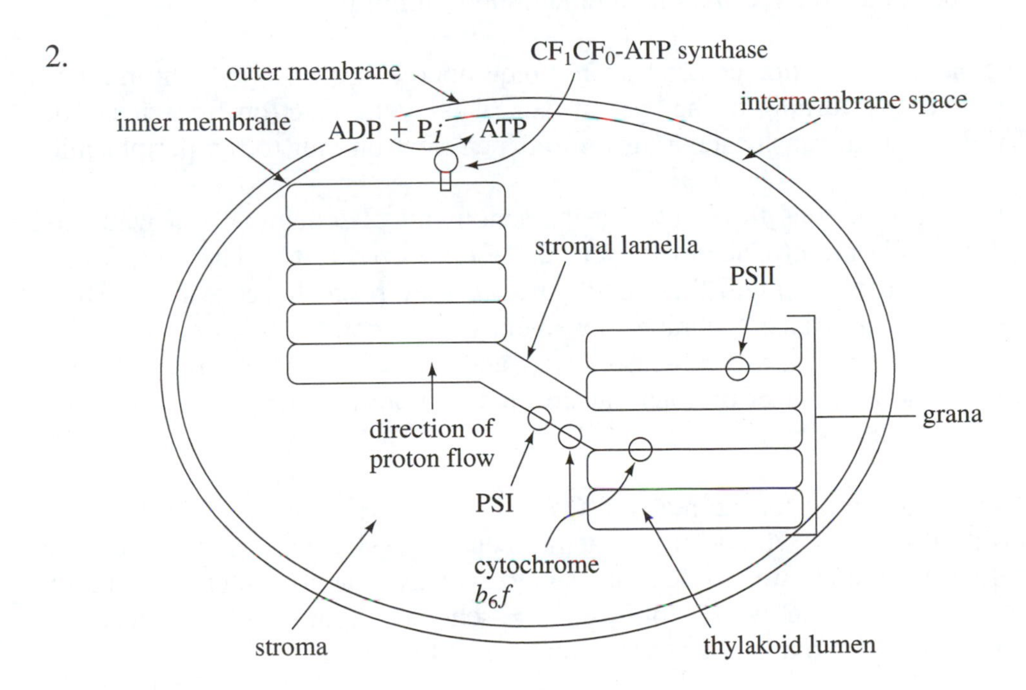

3. Use Planck's law multiplied by Avogadro's number:

$$E = \frac{hc}{\lambda}N$$
$$= (6.626 \times 10^{-34}\ \text{J} \cdot \text{s})\ (2.998 \times 10^{8}\ \text{m} \cdot \text{s}^{-1})(6.0221 \times 10^{23}\ \text{mol}^{-1})/\lambda$$
$$= (0.1196\ \text{J} \cdot \text{m} \cdot \text{mol}^{-1})/\lambda$$

(a) $E = (0.1196\ \text{J} \cdot \text{m} \cdot \text{mol}^{-1})/(4 \times 10^{-7}\ \text{m}) = 3.0 \times 10^{5}\ \text{J} \cdot \text{mol}^{-1} = 300\ \text{kJ} \cdot \text{mol}^{-1}$
(b) $E = (0.1196\ \text{J} \cdot \text{m} \cdot \text{mol}^{-1})/(5 \times 10^{-7}\ \text{m}) = 2.4 \times 10^{5}\ \text{J} \cdot \text{mol}^{-1} = 240\ \text{kJ} \cdot \text{mol}^{-1}$
(c) $E = (0.1196\ \text{J} \cdot \text{m} \cdot \text{mol}^{-1})/(6 \times 10^{-7}\ \text{m}) = 2.0 \times 10^{5}\ \text{J} \cdot \text{mol}^{-1} = 200\ \text{kJ} \cdot \text{mol}^{-1}$
(d) $E = (0.1196\ \text{J} \cdot \text{m} \cdot \text{mol}^{-1})/(7 \times 10^{-7}\ \text{m}) = 1.7 \times 10^{5}\ \text{J} \cdot \text{mol}^{-1} = 170\ \text{kJ} \cdot \text{mol}^{-1}$

4. The pigments in purple photosynthetic bacteria absorb radiation with longer wavelengths than visible light. This is the most intense radiation in the environments that they inhabit, the murky bottoms of stagnant ponds.

5. The chlorophyll molecules in a reaction center have a slightly lower-energy excited state than that of the antenna chlorophyll molecules. This allows excitation energy to be transferred from the antenna molecules to the reaction center, where photooxidation occurs.

6. (a) The electron ejected from the excited special pair must be transferred away so that it cannot return immediately to the special pair, which would allow the excitation energy to be released as heat (or possibly in a way that would damage the reaction center).
(b) After its reduction by an electron, the quinone (now a semiquinone) must await a second excitation event and electron transfer to become fully reduced to a quinol so that it can transfer both electrons to the membrane-bound quinol pool.

7. Electrons from the bacterial reaction center flow to ubiquinone and then to cytochrome bc_1 and cytochrome c_2 before returning to the special pair. This cyclic electron flow does not yield reduced $NADP^+$, but it translocates four protons per electron pair to the periplasmic space.

In the chloroplast, electrons flow in a linear fashion (the Z-scheme) from water to $NADP^+$, so that 2 NADPH are produced for every 2 H_2O oxidized to O_2. The 4 electrons pass from PSII to cytochrome b_6f and then to PSI before they reach ferredoxin–$NADP^+$ reductase. This results in the transmembrane movement of 12 protons. In cyclic electron flow, electrons cycle from PSI back to cytochrome b_6f and hence no NADPH is produced. Instead, this increases the number of protons translocated without affecting the stoichiometry of the $H_2O \rightarrow O_2$ reaction.

8. The standard reduction potential for the reaction $O_2 + 4\,e^- + 4\,H^+ \rightarrow 2\,H_2O$ is 0.815 V. The reduction potential of the P870 bacterial reaction center is ~0.500 V (Figure 18-10), which is not sufficient to oxidize water (electrons spontaneously flow to centers with more positive reduction potentials). The two-reaction center Z-scheme (Figure 18-12) spans a redox range that allows both the oxidation of water and the reduction of $NADP^+$.

9. The longer wavelength activates only PSI (which contains P700), whereas the shorter wavelength activates both PSI and PSII (which contains P680). Since PSII feeds electrons to PSI, both photosystems must operate together for the redox reactions to proceed most efficiently. When only 700 nm light is available, PSII is unable to extract the electrons from H_2O necessary to form O_2. However, in the presence of 700 nm and 500 nm light, PSII can supply electrons to PSI, which can energize them with 700 nm light, thereby driving O_2 production.

10. DCMU blocks electron flow between PSII and PSI. This would cause the excitation energy of PSII to be dissipated by a mechanism other than photooxidation (since the electrons have nowhere to go). Some of the absorbed energy is released as fluorescence. Plants in which electron flow was blocked by DCMU can be detected by their fluorescence.

11. The *S* states in the OEC are chemical intermediates in the five-step reaction in which H_2O is oxidized to O_2. Dark-adapted chloroplasts are in the S_1 state, since O_2 is generated on the third flash of light (Figure 18-13).

12. The cytochrome b_6f complex, via the Q cycle, pumps the majority of the protons that make up the proton gradient required for ATP synthesis. Eight protons are translocated for every 2 H_2O oxidized; more during cyclic electron flow. The membrane-bound cytochrome b_6f complex resembles mitochondrial Complex III (cytochrome bc_1).

13. The Cu(II)/C(I) half-reaction normally has a redox potential of 0.158 V, but plastocyanin's redox potential is 0.370 V. In the protein, the Cu(II) atom is strained toward the tetrahedral coordination geometry of Cu(I), which promotes its reduction.

14. Both photosystems are membrane-bound protein complexes, contain special pairs of chlorophyll where photooxidation occurs, and have near-symmetrical arrangements of pigment molecules at the reaction center.

    The photosystems differ in overall structure, their redox groups, and the pathway of electron transfer.

15. The reduction potential for P680 should be greater than that of the O_2/H_2O half-reaction (0.815 V; Table 13-3) since electrons flow from H_2O to P680. The standard reduction potential of ferredoxin should be less than that for the $NADP^+$/NADPH half-reaction (−0.320 V) since electrons flow from ferredoxin to $NADP^+$.

16. Under bright sun (high proportion of short-wavelength light), PSII is more active than PSI. As a result, reduced plastoquinone accumulates. This activates a protein kinase to phosphorylate the LHCs, which then move to the stromal lamellae, where they associate with and funnel more light energy to PSI.

    Under shady light (high proportion of long-wavelength light), PSI is more active than PSII. Oxidized plastoquinone therefore accumulates. This leads to dephosphorylation of the LHCs, which move to the grana to associate with and funnel light energy to PSII.

17. 3-Phosphoglycerate is the first stable sugar that incorporates the $^{14}CO_2$. This suggests that $^{14}CO_2$ is added to a 2-carbon compound. Ribulose-5-phosphate levels increase after the removal of $^{14}CO_2$, which instead suggests that ribulose-5-phosphate is the $^{14}CO_2$ acceptor.

18. Shown below are the 13 reactions of the Calvin cycle:

(1) $3\ Ru5P + 3\ ATP \rightarrow 3\ RuBP + 3\ ADP$
(2) $3\ RuBP + 3\ CO_2 \rightarrow 6\ 3PG$
(3) $6\ 3PG + 6\ ATP \rightarrow 6\ BPG + 6\ ADP$
(4) $6\ BPG + 6\ NADPH \rightarrow 6\ GAP + 6\ P_i + 6\ NADP^+$
(5) $2\ GAP \rightleftharpoons 2\ DHAP$
(6) $GAP + DHAP \rightleftharpoons FBP$
(7) $FBP \rightarrow F6P + P_i$
(8) $F6P + GAP \rightleftharpoons Xu5P + E4P$
(9) $E4P + DHAP \rightleftharpoons SBP$
(10) $SBP \rightarrow S7P + P_i$
(11) $S7P + GAP \rightleftharpoons Xu5P + R5P$
(12) $2\ Xu5P \rightleftharpoons 2\ Ru5P$
(13) $R5P \rightleftharpoons Ru5P$

Stage I (Reactions 1–4) is the production phase or energy-requiring phase and consists of carboxylation, phosphorylation, and reduction steps. Stage II (Reactions 5–13) is the rearrangement or recovery phase, which regenerates ribulose-5-phosphate.

Stage I:

$$3\ Ru5P + 3\ CO_2 + 9\ ATP + 6\ NADPH \rightarrow 6\ GAP + 9\ ADP + 6\ P_i + 6\ NADP^+$$

Stage II:

$$5\ GAP \rightarrow 3\ Ru5P + 2\ P_i$$

19. The interconversion of fructose-1,6-bisphosphate (FBP) and fructose-6-phosphate (F6P) is a potential futile cycle. The relevant enzymes are phosphofructokinase (PFK, for glycolysis) and fructose bisphosphatase (FBPase, for the Calvin cycle). To control these enzymes, a redox-sensing protein, thioredoxin, activates FBPase and deactivates PFK in the light, and activates PFK and deactivates FBPase in the dark.

20. Photorespiration is a side reaction of ribulose bisphosphate carboxylase in which O_2 (which competes with CO_2 for binding to the active site) reacts with RuBP to form 3PG and 2-phosphoglycolate. The 2-phosphoglycolate is eventually converted back to 3PG by a multistep pathway that consumes NADH and ATP and yields CO_2. Photorespiration therefore wastes some of the free energy captured in the photosynthetic light reactions and "unfixes" some of the CO_2 fixed by photosynthesis.

21. C_4 plants have an advantage when CO_2 is relatively scarce, since they concentrate CO_2 in mesophyll cells. However, this process consumes 2 ATP equivalents. The more energy-efficient C_3 plants therefore have an advantage when CO_2 is not limiting. Thus, a higher atmospheric concentration of CO_2 would favor C_3 plants over C_4 plants.

Chapter 19 Lipid Metabolism

As we saw in Chapter 9, cells contain a diverse array of lipids. The major functions of lipids include forming a barrier to the extracellular environment, maintaining membrane fluidity, and serving as an important energy store, principally in the form of triacylglycerols. This chapter focuses on the diverse metabolic pathways of cellular lipids. First, the means by which ingested dietary lipids are degraded and assimilated is considered. The chapter then presents the oxidative pathways by which long acyl chains are converted to successive two-carbon acetyl-CoA fragments, which are either further oxidized via the citric acid cycle and the electron-transport system, converted to ketone bodies, or used in biosynthesis. Next, mechanisms of fatty acid biosynthesis by successive condensations of two-carbon fragments are presented. The steps leading to the biosynthesis of membrane phospholipids and glycolipids are outlined, followed by the pathway of cholesterol biosynthesis and its regulation. The reader should contemplate the varied metabolic reactions that result in the formation or degradation of lipids in the context of the nutritive, structural, and regulatory roles that lipids fulfill in the cell.

Essential Concepts

Lipid Digestion, Absorption, and Transport

1. The enzymatic digestion of triacylglycerols occurs at the lipid–water interface and is aided by the presence of bile acids, which help solubilize the lipids. Bile acids are cholesterol derivatives that are synthesized in the liver as taurine or glycine derivatives, stored in the gall bladder, and released into the small intestine.

2. Pancreatic lipase hydrolyzes triacylglycerols first to diacylglycerols and then to 2-monoacylglycerols plus free fatty acids. Binding of the enzyme to triacylglycerol at the lipid–water interface requires colipase. The interaction of this protein with lipase produces a hydrophobic surface which promotes binding of the protein complex to the lipid. Phospholipase A_2 also degrades phospholipids to lysophospholipids and fatty acids at a lipid–water interface.

3. Micelles containing bile acids promote the absorption of triacylglycerol and phospholipid hydrolysis products by the intestinal mucosa. Inside intestinal cells, fatty acids are complexed to a fatty acid–binding protein that, in effect, increases the solubility of these hydrophobic compounds.

4. Triacylglycerols are resynthesized in intestinal cells and incorporated into chylomicrons. These lipoproteins enter the bloodstream via the lymphatic system and eventually provide triacylglycerols to peripheral tissues, chiefly skeletal muscle and adipose tissue. The delivery process converts chylomicrons to much smaller chylomicron remnants, which are taken up by the liver.

5. The liver synthesizes other lipoproteins, including very low density lipoproteins (VLDL), which transport triacylglycerols and cholesterol from the liver to skeletal muscle and adipose tissue. In the capillaries, triacylglycerols are degraded by lipoprotein lipase to yield

fatty acids (which the cells can either oxidize or reincorporate into triacylglycerols) and glycerol (which can be transformed to the glycolytic intermediate dihydroxyacetone phosphate). As they lose their triacylglycerol component, VLDL are converted first to intermediate density lipoproteins (IDL) and then to low density lipoproteins (LDL).

6. Cholesterol is removed from cell surface membranes and carried as cholesteryl esters through the bloodstream to the liver by high density lipoproteins (HDL). There, the cholesteryl esters are transferred to VLDL.

Fatty Acid Oxidation

7. When energy needs dictate, triacylglycerols stored in adipose tissue are broken down (mobilized) by hormone-sensitive lipase. Released free fatty acids are transported in complex with serum albumin to the liver and other tissues.

8. Before being degraded by oxidation, fatty acids are first activated by the formation of an acyl-CoA in an ATP-dependent reaction catalyzed by thiokinase.

9. Since β oxidation takes place in the mitochondrial matrix, the acyl groups must cross the inner mitochondrial membrane, which is impermeable to fatty acyl-CoA derivatives. Therefore, the acyl group is transferred to carnitine by carnitine palmitoyl transferase I. The resulting acyl-carnitine readily crosses the membrane via a carrier protein. Once in the mitochondrial matrix, the acyl group is transferred back to a CoA molecule by carnitine palmitoyl transferase II, and the liberated carnitine crosses the membrane back to the cytosol.

10. Fatty acyl groups are degraded by a process called β oxidation, in which successive two-carbon fragments are removed as acetyl-CoA units.
    (a) Acyl-CoA dehydrogenase catalyzes formation of a *trans*-2,3 (α,β) double bond. The enzyme's bound FAD is thereby reduced to $FADH_2$.
    (b) Enoyl-CoA hydratase catalyzes the hydration of the double bond to produce a 3-L-hydroxyacyl-CoA.
    (c) 3-L-Hydroxyacyl-CoA dehydrogenase catalyzes the formation of a β-ketoacyl-CoA with the reduction of NAD^+ to NADH.
    (d) Thiolase catalyzes the thiolysis of the C2—C3 bond, releasing acetyl-CoA and forming a new acyl-CoA which is two carbons shorter than the starting substrate.

    This sequence of reactions is repeated until the acyl-CoA has been converted entirely to acetyl-CoA. For oxidation of palmitoyl-CoA, the sequence occurs 7 times to yield 8 acetyl-CoA.

11. The oxidation of fatty acids is highly exergonic. For example, palmitate's 8 acetyl-CoA can enter the citric acid cycle, and the $FADH_2$ and NADH generated by β oxidation and the citric acid cycle can be reoxidized by the electron-transport chain, which yields a total of 129 ATP.

12. The oxidation of unsaturated fatty acids requires additional enzymatic reactions to accommodate the double bond at C9 in monounsaturated fatty acids (e.g., oleic acid) and at three-carbon intervals in polyunsaturated fatty acids (e.g., linoleic acid). When a *cis*-3,4 (β,γ) double bond is encountered after several rounds of β oxidation, enoyl-CoA isomerase converts it to a *trans*-2,3 double bond. Oxidation of a polyunsaturated fatty acid also requires a re-

action catalyzed by 2,4-dienoyl-CoA reductase, which removes a double bond at the expense of NADPH.

13. The final round of β oxidation of odd-chain fatty acids yields propionyl-CoA. This three-carbon compound is converted to succinyl-CoA, a citric acid cycle intermediate, by three reactions:
    (a) Propionyl-CoA carboxylase catalyzes an ATP-dependent carboxylation reaction that requires the coenzyme biotin and produces (*S*)-methylmalonyl-CoA.
    (b) Methylmalonyl-CoA racemase converts (*S*)-methylmalonyl-CoA to (*R*)-methylmalonyl-CoA.
    (c) Methylmalonyl-CoA mutase transforms (*R*)-methylmalonyl-CoA into succinyl-CoA in a reaction that requires the coenzyme 5′-deoxyadenosylcobalamin, which is derived from cobalamin (vitamin B_{12}).

14. The methylmalonyl-CoA mutase reaction rearranges the substrate's carbon skeleton. The reaction mechanism features homolytic cleavage of the C—Co bond in the coenzyme so that the C and Co atoms each retain one electron. Such homolytic cleavage is rare in biological systems; in biochemical reactions, bonds are usually broken by heterolytic cleavage in which one of the atoms acquires both electrons. The Co ion therefore functions as a generator of free radicals, which are essential for the reaction.

15. The peroxisome also carries out β oxidation. In animals, peroxisomes oxidize very long acyl chains (>22 carbons). These are shortened in the peroxisome, and β oxidation is then completed in the mitochondrion. Peroxisomal oxidation differs from oxidation in mitochondria in two ways:
    (a) No carnitine is required.
    (b) The first step of acyl-CoA oxidation by acyl-CoA oxidase involves transfer of electrons to O_2 with formation of hydrogen peroxide (H_2O_2) and a *trans*-2,3-enoyl-CoA. Thus, peroxisomal oxidation of fatty acids yields less energy than mitochondrial oxidation.

Ketone Bodies

16. Acetyl-CoA, in addition to undergoing oxidation via the citric acid cycle, can also undergo ketogenesis to form acetoacetate, D-β-hydroxybutyrate, and acetone. These water-soluble compounds are collectively called ketone bodies. Acetoacetate and D-β-hydroxybutyrate are important sources of metabolic energy under certain circumstances.

17. Ketogenesis, the formation of ketone bodies from acetyl CoA, occurs as follows:
    (a) Two acetyl-CoA units condense to form acetoacetyl-CoA in a reversal of the thiolase reaction.
    (b) Hydroxymethylglutaryl-CoA synthase catalyzes the condensation of acetoacetyl-CoA with a third acetyl-CoA unit to form β-hydroxy-β-methylglutaryl-CoA (HMG-CoA).
    (c) HMG-CoA lyase cleaves HMG-CoA to form acetyl-CoA and acetoacetate.
    (d) β-Hydroxybutyrate dehydrogenase can reduce acetoacetate to β-hydroxybutyrate in an NADH-dependent reaction.
    (e) Acetoacetate also undergoes spontaneous decarboxylation to form acetone.

18. Acetoacetate and β-hydroxybutyrate formed in the liver are transported in the bloodstream to peripheral tissues and there are converted into two acetyl-CoA units. Succinyl-CoA supplies the CoA for this process and is therefore not utilized in the citric acid cycle to form GTP and succinate.

Fatty Acid Biosynthesis

19. Although fatty acid biosynthesis involves the condensation of successive acetyl-CoA units, this metabolic pathway is distinct from β oxidation in several respects:
    (a) It is a reductive process.
    (b) It takes place in the cytosol.
    (c) It utilizes NADPH as hydrogen donor.
    (d) It uses the C_3 dicarboxylic acid derivative, malonyl-CoA, as its C_2 donor.
    (e) The growing acyl chain is attached to acyl-carrier protein (ACP) rather than to CoA.
    (f) It employs entirely different enzymes and is independently regulated.

20. In order for fatty acid biosynthesis to proceed, sufficient amounts of both acetyl-CoA and NADPH must be available in the cytosol. Acetyl-CoA, which is generated by pyruvate dehydrogenase in the mitochondrion, cannot cross the inner mitochondrial membrane to reach the cytosol. Instead, transport occurs by means of the tricarboxylate transport system in which acetyl-CoA reacts with oxaloacetate to form citrate, which readily crosses the membrane via a transporter. Once in the cytosol, citrate is converted to pyruvate in a series of reactions that liberates acetyl-CoA and also generates NADPH in a 1:1 ratio. Pyruvate then enters the mitochondrion and is converted to oxaloacetate.

21. The first committed step in fatty acid biosynthesis is catalyzed by acetyl-CoA carboxylase, a biotin-dependent enzyme, which converts acetyl-CoA to malonyl-CoA. In the first reaction step, bicarbonate becomes covalently attached to the biotin prosthetic group. This "activated" CO_2 is then transferred from biotin to acetyl-CoA, forming malonyl-CoA.

22. In eukaryotic cells, acetyl-CoA carboxylase is regulated both allosterically and by covalent modification. Citrate allosterically increases the V_{max} of the enzyme, whereas long-chain acyl-CoAs inhibit the reaction. AMP-dependent kinase phosphorylates Ser 79, thereby inactivating the enzyme. Glucagon acts through a cAMP-dependent pathway to inhibit the enzyme, possibly by preventing its dephosphorylation. In contrast, insulin enhances enzyme activity by promoting its dephosphorylation.

23. Fatty acid synthesis from acetyl-CoA and malonyl-CoA takes place by successive cycles of six enzymatic reactions. In *E. coli,* these reactions are catalyzed by separate enzymes. In eukaryotic cells, the process occurs within fatty acid synthase, a protein whose sequence contains all the required enzyme activities and an ACP domain. Fatty acid synthase catalyzes the following reactions:
    (a) In the first two reactions, the synthase is primed by transferring an acetyl group from acetyl-CoA to a Cys SH group and by transferring a malonyl group from malonyl-CoA to the terminal SH of the phosphopantetheine prosthetic group of ACP.
    (b) In the third reaction, malonyl-ACP is decarboxylated and condenses with the acetyl group to form a β-ketoacyl-ACP.

(c) In reactions 4 to 6, the β-keto group is reduced by NADPH to form a hydroxyl group, the hydroxyl group is eliminated in a dehydration reaction to form a double bond, and a second reduction by NADPH reduces the double bond to produce an alkyl group.

(d) The four-carbon acyl chain is then transferred from the ACP to the enzyme Cys SH. The cycle starts anew by the transfer of another malonyl group to the now vacant ACP site in the synthase.

Seven cycles are required to synthesize a C_{16} acyl chain (palmitate), a process that consumes 8 acetyl-CoA and 14 NADPH (which are provided by glucose oxidation via the pentose phosphate pathway). The completed saturated acyl chain is released by thioester cleavage catalyzed by the seventh enzyme activity associated with the synthase.

24. Mammalian fatty acid synthase consists of two identical monomers in a head-to-tail association such that the Cys-linked acyl group in one polypeptide is located close to the phosphopantetheine moiety in the other. This allows the synthase dimer to simultaneously synthesize two acyl chains.

25. Palmitate may be lengthened and desaturated by elongases and desaturases, respectively. However, because animals, unlike plants, cannot introduce a double bond beyond C9 in an acyl chain, linoleic acid (9,12-*cis*-octadecadienoic acid) must be obtained from the diet. It is therefore called an essential fatty acid.

26. The synthesis of triacylglycerols begins with successive acylations of glycerol-3-phosphate to yield lysophosphatidic acid and then phosphatidic acid. Dephosphorylation produces 1,2-diacylglycerol, which then accepts another fatty acyl group. An alternative pathway involves acylation of dihydroxyacetone phosphate, followed by an NADPH-dependent reduction to produce lysophosphatidic acid.

Regulation of Fatty Acid Metabolism

27. Triacylglycerol metabolism, like that of glycogen, is important for the well-being of the whole organism and is regulated by hormones. Fatty acid oxidation is controlled by the rate of triacylglycerol hydrolysis in adipose tissue by hormone-sensitive triacylglycerol lipase. The lipase is stimulated by glucagon through cAMP-dependent phosphorylation, which also inhibits acetyl-CoA carboxylase, so fatty acid oxidation is enhanced and fatty acid synthesis is inhibited.

28. Insulin opposes the effects of glucagon by reducing cAMP levels, thereby inactivating triacylglycerol lipase and stimulating fatty acid synthesis. The ratio of glucagon to insulin therefore controls the status of fatty acid metabolism.

29. These mechanisms of short-term regulation are complemented by long-term hormonal regulation of fatty acid metabolism, which alters the levels of key enzymes such as acetyl-CoA carboxylase and triacylglycerol lipase.

Membrane Lipid Synthesis

30. The formation of choline- and ethanolamine-containing glycerophospholipids involves three enzymatic steps:
    (a) The phosphorylation of the nitrogen-containing base.
    (b) Activation of the phosphocholine or phosphoethanolamine by CTP to form CDP–choline or CDP–ethanolamine.
    (c) Transfer of the activated base to 1,2-diacylglycerol to form the phospholipid.

31. The synthesis of phosphatidylglycerol and phosphatidylinositol involves activation of phosphatidic acid by reaction with CTP to form CDP–diacylglycerol. The activated lipid is then transferred to glycerol-3-phosphate or inositol. Cardiolipin is synthesized from two phosphatidylglycerol.

32. Enzymes that acylate glycerol-3-phosphate show a preference for introducing a saturated fatty acid in position 1 and an unsaturated fatty acid in position 2. However, there must be additional reactions, catalyzed by phospholipases and acyltransferases, that result in the exchange of acyl groups, to account for the fatty acid compositions of all membrane phospholipids.

33. In sphingolipid synthesis, ceramide (*N*-acylsphingosine) is formed in four reactions from palmitoyl-CoA and serine. Phosphatidylcholine donates its phosphocholine group to ceramide to produce sphingomyelin. Ceramide can also be glycosylated, with UDP–glucose or UDP–galactose serving as the sugar donor, to form cerebrosides. Additional glycosylation of cerebrosides generates more complex sphingoglycolipids such as globosides and gangliosides.

Cholesterol Metabolism

34. All 27 carbon atoms of cholesterol are derived from acetate. The major stages in cholesterol formation are:
    (a) Acetate is converted to hydroxymethylglutaryl-CoA (HMG-CoA) and then via mevalonate to an isoprene unit, isopentenyl pyrophosphate.
    (b) Condensation of six isoprene units forms squalene, a linear 30-carbon compound.
    (c) Squalene is oxidized and cyclized to form lanosterol.
    (d) Further modification and removal of three carbons yields cholesterol.

35. After its synthesis in the liver, cholesterol may be transformed into bile acids or converted to cholesteryl esters. Both endogenously synthesized and dietary cholesterol are esterified, packaged in VLDL, and transported through the bloodstream. Peripheral tissues take up LDL (which are derived from VLDL) by receptor-mediated endocytosis, after which the acyl groups of cholesteryl esters are removed by hydrolysis, yielding free cholesterol. The cholesterol may become a cell membrane constituent, it may be re-esterified for intracellular storage, or it may be transported to the liver by HDL.

36. Cholesterol is essential for cell membrane integrity, yet excess cholesterol may be harmful to the organism. Thus, its biosynthesis, utilization, and cellular distribution are carefully controlled. Cholesterol metabolism is regulated in two ways:
    (a) Cholesterol biosynthesis is controlled by HMG-CoA reductase, which catalyzes the rate-limiting conversion of HMG-CoA to mevalonate. In the short term, this enzyme

is inactivated by phosphorylation (catalyzed by AMP-dependent kinase, the same enzyme that inactivates acetyl-CoA carboxylase) and activated by dephosphorylation. Long-term regulation involves changes in the level of enzymes in inverse proportion to the concentrations of mevalonate and cholesterol-containing LDL.

(b) Cholesterol transport and removal from blood is governed largely by the activity of LDL receptors on the liver cell surface, which in turn depends on the number of LDL receptors and hence on the rate of receptor synthesis.

37. High blood cholesterol results from the genetic disease familial hypercholesterolemia (which is characterized by an absence of LDL receptors) or by high dietary cholesterol intake (which tends to repress LDL receptor synthesis).

Guide to Study Exercises (text p. 609)

1. Bile acids (also called bile salts), which are amphipathic molecules, help solubilize dietary lipids so that they become accessible to intestinal lipases. Micelles containing bile acids also take up lipid-soluble vitamins and the products of lipid digestion so that these hydrophobic molecules can pass through the aqueous solution to reach the intestinal cell surface, where they are absorbed. (Section 19-1A)

2. Lipoproteins package hydrophobic and amphipathic lipids into water-soluble particles for transport in the bloodstream. Chylomicrons transport dietary lipids from the intestine through the lymphatic system to the bloodstream. They deliver triacylglycerols to skeletal muscle and adipose tissue, and cholesterol to the liver. The liver synthesizes VLDL particles, which also contain triacylglycerols and cholesterol. The triacylglycerols are degraded by lipoprotein lipase in the capillaries so that free fatty acids can be absorbed by cells. As it gives up its triacylglycerols and becomes smaller and denser, a VLDL becomes an IDL and then an LDL before being taken up by the liver. HDL particles transport cholesterol from the tissues to the liver. They are assembled in the plasma and contain mostly cholesteryl esters, which they transfer to VLDL. (Section 19-1B)

3. Fatty acids are first activated by linking them to CoA via a "high-energy" thioester bond, whose formation consumes the free energy of one phosphoanhydride bond of ATP. After transport into the mitochondrion, the acyl-CoA is degraded, two carbons at a time, in a process called β oxidation. An α,β double bond forms by dehydrogenation, and the electrons are transferred to the mitochondrial electron-transport chain. Next, water is added across the double bond to form a 3-hydroxyacyl-CoA. NAD^+-dependent dehydrogenation then yields a β-ketoacyl-CoA and NADH. Finally, the C_α—C_β bond is cleaved by attack of a second CoA, eliminating acetyl-CoA and producing an acyl-CoA two carbons shorter than the original substrate. These last four reactions are repeated until the entire acyl chain has been degraded to acetyl units. (Section 19-2)

4. Unsaturated fatty acids are degraded by the β oxidation enzymes until their double bonds prevent their binding to these enzymes. The acyl groups must then be enzymatically transformed to substrates of the β oxidation enzymes. When a β,γ double bond is encountered, it is isomerized to an α,β double bond, which is a substrate for enoyl-CoA hydratase. When a 2,4-dienoyl-CoA (with two consecutive double bonds) is encountered, the γ,δ double bond is eliminated by NADPH-dependent reduction.

Fatty acids that contain an odd number of carbon atoms yield the C_3 derivative propionyl-CoA in the final round of β oxidation. The propionyl group is carboxylated to a C_4 methylmalonyl group in an ATP-dependent reaction. Next, the configuration of the methylmalonyl group is inverted by a racemase, and then the carbon skeleton is rearranged to a succinyl group by methylmalonyl-CoA mutase in a reaction that requires coenzyme B_{12}. (Sections 19-2D and E)

5. Ketone bodies are synthesized from acetyl-CoA in liver mitochondria. First, 2 acetyl-CoA are condensed, in a reaction that is the reversal of the last step of β oxidation, to acetoacetyl-CoA. A third acetyl-CoA is added, generating the C_6 group-containing β-hydroxy-β-methylglutaryl-CoA (HMG-CoA). This compound is degraded by HMG-CoA lyase to produce the ketone body acetoacetate and acetyl-CoA. Acetoacetate can be reduced by NADH to produce the ketone body β-hydroxybutyrate, or it may be nonenzymatically decarboxylated to acetone + CO_2.

   Acetoacetate and β-hydroxybutyrate travel from the liver to tissues to be used as alternative fuels to glucose. β-Hydroxybutyrate is oxidized by NAD^+ to produce acetoacetate. The acetoacetate is then linked to CoA donated by succinyl-CoA. A free CoA group then attacks the acetoacetyl-CoA to produce 2 acetyl-CoA. (Section 19-3)

6. A fatty acid, in the form of acyl-CoA, is transported into the mitochondrion by a shuttle system. The acyl portion of acyl-CoA is transferred to carnitine, and the acyl-carnitine is transported across the inner mitochondrial membrane by a carrier protein. In the matrix, the acyl group is transferred back to CoA from the mitochondrial pool of CoA to produce the original acyl-CoA, and carnitine returns to the cytosol.

   Acetyl-CoA produced in the matrix is shuttled back to the cytosol, where it can be used for fatty acid synthesis. First, acetyl-CoA reacts with mitochondrial oxaloacetate to produce citrate (the first step of the citric acid cycle). Citrate is transported across the inner mitochondrial membrane by a transport protein. In the cytosol, ATP-citrate lyase catalyzes the conversion of citrate back to oxaloacetate and acetyl-CoA, in a reaction that consumes ATP. The shuttle system is regenerated when cytosolic oxaloacetate is reduced to malate and then oxidatively decarboxylated to pyruvate (this reaction sequence has the net effect of converting a reducing equivalent from NADH to NADPH, the form required for fatty acid synthesis). After transport back into the matrix, the pyruvate is carboxylated to regenerate oxaloacetate. (Sections 19-2B and 4A)

7. Fatty acid oxidation and synthesis are both four-step cyclic pathways that proceed in increments of C_2 units, and many of their intermediates are chemically similar. The differences are: (a) Oxidation occurs in the mitochondrion and synthesis in the cytosol; (b) the acyl group is linked to CoA for oxidation and to ACP for synthesis; (c) FAD and NAD^+ are electron acceptors in oxidation, whereas NADPH is the electron donor in synthesis; and (d) synthesis but not oxidation involves a C_3 unit, malonyl-CoA (this is critical for the regulation of fatty acid synthesis). (Section 19-4 and Figure 19-19)

8. Fatty acid metabolism is regulated by controlling the availability of fatty acids and by controlling the activity of acetyl-CoA carboxylase. The concentration of fatty acids in the blood depends on the rate of triacylglycerol hydrolysis by hormone-sensitive lipase. Glucagon, epinephrine, and norepinephrine (which signal low glucose levels) lead to activation of the lipase and thereby increase the supply of fatty acids for energy production by β oxidation.

At the same time, these hormones, via the cAPK pathway, inactivate acetyl-CoA carboxylase, which prevents fatty acid synthesis from acetyl-CoA. Insulin, which reflects high glucose levels, reverses the effects of the cAPK pathway. This results in inhibition of hormone-sensitive lipase (which decrease the mobilization of fatty acids) and activates acetyl-CoA carboxylase (which promotes fatty acid synthesis).

Long-term regulation of fatty acid metabolism is accomplished through changes in enzyme concentrations. Insulin stimulates the synthesis of acetyl-CoA carboxylase and fatty acid synthase, whereas starvation inhibits the synthesis of these enzymes. (Section 19-5)

9. Triacylglycerols can be synthesized from acyl-CoA and glycerol-3-phosphate to produce first a lysophosphatidic acid. Triacylglycerol synthesis that begins with dihydroxyacetone phosphate rather than glycerol-3-phosphate requires a reduction step. Addition of a second acyl group to lysophosphatidic acid and removal of the phosphate group yields diacylglycerol. Addition of a third acyl group yields a triacylglycerol.

   Diacylglycerols can be converted to glycerophospholipids in two ways. Phosphatidylethanolamine and phosphatidylcholine are synthesized by adding the respective head group, in the form of a CDP adduct, to 1,2-diacylglycerol. Phosphatidylserine is produced by exchanging the ethanolamine group of phosphatidylethanolamine for serine. In phosphatidylinositol and phosphatidylglycerol synthesis, the diacylglycerol group is activated by linkage to CDP before being linked to inositol or glycerol-3-phosphate.

   Sphingolipid synthesis begins with the condensation of palmitoyl-CoA with serine to yield 3-ketosphinganine. Reduction and transfer of an acyl group from acyl-CoA produce dihydroceramide, which is then oxidized to produce the double bond in ceramide. A head group (phosphocholine in sphingomyelin or a glycosyl unit in cerebrosides) is then added. Additional glycosylation yields the elaborate head groups of gangliosides and globosides. (Sections 19-4E and 19-6)

10. Cholesterol is synthesized entirely from acetyl-CoA. HMG-CoA, formed from acetyl-CoA by cytosolic counterparts of the mitochondrial enzymes that synthesize ketone bodies, is converted to the isoprenoid derivative isopentenyl pyrophosphate in a series of four reactions that consume 2 NADPH and 3 ATP and include a decarboxylation. Two C_5 isopentenyl pyrophosphate (one isomerized to dimethylallyl pyrophosphate) condense to form the C_{10} compound geranyl phosphate. Addition of another isopentenyl group yields the C_{15} compound farnesyl pyrophosphate. Two of these molecules condense to form the linear C_{30} hydrocarbon squalene. Squalene epoxidase oxidizes squalene to produce an epoxide, then squalene oxidocyclase catalyzes a series of cyclizations that convert the epoxide to the four-ring lanosterol. Further oxidation and the elimination of three carbons, in 19 steps, yields cholesterol. (Section 19-7A)

11. Serum cholesterol occurs in the form of lipoproteins. These are removed from the circulation via receptor-mediated endocytosis when a lipoprotein binds to the LDL receptor in liver cells. Consequently, the number of LDL receptors and therefore the rate of uptake of lipoproteins regulates the level of circulating cholesterol. (Section 19-7C)

Questions

1. From a chemical perspective, why is the energy content of fats so much greater than that of carbohydrates or proteins?

Lipid Digestion, Absorption, and Transport

2. Why do individuals who have their gall bladders removed sometimes encounter difficulties in digesting fats?

3. How do pancreatic lipase and phospholipase A_2 differ in the mechanism by which they promote hydrolysis of glycerolipids at interfaces?

4. Match each term on the left with its description on the right.

______ Bile acid
______ Intestinal fatty acid–binding protein
______ Albumin
______ Phospholipase A_2
______ Colipase
______ Chylomicrons

A. Helps bind lipase to the lipid–water interface
B. Hydrolyzes phospholipids to yield lysophospholipids and free fatty acids
C. Forms micelles that take up nonpolar lipid degradation products and transports them through the intestinal wall
D. Transports lipid digestion products through the lymphatic system and then the bloodstream to the tissues
E. Transports through the bloodstream free fatty acids released from adipose tissue stores
F. Forms complexes with free fatty acids to shield intestinal cells from their detergent-like effects

5. Determine the order of the following events involving lipoprotein-mediated transport of dietary triacylglycerols and cholesterol.

______ Triacylglycerols are removed from circulating VLDL by lipoprotein lipase.
______ Chylomicrons are transported through the lymphatic system and enter the bloodstream.
______ Chylomicrons are formed in the intestinal mucosa.
______ Cells take up cholesterol via receptor-mediated LDL endocytosis.
______ Chylomicrons are degraded by lipoprotein lipase to chylomicron remnants.
______ LDL components are rapidly degraded by lysosomal enzymes.
______ VLDL are synthesized in liver.

Fatty Acid Oxidation

6. What products would be isolated from urine when dogs are fed (a) phenylheptanoic acid and (b) phenyloctanoic acid? How does this experiment, originally performed by Knoop, shed light on the process of fatty acid oxidation?

7. A patient develops an enlarged fatty liver and low blood glucose. These symptoms can be partially overcome only when massive amounts of carnitine are included in the diet. What enzyme defect might be responsible for this problem?

8. Which of the following statements is(are) correct? In the β oxidation of fatty acids,
   (a) The activation of fatty acids by acyl-CoA synthetase is driven by the hydrolysis of pyrophosphate.
   (b) The reaction catalyzed by acyl-CoA dehydrogenase uses FAD as electron acceptor.
   (c) The reaction catalyzed by enoyl-CoA hydratase produces a 3-D-hydroxyacyl-CoA.

9. Calculate the yield of ATP when one mole of stearic acid is completely oxidized to CO_2 and H_2O.

10. In the β oxidation of oleic acid, the presence of the double bond at the ________ position necessitates modification of the β oxidation cycle. The modification occurs after the ______ round of β oxidation because the ________ enoyl-CoA is not a substrate for _______. The problem is overcome by the action of the enzyme ____________ which converts the ________ bond to a __________ bond so that β oxidation can continue.

11. In the β oxidation of linoleic acid, a second difficulty presents itself because of the additional double bond in the ________ position. In this case, after the fifth round of oxidation, a ________ CoA is formed, which is a poor substrate for _______. In mammals, two reactions are necessary for β oxidation to resume. In the first, NADPH-dependent _________ reductase reduces one double bond to form a __________, and then the action of a _______ isomerase yields a ________ CoA that can participate in β oxidation.

12. Describe the role of cobalamin (vitamin B_{12}) in the methylmalonyl-CoA mutase reaction.

13. Explain why succinyl-CoA arising from the oxidation of odd-chain fatty acids cannot be directly oxidized by the citric acid cycle. How can it be further degraded?

14. Why does the β oxidation of fatty acids in peroxisomes yield less ATP than the corresponding process in mitochondria?

Ketone Bodies

15. Infants have high levels of ketone bodies in their blood and abundant 3-ketoacyl-CoA transferase in their tissues (except in liver) prior to weaning. What nutritional advantage does this confer?

16. Calculate the ATP that is produced when linoleic acid (9,12-octadecadienoic acid; 18:2) is (a) oxidized to CO_2 and H_2O or (b) converted to the ketone body acetoacetate in the liver and then oxidized to CO_2 and H_2O in the peripheral tissues.

17. Write equations for ketone body synthesis and degradation. What is the net result of combined synthesis and degradation?

Fatty Acid Biosynthesis

18. A rat liver cytosol preparation is incubated with acetate and all cofactors necessary for the biosynthesis of palmitate. Which carbon atoms of palmitate will be isotopically labeled when the preparation contains (a) $H^{14}CO_3^-$ or (b) $^{14}CH_3COO^-$?

19. What are the advantages of the multifunctional "head-to-tail" dimer structure of animal fatty acid synthase in the formation of long-chain fatty acids?

20. Which of the following statements is(are) correct? The transport of acetyl-CoA from the mitochondrial matrix to the cytosol for fatty acid biosynthesis:
    (a) is necessary because the inner mitochondrial membrane is impermeable to acetyl-CoA.
    (b) uses the tricarboxylate transport system, which operates in both directions.
    (c) can generate equal numbers of acetyl-CoA and NADPH molecules in the cytosol.

Regulation of Fatty Acid Metabolism

21. Match each term on the left with its function in the short-term regulation of fatty acid metabolism.

______ Citrate	A. Activates acetyl-CoA carboxylase
______ cAMP-dependent phosphorylation	B. Inhibits carnitine palmitoyl transferase
______ Palmitate	C. Activates hormone-sensitive lipase
______ Malonyl-CoA	D. Inhibits acetyl-CoA carboxylase

Membrane Lipid Synthesis

22. Match each term on the left with its metabolic role on the right.

______ CDP–diacylglycerol	A. Precursor of cardiolipin
______ Ceramide	B. Precursor of phosphatidylcholine
______ Phosphatidylglycerol	C. Contains a vinyl ether linkage
______ Plasmalogen	D. Precursor of sphingomyelin
______ CDP–Choline	E. Precursor of phosphatidylinositol

23. When tissues are incubated *in vitro* with ^{32}P-labeled P_i, the incorporation of radioactivity into phospholipids can be readily demonstrated. High concentrations of the drug propranolol dramatically alter the pattern of phospholipid labeling, such that radioactivity in phosphatidylcholine and phosphatidylethanolamine markedly declines while that in phosphatidylinositol and phosphatidylglycerol is greatly elevated. From this information and your knowledge of phospholipid biosynthetic pathways, determine which enzymatic reaction is principally affected by propranolol and explain how the drug alters phospholipid biosynthesis.

24. A metabolic disease may result from an enzyme deficiency that prevents the synthesis of an essential metabolite or that prevents the normal breakdown of a metabolite. Which type of defect is exhibited in Tay–Sachs disease and what enzyme is affected?

Cholesterol Metabolism

25. What are the two principal ways in which the cholesterol needs of many tissues are met?

26. Place the following intermediates in cholesterol biosynthesis in the correct order: squalene, farnesyl pyrophosphate, 2,3-oxidosqualene, hydroxymethylglutaryl-CoA (HMG-CoA), lanosterol, mevalonate, geranyl pyrophosphate.

27. Explain how the regulation of HMG-CoA reductase, the principal control site for cholesterol synthesis, can conserve cellular ATP.

28. Which of the following statements is(are) correct? Inhibitors of HMG-CoA reductase are used to decrease serum cholesterol levels. Such inhibitors would also:
    (a) reduce the intracellular level of mevalonate.
    (b) reduce the synthesis of LDL receptors.
    (c) reduce the synthesis of ubiquinone.

Answers to Questions

1. The energy content of fats is greater because the carbon atoms of fatty acids are more reduced and therefore release more free energy upon oxidation than the carbons of carbohydrates or proteins. In addition, fats can be stored in large amounts in anhydrous form. In contrast, proteins cannot be stored in large amounts, and carbohydrates require a large volume for hydration.

2. The gall bladder stores and secretes bile acids, which are essential for emulsifying triacylglycerols and thereby promoting their hydrolysis. In addition, bile acids and released fatty acids are components of mixed micelles that are absorbed across the intestinal wall.

3. Both pancreatic lipase and phospholipase A_2 hydrolyze their substrates at the lipid–water interface, where the enzymes undergo interfacial activation. Pancreatic lipase binds to the interface only when complexed with pancreatic colipase. In the presence of a lipid micelle, the lipase undergoes a structural change that exposes the active site, forms an oxyanion hole, and, in conjunction with colipase, creates a large hydrophobic surface near the active site that helps to bind the lipase–colipase complex to the lipid.
   Phospholipase A_2 does not alter its conformation but instead has a hydrophobic channel that enables a micellar phospholipid to gain access to the enzyme active site without having to pass through the aqueous phase.

4.
   C Bile acid
   F Intestinal fatty acid–binding protein
   E Albumin
   B Phospholipase A_2
   A Colipase
   D Chylomicrons

5.
   5 Triacylglycerols are removed from circulating VLDL by lipoprotein lipase.
   2 Chylomicrons are transported through the lymphatic system and enter the bloodstream.
   1 Chylomicrons are formed in the intestinal mucosa.
   6 Cells take up cholesterol via receptor-mediated LDL endocytosis.
   3 Chylomicrons are degraded by lipoprotein lipase to chylomicron remnants.
   7 LDL components are rapidly degraded by lysosomal enzymes.
   4 VLDL are synthesized in liver.

6. (a) Hippuric acid (a benzoic acid derivative).
   (b) Phenylaceturic acid (a phenylacetic acid derivative). This experiment suggested that fatty acids are degraded by oxidation and removal of successive two-carbon fragments.

7. The patient probably has a carnitine palmitoyl transferase I deficiency. As a result, acyl-CoA transport across the mitochondrial membrane is inadequate. Fatty acids released from adipose tissue stores would accumulate in the liver. Glucose levels would fall because it would be the primary metabolic fuel in the absence of oxidizable fatty acids. Treatment with high doses of carnitine would elevate tissue carnitine levels and hence enable some acyl-carnitine to form and enter the mitochondrion so that the acyl group could be oxidized by β oxidation.

8. (a) and (b) are correct. (c) is incorrect because the product is 3-L-hydroxyacyl-CoA.

9. 146 ATP. Stearic acid (an 18:0 fatty acid) is first activated by conversion to stearoyl-CoA, with the consumption of 2 equivalents of ATP. Stearoyl-CoA is then degraded by eight rounds of β oxidation to form 9 acetyl-CoA, 8 $FADH_2$, and 8 NADH. Oxidation of each acetyl-CoA by the citric acid cycle yields GTP, NADH, and $FADH_2$. Reoxidation of the $FADH_2$ and NADH, with the transfer of electrons to O_2 to form H_2O, yields ATP by oxidative phosphorylation. The stoichiometry of ATP production can be summarized as follows:

Process	*Reduced coenzymes formed*	*ATPs formed*
Fatty acid activation		−2
8 rounds of β oxidation	8 $FADH_2$	16
	8 NADH	24
9 rounds of citric acid cycle		9 (GTP)
	27 NADH	81
	9 $FADH_2$	18
Total		146

10. The missing terms are: cis Δ^9; third; cis Δ^3; enoyl-CoA hydratase; enoyl-CoA isomerase; cis Δ^3; trans Δ^2.

11. The missing terms are: cis Δ^{12}; 2,4-dienoyl-; enoyl-CoA hydratase; 2,4-dienoyl-CoA; *trans*-Δ^3-enoyl-CoA; 3,2-enoyl-CoA; *trans*-Δ^2-enoyl.

12. Cobalamin is essential for converting (*R*)-methylmalonyl-CoA to succinyl-CoA. The cobalt alternates between the Co(III) and Co(II) states and therefore functions as a free radical generator. The weak carbon—cobalt(III) bond of the coenzyme undergoes homolytic cleavage such that the carbon and cobalt each acquire one electron. This yields a deoxyadenosyl radical that can abstract a hydrogen atom from methylmalonyl-CoA, which then rearranges to form succinyl-CoA.

13. Since citric acid cycle intermediates function as catalysts and are continuously regenerated, the net oxidation of succinyl-CoA can occur only if it is converted to another compound that is oxidized by the operation of the cycle, such as pyruvate or acetyl-CoA. This is accomplished by transforming succinyl-CoA to malate and then transporting the malate to the cytosol, where it is oxidatively decarboxylated to pyruvate and CO_2 in a reaction catalyzed by malic enzyme. Pyruvate can then be transported into the mitochondrion and oxidized via pyruvate dehydrogenase and the citric acid cycle.

14. The first step in β oxidation in peroxisomes is catalyzed by acyl-CoA oxidase rather than acyl-CoA dehydrogenase as occurs in mitochondria. In the acyl-CoA oxidase reaction, electrons from acyl-CoA are transferred directly to O_2 to form H_2O_2 and therefore do not pass through the electron-transport chain with concomitant formation of ATP. For this reason, peroxisomal β oxidation yields less ATP than mitochondrial β oxidation.

15. The large intake of milk means that fat supplies a major portion of the calories consumed by infants prior to weaning. The oxidation of fatty acids produces abundant acetyl-CoA which is partly converted to ketone bodies in the liver. The ketone bodies, like glucose, are water-soluble and easily transported. They are therefore readily available as metabolic fuels to support the rapid growth and development of the infant.

16. (a) The net production of ATP from the complete oxidation of linoleic acid can be calculated as in Problem 9. The total yield is 141 ATP. This is 5 less than for the oxidation of stearate because the first double bond of linoleate does not need an FAD to reduce it ($FADH_2$ is equivalent to 2 ATP) and the reduction of its second double bond consumes an NADPH (which is equivalent to ~3 ATP).
    (b) If the 9 acetyl-CoA generated by the β oxidation of linoleic acid were instead converted to ketone bodies, 4.5 molecules of acetoacetate would be formed. The transformation of acetoacetate back into acetyl-CoA does not directly consume ATP. However, it involves the consumption of 4.5 succinyl-CoA, which would otherwise provide the energy for a substrate level phosphorylation. Therefore, the net yield of ATP for this metabolic route can be considered to be 141 − 4.5 = 136.5 ATP.

17. *Ketogenesis:*

$$2\ \text{Acetyl-CoA} + H_2O \rightarrow \text{acetoacetate} + 2\ \text{CoASH}$$

Ketone body breakdown:

$$\text{Acetoacetate} + \text{succinyl-CoA} + \text{CoASH} \rightarrow 2\ \text{acetyl-CoA} + \text{succinate}$$

Combined synthesis and breakdown:

$$\text{Succinyl-CoA} + H_2O \rightarrow \text{succinate} + \text{CoASH}$$

18. (a) $H^{14}CO_3^-$ is incorporated into malonyl-CoA by the action of acetyl-CoA carboxylase, but the labeled carboxyl group is subsequently eliminated as $^{14}CO_2$ during condensation of malonyl-ACP with the growing fatty acyl chain. Therefore, the biosynthesized palmitate will be unlabeled.

(b) The radioactive methyl carbon in acetate will label each even-numbered carbon atom in the completed fatty acid.

19. The proximity of multiple enzyme activities involved in fatty acid synthesis on a single polypeptide chain may enhance the efficiency of the process. Moreover, the head-to-tail orientation of the two synthase subunits may enable two fatty acyl chains to be synthesized simultaneously: At each end of the dimer, an acyl group attached to the enzyme–Cys residue on one subunit is extended by addition of an acetyl group from malonyl-ACP on the other subunit.

20. (a) and (c) are correct. (b) is incorrect because the tricarboxylate transport system is irreversible, that is, unidirectional. This is because both the ATP-citrate lyase reaction (in the cytosol) and the pyruvate carboxylase reaction (in the matrix) involve ATP hydrolysis. Thus, the net result of one turn of the cycle is the hydrolysis of 2 ATP to 2 ADP + 2 P_i and the transport of acetyl-CoA from the matrix to the cytosol (assuming that the NADH and NADPH in the malate dehydrogenase and malic enzyme reactions are equivalent).

21. __A__ Citrate
__C__ cAMP-dependent phosphorylation
__D__ Palmitate
__B__ Malonyl-CoA

22. __E__ CDP–diacylglycerol
__D__ Ceramide
__A__ Phosphatidylglycerol
__C__ Plasmalogen
__B__ CDP–Choline

23. Propranolol primarily inhibits phosphatidic acid phosphatase, which converts phosphatidic acid to 1,2-diacylglycerol. Since diacylglycerol is an intermediate in the synthesis of phosphatidylcholine and phosphatidylethanolamine, propranolol decreases the production of these lipids. Inhibition of phosphatidic acid phosphatase also increases the concentration of phosphatidic acid, which is then converted in greater amounts to phosphatidylinositol and phosphatidylglycerol.

24. Tay–Sachs disease results from a defect in the enzyme hexosaminidase A, which breaks down ganglioside G_{M2} to ganglioside G_{M3}.

25. Cholesterol can be obtained either by synthesis *de novo* from acetyl-CoA or from circulating lipoproteins, primarily LDL, that enter cells by receptor-mediated endocytosis.

26. The order is: HMG-CoA, mevalonate, geranyl pyrophosphate, farnesyl pyrophosphate, squalene, 2,3-oxidosqualene, lanosterol.

27. HMG-CoA reductase catalyzes the first unique step in the synthesis of cholesterol, the conversion of HMG-CoA to mevalonate. This reaction is followed by three reactions that consume ATP. By regulating the activity of HMG-CoA reductase so that it operates only when cholesterol is needed, the cell avoids wasting metabolic energy on the production of cholesterol precursors.

28. (a) and (c) are correct. (b) is incorrect because inhibition of cholesterol biosynthesis tends to increase the synthesis of LDL receptors.

Chapter 20 Amino Acid Metabolism

This chapter surveys nitrogen metabolism, with its key focus on amino acid metabolism. The chapter begins with a brief discussion of protein turnover and pathways of protein degradation, including the lysosomal and ubiquitin-dependent pathways. The following sections describe the mechanisms by which amino acids are catabolized, beginning with amino acid deamination by transamination and oxidative deamination. Next is the urea cycle, a pathway that transfers an ammonium ion and the amino group of aspartate through a series of intermediates to arginine, which is cleaved to generate urea and regenerate the amino acid ornithine. The chapter then discusses the degradation of amino acids, which yield common intermediates that can be used for gluconeogenesis (from glucogenic amino acids) or fatty acid synthesis (from ketogenic amino acids). Many of these degradative pathways involve enzymatic reactions mediated by the coenzyme pyridoxal-5′-phosphate (PLP), which catalyzes transamination reactions, decarboxylation reactions, and reactions that break the C_α—C_β covalent bond of amino acids. This coenzyme, like thiamine pyrophosphate, is a resonance-stabilized electron sink that stabilizes carbanions. Genetically inherited defects in amino acid metabolism are discussed in Box 20-1.

The sections in the final part of the chapter include the anabolic reactions of amino acid metabolism. Amino acid biosynthetic reactions are divided into two groups, those synthesizing the nonessential (for mammals) and essential amino acids. Within each group, the derivation of amino acids from common precursors is emphasized. The chapter then examines the synthesis of heme, which is assembled in a modular fashion beginning with glycine and succinyl-CoA. Disorders of heme metabolism are discussed in the text and in Box 20-2. The following section describes the conversion of select amino acids to hormones and neurotransmitters. This chapter concludes with a discussion of how microorganisms fix N_2 into NH_3 by an energetically costly process.

Essential Concepts

Intracellular Protein Degradation

1. Proteins are continuously synthesized and degraded into amino acids. This serves three functions:
   (a) Proteins serve as a form of long-term energy storage that is utilized to provide gluconeogenic precursors and ketone bodies.
   (b) Degradation eliminates abnormal and damaged proteins, whose accumulation is potentially hazardous to cell function.
   (c) The regulation of metabolic activity under changing physiological conditions requires the degradation of one set of regulatory proteins and its replacement by a new set.

2. Proteins have highly variable half-lives, ranging from a few minutes to several days. The half-life of a protein varies with the identity of its N-terminal residue, a phenomenon which is referred to as the N-end rule. Other signals are contained in the protein's amino acid sequence. For example, proteins with segments rich in Pro (P), Glu (E), Ser (S), and Thr (T), which are called PEST sequences, have very short half-lives.

3. Intracellular proteins are degraded either in lysosomes or, after ubiquitination, in proteasomes.
   (a) Lysosomes degrade proteins (and other biological molecules) that are taken up by endocytosis, and they recycle cellular proteins that are enclosed in vacuoles. In well-nourished cells, lysosomal protein degradation is nonselective, but in starved cells, this pathway is more selective.
   (b) Intracellular proteins may be tagged for degradation by covalently linking them to a small protein called ubiquitin. Ubiquitination requires three enzymes: ubiquitin-activating enzyme (E1), ubiquitin-conjugating enzyme (E2), and ubiquitin-protein ligase (E3). E1 forms a thioester bond with ubiquitin in a reaction that consumes one ATP. E2 transfers this activated ubiquitin from E1 to itself, forming another thioester bond. E3 transfers activated ubiquitin to the Lys ε-amino group of a previously selected target protein.

4. Ubiquitinated proteins are degraded in an ATP-dependent manner inside a large multiprotein complex called the 26S proteasome. The complex has a barrel-like core (20S subunit) that degrades proteins into ~8 residue fragments, which diffuse out of the proteasome and are subsequently hydrolyzed by cytosolic peptidases.

Amino Acid Deamination

5. The degradation of an amino acid usually begins with its transamination by a pyridoxal-5′-phosphate (PLP)-containing enzyme, in which the amino group is transferred to an α-keto acid (usually α-ketoglutarate). Transamination utilizing α-ketoglutarate generates glutamate, which can be subsequently oxidatively deaminated by glutamate dehydrogenase to regenerate α-ketoglutarate and release a free ammonium ion. High concentrations of ammonia are toxic, so the steady-state level of ammonia is kept low.

6. PLP functions as an electron sink in the resonance stabilization of a C_α carbanion derived from an amino acid. This carbanion is an intermediate in amino acid transamination, decarboxylation, and the breakage of the $C_\alpha—C_\beta$ bond.

7. Glutamate dehydrogenase catalyzes the oxidative deamination of glutamate to yield α-ketoglutarate and ammonium ion.

The Urea Cycle

8. Living organisms excrete excess nitrogen either as free ammonia (aquatic organisms only), urea (most terrestrial vertebrates), or uric acid (birds and terrestrial reptiles). The overall reaction of the urea cycle is

$$NH_3 + HCO_3^- + \text{Aspartate} \xrightarrow[\;3\,ADP + 2\,P_i + AMP + PP_i\;]{3\,ATP} \text{Urea} + \text{Fumarate}$$

9. The urea cycle takes place in the liver of adult vertebrates. Of the five reactions in the urea cycle, two are mitochondrial and three are cytosolic.
   (a) Ammonia (a product of the glutamate dehydrogenase reaction) reacts with HCO_3^- and two ATP to form the "activated" product carbamoyl phosphate in a reaction catalyzed by mitochondrial carbamoyl phosphate synthetase.
   (b) Transfer of the carbamoyl group to ornithine results in the formation of citrulline, which is transported out of the mitochondrion to the cytosol.
   (c) Condensation of aspartate with citrulline in an ATP-dependent reaction forms argininosuccinate.
   (d) Argininosuccinate is cleaved into fumarate and arginine.
   (e) In the final reaction, hydrolytic cleavage of arginine yields urea and regenerates ornithine, which is transported back into the mitochondrion.

10. The urea cycle is regulated by the activity of carbamoyl phosphate synthetase I, which is allosterically activated by *N*-acetylglutamate. Increasing concentrations of glutamate, resulting from transamination during protein degradation, stimulate the synthesis of *N*-acetylglutamate, thereby boosting urea cycle activity to increase the excretion of amino groups as urea.

Breakdown of Amino Acids

11. The carbon skeletons of the 20 "standard" amino acids can be completely oxidized to CO_2 and H_2O. They can also be used to synthesize glucose or fatty acids:
   (a) Glucogenic amino acids are broken down to pyruvate, α-ketoglutarate, succinyl-CoA, fumarate, and oxaloacetate and can be used as glucose precursors.
   (b) Ketogenic amino acids can be broken down to the ketone body acetoacetate or to acetyl-CoA, which can be used for fatty acid synthesis.

12. The amino acids can be grouped according to their common degradative products.
   (a) Alanine, cysteine, glycine, serine, and threonine are degraded to pyruvate.
   (b) Asparagine and aspartate are degraded to oxaloacetate.
   (c) Arginine, glutamate, glutamine, histidine, and proline are degraded to α-ketoglutarate.
   (d) Isoleucine, methionine, and valine are degraded to succinyl-CoA.
   (e) Leucine and lysine are degraded to acetoacetate and acetyl-CoA.
   (f) Tryptophan is degraded to alanine and acetoacetate.
   (g) Phenylalanine and tyrosine are degraded to fumarate and acetoacetate.

13. Important cofactors involved in the degradation of amino acids are PLP, tetrahydrofolate (THF), biopterin, and *S*-adenosylmethionine (SAM).
   (a) PLP mediates the catalytic breakdown of the $C_\alpha—C_\beta$ covalent bond in threonine to generate acetaldehyde and glycine.
   (b) THF is composed of the following functional groups: 2-amino-4-oxo-6-methylpterin, *p*-aminobenzoic acid, and several glutamates. THF is derived from the vitamin folic acid, which must be reduced to form THF. THF serves as a one-carbon carrier in several reactions involved in amino acid catabolism. The carbon unit may exist in one of several oxidation states ranging from a relatively oxidized formyl group to a methyl group.
   (c) Biopterin is a redox cofactor involved in the oxidation of phenylalanine to tyrosine. Biopterin contains a pteridine ring that is similar to the isoalloxazine ring of the flavin coenzymes.

(d) The reaction of ATP with methionine yields SAM, a potent methylating reagent in several reactions in which the transfer of a methyl group from THF is not favored.

14. The catabolism of the branched-chain amino acids (isoleucine, leucine, and valine), lysine, and tryptophan employs enzymatic conversions resembling those in the β oxidation of fatty acids and in the oxidative decarboxylation of α-keto acids such as pyruvate and α-ketoglutarate.

Amino Acid Biosynthesis

15. Amino acids that cannot be synthesized by mammals and that must therefore be obtained from their diet are called essential amino acids. Nonessential amino acids can be synthesized from common intermediates:
    (a) Alanine, aspartate, and glutamate are formed by one-step transamination reactions. Asparagine and glutamine are formed by amidation of aspartate and glutamate. The activation of glutamate by glutamine synthetase, which occurs prior to its amidation, is a key regulatory point in bacterial nitrogen metabolism.
    (b) Glutamate gives rise to proline and arginine (which is also considered to be an essential amino acid because, when synthesized, it is largely degraded to urea).
    (c) 3-Phosphoglycerate is the precursor of serine, which can be converted to cysteine and glycine. Glycine can also be produced from CO_2, NH_4^+, and N^5,N^{10}-methylene-THF.

16. The essential amino acids can be categorized into four groups based on their synthetic pathways:
    (a) The aspartate family. Aspartate serves as the precursor for the synthesis of lysine, threonine, and methionine. This pathway also produces homoserine and homocysteine.
    (b) The pyruvate family. Pyruvate serves as a precursor for the synthesis of valine, leucine, and isoleucine.
    (c) Aromatic amino acids. Phosphoenolpyruvate and erythrose-4-phosphate serve as the precursors for tyrosine, phenylalanine, and tryptophan. Serine is also required for the synthesis of tryptophan.
    (d) Histidine. 5-Phosphoribosyl-α-pyrophosphate is a precursor for the synthesis of histidine.

Other Products of Amino Acid Metabolism

17. Heme synthesis occurs in a modular fashion similar to the synthesis of cholesterol. Heme synthesis begins in the mitochondrion with the condensation of glycine and succinyl-CoA. The product, δ-aminolevulinic acid (ALA), diffuses into the cytosol, where two ALA condense to form porphobilinogen (PBG). Four PBG then condense to form the porphyrin uroporphyrinogen III. Decarboxylation reactions form coproporphyrinogen III, which diffuses into the mitochondrion and is converted to heme.

18. Heme synthesis occurs in erythroid cells and in the liver. In the liver, heme acts by feedback inhibition to regulate ALA synthase activity. In reticulocytes (immature red blood cells), heme stimulates the synthesis of globin and possibly itself. Once heme synthesis is "switched on," it appears to occur at its maximal rate.

19. Red blood cells survive in the bloodstream for about 120 days and then are removed by the spleen and destroyed. Heme catabolism results in the formation of bilirubin, which is trans-

ported to the gall bladder for secretion into the intestinal tract. Bilirubin is converted by microbes in the colon into urobilinogen then into stercobilin, which gives the feces their red-brown pigment. Some urobilinogen is reabsorbed and converted in the kidneys to urobilin, which gives urine its characteristic yellow color.

20. Several hormones and neurotransmitters are derived from amino acids. PLP-mediated decarboxylation forms histamine from histidine and γ-aminobutyric acid (GABA) from glutamate. Decarboxylation and hydroxylation of tryptophan yield serotonin. Hydroxylated tyrosine is the precursor of L-DOPA and melanin. PLP-dependent decarboxylation of L-DOPA yields dopamine, which is hydroxylated to form norepinephrine. Donation of a methyl group from SAM to norepinephrine yields epinephrine.

21. Nitric oxide (NO) is produced by a five-electron oxidation of arginine. NO, a relatively stable gaseous free radical, is synthesized by vascular endothelial cells and induces vasodilation. A variety of cells, including neuronal cells and leukocytes, synthesize NO, which acts as a signaling molecule and is involved in the generation of antimicrobial hydroxyl radicals. Sustained NO release is implicated in endotoxic shock and neuronal damage.

Nitrogen Fixation

22. N_2, an extremely stable gaseous molecule, can be reduced, or fixed, by species of bacteria called diazatrophs, such as those of the genus *Rhizobium,* which live symbiotically in the root nodules of leguminous plants. Diazatrophs convert N_2 to NH_3 by the following net reaction

$$N_2 + 8\ H^+ + 8\ e^- + 16\ ATP + 16\ H_2O \rightarrow 2\ NH_3 + H_2 + 16\ ADP + 16\ P_i$$

The ammonia formed is added to α-ketoglutarate or glutamate by glutamate dehydrogenase or glutamine synthetase, respectively.

23. Nitrogenase, which catalyzes the reduction of N_2 to NH_3, is a two-subunit enzyme. Its Fe-protein contains one [4Fe–4S] cluster and two ATP-binding sites, and its MoFe-protein contains a P-cluster (a [4Fe–4S] cluster linked to a [4Fe–3S] cluster) and a FeMo cofactor (a [4Fe–3S] cluster and a [Mo–3Fe–3S] cluster bridged by 3 sulfide ions). The electrons to reduce N_2 come from oxidative or photosynthetic reactions, depending on the species. Although electron transfer must occur at least 8 times per N_2 molecule fixed (requiring a total of 16 ATP), the actual physiological cost of N_2 fixation is as high as 20–30 ATP.

Guide to Study Exercises (text p. 661)

1. Eukaryotic intracellular proteins are broken down by a pathway that begins with the covalent attachment of ubiquitin to the proteins. Ubiquitin is "activated" by the formation of a thioester bond between its terminal carboxyl group and an SH group in ubiquitin-activating enzyme (E1). E1 then transfers its ubiquitin to an SH group in a ubiquitin-conjugating enzyme (E2). Ubiquitin-protein ligase (E3) then selects a protein to be degraded, often according to its N-terminal residues, and transfers the ubiquitin from the E2–ubuiquitin complex to an ε-amino group in the protein. More than one ubiquitin may be attached to the

protein, and additional ubiquitin molecules may be added through isopeptide bonds between the Lys 48 of one ubiquitin and the C-terminus of another. The ubiquitin-tagged protein enters the interior of a proteasome, a macromolecular complex whose subunits are arranged in stacked rings to form a barrel shape. The proteasome's seven different types of β subunits are proteases with different specificities. In an ATP-dependent process, they degrade the ubiquitinated protein into ~8-residue fragments, which diffuse out of the proteasome. The ubiquitin molecules are recycled rather than degraded. (Sections 20-1B and C)

2. Transaminases remove the amino group of an amino acid with the assistance of pyridoxal-5′-phosphate (PLP), which is attached to the enzyme via a Schiff base linkage to a Lys residue. The amino group of the amino acid attacks this Schiff base, displacing the enzyme Lys group so that the PLP forms a Schiff base with the amino acid. Tautomerization of the bound amino acid (as catalyzed by the freed Lys residue) generates a ketimine that can be hydrolyzed to yield an α-keto acid and pyridoxamine phosphate (PMP; PLP with its aldehyde oxygen replaced by the amino group from the amino acid). The electron-withdrawing nature of PLP facilitates deamination by promoting removal of the amino acid's α proton for the tautomerization step. The resulting carbanion is stabilized by the resonance of the PLP group. (Section 20-2A)

3. The urea cycle consists of five enzyme-catalyzed steps in which the carbon of HCO_3^- and the nitrogens of NH_3 and aspartate are combined to form urea:
   (a) Mitochondrial carbamoyl phosphate synthetase, in a reaction that consumes two ATP, combines NH_4^+ and HCO_3^- to form carbamoyl phosphate.
   (b) Ornithine transcarbamoylase transfers the carbamoyl moiety to the amino acid ornithine to yield citrulline, which is then transported to the cytosol.
   (c) Argininosuccinate synthetase condenses citrulline with aspartate, thereby consuming two ATP equivalents.
   (d) The resulting argininosuccinate eliminates fumarate, leaving arginine, in a reaction catalyzed by argininosuccinase. The fumarate can be transformed back to aspartate, a substrate for the preceding reaction, by the citric acid cycle reactions fumarate → malate → oxaloacetate followed by a transamination reaction that converts the oxaloacetate to aspartate.
   (e) Arginase catalyzes the hydrolysis of arginine to produce urea and regenerate ornithine, which is transported back into the mitochondrion to serve as a substrate for the second step of the urea cycle. (Section 20-3A)

4. The carbon skeletons of amino acids are degraded to compounds that directly or indirectly enter the citric acid cycle for complete oxidation to CO_2 and H_2O. Amino acids whose skeletons yield pyruvate or citric acid cycle intermediates are known as glucogenic amino acids since they support the net synthesis of glucose. Amino acids that are broken down to acetyl-CoA or acetoacetate are said to be ketogenic since they can be oxidized as ketone bodies but cannot be converted to glucose. (Section 20-4 and Figure 20-11)

5. The nonessential amino acids can be synthesized from common metabolites of glycolysis and the citric acid cycle in all organisms. Pyruvate gives rise to alanine by transamination. Similarly, transamination converts oxaloacetate to aspartate, and α-ketoglutarate to glutamate; these amino acids can then be amidated to produce asparagine and glutamine. Glutamate gives rise to proline in four steps, and to ornithine, which is converted to arginine by

the reactions of the urea cycle. However, because the urea cycle requires so much arginine, this amino acid is best considered an essential amino acid. The glycolytic intermediate 3-phosphoglycerate yields serine, which can then be converted to glycine via transfer of one carbon to tetrahydrofolate (THF). Glycine can also be synthesized from NH_4^+, CO_2, and a carbon unit donated by THF. Cysteine is a product of the degradation of the essential amino acid methionine, and tyrosine is a product of phenylalanine breakdown. (Section 20-5A)

6. N_2 contains an extremely unreactive triple bond, and therefore its fixation (conversion to a biologically useful form) requires the input of considerable free energy. The electrons that reduce N_2 to 2 NH_3 are produced by oxidative metabolism or by photosynthetic electron transport. Two ATP are also required to decrease the reduction potential of the Fe-protein of nitrogenase so that it can donate an electron to N_2. Although a minimum of 6 electrons and 12 ATP are required to fix N_2, side reactions that reform N_2 from N_2H_2 boost the cost of the reaction to at least 8 electrons and 16 ATP per N_2 fixed. (Section 20-7)

Questions

Intracellular Protein Degradation

1. Indicate whether the following statements are true:
   (a) Proteins with the sequence Lys-Phe-Glu-Arg-Gln are selectively degraded by proteasomes.
   (b) Proteins containing sequences rich in Pro, Glu, Ser, and Thr often have short half-lives.
   (c) The addition of ubiquitin protects segments of a protein from proteolysis.
   (d) Lysosomal proteases degrade only extracellular proteins that enter the cell by endocytosis.
   (e) The ubiquitin-transfer reactions catalyzed by E2 and E3 do not require the input of free energy in the form of ATP.

Amino Acid Deamination

2. Draw the chemical structures for the products of transamination reactions involving α-ketoglutarate and (a) threonine, (b) isoleucine, and (c) glycine.

3. In what form is PLP found in aminotransferases prior to reacting with an amino acid? In what form is this coenzyme found after the release of the α-keto acid?

4. Aminotransferase reactions occur via a Ping Pong mechanism (see p. 334). Use Cleland notation to describe the reaction shown in Figure 20-7, including substrates, intermediates, and products.

5. In which direction would you expect flux through the glutamate dehydrogenase reaction in starved individuals? Why?

The Urea Cycle

6. Write the overall equation for the formation of urea. What is the total free energy cost (in terms of "high-energy" phosphoanhydride bonds) per mole of urea synthesized?

7. Which reactions of the urea cycle take place in the mitochondrion? Which step is the first committed step?

8. Complete the diagram below, which shows the flux of alanine's amino group from its entry into the liver to its exit as urea.

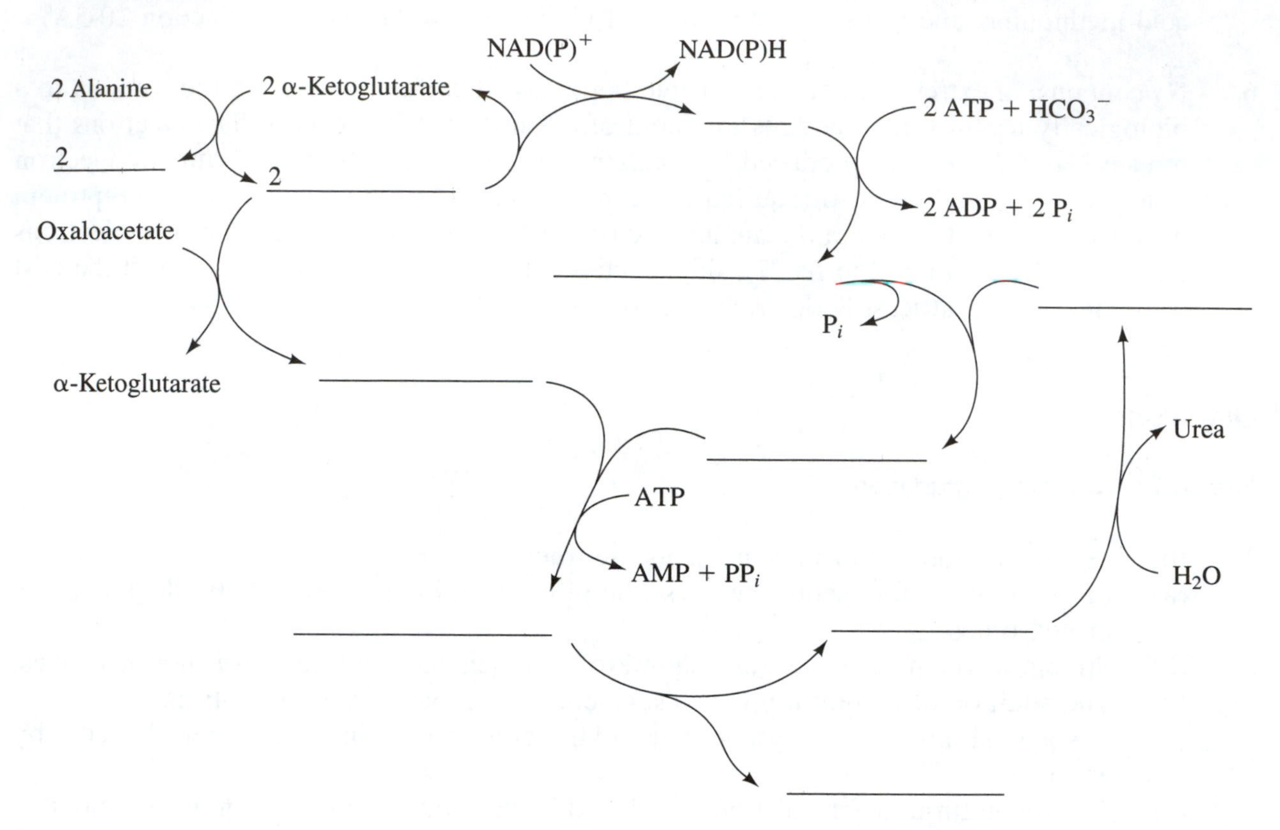

Breakdown of Amino Acids

9. Which amino acids yield citric acid cycle intermediates upon transamination?

10. Reactions that involve PLP can be classified according to which bond to the alpha carbon is broken. For the amino acid–PLP adduct shown below, indicate what kind of reaction involves cleavage of bonds *a*, *b*, and *c* and provide an example of each.

Amino acid–PLP Schiff base (aldimine)

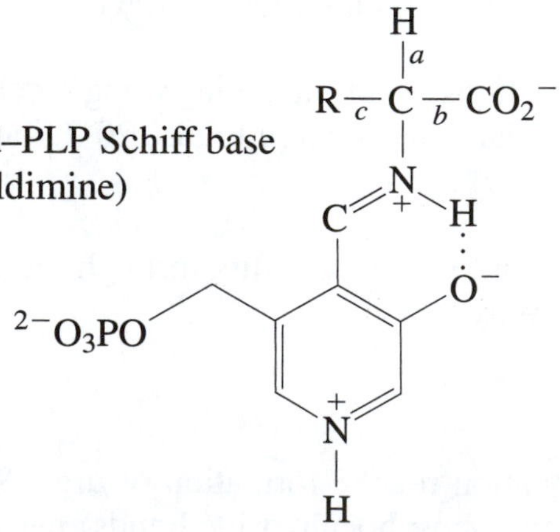

11. Many glucogenic amino acids are broken down to a citric acid cycle intermediate, which merely increases the catalytic activity of the cycle. How does the cell obtain a net yield of glucose from such amino acids?

12. Examine Figure 20-19 for the degradation of isoleucine, valine, and leucine.
    (a) Which reaction is analogous to the reaction catalyzed by the pyruvate dehydrogenase complex?
    (b) Which reaction is analogous to the reaction catalyzed by acyl-CoA dehydrogenase in fatty acid oxidation?

13. Some forms of maple syrup urine disease are amenable to treatment with large doses of dietary thiamine. What metabolic defect is being treated in these cases?

14. Which coenzymes mediate one-carbon transfer reactions?

15. How do sulfonamides selectively inhibit bacterial growth? Why aren't animals also adversely affected by these antibiotics?

16. *S*-Adenosylmethionine (SAM) is a key methylating agent in several physiological processes, but its regeneration depends on the presence of another methylating agent. How is SAM regenerated?

17. Calculate the free energy yield for the complete oxidation of the carbon skeletons of (a) aspartate, (b) glutamine, and (c) lysine. Express your answer in ATP equivalents and assume that NADPH is equivalent to NADH.

Amino Acid Biosynthesis

18. Rationalize the significance of α-ketoglutarate as an activator of glutamine synthetase.

19. The four intermediates of the urea cycle are α-amino acids.
    (a) Which one of these is considered to be an essential amino acid in children?
    (b) Outline a pathway by which adults can synthesize this amino acid from glucose.

20. Chorismate is the seventh intermediate in the synthesis of tryptophan from phosphoenolpyruvate and erythrose-4-phosphate (Figure 20-32). Why is it considered a branch point in amino acid biosynthesis?

21. What is the advantage of channeling? Why is it especially important for indole in the tryptophan synthase reaction?

Other Products of Amino Acid Metabolism

22. Porphyrin is synthesized in a modular fashion, much like cholesterol. How many molecules of succinyl-CoA are required?

23. How many carbons of heme would be labeled with ^{14}C if the starting material included (a) glycine with a ^{14}C-labeled carboxyl group, and (b) succinyl-CoA with a ^{14}C-labeled carboxyl group?

24. Heme oxygenase converts heme to the linear ______________________ and releases ______________________ and ______________________.

25. Bilirubin serves as an antioxidant (i.e., it undergoes oxidation to protect other cell components from oxidation). How is bilirubin regenerated after its oxidation?

Nitrogen Fixation

26. In mitochondrial respiration, electrons transferred from NADH and $FADH_2$ reduce O_2 to H_2O. Why can't these same coenzymes participate in the reduction of N_2 to NH_3?

Answers to Questions

1. (a) False, (b) true, (c) false, (d) false, (e) true.

2. (a)

$$\underset{\textbf{Threonine}}{H{-}C(COO^-)(NH_3^+){-}C(H)(OH){-}CH_3} + \underset{\boldsymbol{\alpha}\textbf{-Ketoglutarate}}{^-OOC{-}CH_2{-}CH_2{-}C({=}O){-}COO^-}$$

$$\downarrow$$

$$^-OOC{-}C({=}O){-}C(H)(OH){-}CH_3 + \underset{\textbf{Glutamate}}{^-OOC{-}CH_2{-}CH_2{-}CH(NH_3^+){-}COO^-}$$

(b)

$$\underset{\textbf{Isoleucine}}{H{-}C(COO^-)(NH_3^+){-}CH(CH_3){-}CH_2{-}CH_3} + \underset{\alpha\textbf{-Ketoglutarate}}{{}^-OOC{-}CH_2{-}CH_2{-}C({=}O){-}COO^-}$$

$$\downarrow$$

$$^-OOC{-}C({=}O){-}CH(CH_3){-}CH_2{-}CH_3 + \underset{\textbf{Glutamate}}{{}^-OOC{-}CH_2{-}CH_2{-}CH(NH_3^+){-}COO^-}$$

(c)

$$\underset{\textbf{Glycine}}{H{-}C(COO^-)(NH_3^+){-}H} + \underset{\alpha\textbf{-Ketoglutarate}}{{}^-OOC{-}CH_2{-}CH_2{-}C({=}O){-}COO^-}$$

$$\downarrow$$

$$^-OOC{-}C({=}O){-}H + \underset{\textbf{Glutamate}}{{}^-OOC{-}CH_2{-}CH_2{-}CH(NH_3^+){-}COO^-}$$

3. Prior to reacting with an amino acid, PLP is covalently bound to an enzyme Lys side chain via an imine linkage (Schiff base). With the release of the α-keto acid, PLP is converted to PMP, which is no longer covalently attached to the enzyme.

4.

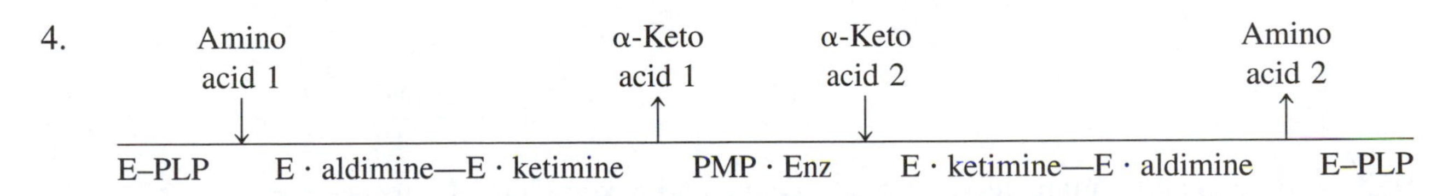

5. Flux in this normally close-to-equilibrium reaction would favor the production of α-ketoglutarate. Starved individuals break down protein (mostly muscle protein) and oxidize the resultant amino acids for energy. The α-ketoglutarate would be needed as an amino-group acceptor in the transamination of amino acids to be used as fuel.

6. $NH_3 + HCO_3^- + \text{aspartate} + 3\ ATP + H_2O \rightarrow$
$$\text{Urea} + \text{fumarate} + 2\ ADP + AMP + 4\ P_i$$

The total free energy cost is four ATP equivalents, since the PP_i produced in Step 3 is hydrolyzed by inorganic pyrophosphatase. However, if the reactions reconverting fumarate to aspartate and the glutamate dehydrogenase reaction that supplies the ammonia to the urea cycle are taken into account, the overall reaction is

$$2\ \text{Amino acid} + 2\ NAD^+ + HCO_3^- + 3\ ATP + H_2O \rightarrow$$
$$2\ \alpha\text{-Keto acid} + 2\ NADH + \text{urea} + 2\ ADP + AMP + 4\ P_i$$

and, assuming each NADH is energetically equivalent to 3 ATP, the free energy is actually a profit of 2 ATP.

7. Carbamoyl phosphate synthetase and ornithine transcarbamoylase are mitochondrial enzymes. The reaction catalyzed by carbamoyl phosphate synthetase is the first committed step.

8.

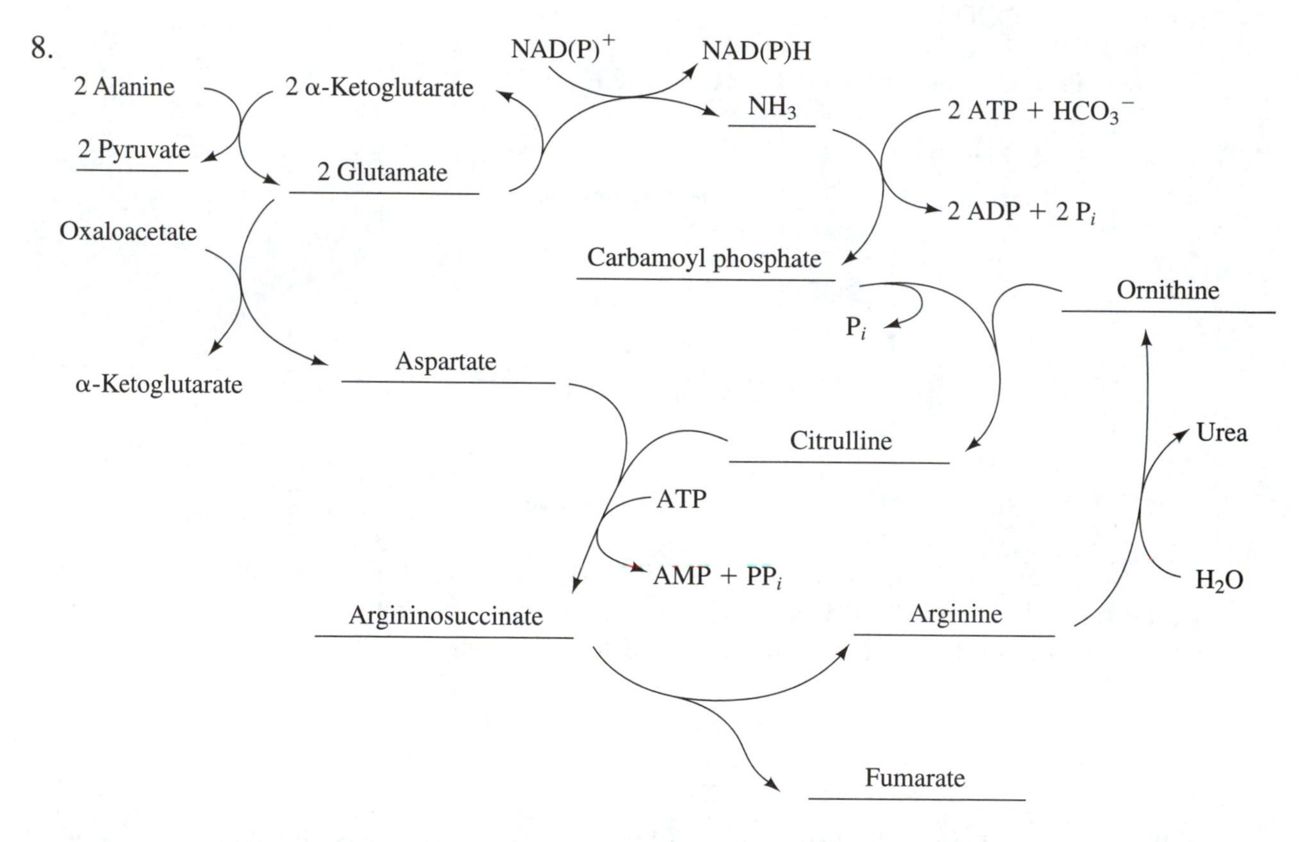

9. Aspartate and glutamate yield oxaloacetate and α-ketoglutarate, respectively.

10. Bond *a*: Transamination reaction such as occurs in the transfer of glutamic acid's amino group to oxaloacetate to yield α-ketoglutarate and aspartate.
Bond *b*: The decarboxylation of α-amino acids such as occurs in the synthesis of physiologically active amines.
Bond *c*: A C—C bond cleavage such as occurs in the serine hydroxymethyltransferase reaction.

11. All citric acid cycle intermediates can form oxaloacetate, which can then be decarboxylated and phosphorylated by PEP carboxykinase. The resulting PEP can then be converted to glucose by gluconeogenesis.

12. (a) Reaction 2
    (b) Reaction 3

13. In such cases, the defect is poor binding of thiamine pyrophosphate (TPP) to the E1 subunit of branched-chain α-keto acid dehydrogenase. The administration of large doses of the vitamin precursor of TPP improves the catalytic efficiency of the defective enzyme, which breaks down isoleucine, valine, and leucine.

14. Biotin, *S*-adenosylmethionine, and tetrahydrofolate.

15. Sulfonamides are structural analogs of the *p*-aminobenzoate group of folate and thereby interfere with folate synthesis. Unlike bacteria and plants, most animals cannot synthesize folate and hence are unaffected by these drugs.

16. After SAM donates its methyl group, the resulting *S*-adenosylhomocysteine is hydrolyzed to adenosine and homocysteine. Homocysteine is then methylated by N^5-methyltetrahydrofolate to form methionine, which is then adenylated by ATP to regenerate SAM (see Figure 20-16).

17. (a) 15 ATP. Aspartate is converted to oxaloacetate by transamination. Oxaloacetate is decarboxylated by PEP carboxykinase and the resulting PEP is converted to pyruvate and then to acetyl-CoA for entry into the citric acid cycle.

Reaction	***Reduced coenzymes produced***	***ATP equivalents***
PEP carboxykinase		−1 (GTP)
Pyruvate kinase		1
Pyruvate dehydrogenase	1 NADH	3
Citric acid cycle		1 (GTP)
	3 NADH	9
	1 $FADH_2$	2
Total		15

    (b) 27 ATP. Glutamine is converted to glutamate and then oxidized to α-ketoglutarate (Figure 20-15), which is then converted to oxaloacetate. Oxidation of oxaloacetate proceeds as outlined in part *a*. (Alternatively, α-ketoglutarate can be converted to malate by the citric acid cycle. Malic enzyme converts malate to pyruvate in a reaction that produces NADPH. The net ATP yield is the same.)

Reaction	*Reduced coenzymes produced*	*ATP equivalents*
Glutamate dehydrogenase	1 NADPH	3
α-Ketoglutarate dehydrogenase	1 NADH	3
Succinyl-CoA synthetase		1 (GTP)
Succinate dehydrogenase	1 $FADH_2$	2
Malate dehydrogenase	1 NADH	3
PEP carboxykinase		−1 (GTP)
Pyruvate kinase		1
Pyruvate dehydrogenase	1 NADH	3
Citric acid cycle		1 (GTP)
	3 NADH	9
	1 $FADH_2$	2
Total		27

(c) 34 ATP. Lysine is converted to acetoacetate in 11 reactions (Figure 20-20), several of which yield reduced coenzymes. The acetoacetate is converted to 2 acetyl-CoA at the expense of 1 GTP (since a CoA group is donated by succinyl-CoA).

Reaction	*Reduced coenzymes produced*	*ATP equivalents*
Lysine degradation		
Reaction 1	−1 NADPH	−3
Reaction 2	1 NADH	3
Reaction 3	1 NAD(P)H	3
Reaction 5	1 NADH	3
Reaction 6	1 $FADH_2$	2
Reaction 9	1 NADH	3
3-Ketoacyl-CoA transferase		−1 (succinyl-CoA)
2 Citric acid cycle		2 (GTP)
	6 NADH	18
	2 $FADH_2$	4
Total		34

18. The oxidative deamination of glutamate by glutamate dehydrogenase yields ammonia and α-ketoglutarate. Hence rising levels of α-ketoglutarate signal rising levels of free ammonia. The activation of glutamine synthetase by α-ketoglutarate increases the production of glutamine to help prevent the accumulation of ammonia, which is toxic in high concentrations.

19. (a) Arginine
(b) Glucose is degraded to 2 pyruvate by glycolysis. One pyruvate is converted to oxaloacetate by the pyruvate carboxylase reaction. The other pyruvate, converted to acetyl-CoA by pyruvate dehydrogenase, combines with the oxaloacetate to form citrate, then isocitrate and then α-ketoglutarate (the first three reactions of the citric acid cycle). Glutamate dehydrogenase, which operates close to equilibrium, converts α-ketoglutarate and NH_3 to glutamate. Three additional reactions convert glutamate

to ornithine (Figure 20-28). Since ornithine is a urea cycle intermediate, the reactions of the urea cycle convert it to the other cycle intermediates, including arginine.

20. Chorismate is also the precursor of phenylalanine and tyrosine, whose biosynthesis branches from the tryptophan pathway at chorismate.

21. Channeling, the transfer of an intermediate directly from one active site to another, increases the efficiency of the overall reaction by preventing the loss of the intermediate, which could then diffuse away or be degraded. Channeling is important for indole, a nonpolar molecule that might otherwise be lost from the cell.

22. Eight succinyl-CoA units are required to synthesize the porphyrin ring:

$$8 \text{ Succinyl-CoA} + 8 \text{ glycine} \rightarrow 8 \text{ ALA} \rightarrow 4 \text{ BPG} \rightarrow \text{uroporphyrinogen III}$$

23. (a) None. The carboxyl group of glycine is lost as carbon dioxide during condensation of glycine with succinyl-CoA.
    (b) Two. Uroporphyrinogen III contains all eight succinyl-derived carboxyl groups. However, subsequent steps remove six of these (Figure 20-35).

24. Heme oxygenase converts heme to the linear biliverdin and releases Fe^{3+} and CO.

25. Oxidized bilirubin is biliverdin, which is reduced by NADPH to regenerate bilirubin.

26. The reduction potentials of NADH (−0.315 V) and $FADH_2$ (−0.219 V) are not sufficiently negative for these coenzymes to reduce N_2 ($\mathcal{E}^{\circ\prime} = -0.340$ V).

Chapter 21 Mammalian Fuel Metabolism: Integration and Regulation

This chapter summarizes and integrates the major pathways involved in fuel metabolism. It first examines the variations of fuel metabolism in the brain, muscle, adipose tissue, and liver. Each of these organs exhibits specialized needs and functions, which influence the regulation of fuel metabolism in that organ. The discussion then turns to the ways in which fuel metabolism is interconnected between the liver and skeletal muscle via the Cori cycle and the glucose–alanine cycle. Fuel use in different organs depends in part on tissue-specific glucose transporter isoforms. Fuel metabolism is regulated principally by the action of hormones, which bind to specific receptor proteins. The binding of a hormone to its receptor induces a conformational change that transduces the signal into new biochemical activity inside the cell. Three signal transduction pathways regulate fuel metabolism: the adenylate cyclase signaling system, the receptor tyrosine kinase system, and the phosphoinositide pathway. Signal transduction pathways are the target of a variety of drugs and toxins, as is discussed in Box 21-1. In addition, Box 21-2 discusses the alterations of signaling pathways in cancers, which often result from mutations in genes (oncogenes) that encode signaling proteins. This chapter concludes with a discussion of disturbances in fuel metabolism, in particular, starvation, diabetes, and obesity.

Essential Concepts

Organ Specialization

1. The biochemical pathways for the synthesis and breakdown of the three major metabolic fuels (carbohydrates, fatty acids, and protein) converge on a small number of metabolites: pyruvate, acetyl-CoA, and citric acid cycle intermediates. In animals, flux through these pathways is tissue-specific. In mammals, the liver is one of the few organs that can carry out most of the biochemical pathways involved in energy metabolism (see Figure 21-1).

2. Some organs do not synthesize fuel molecules but only catabolize these molecules. Muscle and brain are two such organs. The brain's exclusive fuel source is normally glucose, and because the brain stores no glycogen, it relies on glucose present in the bloodstream. Prolonged fasting results in a gradual switch from glucose catabolism to ketone body catabolism in the brain. Skeletal and heart muscles rely on fatty acids when resting and use glucose from glycogen or from the blood upon exertion. The heart relies almost exclusively on aerobic metabolism.

3. Adipose tissue is largely involved in the long-term storage of fatty acids (in the form of triacylglycerols). It releases fatty acids into the bloodstream in response to hormones (e.g., epinephrine). The synthesis of triacylglycerols requires the catabolism of glucose to glycerol-3-phosphate, to which fatty acids are esterified.

4. The liver maintains homeostatic levels of circulating fuel molecules for use by all tissues. The liver metabolizes excess glucose by converting it to glucose-6-phosphate (G6P). This reaction is catalyzed by glucokinase (an isozyme of hexokinase) whose relatively high K_M

(~5 mM) makes it sensitive to changes in glucose concentration over the physiological range (~5 mM). In contrast, hexokinase ($K_M < 0.1$ mM) is saturated with glucose at these glucose levels.

5. In the liver, G6P has several fates:
   (a) Polymerization into glycogen.
   (b) Conversion to glucose, which then equilibrates with the glucose in the bloodstream.
   (c) Catabolism into acetyl-CoA, which can be further catabolized for the production of ATP, or diverted to the anabolic routes for the synthesis of fatty acids and cholesterol.
   (d) Conversion via the pentose phosphate pathway to ribose-5-phosphate, an important precursor for nucleotide synthesis, and NADPH for use in biosynthetic reactions.

6. The liver can synthesize and degrade triacylglycerols based on the body's metabolic needs. It also degrades amino acids to a variety of intermediates, which are then used for fatty acid synthesis (from the ketogenic amino acids) and gluconeogenesis (from the glucogenic amino acids).

Interorgan Metabolic Pathways

7. Some metabolic pathways, called interorgan pathways, are segregated between tissues that exchange metabolites (if such a pathway occurred entirely within a single organ, it would constitute a futile cycle). The Cori cycle is defined by the exchange of glucose and lactate between muscle and liver. The lactate generated by anaerobic glycolysis in muscle permits the muscle to anaerobically generate ATP required for contraction, while postponing the aerobic production of ATP in the liver necessary to resynthesize glucose.

8. Another interorgan pathway is the glucose–alanine cycle. Glycolytically produced pyruvate serves as an amino-group acceptor in transamination reactions in the muscle during protein degradation. The resultant alanine is transported to the liver, where it is reconverted to pyruvate for gluconeogenesis, with the released amino group appearing in urea. The glucose–alanine cycle can be viewed as a mechanism to transport nitrogen to the liver from the muscle.

9. Glucose transporters (the GLUT family) occur in several isoforms, some of which are expressed in a tissue-specific fashion:
   (a) GLUT1 is the passive glucose transporter in erythrocytes and most other tissues except liver.
   (b) GLUT2 occurs primarily in the liver and pancreatic β cells. Its high K_M for glucose (~60 mM) allows glucose flux in and out of these cells to vary linearly with glucose concentration, thereby allowing these cells to equilibrate their cytosolic glucose with blood glucose.
   (c) GLUT3 is expressed in the central nervous system and may operate in concert with GLUT1.
   (d) GLUT4 is an insulin-sensitive glucose transporter in muscle and adipose tissue. Stimulation by insulin results in translocation of GLUT4 from intracellular vesicles to the cell surface via exocytosis.
   (e) GLUT5 occurs primarily in the small intestine.
   (f) GLUT7 occurs in the endoplasmic reticulum of liver cells.

Mechanisms of Hormone Action: Signal Transduction

10. Hormones are chemical signals that reach their target cells via the bloodstream. Each hormone binds to a specific receptor, and its binding results in the transduction of the hormone signal into a biochemical response inside the cell. Cell-surface receptors are thought to behave like allosteric proteins, alternating between two distinct conformations, one with bound ligand and one without. Many factors may modulate the response of a cell to a specific hormone, such as the number of receptors and the activities of other receptors.

11. Energy metabolism is regulated largely by pancreatic and adrenal hormones. The pancreas secretes the polypeptide hormones glucagon and insulin, which act antagonistically to each other to regulate blood glucose concentration. The adrenal medulla secretes the catecholamines epinephrine and norepinephrine, which bind to α- and β-adrenergic receptors. These receptors often respond to the same ligand in opposite ways in different tissues.

12. Three of the best-studied signal transduction systems are the adenylate cyclase system, the receptor tyrosine kinase system, and the phosphoinositide pathway. In each case, reception of a signal at the cell surface is transduced to the cytoplasm. In the adenylate cyclase system and the phosphoinositide pathway, signal transduction leads to the rapid production of intracellular second messengers, which leads to activation of many target proteins. The tyrosine kinase system usually activates a protein kinase cascade. In all systems, the initial hormone signal is amplified, but its duration can be limited.

13. Signal transduction via the adenylate cyclase system involves activation of a heterotrimeric G-protein. Binding to a hormone–receptor complex causes the G-protein to release its bound GDP and bind GTP. This in turn induces the G-protein to dissociate into G_α and $G_{\beta\gamma}$ components, each of which can bind to and activate target proteins. The α subunit of a stimulatory G-protein ($G_{s\alpha}$) activates adenylate cyclase, whereas the α subunit of an inhibitory G-protein ($G_{i\alpha}$) inhibits the enzyme. The α subunit is deactivated by its hydrolysis of GTP, which promotes the reassociation of G_α with $G_{\beta\gamma}$.

14. The second messenger cyclic AMP (cAMP), the product of the adenylate cyclase reaction, activates cAMP-dependent protein kinase (cAPK). cAMP binds to the regulatory subunits of cAPK so as to release the cAPK catalytic subunits in active form. The effects of cAMP are attenuated by the action of a phosphodiesterase that hydrolyzes cAMP to AMP, and by protein phosphatases that reverse the phosphorylation catalyzed by cAPK. Many drugs and toxins interfere with components of the adenylate cyclase signaling system.

15. Proteins called growth factors stimulate proliferation and/or differentiation of target cells. Many growth factor receptors have protein kinase activity that specifically phosphorylates tyrosine residues on target proteins. These receptors, called receptor tyrosine kinases (RTKs), usually dimerize upon binding growth factor, which leads to autophosphorylation of the receptor protein. This phosphorylation stimulates phosphorylation of target proteins, many of which bind to the RTK via a protein domain called the Src homology 2 (SH2) domain, which recognizes the RTK phospho-Tyr residues.

16. Some RTKs interact with target proteins that activate the monomeric G-protein Ras, which in turn activates a protein kinase cascade. In the best-known pathway, activation of Ras leads to the activation of Raf, which is a Ser/Thr protein kinase that phosphorylates and activates MEK. MEK is both a Tyr kinase and a Ser/Thr kinase that activates MAP kinase (MAPK), another Ser/Thr kinase with many targets including other kinases and transcription factors.

17. Many genes that encode elements of growth-factor signaling pathways are known as proto-oncogenes, because their mutation may yield an oncogene. The product of an oncogene may disrupt normal cell growth and lead to cancer. Many cancer-causing viruses bear oncogenes. For example, the v-*erbB* oncogene, first identified in an animal virus, codes for the epidermal growth factor receptor without its ligand-binding domain, which results in constitutively active phosphorylation of protein substrates by this RTK. Other oncogenes include mutated forms of Ras and the transcription factors Fos and Jun.

18. The phosphoinositide signaling pathway produces two second messengers: inositol trisphosphate (IP_3) and diacylglycerol (DG), which are hydrolysis products of the plasma membrane phospholipid phosphatidylinositol-4,5-bisphosphate (PIP_2). Activation of a heterotrimeric G-protein by a ligand–receptor complex activates phospholipase C (PLC), which hydrolyzes PIP_2. The resulting IP_3 binds to IP_3-gated Ca^{2+} channels in the endoplasmic reticulum, thereby releasing Ca^{2+} into the cytosol. This Ca^{2+} can then bind to calmodulin and activate calmodulin-sensitive kinases. DG activates protein kinase C (PKC), which activates other kinases and proteins.

19. Different signal transduction pathways may antagonize each other or act synergistically via cross talk. For instance, some RTKs activate a PKC isoform that contains SH2 domains.

Disturbances in Fuel Metabolism

20. Starvation requires the mobilization of stored fuel molecules. Early in a fast, the liver accelerates glycogenolysis and increases gluconeogenesis from available materials. This makes glucose available continuously to the central nervous system. Skeletal muscle protein becomes a source of glucose via glucogenic amino acids released upon proteolysis. After several days, the liver shifts its metabolic machinery toward the breakdown of fatty acids in order to produce ketone bodies. The brain eventually adapts to use ketone bodies rather than glucose as a primary metabolic fuel.

21. Diabetes mellitus is a pathological condition that results from insufficient secretion of insulin or inefficient stimulation of its target cells. There are two major forms of diabetes:
    (a) Insulin-dependent or juvenile-onset diabetes mellitus (type I) results from an autoimmune response that destroys pancreatic β cells, which secrete insulin.
    (b) Non-insulin-dependent or maturity-onset diabetes mellitus (type II) may be caused by a decrease in the number of insulin receptors or their ability to bind insulin. This leads to the clinical condition called insulin resistance, in which insulin levels are far higher than normal but target tissues exhibit little or no response to the hormone. Type II diabetes may also be caused by abnormal insulin production resulting from defects in glucokinase in pancreatic β cells and from the age-dependent decline in mitochondrial oxidative capacity (see Box 21-3).

22. Obesity is a condition that arises when triacylglycerol intake exceeds its utilization. In mice, the 16-kD protein leptin appears to suppress appetite. Mice homozygous for the absence of the *obese* gene, which encodes leptin, eat continuously and become obese. In humans, the causes of obesity appear to be more complicated but may also involve abnormal activity of the *obese* gene.

Guide to Study Exercises (text p. 692)

1. The metabolism of fuel molecules varies by organ and depends on the organism's nutritional state. The brain, for example, consumes fuels but does not generate or store them. This organ requires a steady supply of glucose, which is completely oxidized to produce ATP. During starvation, the brain gradually adapts to using ketone bodies supplied by the liver.

   Muscle uses glucose, fatty acids, and ketone bodies as fuels. Its own glycogen store supplies glucose-6-phosphate (G6P), which is oxidized to produce ATP for contraction. At maximum exertion (anaerobic conditions), glycolysis proceeds faster than the citric acid cycle, so G6P is degraded to lactate, which is transported to the liver for further oxidation or gluconeogenesis. Some muscles, such as the heart, are entirely aerobic.

   The primary function of adipose tissue is the storage and mobilization of fatty acids. Circulating fatty acids are esterified to glycerol derived from glucose to form triacylglycerols. The triacylglycerols can be hydrolyzed to release fatty acids for use as fuels by other tissues.

   The liver carries out a variety of metabolic reactions so that it can maintain appropriate levels of fuels in the bloodstream. Excess circulating glucose is taken up by the liver and converted to G6P by the action of glucokinase. The G6P can be used to synthesize glycogen or catabolized by the pentose phosphate pathway or glycolysis, depending on the need for NADPH, R5P, and acetyl-CoA. When the blood glucose level is low, liver glycogen is broken down to supply glucose to other tissues. Similarly, excess fatty acids are incorporated into triacylglycerols and disseminated via the bloodstream in the form of lipoproteins. The liver degrades fatty acids to ketone bodies (the liver itself does not use ketone bodies for fuel). The liver also makes amino acids available as fuel, either by degrading them to other metabolites or by converting them to glucose. This is especially important when other fuels are not available. (Section 21-1)

2. Glucokinase is the liver isozyme of hexokinase, the enzyme that phosphorylates glucose to produce G6P. Because it has a high K_M (~5 mM), it does not become saturated at physiological glucose concentrations (~5 mM). Instead, it operates with a velocity roughly proportional to the glucose concentration. Therefore, at low [glucose], glucokinase is less active than hexokinase ($K_M < 0.1$ mM) so that other tissues can direct glucose toward glycolysis by phosphorylating it. At high [glucose], glucokinase activity is high so that the liver can metabolize the excess glucose. (Section 21-1D)

3. The Cori cycle, in which muscle lactate is converted to glucose in the liver, operates when the rate of muscle glycolysis exceeds the cells' oxidative capacity. Under conditions of high exertion, glucose is rapidly catabolized to produce ATP to power muscle contraction. The resulting lactate is transported from muscle to the liver for use in gluconeogenesis. The newly synthesized glucose then returns to the muscle, where it may again be catabolized by glycolysis.

In the glucose–alanine cycle, muscle-produced alanine is transported to the liver for use in gluconeogenesis. When other fuels are scarce, proteins (which are present in high concentrations in muscle tissue) are broken down. The resulting amino acids donate their amino groups to pyruvate in transamination reactions, producing alanine. Alanine travels to the liver, where it is converted back to pyruvate for gluconeogenesis. The amino groups that originate in muscle proteins are thereby transferred to the liver, where they are disposed of as urea. (Sections 21-2A and B)

4. Different tissues contain different glucose transporters depending on how they metabolize glucose. Most cells take up glucose passively, via GLUT1, according to its concentration in the blood. GLUT2, which has a high K_M, allows glucose to enter the liver at a rate proportional to its concentration and does not become saturated even at high [glucose]. GLUT3 facilitates glucose uptake in brain, a tissue that relies heavily on glucose. GLUT4 responds to insulin to rapidly increase the entry of glucose into muscle and adipose tissue, which are specialized to store metabolic fuels in times of plenty. GLUT5 may facilitate the exit of glucose from intestinal cells to the bloodstream or may mediate the uptake of fructose. In the liver, GLUT7 transports glucose from the lumen of the endoplasmic reticulum (where glucose-6-phosphatase is located) to the cytosol. Only the liver converts G6P to glucose, so GLUT7 occurs only in the liver. (Section 21-2C)

5. The adenylate cyclase signaling system consists of an extracellular hormone, a transmembrane receptor protein, a heterotrimeric G-protein that is tethered to the cytoplasmic face of the plasma membrane, and the membrane-bound enzyme adenylate cyclase. Hormone binding to the receptor induces the associated G-protein to release GDP and bind GTP, which in turn causes the G-protein to dissociate into its G_α and $G_{\beta\gamma}$ components. G_α activates adenylate cyclase, which converts ATP to the second messenger cAMP. cAMP can then activate its intracellular targets, such as cAMP-dependent protein kinase. This signaling system can also operate in a negative fashion when the hormone binds to an inhibitory receptor, whose associated G-protein then inhibits the activity of adenylate cyclase.

   The receptor tyrosine kinase (RTK) signaling system consists of an extracellular hormone (many of which are called growth factors) and a monomeric transmembrane receptor protein whose intracellular domain is a tyrosine kinase. Hormone binding to its receptor causes receptor dimerization, which activates the kinase in a way that allows autophosphorylation. The phosphorylated RTK can then activate other proteins by binding to them and/or phosphorylating their Tyr residues.

   Signaling via the phosphoinositide system requires an extracellular hormone, a transmembrane receptor protein, a heterotrimeric G-protein, the membrane-bound enzyme phospholipase C, and protein kinase C. Ligand binding to the receptor activates the G-protein, whose α subunit, in complex with GTP, activates phospholipase C. This enzyme catalyzes hydrolysis of a membrane lipid to release the second messengers inositol-1,4,5-trisphosphate (IP_3) and 1,2-diacylglycerol (DG). The water-soluble IP_3 induces the opening of Ca^{2+} transport channels in the endoplasmic reticulum membrane, leading to an increase in cytosolic [Ca^{2+}], which also acts as a second messenger. The lipid-soluble DG activates protein kinase C, which then phosphorylates and activates other cellular proteins. (Sections 21-3B, C, and D)

6. A heterotrimeric G-protein consists of three subunits that are anchored to the inside of the plasma membrane through covalently attached lipids. In its inactive form, the G-protein binds GDP. The G-protein associates with the intracellular portion of a receptor that has bound its ligand. This association causes the G-protein to exchange its bound GDP for GTP, whereupon the α subunit of the G-protein, which contains the GTP binding site, dissociates from the $\beta\gamma$ dimer. The $G_\alpha \cdot$ GTP unit can activate other cellular components (e.g., adenylate cyclase or phospholipase C) until its intrinsic GTPase activity hydrolyzes its bound GTP to GDP + P_i. The α subunit then reassociates with $G_{\beta\gamma}$, which may also activate cellular components. (Section 21-3B)

7. The body adapts to starvation by altering fuel metabolism so that stored fuels are mobilized and supplied to tissues that require them. The major energy stores are fats in adipose tissue, protein in muscle, and glycogen in liver and muscle. Glycogen stores last about one day, after which muscle protein breakdown yields amino acids that can serve as gluconeogenic precursors in the liver. The liver thereby generates a constant supply of glucose for tissues such as the brain, which rely primarily on glucose. At the same time, the liver and muscle oxidize fatty acids mobilized from adipose tissue to meet their energy needs. To prevent severe loss of muscle mass, gluconeogenesis from protein-derived amino acids is supplemented by ketone body synthesis. The liver converts the acetyl-CoA product of fatty acid β oxidation to ketone bodies, which are released into the bloodstream. The brain gradually adapts to use ketone bodies as fuel so that during prolonged starvation, most of the body's energy needs are met through metabolism of fats rather than of carbohydrates or protein. (Section 21-4A)

8. Insulin-dependent diabetes mellitus (type I or juvenile-onset diabetes) is characterized by insufficient production of insulin by the pancreatic β cells, usually as a result of autoimmune destruction of these cells. This form of diabetes can be moderated by daily injections of insulin to replace the missing hormone. Non-insulin-dependent diabetes mellitus (type II or maturity-onset diabetes) is the most common form of the disease and may have several causes. In type II diabetes, insulin levels are normal or elevated, but cells that normally respond to insulin are "resistant" due to mutations in the insulin receptor or diminished synthesis of the receptor. This form of the disease can be moderated to some extent by dietary changes. (Section 21-4B)

9. In obese individuals, fuel intake outweighs fuel consumption so that the excess is stored as fat in adipose tissue. With every meal, insulin production is stimulated, leading to chronic overproduction of the hormone. To compensate, insulin-responsive cells gradually lose their sensitivity to insulin, possibly by decreasing the synthesis of insulin receptors, so that blood glucose levels remain high. (Sections 21-4B and C)

Questions

Organ Specialization

1. List the possible products of acetyl-CoA metabolism in animals. Include the pathway in-

volved and, where appropriate, the key regulatory enzyme that commits acetyl-CoA to that pathway.

	Fate	*Pathway*	*Key Enzyme*
1.			
2.			
3.			
4.			
5.			

2. List the possible products of pyruvate metabolism in animals. Include the pathway involved and, where appropriate, the key regulatory enzyme that commits pyruvate to that pathway.

	Fate	*Pathway*	*Key Enzyme*
1.			
2.			
3.			
4.			

3. List the metabolic products of glucose-6-phosphate metabolism. Include the pathways involved and their key regulatory enzymes.

	Fate	*Pathway*	*Key Enzyme*
1.			
2.			
3.			
4.			

4. Match the metabolic pathways below with the cellular compartments in which they occur.

A. Cytosol
B. Inner mitochondrial membrane
C. Mitochondrial matrix
D. Mitochondrial intermembrane space
E. Peroxisome
F. Endoplasmic reticulum

______ Glycolysis
______ Lactic fermentation
______ Pyruvate dehydrogenation
______ Citric acid cycle
______ Oxidative phosphorylation of ADP
______ Pentose phosphate pathway
______ Fatty acid elongation and desaturation
______ Palmitoyl-CoA oxidation
______ Palmitate synthesis
______ Lignoceric (24:0) acid oxidation
______ Amino acid degradation
______ Urea cycle
______ Gluconeogenesis
______ Cholesterol synthesis

5. Match each of the metabolic functions below with the appropriate pathway(s).

A. Glycolysis
B. Pentose phosphate pathway
C. Gluconeogenesis
D. Fatty acid oxidation
E. Glycogen synthesis
F. Amino acid degradation

(a) Provides reducing equivalents for biosynthesis.
(b) Provides glucose during the early phase of a fast, after glycogen stores are depleted.
(c) Provides energy for heart muscle.
(d) Stores metabolic fuel in the liver.

(e) Provides ATP during a rapid burst of skeletal muscle activity.
(f) Generates the nucleotide precursor ribose-5-phosphate.
(g) Provides energy for gluconeogenesis during a fast.
(h) Provides energy immediately after a meal.

6. What is the significance of high levels of hexokinase activity in the brain?

Interorgan Pathways

7. What are the similarities in the Cori cycle and the glucose–alanine cycle? What distinguishes them?

8. Muscle phosphofructokinase is allosterically stimulated by NH_4^+. What is the physiological function of this stimulation?

9. In the liver, the "fight-or-flight" hormone epinephrine stimulates the FBPase activity of the bifunctional enzyme PFK-2/FBPase-2. In muscle, epinephrine stimulates the PFK activity of the enzyme. What is the adaptive significance of this biochemistry?

10. How does an increase in blood glucose affect triacylglycerol metabolism in adipocytes?

11. Why is GLUT1 expressed only at low levels in the liver?

Mechanisms of Hormone Action: Signal Transduction

12. Compare the overall structure and mode of signal transduction by receptor tyrosine kinases and receptors that act via second messengers.

13. During G-protein activation, which subunit exchanges GDP for GTP? What is the immediate consequence of this exchange?

14. How are SH2 and SH3 domains important in the activation of Ras by receptor tyrosine kinases?

15. How does binding of ligand to a receptor tyrosine kinase lead to changes in gene expression?

16. Describe the intracellular signaling cascade mediated by phospholipase C activity. How is phospholipase C activated?

Disturbances in Fuel Metabolism

17. List the order in which the following energy sources are used by skeletal muscle to produce ATP: protein, phosphocreatine, fatty acids, glycogen.

18. List the order in which the liver uses the following substances to provide the body with metabolic fuel during starvation: glycogen, fatty acids, muscle protein, nonmuscle protein. Explain your answer.

19. You obtain a liver homogenate from a fasting mouse and "spike" it with a high concentration of glucose. In what form would you expect glycogen phosphorylase?

20. During a fast to lose weight, is it important to be physically active or might it be better to remain sedentary? Explain.

Answers to Questions

1. Note that animals cannot undertake the net conversion of acetyl units to glucose.

Fate	*Pathway*	*Key Enzyme*
1. CO_2 and H_2O	Citric acid cycle	Citrate synthase
2. Acetoacetate and/or β-hydroxybutyrate	Ketogenesis	Mitochondrial HMG-CoA synthase
3. Fatty acids	Fatty acid synthesis	Acetyl-CoA carboxylase
4. Cholesterol	Cholesterol synthesis	Cytosolic HMG-CoA synthase
5. Amino acids	Pathways utilizing citric acid cycle intermediates.	Varies; often catalyzes an exergonic reaction coupled to ATP hydrolysis, e.g., glutamine synthetase.

2. Note that the conversion of lactate to pyruvate is reversible.

Fate	*Pathway*	*Key Enzyme*
1. CO_2 and H_2O	Pyruvate dehydrogenation/ citric acid cycle	Pyruvate dehydrogenase
2. Oxaloacetate for citric acid cycle	Anaplerotic reactions	Pyruvate carboxylase
3. Glucose	Gluconeogenesis	Phosphofructokinase-2, which alters the levels of F2,6P, a key allosteric regulator of PFK-1 and FBPase-1.
4. Amino acids	Transamination and pathways utilizing citric acid cycle intermediates	Varies; the transamination of pyruvate to alanine is a reversible transamination.

3.

Fate	*Pathway*	*Key Enzyme*
1. Glucose	Glucose transport (liver)	Glucose-6-phosphatase

2. Glycogen	Glycogen synthesis	Glycogen synthase
3. Acetyl-CoA	Glycolysis/pyruvate dehydrogenation	PFK-1 and pyruvate dehydrogenase
4. Ribose-5-phosphate and NADPH	Pentose phosphate pathway	G6P dehydrogenase

4.

A	Glycolysis
A	Lactic fermentation
C	Pyruvate dehydrogenation
C	Citric acid cycle
B	Oxidative phosphorylation of ADP
A	Pentose phosphate pathway
F	Fatty acid elongation and desaturation
C	Palmitoyl-CoA oxidation
A	Palmitate synthesis
E	Lignoceric (24:0) acid oxidation
A, C	Amino acid degradation
A, C	Urea cycle
A, C	Gluconeogenesis
A	Cholesterol synthesis

5. (a) B; (b) C, F; (c) D; (d) E; (e) A; (f) B; (g) D; (h) A, F

6. The brain (and the rest of the central nervous system) has a high energy demand, and under normal conditions, its sole source of metabolic fuel is blood-delivered glucose. Hexokinase has a high affinity for glucose, so that glucose that enters a brain cell via facilitated diffusion can be rapidly converted to glucose-6-phosphate, thereby allowing additional glucose to enter the cell.

7. Both cycles operate under conditions of metabolic stress. In both cycles, metabolites arriving from skeletal muscle are converted by the liver to glucose, which is returned to the bloodstream. The Cori cycle involves the synthesis of glucose from lactate produced by lactic fermentation in the muscle during sustained vigorous exercise. The glucose–alanine cycle involves the synthesis of glucose from alanine produced in the muscle. It is thereby a mechanism of transporting nitrogen from muscle to liver. However, under conditions of starvation, the cycle is broken and the muscle-derived glucose is supplied to other tissues.

8. The degradation of proteins yields amino acids whose further breakdown, in many cases [e.g., glycine, cysteine, serine, and threonine (Figure 20-12); and glutamine (Figure 20-15)], yields ammonia. The ammonia is then transferred to α-ketoglutarate via the glutamate dehydrogenase reaction to yield glutamate, and the resulting α-amino group is transferred to pyruvate by a transaminase to yield alanine. The alanine is then transported to the liver where, via transaminases, glutamate dehydrogenase, and the urea cycle, its α-amino group is transferred to urea for excretion by the kidneys.

   The elevation of the NH_4^+ level in muscle is indicative of increased protein breakdown (e.g., due to starvation). The activation by NH_4^+ of muscle phosphofructokinase (PFK) stimulates glycolysis (PFK is the major flux-determining enzyme) and thereby increases the rate of pyruvate production to accommodate the increased need to synthesize alanine.

9. In the liver, epinephrine decreases the concentration of F2,6P, which is an allosteric activator of PFK-1 and an inhibitor of FBPase. As a result, gluconeogenesis is favored over glycolysis. This allows the liver to provide glucose to other organs, including muscle. In mus-

cle, epinephrine increases the concentration of F2,6P, which promotes glycolytic flux. The muscle can therefore produce ATP at a greater rate, which is necessary for vigorous muscle activity. Thus, although epinephrine has opposite effects in liver and muscle, the hormone acts to provide additional metabolic energy to muscle under conditions of need.

10. Increasing levels of blood glucose favor triacylglycerol synthesis as the rate of glucose uptake by adipocytes increases. The three-carbon backbone of triacylglycerols is glycerol-3-phosphate, which is obtained by the reduction of dihydroxyacetone phosphate. A high level of glycolytic activity in adipocytes results in a sufficiently large pool of glycerol-3-phosphate to favor fatty acid esterification.

11. In the liver, GLUT2 mediates the entry of glucose into cells. GLUT2, like hexokinase, has a relatively high K_M for glucose so that glucose utilization occurs in proportion to the amount of glucose present. In other cells, GLUT1 (which has a relatively low K_M for glucose) permits glucose entry at low [glucose] but becomes saturated at higher [glucose]. If liver cells relied on GLUT1 rather than GLUT2, they would be unable to rapidly take up glucose at high concentrations.

12. Receptor tyrosine kinases are usually monomeric proteins with a single membrane-spanning segment. Ligand binding induces dimerization of the receptor, which leads to autophosphorylation at one or more tyrosine residues in its cytoplasmic domain. These phosphorylated residues provide binding sites for target proteins with SH2 domains, which are generally activated upon binding to the receptor tyrosine kinase.

    Receptors that act via second messengers are transmembrane proteins that usually contain multiple membrane-spanning segments. Ligand binding leads to interactions between the receptor and other plasma membrane proteins, such as G-proteins, that participate in signal transduction. A sequence of such interactions culminates in the production or release of cytoplasmic second messengers such as cAMP, IP_3, or Ca^{2+}.

13. When the G-protein binds to a hormone–receptor complex, the G_α subunit exchanges GDP for GTP. Binding of GTP induces a conformational change in G_α that causes it to dissociate from $G_{\beta\gamma}$. GTP binding also increases the affinity of G_α for adenylate cyclase, which is thereby activated.

14. The SH2 and SH3 domains function in the assembly of a multiprotein signaling complex that links the phosphorylated receptor tyrosine kinase to the monomeric G-protein Ras. For example, the SH2 domain in the Grb2/Sem-5 protein binds the phosphorylated cytoplasmic domain of the receptor tyrosine kinase, while its SH3 domains recognize and bind proline-rich sequences in Sos protein. This complex induces Ras to release GDP and bind GTP in order to become active (see Figure 21-16).

15. Ligand binding to a receptor tyrosine kinase leads to the activation of Ras via the formation of a complex of SH2- and SH3-containing proteins. Activated Ras triggers a kinase cascade consisting of Raf, MEK, and MAPK, which are activated in series. Activated MAPK migrates to the nucleus, where it phosphorylates a variety of proteins, including transcription factors. The activated transcription factors stimulate gene expression.

16. Activation of phospholipase C leads to the hydrolysis of phosphatidylinositol-4,5-bisphosphate (PIP_2) to inositol-1,4,5-trisphosphate (IP_3) and 1,2-diacylglycerol (DG). Both of these

products are second messengers. IP_3 moves through the cytosol and binds to the IP_3 receptor, a Ca^{2+} channel on the endoplasmic reticulum. Binding of IP_3 to its receptor triggers the release of Ca^{2+} from the ER, increasing intracellular Ca^{2+}. The Ca^{2+} can stimulate many cellular activities either by interacting directly with target proteins or through the formation of the Ca^{2+}-calmodulin complex, which regulates many other target proteins. The DG is retained within the membrane, where it activates protein kinase C. This enzyme then phosphorylates and activates additional target proteins.

Members of the phospholipase C family can be activated in two ways. Some forms are activated by interacting with a $G_{\alpha} \cdot GTP$ complex derived from a G-protein following ligand binding to a G-protein-linked receptor. Other isoforms are activated by binding, via their SH2 domains, to a receptor tyrosine kinase that has undergone ligand binding, dimerization, and autophosphorylation.

17. Phosphocreatine, glycogen, fatty acids, protein.

18. The order of usage of these metabolic fuels is glycogen, muscle protein, fatty acids, and nonmuscle protein. Glucose is the primary metabolic fuel of muscle and the central nervous system. After the first day of a fast, the glycogen stores of the liver (and muscle) are exhausted. In order to maintain circulating glucose levels, muscle protein is broken down into amino acids, many of which are glucogenic. Extended fasting alters the metabolic machinery of the liver to favor fatty acid degradation into ketone bodies, thereby conserving critical muscle mass. Severe, extended starvation eventually leads to loss of non-muscle protein, which can severely compromise organ function and can be life-threatening.

19. During a fast, glucagon stimulates glycogen degradation via cAPK, which phosphorylates and activates glycogen phosphorylase (glycogen phosphorylase *a*). The addition of a large amount of glucose shifts the $T \rightleftharpoons R$ equilibrium toward the T state. Stabilization of the T state by glucose binding promotes dephosphorylation of Ser 14, which converts the enzyme to the less active glycogen phosphorylase *b*.

20. A certain level of activity facilitates the transition from metabolism of glucose to metabolism of fatty acids. In a sedentary fasting individual, inactive muscle would provide more metabolic fuel via protein degradation than active muscle; hence, a completely sedentary individual might experience significant loss of muscle mass without a major loss of adipose tissue mass, the primary target for weight loss. Therefore, it is important to be active rather than sedentary while dieting.

Chapter 22 Nucleotide Metabolism

This chapter presents nucleotide metabolism, including the synthesis of purine and pyrimidine ribonucleotides, the conversion of ribonucleotides to deoxyribonucleotides, and the degradation of nucleotides. As you saw in Chapter 3, nucleotides are composed of a purine or pyrimidine which is linked to C1′ of either ribose or deoxyribose. The pentose in turn is esterified through C5′ to one or more phosphate groups. Nucleoside triphosphates are precursors of nucleic acids and also, through the hydrolysis of one or both of their phosphoanhydride bonds, provide the free energy that drives many biochemical reactions. This chapter begins with a detailed discussion of the synthesis of purine nucleotides, describing the source of each of the carbons and nitrogens of the purine ring. The chapter then discusses the intricate system of negative and positive feedback regulation that balances the synthesis of ATP and GTP. Most cells do not synthesize purines *de novo*; instead they are recycled via salvage pathways. The following section describes the synthesis of pyrimidines, in which negative feedback mechanisms balance the amounts of UTP and CTP in the cell. Next, the chapter discusses the formation of deoxyribonucleotides by the action of ribonucleotide reductase. This enzyme has several unique features, including its mechanism of action and its intricate allosteric regulatory mechanisms. The formation of deoxyUTP provides the precursor for deoxythymidylate. The key enzyme involved here is thymidylate synthase, which is a target of cancer chemotherapy, as described in Box 22-1. Finally, the chapter considers the degradation of nucleotides, which is in some respects a continuation of nitrogen metabolism discussed in Chapter 20. The catabolism of purines leads to the formation of uric acid, and the catabolism of pyrimidines leads to the formation of β-alanine and β-aminoisobutyrate. The fate of uric acid varies among animal groups depending on their strategies for excreting excess nitrogen. Heritable abnormalities in nucleotide metabolism lead to a variety of diseases, including immune system defects, as discussed in Box 22-2.

Essential Concepts

Synthesis of Purine Ribonucleotides

1. Early investigations of the nucleotide biosynthetic pathway identified the precursors that supply the carbon and nitrogen atoms of the purine ring: N1 from aspartate; C2 and C8 from formate; N3 and N9 from glutamine; C4, C5, and N7 from glycine; and C6 from bicarbonate.

2. The purines are synthesized as ribonucleotides rather than as free bases. The purine nucleotide biosynthetic pathway produces inosine 5′-monophosphate (IMP) in 11 steps:
   (a) Ribose is activated by reaction with ATP to form 5-phosphoribosyl-α-pyrophosphate (PRPP).
   (b) The amide nitrogen of glutamine (which becomes N9) displaces the pyrophosphate group of PRPP, inverting the configuration at C1′ to form β-5-phosphoribosylamine.
   (c) A molecule of glycine is added, providing C4, C5, and N7 and forming glycinamide ribotide (GAR).
   (d) The free amino group of GAR reacts with the one-carbon carrier N^{10}-formyltetrahydrofolate to add C8 and give formylglycinamide ribotide (FGAR).

(e) The amide nitrogen of a second glutamine is added to FGAR in an ATP-dependent reaction, which provides N3 and yields formylglycinamidine ribotide (FGAM).
(f) The purine imidazole ring is closed in a reaction requiring ATP and producing 5-aminoimidazole ribotide (AIR).
(g) Purine C6 is incorporated into AIR in a carboxylation reaction to give carboxyaminoimidazole ribotide (CAIR).
(h) Aspartate contributes N1 by forming an amide bond with C6 to yield 5-aminoimidazole-4-(*N*-succinylocarboxamide) ribotide (SACAIR).
(i) Fumarate is cleaved from SACAIR to produce 5-aminoimidazole-4-carboxamide ribotide (AICAR). Note that reactions (h) and (i) resemble reactions of the urea cycle.
(j) The last purine ring atom, C2, is added through formylation by N^{10}-formyltetrahydrofolate to form 5-formaminoimidazole-4-carboxamide ribotide (FAICAR).
(k) In the final step, the larger ring is closed to form IMP.

3. IMP is rapidly transformed into AMP and GMP. To form AMP, the amino group of aspartate displaces the carbonyl oxygen at C6 of the purine base, which is followed by removal of fumarate. To form GMP, IMP is oxidized in an NAD^+-dependent reaction to yield xanthosine monophosphate (XMP), which then accepts a glutamine amide nitrogen to produce the guanine nucleotide. Note that AMP synthesis from IMP requires cleavage of GTP to GDP + P_i, whereas GMP formation requires ATP cleavage to AMP + PP_i.

4. Nucleoside monophosphates are converted to di- and triphosphate derivatives in two steps. First, a specific nucleoside monophosphate kinase catalyzes phosphorylation of a nucleoside monophosphate by a nucleoside triphosphate to yield two nucleoside diphosphates. Second, a nucleoside diphosphate is phosphorylated to the corresponding triphosphate by the action of nucleoside diphosphate kinase.

5. Purine nucleotide biosynthesis is carefully regulated. IMP production is controlled at its first two steps. Ribose phosphate pyrophosphokinase, which generates PRPP, is inhibited through negative feedback by ADP and GDP. Amidophosphoribosyl transferase, which produces phosphoribosylamine, is inhibited by adenine nucleotides and by guanine nucleotides, which bind at separate inhibitory sites. PRPP also allosterically stimulates amidophosphoribosyl transferase.

6. Additional regulation occurs immediately beyond the IMP branch point, because AMP and GMP inhibit their own synthesis. Moreover, a balance in the synthesis of the two purine nucleotides is achieved because GTP is necessary for AMP synthesis and ATP is required for GMP formation (see Figure 22-4).

7. Purines released from the degradation of nucleic acids are re-utilized via metabolic salvage pathways. In mammals, distinct enzymes catalyze the reactions of adenine and guanine with PRPP to form AMP and GMP, respectively. Hypoxanthine–guanine phosphoribosyltransferase (HGPRT) uses guanine as a substrate but also transforms the purine base hypoxanthine into IMP. An inherited deficiency in HGPRT results in Lesch–Nyhan syndrome, a disease characterized by bizarre behavioral abnormalities.

Synthesis of Pyrimidine Ribonucleotides

8. Pyrimidine biosynthesis utilizes only three precursors. Aspartate contributes N1 and C4, C5, and C6 of the pyrimidine ring, whereas C2 is derived from bicarbonate and N3 is donated by glutamine.

9. In pyrimidine nucleotide formation, the ring is completed before being linked to ribose-5-phosphate. The six steps of the UMP biosynthetic pathway are as follows:
   (a) Bicarbonate and the amide nitrogen of glutamine combine to form carbamoyl phosphate in a reaction that requires 2 ATP and is catalyzed by carbamoyl synthetase II. This enzyme is distinct from carbamoyl synthetase I, which forms carbamoyl phosphate during the urea cycle, a process in which ammonia is the nitrogen donor.
   (b) Carbamoyl phosphate reacts with aspartate to yield carbamoyl aspartate. The reaction is catalyzed by aspartate transcarbamoylase, a key enzyme for pyrimidine biosynthesis regulation in *E. coli* but not in animals.
   (c) Carbamoyl aspartate is converted to the ring compound dihydroorotate by elimination of water.
   (d) Dihydroorotate is dehydrogenated to form orotate, with a mitochondrial quinone serving as the ultimate electron acceptor.
   (e) Orotate reacts with PRPP to produce orotidine-5′-monophosphate (OMP). The same enzyme can also salvage uracil and cytosine by catalyzing their conversion to UMP and CMP, respectively.
   (f) OMP undergoes decarboxylation to form UMP.

10. In microorganisms, the six reactions of UMP biosynthesis are carried out by separate enzymes. However, in animals, several of the reactions are catalyzed by polypeptides that possess more than one enzyme activity. This is also true for certain steps in purine formation. In these instances, pathway intermediates are not liberated from the multifunctional proteins but are channeled directly to the next active site.

11. As in purine nucleotide biosynthesis, UTP is formed from UMP by the successive actions of a nucleoside monophosphate kinase and nucleoside diphosphate kinase.

12. CTP is synthesized through amination of UTP by glutamine (in animals) or ammonia (in bacteria) in an ATP-dependent reaction.

13. The regulation of pyrimidine biosynthesis in bacteria is accomplished by the allosteric modulation of aspartate transcarbamoylase, such that the binding of ATP to the enzyme stimulates activity and the binding of CTP or UTP inhibits it. Pyrimidine formation in animals is regulated by changes in carbamoyl synthetase II activity. This enzyme is inhibited by UDP and UTP and stimulated by ATP and PRPP. In addition, decarboxylation of OMP is diminished by elevated UMP levels.

Formation of Deoxyribonucleotides

14. Deoxyribonucleotides are synthesized by the reduction of the hydroxyl group at C2′ of the corresponding ribonucleotide. Enzymes that catalyze these reactions are called ribonucleotide reductases.

15. Ribonucleotide reductase in eukaryotes and some prokaryotes reduces a ribonucleoside diphosphate (NDP) to a deoxyribonucleoside diphosphate (dNDP). The enzyme is a tetramer that can be dissociated into inactive homodimers, R1 and R2. R1 contains a substrate-binding site and two independent allosteric sites that control enzymatic activity and substrate specificity. Each subunit of R2 has an Fe(III) prosthetic group that interacts with Tyr 122 to form a tyrosyl free radical, which plays a critical role in the reaction mechanism. Five cysteine residues in each subunit of R1 also participate, some acting as electron donors to reduce the substrate and generate a disulfide bond.

16. Following NDP reduction, the ribonucleotide reductase disulfide bond is reduced via disulfide interchange with the small protein thioredoxin, which regenerates the active enzyme. Oxidized thioredoxin is subsequently reduced by thioredoxin reductase, which obtains electrons from NADPH. This coenzyme is thus the ultimate reducing agent in the conversion of an NDP to a dNDP.

17. The production of balanced amounts of deoxyribonucleotides needed for normal DNA synthesis is achieved by elaborate feedback control of ribonucleotide reductase (see Figure 22-12). For example, when ATP binds to the activity site of R1, the enzyme is stimulated to reduce NDPs. In contrast, binding of dATP to this site inhibits enzyme activity toward all substrates. The nature of the substrates reduced is determined by allosteric regulation at the substrate specificity site of R1, which has the following characteristics:
    (a) Binding of ATP or dATP stimulates reduction of CDP and UDP.
    (b) Binding of dTTP stimulates GDP reduction and inhibits CDP and UDP reduction.
    (c) Binding of dGDP enhances ADP reduction, but blocks CDP, UDP, and GDP reduction.

18. Nucleoside diphosphate kinase catalyzes the ATP-dependent phosphorylation of dNDPs to produce the dNTPs needed for DNA synthesis.

19. In the synthesis of thymine deoxyribonucleotides, dUTP is first hydrolyzed to dUMP and then methylated by thymidylate synthase to form dTMP. The methyl donor in this reaction is N^5,N^{10}-methylenetetrahydrofolate, which is simultaneously oxidized to dihydrofolate. Dihydrofolate is reduced to tetrahydrofolate with NADPH by dihydrofolate reductase. Tetrahydrofolate then accepts a hydroxymethyl group from serine to regenerate the methyl donor. Finally, dTMP is phosphorylated to dTTP. Drug-induced inhibition of dTMP synthesis is an important tool in cancer therapy (see Box 22-1).

Nucleotide Degradation

20. The breakdown of purine nucleotides leads to production of uric acid. Degradation involves the dephosphorylation of the nucleotides to nucleosides, followed by the hydrolysis or phosphorolysis of the nucleosides to yield the free purine base and ribose or ribose-1-phosphate. The bases may be reincorporated into purine nucleotides via the salvage reactions. Adenosine is deaminated to form inosine prior to further degradation.

21. Both hypoxanthine, which is derived by phosphorolysis of inosine, and guanine are oxidized sequentially to xanthine and then uric acid. Xanthine oxidase catalyzes the two reactions that convert hypoxanthine to uric acid, via a complex series of electron-transfer events.

22. Uric acid is the final product of purine degradation for primates, birds, terrestrial reptiles, and some insects. Excretion of uric acid is advantageous in that it requires little accompanying water. Other organisms oxidize uric acid to allantoin, allantoic acid, urea, or ammonia prior to excretion.

23. A variety of factors may lead to abnormally high levels of uric acid, which can result in gout.

24. Pyrimidine nucleotide catabolism occurs through reactions similar to those for purine nucleotides to yield the free bases uracil and thymine. These are further broken down to the amino acids β-alanine and β-aminoisobutyrate, which can be further metabolized to furnish energy-yielding compounds.

Guide to Study Exercises (text p. 720)

1. See Section 3-1 and Table 3-1.

2. The synthesis of purine nucleotides in general is more elaborate than the synthesis of pyrimidine nucleotides.
   (a) The precursors of pyrimidine nucleotides are ribose-5-phosphate, HCO_3^-, aspartate, and glutamine. Purine nucleotide synthesis requires all these plus glycine and the formyl group of N^{10}-formyltetrahydrofolate.
   (b) The synthesis of the purine nucleotide IMP requires 7 ATP equivalents. AMP synthesis requires one additional ATP, and GMP synthesis requires two additional ATP. Four ATP equivalents are required to synthesize the pyrimidine nucleotide UMP (two of these are consumed in the activation of R5P to PRPP). Three more ATP are required to convert UMP to CTP.
   (c) The ribose portion of purine nucleotides is acquired in the first step of the pathway, the activation of R5P to PRPP, before the first purine ring component (the amide nitrogen of glutamine) is introduced. The ribose group, as PRPP, does not enter the pyrimidine nucleotide synthetic pathway until the fifth step, after the orotate ring has formed.
   (d) Eleven steps are required to synthesize IMP. AMP and GMP synthesis each require two additional steps. Six reactions synthesize UMP. Three additional steps convert UMP to CTP. (Sections 22-1A and B and 22-2A and B)

3. PRPP is one of the reactants for purine nucleotide synthesis. It is also required for pyrimidine nucleotide synthesis and the salvage of bases. The activation of R5P to PRPP is inhibited by ADP and GDP (feedback inhibition by products of purine nucleotide biosynthesis), and PRPP allosterically activates amidophosphoribosyl transferase, the pathway's flux-controlling step (feed-forward activation). The level of PRPP determines the rate of OMP production in pyrimidine nucleotide synthesis. In animals but not bacteria, PRPP activates carbamoyl phosphate synthetase II, which catalyzes the first reaction of the pyrimidine nucleotide biosynthetic pathway. (Sections 22-1C and 22-2C)

4. Folate cofactors are required to synthesize purine nucleotides and deoxythymidine. N^{10}-formyl-THF is a substrate for Reactions 4 and 10 in the IMP synthetic pathway. N^5,N^{10}-methylene-THF is the methyl-group donor in the synthesis of dTMP from dUMP. (Sections 22-1A and 22-3B)

5. Antifolates are substances that inhibit reactions involving folate cofactors. Drugs such as methotrexate, aminopterin, and trimethoprim bind more tightly than folate to folate-requiring enzymes. Antifolates are particularly effective inhibitors because mammals, which cannot synthesize folate, have a limited supply of the folate cofactors. Thus, rapidly growing cells, such as cancer cells, which need a steady supply of dTMP for DNA synthesis, are more greatly affected by antifolates than are more slowly growing cells. (Box 22-1)

6. Lesch–Nyham syndrome results from a severe deficiency of hypoxanthine–guanine phosphoribosyltransferase (HGPRT), which salvages guanine and hypoxanthine bases by linking them to PRPP to form GMP and IMP. Defective HGPRT causes PRPP to accumulate, which stimulates purine nucleotide synthesis. The resulting purines are broken down to uric acid.

    A deficiency of the bifunctional enzyme that catalyzes the conversion of orotate to OMP and then UMP results in orotic aciduria. The excess orotic acid is excreted rather than converted to UMP and CTP.

    SCID (severe combined immunodeficiency disease) results from a deficiency of adenosine deaminase (ADA), the enzyme that converts adenosine and deoxyadenosine to inosine for further degradation. The enzyme defect increases the concentration of dATP, which inhibits ribonucleotide reductase. This prevents the synthesis of other dNTPs and blocks cell division. Cells of the immune system are particularly vulnerable to ADA deficiency in part because they normally have high ADA activity.

    Gout is caused by an increase in the level of uric acid (a purine degradation product), which tends to precipitate as crystals of sodium urate in the joints and kidneys. The excess uric acid may be a consequence of impaired uric acid excretion by the kidneys, HGPRT deficiency, or another defect that increases uric acid production. (Sections 22-1D, 22-2C, Box 22-2, and 22-4B)

7. A large portion of nucleotides synthesized by the cell is intended for DNA synthesis, which requires balanced concentrations of purines and pyrimidines. On the other hand, RNA synthesis and other activities call for ribonucleotides rather than deoxyribonucleotides.
    (a) The cell balances the supplies of purine and pyrimidine nucleotides by several mechanisms that rely primarily on feedback inhibition. First, the supply of ATP and GTP is balanced by a system in which each of these nucleotides serves as a substrate for the other's synthesis. High concentrations of AMP and GMP inhibit their own synthesis, and adenine and guanine nucleotides both inhibit the synthesis of their precursor, IMP.

        The synthesis of pyrimidine nucleotides depends on the supply of PRPP, which is produced in the first step of purine synthesis. Pyrimidine nucleotide synthesis is inhibited by its products CTP (in bacteria) and uridine nucleotides (in mammals). ATP (a purine nucleotide) activates carbamoyl phosphate synthetase II in mammals, and aspartate transcarbamoylase in bacteria, thereby coordinating the rates of synthesis of purines and pyrimidines. Finally, the concentrations of various NDPs and NTPs are equilibrated through the action of kinases such as nucleoside diphosphate kinase.

(b) The balance between ribose and deoxyribose nucleotides is maintained by regulating the activity of ribonucleotide reductase. ATP stimulates the reduction of CDP and UDP. dUDP is converted to dTTP, which inhibits further CDP and UDP reduction but stimulates GDP reduction. dGTP inhibits reduction of CDP, UDP, and GDP but stimulates reduction of ADP. Eventually, dATP inhibits all NDP reduction. (Sections 22-1C, 22-2C, and 22-3B)

8. Nucleotide degradation gives rise to free bases and ribose or ribose-1-phosphate. Some of the bases are salvaged but most are further degraded. Purines are broken down to xanthine, which is then oxidized to uric acid in a reaction that also produces H_2O_2. The deamination of adenine and guanine also produces NH_4^+. Depending on the species, uric acid may be further degraded to allantoin, allantoate, urea, or NH_4^+.

   Pyrimidine degradation leads to uracil (or thymine), which is reduced, decarboxylated, and deaminated before being linked to CoA as malonyl-CoA (a precursor in fatty acid synthesis) or methylmalonyl-CoA (which can be converted to succinyl-CoA). (Section 22-4)

Questions

Synthesis of Purine Ribonucleotides

1. Which of the following statements is(are) true? The enzyme ribose phosphate pyrophosphokinase:
   (a) catalyzes the transfer of pyrophosphate from 5-phosphoribosyl-α-pyrophosphate (PRPP) to activate 5-phosphoribosylamine.
   (b) catalyzes the formation PRPP from ribose-5-phosphate and ATP.
   (c) transfers a pyrophosphoryl group from ATP to the C1 of ribose.
   (d) produces pyrophosphate as a reaction product.

2. Which of the following compounds contribute atoms in the synthesis of the purine ring?

NH_4^+	N^5,N^{10}-methylenetetrahydrofolate
S-adenosylmethionine	Aspartate
Glutamine	Fumarate
N^{10}-formyltetrahydrofolate	CO_2
Serine	Glycine

3. In the synthesis of AMP from inosine monophosphate, the amino group that replaces the 6-keto group is obtained from __________ with release of __________, whereas GMP formation involves an oxidation reaction in which the electron acceptor is __________, followed by transfer of an amino group from __________.

4. Which of the following statements is(are) true? Purine nucleotide biosynthesis is regulated in part by:
   (a) feedback inhibition of amidophosphoribosyl transferase by CTP and UMP.
   (b) feedback inhibition of amidophosphoribosyl transferase by PRPP.
   (c) feedback inhibition of amidophosphoribosyl transferase by ADP and GDP.
   (d) feedback inhibition of ribose phosphate pyrophosphokinase by ADP and GDP.

5. In most tissues, purine bases are salvaged rather than synthesized *de novo*. What is the advantage of this metabolic strategy?

Synthesis of Pyrimidine Ribonucleotides

6. Identify the compound shown below.

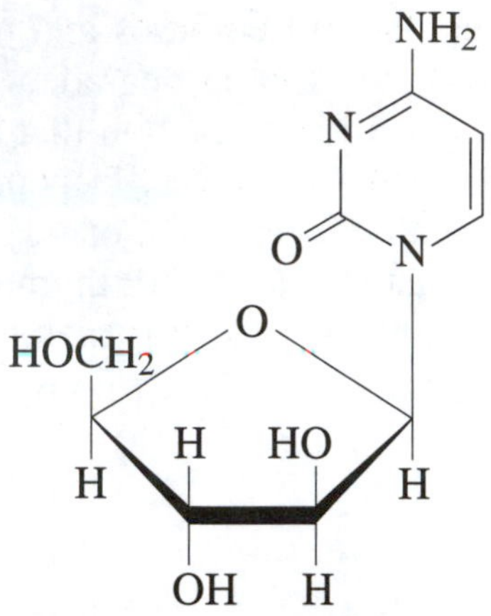

7. Determine which of the following compounds participate in CTP biosynthesis and place them in the correct order:

Orotic acid	Aspartic acid
Glutamine	N^{10}-formyltetrahydrofolate
Uracil	HCO_3^-
Cytidine	Uridylic acid
Dihydroorotate	PRPP

8. In what way is UMP synthesis similar to fatty acid synthesis in mammals?

9. Why is the administration of uridine an effective treatment for human orotic aciduria? Why is it preferable to use uridine instead of UMP?

Formation of Deoxyribonucleotides

10. The pentose phosphate pathway converts glucose to the ribose-5-phosphate needed for DNA synthesis. What other pentose phosphate pathway product is required for DNA synthesis?

11. Match each component of the ribonucleotide reductase reaction on the left with its function on the right.

______ dATP	A. Location of tyrosyl radical
______ R1 subunit	B. Stimulates GDP reduction and inhibits CDP and UDP reduction
______ R2 subunit	C. Physiological reducing agent of the enzyme
______ dTTP	D. Inhibits reduction of all NDPs
______ dGTP	E. Location of substrate specificity site
______ thioredoxin	F. Stimulates ADP reduction and inhibits CDP, UDP, and GDP reduction

12. Why is 5-fluorouracil a powerful antitumor agent? Why is it considered a suicide substrate?

13. What is the rationale for administering a high dose of methotrexate followed by a massive amount of N^5-formyltetrahydrofolate and thymidine to a cancer patient?

Nucleotide Degradation

14. Adenosine deaminase (ADA) catalyzes the first step in the catabolism of adenine nucleotides. Why does an absence of ADA seriously compromise the immune system?

15. Summarize the metabolic pathway by which ^{14}C initially present at position 5 of uracil could appear in long-chain fatty acids.

16. Some terrestrial organisms secrete excess nitrogen as uric acid rather than urea. Why is this an advantage?

Answers to Questions

1. (b) and (c) are correct

2. Glutamine, N^{10}-formyltetrahydrofolate, aspartate, CO_2, and glycine

3. Aspartate; fumarate; NAD^+; glutamine

4. (c) and (d) are correct

5. The synthesis of the purine ring is energetically expensive (7–9 ATP equivalents). In contrast, the salvage pathways, in which a pre-existing purine base reacts with PRPP to reform a nucleotide, requires only the energy of the activated PRPP.

6. This molecule is cytosine arabinoside, an analog of cytidine in which arabinose replaces ribose.

7. HCO_3^- and glutamine; aspartic acid; dihydroorotate; orotic acid; PRPP; uridylic acid (UMP), glutamine (second use)

8. In both pathways in mammals, multiple enzyme activities are located on a single polypeptide chain. This increases the rate of the multistep process and allows intermediates to be channeled from one active site to the next without being lost to the surrounding medium.

9. Orotic aciduria is caused by a deficiency of the bifunctional enzyme that converts orotic acid to UMP. Uridine can be phosphorylated to UMP, thereby bypassing the metabolic block and acting as precursor for both uridine and cytidine nucleotides. Uridine is preferable to UMP because, being an uncharged molecule, it can readily cross cell membranes.

10. The pentose phosphate pathway also produces NADPH, which is the ultimate electron donor for the reduction of NDPs to dNDPs via thioredoxin reductase, thioredoxin, and ribonucleotide reductase.

11. __D__ dATP
__E__ R1 subunit
__A__ R2 subunit
__B__ dTTP
__F__ dGTP
__C__ thioredoxin

12. 5-Fluorouracil is readily taken up by cancer cells and converted to 5-fluorodeoxyuridylate (FdUMP), which is an inactivator of thymidylate synthase. The rapid proliferation of tumor cells requires a constant supply of dTTP for DNA synthesis. Most slow-growing normal cells are relatively insensitive to 5-fluorouracil treatment. FdUMP only inactivates thymidylate synthase after going partially through the catalytic cycle. Since the enzyme actually generates the species that inactivates it, FdUMP is termed a mechanism-based inhibitor or a suicide substrate.

13. Methotrexate is a dihydrofolate (DHF) analog that binds avidly to dihydrofolate reductase and consequently prevents the reduction of dihydrofolate to tetrahydrofolate (THF). Rapidly growing cancer cells, which require THF for the synthesis of dTTP, are particularly sensitive to methotrexate. A high dose of methotrexate is intended to rapidly kill cancer cells. Subsequent administration of a tetrahydrofolate derivative and thymidine supplies sufficient amounts of folate and a dTTP precursor to enable the more slowly growing normal cells to survive.

14. In the absence of ADA activity, deoxyadenosine accumulates and is phosphorylated. The elevated concentration of dATP inhibits ribonucleotide reductase. Inadequate production of deoxyribonucleotides inhibits DNA synthesis and hence cell proliferation. Lymphocytes exhibit especially active phosphorylation of deoxyadenosine and are therefore particularly vulnerable to ADA deficiency.

15. In the catabolism of pyrimidines, uracil is degraded in several steps to malonyl-CoA (Figure 22-23), which is a substrate for fatty acid synthesis (Section 19-4C). C5 of the pyrimidine ring becomes the methylene carbon of malonyl-CoA and can therefore be incorporated into fatty acyl chains.

16. Uric acid excretion requires little water since it is eliminated as insoluble crystals. Urea is water-soluble and requires large volumes of water to be excreted. Thus uric acid excretion is advantageous over urea excretion in environments where water is scarce.

Chapter 23 Nucleic Acid Structure

This chapter takes a closer look at the structures of nucleic acids, which were first introduced in Chapter 3. This chapter begins by examining the structure of DNA in some detail, focusing on the different helical conformations and the limited flexibility of DNA that results from restrictions on rotation of various covalent bonds. Next, you are introduced to supercoiling, a structural feature of DNA that has important consequences for the biological activity of DNA. The forces that stabilize double helical structures are considered next. Knowledge of these forces is central to understanding the techniques used to isolate, analyze, and manipulate nucleic acids.

The chapter then turns to the interactions of DNA with proteins, giving as examples the restriction endonucleases and well-characterized prokaryotic and eukaryotic sequence-specific regulators of transcription. These transcription factors include the repressor from bacteriophage 434, the *E. coli trp* repressor, and the *E. coli met* repressor. The eukaryotic transcription factors that are discussed are proteins that contain the zinc finger DNA-binding motif and the leucine zipper dimerization motif (including those containing the helix–loop–helix motif near the DNA-binding region). In all cases, your attention is drawn to the common molecular interactions found in all of these proteins.

Finally, the structure of eukaryotic chromosomes is considered. This section explores how the histone proteins interact with DNA to generate higher-order structures that condense DNA over 50,000-fold. First, the structure of the nucleosome is discussed. This is followed by descriptions of the 300-Å filament of chromatin and the organization of highly condensed metaphase chromosomes.

Essential Concepts

1. DNA and RNA are the two kinds of nucleic acids that store genetic information and make it available to the cell. These molecules must therefore have the following properties:
   (a) Genetic information must be stored in a form that is stable and manageable in size.
   (b) Genetic information must be decoded by transcription and translation, which together convert nucleic acid sequences into protein sequences.
   (c) The information in DNA or RNA must be accessible to proteins and other nucleic acids.
   (d) Replication, in the case of DNA, must be template-driven so that each daughter cell receives the same genetic information.

The DNA Helix

2. DNA occurs in three major structural forms called A-DNA, B-DNA, and Z-DNA. B-DNA is the most common form and has the structural form first described by Watson and Crick.

3. The key structural features of B-DNA include:
   (a) The two antiparallel strands wind in a right-handed manner around a common axis, producing a double helix that is about 20 Å in diameter.
   (b) Pairs of nucleotide bases are nearly perpendicular to the axis of the helix. The base pairs are in the interior of the double helix, while the sugar–phosphate backbone is on the outside, thus giving the appearance of a spiral staircase.

(c) The "ideal" B-DNA double helix has ten base pairs (bp) per turn with a pitch of about 34 Å.

(d) The double helix contains a wide and deep major groove and a narrow and deep minor groove.

4. A-DNA is a wider and flatter right-handed helix than B-DNA. The planes of its base pairs are tilted about 20° to the helix axis. Its major groove is deep and narrow, while its minor groove is shallow and wide. A-DNA forms under dehydrating conditions.

5. Z-DNA is a left-handed DNA helix that contains a deep minor groove but no discernible major groove. It forms *in vitro* in nucleic acids containing alternating purines and pyrimidines, but its existence *in vivo* is uncertain.

6. RNA is single-stranded but can form regions of double helix by folding back on itself. 2′-OH groups preclude B-form structures; instead, double helical regions assume conformations resembling A-DNA. DNA–RNA hybrids also show A-form conformations.

7. Segments of B-DNA deviate from the ideal conformation, often in a sequence-dependent manner, which may be important for sequence-specific recognition of DNA by proteins involved in regulating specific gene activity.

8. The conformational flexibility of DNA is limited by steric hindrance at the glycosidic bond between the nitrogenous base and the ribose moiety. Steric hindrance also induces a specific pucker in the ribose and restricts rotation in the bonds of the sugar–phosphate backbone.

9. The twisting of a double helix that is often observed in covalently closed, circular, double-stranded DNA is called supercoiling or superhelicity. A key topological property of a closed circular double helix is that the number of supercoils cannot be altered without first cutting at least one strand of DNA. In the mathematical relationship

$$L = T + W$$

L is the linking number (the number of times that one strand of the double helix winds around the other), T is the twist (the number of complete revolutions that one strand makes around the axis of the double helix), and W is the writhing number (the number of turns the duplex axis makes around the superhelix axis, a measure of the supercoiling of the circular DNA). When $W < 0$, the DNA is negatively supercoiled; when $W > 0$, the DNA is positively supercoiled. Naturally occurring DNA is negatively supercoiled, which promotes strand separation for processes such as DNA replication and transcription.

10. Topoisomerases alter DNA supercoiling by making transient single-strand breaks (Type I) or double-strand breaks (Type II). Prokaryotic Type I topoisomerases relax negative supercoils by increasing the linking number in increments of one. Eukaryotic Type I topoisomerases can relax positive supercoils as well. Type II topoisomerases of both prokaryotes and eukaryotes relax negative and positive supercoils in an ATP-dependent manner. Only the prokaryotic version (also called DNA gyrase) can introduce negative supercoils. Negative supercoils in eukaryotes result primarily from packaging DNA into nucleosomes.

Forces Stabilizing Nucleic Acid Structures

11. Like the denaturation of proteins, the denaturation of DNA is cooperative; i.e., the unwinding of one part of DNA destabilizes the remaining double helix. This can be measured by observing the increase in ultraviolet light absorbance in going from double-stranded to single-stranded DNA. The midpoint of this "melting" curve is called the melting temperature, T_m.

12. The T_m of double-stranded DNA depends on the solvent, the kind and concentration of ions in solution, the pH, and the mole fraction of G · C base pairs.

13. While base pairing provides specificity to the structure of DNA, it contributes little to the stability of DNA. Hydrophobic interactions (which tend to cause free base pairs to aggregate) and van der Waals interactions between base pairs (called stacking interactions) contribute the most to the stability of double-stranded nucleic acids. Stacking interactions between G · C base pairs are stronger than those between A · T base pairs. Hence, the greater stability of GC-rich DNA is a reflection not of the greater number of hydrogen bonds but of greater stacking energies.

14. Most RNA is single-stranded and adopts a much wider variety of shapes than does DNA. These varied shapes are due to various sections of RNA forming double-stranded regions via specific base pairing. For example, ribosomal RNA (rRNA) is about 46% double-stranded. The compact shape of transfer RNA (tRNA) is a result of base pairing and extensive stacking interactions.

15. RNA also has catalytic activity. An RNA found in certain plant viruses, called the hammerhead ribozyme, catalyzes the cleavage of a specific RNA during posttranscriptional processing. Synthetic versions of ribozymes can catalyze many other reactions, including phosphoryl group transfer, isomerization at C—C covalent bonds, and hydrolytic reactions.

Fractionation of Nucleic Acids

16. DNA and RNA in cells are invariably associated with proteins, so nucleic acid preparations must be deproteinized as part of their purification. Deproteinization can be accomplished by treating cell lysates with detergents, chaotropic agents, or proteases. The nucleic acid can then be recovered by precipitation with ethanol.

17. Double-stranded DNA can be separated from single-stranded nucleic acid, nucleotides, and other soluble molecules by chromatography using a column of hydroxyapatite, a form of calcium phosphate.

18. Eukaryotic messenger RNA (mRNA) can be separated from total RNA (of which it constitutes no more than 5% by mass), DNA, nucleotides, and other soluble molecules by affinity chromatography. Most eukaryotic mRNAs contain a poly(A) sequence at their 3′ ends, which is not found in rRNA or tRNA. Poly(U) or poly(T) covalently attached to a cellulose or agarose matrix can bind eukaryotic mRNA by complementary base pairing under high salt conditions; the mRNA is subsequently recovered by exposing the matrix to a low salt buffer.

19. Small nucleic acids (<1000 nucleotides) can be easily resolved by polyacrylamide gel electrophoresis. Larger nucleic acids (up to 100,000 bp) can be separated using less cross-linked agarose gels. For extremely large DNAs (up to 10^7 bp), pulsed-field gel electrophoresis is used. Nucleic acids can be visualized in these gels by adding planar aromatic cations such as ethidium bromide, acridine orange, or proflavin. These dyes are called intercalation agents because they intercalate between the stacked bases, where they exhibit much greater fluorescence than do the free dyes.

20. Equilibrium density gradient centrifugation in CsCl can be used to separate double-stranded DNA from the denser single-stranded DNA and RNA. Different-sized RNAs can be separated by zonal ultracentrifugation through a sucrose gradient.

DNA–Protein Interactions

21. Many proteins bind DNA nonspecifically, primarily through interactions between the protein's functional groups and the sugar–phosphate backbone of DNA. Histones and certain DNA replication proteins are examples of such proteins.

22. Sequence-specific DNA-binding proteins usually interact with base pairs in the major groove of DNA by hydrogen bonding, either directly or indirectly via intervening water molecules. Ionic interactions with the sugar–phosphate backbone also occur, probably to facilitate contact between these proteins and DNA, after which the DNA can be "scanned" for a specific binding site. Dissociation constants for sequence-specific DNA-binding proteins are 10^3 to 10^7 times lower than those for nonspecific DNA-binding proteins.

23. Prokaryotic sequence-specific DNA-binding proteins recognize base pairs directly or through indirect readout, in a manner that depends on the sequence-dependent conformation and/or flexibility of the DNA backbone. Many DNA-binding proteins recognize palindromic (or nearly palindromic) sequences.

24. A large class of eukaryotic sequence-specific DNA-binding proteins are transcription factors. Many of these proteins form dimers and promote transcription at specific sites.

25. Many eukaryotic transcription factors contain structural motifs called zinc fingers. Zinc fingers are compact ~30 residue modules containing one or two Zn^{2+} ions liganded by His and/or Cys residues.

26. Another type of eukaryotic transcription factor contains a structural motif called the leucine zipper. The leucine zipper is a seven-residue pseudorepeating unit $(a\text{-}b\text{-}c\text{-}d\text{-}e\text{-}f\text{-}g)_n$ in which a hydrophobic strip along one side of the α helix formed by these residues promotes dimerization. The DNA-binding domain of such proteins is usually quite separate from the dimerization domain.

Eukaryotic Chromosome Structure

27. Prokaryotic genomes typically consist of a single circular DNA molecule ranging from several hundred thousand bp to several million bp.

28. The 23 chromosomes of the 3-billion-bp human genome have a total extended length of nearly one meter. Individual human chromosomes, with extended lengths between 1.6 and 8.2 cm, are condensed to varying degrees in different cells at different times, down to 1.3 to 10 μm long during mitosis.

29. Eukaryotic DNA is packaged into units called nucleosomes. Each nucleosome contains an octamer of four different basic proteins called histones $[(H2A)_2(H2B)_2(H3)_2(H4)_2]$ and 146 bp of DNA. Approximately 55 bp of DNA (called linker DNA) links two nucleosomes and is associated with a fifth histone, H1.

30. Histones are basic proteins with a large proportion of Arg and Lys residues, which interact with the phosphate oxygens of the sugar–phosphate backbone via salt bridges, hydrogen bonds, and helix dipoles.

31. The winding of DNA in nucleosomes to form the beaded chain seen in electron micrographs reduces its length by sevenfold. Further reduction of length is achieved by coiling the beaded chain into an ~300-Å-diameter filament. Higher-order structures, requiring nonhistone proteins, are less well characterized. The most condensed structure is the metaphase chromosome, which has a central scaffold of fibrous protein that anchors loops of DNA and has an overall diameter of ~1.0 μm.

32. Prokaryotic DNA is also packaged into nucleosome-like structures with highly basic proteins that functionally resemble histones.

Key Equation

$$L = T + W$$

Guide to Study Exercises (text p. 771)

1. See Table 23-1.

2. RNA contains the pentose ribose, whereas DNA has 2′-deoxyribose. Because the 2′-OH group prevents RNA from assuming a B-DNA-like conformation, it adopts a shallower, wider, A-DNA-like helix. Furthermore, because RNA is usually single-stranded, it tends to fold back on itself to satisfy hydrogen-bonding requirements, rather than forming a more linear rodlike molecule typical of double-stranded DNA. Consequently, RNA molecules, which are generally smaller than DNA molecules, exhibit greater conformational variety than DNA and can therefore perform a wider variety of functions than can DNA. (Sections 23-1A and 23-2E)

3. Type I topoisomerases create a single-strand break in DNA through which a DNA strand (or double-stranded DNA) is passed before the nick is sealed. The nicking and closing reactions require no additional source of free energy since the energy of the phosphodiester bond is conserved in the formation of a covalent enzyme–DNA intermediate. Repeated nicking and closing causes supercoiled DNA to fully relax. Prokaryotic Type I topoisomerase

can relax only negatively supercoiled DNA; the eukaryotic enzyme can relax both positively and negatively supercoiled DNA.

Type II topoisomerases cleave both strands of a double-stranded DNA, pass double-stranded DNA through the break, and then reseal the break. These reactions require ATP, whose binding and subsequent hydrolysis provide the free energy that drives enzyme conformational changes required for catalysis. Prokaryotic Type II topoisomerases relax both negative and positive supercoils and can induce negative supercoils into DNA. Eukaryotic Type II topoisomerases can only relax supercoils. (Section 23-1C)

4. During denaturation, the base pairs in nucleic acids separate, yielding single strands that have no fixed conformation. This flexibility results in a decrease in viscosity in the case of a DNA solution. UV absorbance increases by ~40% due to the disruption of electronic interactions among neighboring bases. DNA denaturation typically occurs over a narrow temperature range, because denaturation is a cooperative process in which melting in one part of the molecule destabilizes the rest of the molecule. RNA denaturation is not highly cooperative, because RNA molecules tend to contain many relatively short double-stranded regions that melt independently.

   During renaturation, hydrogen bonds form between complementary regions between strands (in DNA) or between segments of the same strand (in RNA). The temperature must be high enough (only ~25°C below T_m) to provide enough thermal energy to allow short base-paired regions to separate and reanneal until the proper base-pairing interactions are established along the length of the molecule. (Section 23-2A)

5. Nucleic acids are stabilized primarily by Watson–Crick base pairing and van der Waals interactions between stacked bases, particularly G and C bases. Divalent cations, which shield the negative charges of adjacent phosphates, also stabilize nucleic acids. (Section 23-2)

6. Some proteins interact with DNA nonspecifically, that is, through recognition of its overall geometry and sugar–phosphate backbone, without regard to its sequence of bases. Other proteins interact with DNA in a sequence-specific manner, by making contacts with functional groups on the bases or by recognizing subtle base-specific variations in backbone conformation. DNA–protein interactions include hydrogen bonds, often with bridging water molecules, and ionic interactions. Both the protein and the nucleic acid may change conformation on binding. (Section 23-4)

7. A zinc finger unit of a eukaryotic transcription factor binds a short segment of DNA. Typically, multiple zinc fingers are arranged in tandem so that the protein recognizes an extended sequence of bases. The transcription factor often wraps around the DNA, following the surface grooves. Most prokaryotic repressors have a homodimeric structure and thereby recognize DNA sequences that have perfect or nearly perfect palindromic symmetry. Often the repressor interacts with bases in the major groove that are one turn apart. (Sections 23-4B and C)

8. Eukaryotic DNA is dramatically condensed in chromatin, first by being wrapped around a histone octamer to form a nucleosome core particle, which is brought closer to its neighbors through interactions with histone H1. Nucleosomes are then folded in a zigzag fashion to form a 300-Å-diameter solenoid (coil). These filaments are attached as loops to a protein scaffold, producing a metaphase chromosome with a diameter of 1.0 μm. (Section 23-5)

Questions

The DNA Helix

1. What form of DNA might you expect to see in desiccated (but viable!) brine shrimp eggs? Why?

2. Double-stranded DNA is relatively stiff, whereas single-stranded DNA is a flexible coil. What factors influence the structure of single-stranded versus double-stranded DNA?

3. How does the ribose pucker affect DNA structure?

4. A sample of a circular plasmid is digested with Type I topoisomerase and analyzed by agarose gel electrophoresis followed by staining with ethidium bromide. Fifteen bands of DNA are visible. What do these bands represent and how do their linking numbers differ?

Forces Stabilizing Nucleic Acid Structures

5. Why is DNA not susceptible to hydrolysis by NaOH?

6. High concentrations of denaturing agents such as urea or formamide ($HCONH_2$) tend to favor a rodlike conformation for single-stranded DNA, rather than a flexible coil. What molecular interactions promote this behavior?

7. Explain how the following affect the T_m of double-stranded DNA:
   (a) Increasing the monovalent salt concentration.
   (b) Decreasing the pH.
   (c) Increasing the pH.
   (d) Increasing the concentration of formamide.

8. Would the UV absorbance of a solution containing partially stacked poly(A) increase or decrease when [Na^+] increases?

9. What forces stabilize the tertiary structure of tRNA?

Fractionation of Nucleic Acids

10. Ice-cold ethanol is used to precipitate DNA. What is the significance of the temperature of the ethanol? Is this more important for shorter or longer DNA strands?

11. Samples of a circular plasmid are incubated with increasing concentrations of ethidium bromide and then analyzed on an agarose gel. Which gel below shows the results of this procedure?

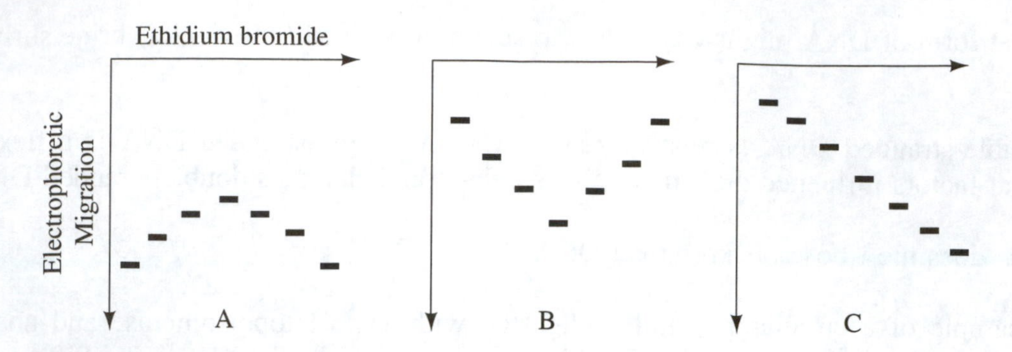

12. What method of nucleic acid fractionation, other than gel electrophoresis, might be useful in separating a native supercoiled plasmid from genomic bacterial DNA or relaxed circular plasmid?

13. In one procedure for isolating RNA, the contents of a cell are homogenized (forcefully mixed) with detergents and chaotropic agents and then subjected to ultracentrifugation in a CsCl equilibrium density gradient. What is the purpose of the detergents and chaotropic agents? Where in the CsCl gradient do you expect to find the RNA and DNA (the protein forms a band at the top of the gradient)?

14. You have obtained 100–1000 bp DNA fragments from a dinosaur bone and have purified them by CsCl equilibrium density ultracentrifugation. Shown below is the banding pattern of DNA. Which band is the richest in G + C? Which band most likely represents mitochondrial DNA?

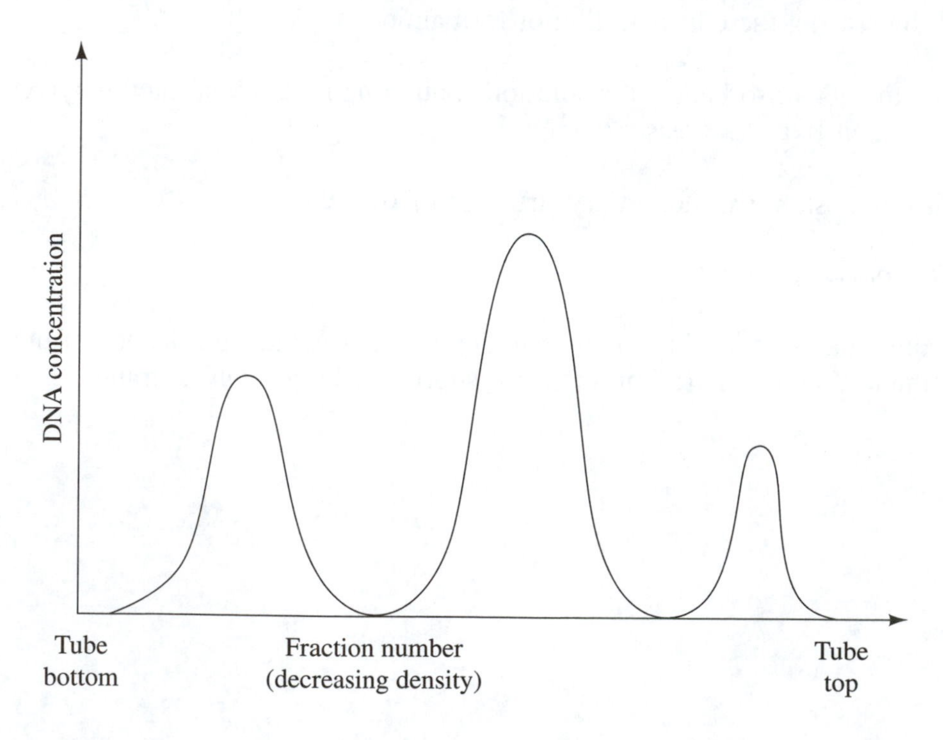

DNA–Protein Interactions

15. Specific DNA-binding proteins mainly contact DNA through hydrogen bonds in the major groove of B-DNA.
    (a) Why might sequence-specific binding be more common in the major groove than in the minor groove?
    (b) What hydrogen-bond contacts can proteins make with the bases in the major groove? Are these different from those in the minor groove?

16. In what ways do HTH proteins and the *met* repressor represent two general modes of DNA–protein interaction?

Eukaryotic Chromosome Structure

17. What insight into the structure of the eukaryotic chromosome was obtained from nuclease digestion studies?

18. Is histone H1 present in the electron micrograph shown in Figure 23-43? What is the relationship of H1 to the nucleosome structure?

19. Estimate the packing ratio of the DNA in human metaphase chromosomes, which have a total length of 200 μm and contain 6×10^9 bp.

20. Limited digestion of chromatin with a bacterial nuclease yields a 166-bp DNA fragment, whereas limited digestion with pancreatic DNase I yields 10-bp fragments. Explain the difference in these results.

21. Why is it important for the nucleic acid in certain RNA viruses to form a double-stranded A-type helix?

Answers to Questions

1. The A form of DNA, favored under dehydrating conditions, is the most likely form to occur in desiccated cysts or spores or in such things as brine shrimp eggs, which survive for years in a desiccated form.

2. Hydrogen bonding between base pairs and the limited rotation of the bases with respect to the sugar residues impose limitations on the rotation of the bonds in the ribose–phosphate backbone of double-stranded DNA. However, in single-stranded DNA, the ribose–phosphate bonds have greater freedom to rotate, allowing the polymer to take up a greater range of conformations.

3. The ribose pucker governs the relative orientations of the phosphate groups to each sugar residue. Residues in B-DNA have the C2′-*endo* conformation; in A-DNA, they have the C3′-*endo* conformation; and in Z-DNA, purine nucleotides are C3′-*endo* and pyrimidine nucleotides are C2′-*endo*.

4. The fifteen bands represent the native, negatively supercoiled plasmid (migrating the fastest because it is the most compact) and progressively less supercoiled DNA molecules (with the most relaxed DNA migrating the slowest because it is the least compact). Because Type I topoisomerase relaxes DNA by making single-strand cuts, the linking numbers of the 15 DNA bands change by increments of one from bottom to top.

5. DNA does not contain a 2′-OH group that can be deprotonated and then serve as a nucleophile to attack the phosphate group at the 3′ position.

6. Chaotropic agents such as urea and formamide tend to disrupt the structure of water. Hence, in their presence, water's ability to solvate DNA's anionic phosphate groups is reduced. Consequently, the phosphate groups repel one another more strongly than they do in the absence of the chaotropic agents, which induces the DNA to take up an extended rodlike conformation.

7. (a) Monovalent salts increase the T_m because they attenuate the repulsions between the negatively charged phosphate groups.
   (b) A large decrease in pH disrupts hydrogen bonding between base pairs by protonating some of the bases and therefore decreases T_m.
   (c) A large increase in pH disrupts hydrogen bonding between base pairs by deprotonating some of the base pairs and therefore decreases T_m.
   (d) Formamide disrupts water structure and thereby promotes denaturation, leading to a decrease in T_m.

8. An increase in ionic strength reduces the repulsions between phosphate groups and hence promotes base stacking. UV absorbance, which increases when the bases melt apart, would therefore decrease.

9. The tertiary structure of tRNA is stabilized by non-Watson–Crick hydrogen bonding and by stacking interactions.

10. Ethanol decreases the T_m of the DNA as well as decreasing its solubility, so it is important to keep the solution cold to keep the duplex intact. Shorter DNA molecules are more vulnerable to melting and hence strand separation, since T_m depends in part on the length of the DNA.

11. Gel A best represents the expected banding pattern. With increasing ethidium bromide, the plasmid unwinds, becoming progressively less supercoiled until it is a relaxed open circle. The relaxed open circle is less compact than the native plasmid; therefore, it migrates more slowly during electrophoresis. Further increases in ethidium bromide induce positive supercoils, so the plasmid again becomes more compact and migrates faster.

12. CsCl equilibrium density centrifugation would also be useful, since the native supercoiled plasmid is likely to be denser than the other two species of DNA.

13. The detergent and chaotropic agents denature proteins and destroy nucleic acid–protein interactions. The denatured protein collects at the top of the gradient, the DNA forms a band

in the gradient, and the RNA, which is too dense to form a band in the CsCl, collects as a pellet at the bottom of the tube.

14. The band nearest the bottom of the tube contains the DNA with the highest G + C content. Because the DNA fragments are so small, it is not possible to tell whether the DNA is nuclear or mitochondrial. Mitochondrial DNA would band separately from nuclear DNA if it were intact, but not when it is mixed with nuclear DNA in small pieces whose differences in base composition are nearly indistinguishable.

15. (a) The bases are more exposed in the major groove of B-DNA and are therefore more accessible to binding proteins. In addition, more base-specific hydrogen bonding donors and acceptors are exposed in the major groove.
    (b) In the major groove, the groups available for hydrogen bonding are N7 and N6 of adenine, N7 and O6 of guanine, N4 of cytosine, and O4 of thymine. In the minor groove, the groups available for hydrogen bonding are N3 of adenine, N3 and N2 of guanine, O2 of cytosine, and O2 of thymine.

16. Both kinds of protein exhibit a two-fold symmetry that is reflected in the palindromic DNA sequence at their binding sites. The α helices of HTH proteins contact the major groove of DNA directly, or indirectly via H_2O bridges. In the *met* repressor-like proteins, β strands contact the major groove of DNA.

17. Studies using micrococcal nuclease digestion indicated that the nucleosome contains ~200 bp of DNA. Further digestion trims this DNA to ~146 bp, leaving the nucleosome core particle.

18. Histone H1 is probably absent from the preparation shown in Figure 23-43; compare with Figure 23-46*a*. H1 appears to compact the DNA by binding to the ends of the DNA entering and leaving the nucleosome core.

19. The uncondensed DNA would have a length of $(6 \times 10^9 \text{ bp})(0.34 \text{ nm/bp}) = 2.04 \text{ m}$. The packing ratio is therefore $2.04 \text{ m}/200\ \mu\text{m} = \sim 10{,}000$.

20. The bacterial nuclease does not cleave the DNA of the nucleosome; however, DNase I appears to be able to cleave nucleosome-bound DNA, cutting it once per helical turn.

21. The RNA must be packaged efficiently within the viral protein capsid. The formation of a double-stranded A helix compacts the RNA in a regular fashion so that it can fit inside the regular shape of the capsid.

Chapter 24 DNA Replication, Repair, and Recombination

This chapter introduces you to the remarkably complex biochemistry of DNA replication, repair, and recombination. The chapter begins with a review of the classic Meselson and Stahl experiment showing that DNA is replicated semiconservatively; that is, each daughter DNA contains one new strand hydrogen bonded to one old, or parental, strand. The chapter reviews the essential steps in the replication of DNA, then takes a more detailed look at the proteins and enzymes involved in this remarkable process. Our best information comes from the enzymology of prokaryotic DNA replication, so this is the focal point of the discussion. Unique features of eukaryotic DNA replication are covered, particularly the problems presented by replicating a linear DNA molecule. The chapter then looks at the genesis of point mutations and small insertions and deletions of nucleotides during DNA replication. This naturally leads to a discussion of the mechanisms cells use to repair damaged DNA or mistakes incurred during replication. Again, prokaryotic mechanisms are understood best, and hence are the focus of the discussion. The chapter then considers the mechanisms by which prokaryotes engage in DNA recombination. This chapter concludes with an introduction to mobile DNA elements called transposons (or retrotransposons in eukaryotes), which can move to new locations in the genome by recombination during their own replication.

Essential Concepts

Overview of DNA Replication

1. The structure of DNA provides a template-driven mechanism for its replication. Experiments by Meselson and Stahl showed that each polynucleotide strand serves as a template for a new daughter strand. On completion of replication, each daughter strand, which is hydrogen bonded to its template, or parental strand, segregates to one of the daughter cells. This mode of DNA replication is called semiconservative DNA replication.

2. DNA polymerase requires a template, all four deoxynucleoside triphosphates (NTPs), and a primer from which to extend the chain. The polymerization reaction involves the nucleophilic attack of the growing DNA chain's (or a primer's) 3′-OH group on the α-phosphoryl group of a free NTP that is hydrogen bonded to the template. The liberated PP_i is subsequently hydrolyzed, making the polymerization irreversible. Because the 3′ end of the chain grows, polymerization is said to proceed from 5′ to 3′.

3. DNA replication is bidirectional, meaning that DNA synthesis proceeds in both directions from the site of initiation of replication. Circular DNA that contains a bubble where the parental DNA has unwound to allow replication is said to be undergoing θ replication.

4. Since DNA replication can proceed only in the 5′→3′ direction, one of the daughter strands at the replication fork must be synthesized as a series of fragments, so that the DNA is synthesized semidiscontinuously. In the leading strand, which extends 5′→3′ in the direction that the replication fork travels, DNA synthesis is continuous from the site of initiation. In the lagging strand, which extends in the opposite direction, DNA is synthesized as a series

of Okazaki fragments (named after the scientist who first observed their synthesis), which are later joined together by DNA ligase.

5. DNA synthesis *in vivo* begins as the extension of an RNA primer synthesized by primase (or RNA polymerase) at the site of initiation of DNA replication. Primers are subsequently removed by the 5′→3′ exonuclease activity of a DNA polymerase, and the polymerase fills in the gap, which is then sealed by DNA ligase. Ligation is endergonic and requires the free energy of ATP or NAD^+ hydrolysis.

Prokaryotic DNA Replication

6. There are three DNA polymerases in prokaryotes:
   (a) DNA polymerase I, the enzyme discovered first by Arthur Kornberg, removes RNA primers (via its 5′→3′ exonuclease activity) and subsequently fills in the gaps (via its 5′→3′ polymerase). These two reactions together are referred to as nick translation. Pol I also has 3′→5′ exonuclease activity to eliminate misincorporated nucleotides.
   (b) DNA polymerase II is involved in DNA repair.
   (c) DNA polymerase III is the primary DNA replicating enzyme. It is the largest of the three with at least 10 subunits. Pol III has 5′→3′ polymerase activity and 3′→5′ exonuclease activity to eliminate misincorporated nucleotides.

7. The initiation of DNA replication in *E. coli* occurs at a specific sequence called the *oriC* locus, where the DNA is melted apart, unwound, and the leading strand primer is synthesized.
   (a) DNA melting, or local denaturation, requires ATP and is accomplished by a complex of DnaA protein.
   (b) The hexameric protein DnaB (a helicase) further unwinds DNA in an ATP-dependent reaction. As positive supercoils develop, Type II topoisomerase is required to generate negative supercoils.
   (c) The tetrameric single-strand binding protein (SSB) prevents reannealing of the DNA. Numerous SSBs coat single-stranded DNA.
   (d) A complex of proteins called the primosome, which contains DnaB, primase (DnaG protein), and five other proteins, carries out primer synthesis. The primosome displaces SSB in an ATP-dependent process.

8. DNA synthesis of both leading and lagging strands is carried out by the replisome, a complex containing two Pol III enzymes. In order for the replisome to move as a single unit in the 5′→3′ direction, the lagging strand template must loop around once so that the Okazaki fragment and the leading strand can be synthesized in the same direction.

9. Specific DNA sequences (*Ter* sites) signal the termination of DNA synthesis. However, arrest of DNA replication requires the activity of Tus protein, which binds to a *Ter* site and inhibits or displaces DnaB.

10. The fidelity of DNA replication in *E. coli* is very high: ~1 error per 1000 bacteria per generation! Four factors contribute to this high fidelity:
    (a) Cells maintain balanced levels of all four deoxynucleotides.
    (b) Polymerization by Pol I, II, or III occurs only after an incoming nucleotide is properly positioned in the catalytic site and on the template.

(c) The 3′→5′ exonuclease activity of DNA polymerases detects and excises mismatches in the new DNA strand.
(d) DNA repair systems detect and correct residual errors after DNA synthesis is complete.

Eukaryotic DNA Replication

11. Eukaryotes contain at least five DNA polymerases that are distinguished largely by their different sensitivities to chemical inhibitors.

12. Eukaryotic DNA polymerases α and δ appear to share the functions of *E. coli* Pol III, the replicase. DNA polymerase α has 5′→3′ polymerase activity and primase activity but no 3′→5′ exonuclease activity and is only moderately processive. DNA polymerase δ is nearly infinitely processive when it is complexed with a protein called proliferating cell nuclear antigen (PCNA). DNA polymerase δ lacks primase activity but does have 3′→5′ exonuclease activity. Presumably, DNA polymerase δ synthesizes the leading strand, while DNA polymerase α synthesizes the lagging strand. However, other proteins are also required to replicate DNA *in vivo*.

13. DNA polymerase ε is required to repair UV-damaged DNA. The function of DNA polymerase β is unknown, but it may be involved in DNA repair as well. DNA polymerase γ is found in mitochondria (chloroplasts contain a similar enzyme), where it presumably replicates mitochondrial DNA.

14. A viral RNA-directed DNA polymerase is reverse transcriptase, which has no primase or 3′→5′ exonuclease activity but does contain ribonuclease (RNase H) activity. Reverse transcriptase uses a tRNA primer to replicate the viral genome.

15. In yeast, DNA replication begins at DNA sequences called autonomously replicating sequences (ARS). In higher eukaryotes, no analogous sequences have been identified. Many initiation sites appear simultaneously, but the control of DNA replication termination is not understood.

16. Eukaryotic chromosomes are linear, so the RNA primers at the 3′ ends of a chromosome cannot be replaced with DNA (DNA polymerase must extend a DNA strand from 5′ to 3′). This would lead to progressively shorter chromosomes, with the concomitant loss of genetic information, with each cell generation. However, eukaryotic chromosomes contain telomeres, repetitive sequences at the ends of chromosomes. These sequences are added by an enzyme called telomerase. Telomerase, a ribonucleoprotein, contains an RNA molecule that serves as a template for a reverse transcriptase activity that extends the telomere.

17. Telomerase activity is absent in somatic tissues of animals, which exhibit a finite replicative lifespan (about 60–80 divisions after birth in humans). Telomeres appear to shorten as cells age in culture. Hence, progressive loss of telomeres may have an impact on aging in at least some highly proliferative tissues. Many immortal cells (such as cancer cells) contain telomerase activity. Hence, telomerase is an attractive target for antitumor drug development.

Mutation

18. A mutation is a random, heritable alteration of genetic information and can arise in several ways:
    (a) Addition or deletion of one or more nucleotides.
    (b) Modification of a base so that it pairs with a different base during DNA replication.
    (c) Translocation of a segment of DNA from one region of the genome to another.

19. Chemicals that induce mutations, called chemical mutagens, produce two classes of DNA damage:
    (a) Point mutations, in which one base pair replaces another. In a transition, a purine (or pyrimidine) is replaced by another. In a transversion, a purine is replaced by a pyrimidine, or *vice versa*.
    (b) Insertion/deletion mutations, in which one or more nucleotide pairs are inserted or deleted.

20. Adenine and cytosine in cellular DNA are modified by methylation in a sequence-specific manner that affects gene regulation without causing a mutation.

21. Many mutagens are also carcinogens (chemicals that cause cancer). This property is used in the Ames test to assess the carcinogenic potential of chemicals. In the Ames test, a his^- strain of bacteria is exposed to a potential mutagen on an agar plate containing no His. The mutagenicity of a substance is scored as the number of his^+ colonies that appear, minus the few spontaneous ones that appear in the absence of the putative mutagen.

DNA Repair

22. DNA repair mechanisms range from simple one-enzyme systems (such as recognition of alkylated bases or pyrimidine dimers) to more elaborate multienzyme systems (such as nucleotide excision repair, recombination repair, and the SOS response).

23. Simple single-enzyme systems correct specific alkylated bases: e.g. O^6-methylguanine residues are recognized by O^6-methylguanine–DNA methyltransferase, which transfers the aberrant methyl group to a Cys residue on the enzyme. Pyrimidine dimers, which form from the absorption of UV light by two adjacent pyrimidine residues, are repaired by light-activated DNA photolyases.

24. Nucleotide excision repair in *E. coli* is carried out by several enzymes:
    (a) UvrABC endonuclease excises an oligonucleotide containing the damaged bases.
    (b) Pol I fills in the single-stranded region.
    (c) DNA ligase catalyses the formation of a phosphodiester linkage to restore the intact DNA molecule.

25. Other damaged bases may be removed by DNA glycosylases and the DNA restored by an endonuclease, polymerase, and a ligase.

26. A complex DNA repair system in *E. coli* is the SOS response, which is activated when DNA damage prevents replication. SOS repair is error prone and therefore mutagenic.

27. RecA, a key protein in genetic recombination, is also involved in recombination repair (postreplicative repair). During DNA replication, DNA synthesis halts at sites of DNA damage and resumes downstream from the damage. The gap in the daughter strand is then repaired by recombination with the homologous segment from the other normal daughter strand.

Recombination

28. General recombination occurs between two sections of DNA with extensive homology. The aligned strands of DNA are cleaved, and the strands cross over to form a four-branched structure called a Holliday junction. The junction can dissociate into two new duplexes in two ways, both yielding new DNA molecules.

29. Site-specific recombination in *E. coli* is mediated by RecA, which polymerizes on single-stranded gaps in DNA duplexes. RecA partially unwinds the duplexes and exchanges two strands in an ATP-dependent reaction. All recombination events require additional proteins to further unwind DNA, produce nicks, maintain proper supercoiling, drive branch migration, and seal nicks.

30. Transposons are DNA segments in prokaryotes and eukaryotes that can move from one site in the genome to another by a mechanism that involves their replication. The simplest transposons are insertion elements (IS), which contain a transposase gene that mediates the recombination reactions involved in transposition. This recombination occurs between short target sequences located at the ends of the transposon. More complex transposons contain genes required for transposition as well as other genes, such as antibiotic resistance genes.

31. Transposons promote inversions, deletions, and rearrangements of host DNA and thereby affect gene expression and permit chromosome evolution. Eukaryotic transposons resemble the DNA from retroviruses rather than bacterial transposons; hence they are called retrotransposons. Transposition of a retrotransposon begins with its transcription to RNA, followed by synthesis of DNA from the RNA template by reverse transcriptase; the DNA then inserts randomly into the host genome.

Guide to Study Exercises (text p. 811)

1. DNA replication is semiconservative since the parental DNA separates into two strands and one parental strand is conserved in each of the two daughter molecules. Synthesis of new DNA occurs at the point where the parental strands separate, called a replication fork. Replication initiates at a point along the chromosome, forming two replication forks so that replication proceeds in opposite directions along the length of DNA, that is, bidirectionally. Replication is semidiscontinuous because one strand (the leading strand) is synthesized continuously from start to finish (origin to termination site) by DNA polymerase moving along the template DNA strand. The other strand (the lagging strand) is synthesized discontinuously as Okazaki fragments because the DNA polymerase, which synthesizes in the $5'\rightarrow3'$ direction, must repeatedly "back up" and restart as template DNA is newly exposed at the replication fork. (Section 24-1)

2. RNA is essential for lagging-strand synthesis because each Okazaki fragment begins with the synthesis of an RNA primer that DNA polymerase then extends. In eukaryotes, shortening of chromosome ends during replication is prevented by the addition of telomeric sequences that serve as templates for the synthesis of the final Okazaki fragment of the lagging strand. The telomeric DNA is added by telomerase, a ribonucleoprotein whose RNA component acts as a template for the addition of nucleotides at the chromosome end. (Sections 24-1 and 24-3C)

3. The 5′→3′ exonuclease activity of *E. coli* DNA polymerase I excises up to 10 nucleotides from the 5′ end of the single-stranded nick between two Okazaki fragments, thereby removing the RNA primer. The polymerase activity replaces these nucleotides with deoxyribonucleotides that are added to the 3′ end of the nick. The 3′→5′ exonuclease activity allows DNA polymerase I to edit its mistakes by excising a mispaired nucleotide at the growing end of a DNA chain. (Section 24-2A)

4. The two β subunits of *E. coli* DNA polymerase III form a sliding clamp around the DNA that helps anchor the polymerase to the DNA and increase its processivity. In eukaryotes, the three PCNA subunits form a ring whose structure closely resembles that of the prokaryotic β clamp and which enhances the processivity of DNA polymerase δ. (Sections 24-2C and 24-3A)

5. ATP hydrolysis provides the free energy for certain steps of DNA replication: (a) Melting open the parental DNA helix at the origin of replication (e.g., by *E. coli* DnaA); (b) Unwinding DNA at the replication fork for primer synthesis (e.g., by *E. coli* DnaB, PriA, and Type II topoisomerase); (c) Reloading the β clamp for synthesis of a new Okazaki fragment, catalyzed by the γ complex of *E. coli* polymerase III; (d) Sealing of single-strand nicks by ATP-requiring DNA ligases. Reactions that do not require ATP hydrolysis include the polymerase reaction itself, in which the addition of a nucleotide to the growing DNA chain is accompanied by the elimination of PP_i, whose subsequent hydrolysis to 2 P_i drives the polymerization reaction. Another non-ATP-dependent reaction is catalyzed by NAD^+-dependent DNA ligases. (Sections 24-1, 24-2B and C)

6. The high fidelity of DNA replication results from (a) balanced levels of dNTPs that prevent misincorporation of an overabundant dNTP or substitution for a scarce dNTP; (b) conformational changes in DNA polymerase that ensure proper base pairing between the incoming nucleotide and the template; (c) the 3′→5′ exonuclease activity that detects and excises misincorporated nucleotides; and (d) enzymes that detect and repair DNA damage. (Section 24-2E)

7. Different enzymes are required to replicate DNA in prokaryotes and eukaryotes. In prokaryotes, replication of the circular chromosome begins at a single point (*oriC* in *E. coli*); in much larger eukaryotic genomes, replication initiates at multiple origins. Prokaryotic Okazaki fragments are 1000–2000 nt long; eukaryotic Okazaki fragments are 100–200 nt long. Bacterial DNA polymerase catalyzes the synthesis of 1000 nt per second; eukaryotic DNA polymerase synthesizes 50 nt per second. Replication of the circular bacterial chromosome terminates where two replication forks meet (at *Ter* sites in *E. coli*); eukaryotes require special mechanisms for replicating the ends of their linear chromosomes. (Sections 24-1, 24-2, and 24-3)

8. The ends of eukaryotic chromosomes consist of at least 1000 tandem G-rich sequences on the 3′-ending strand, which extends to form a 12- to 16-nt single-strand overhang. This extension serves as a template for the primer that initiates synthesis of the final Okazaki fragments of the lagging strand during DNA replication. (Section 24-3C)

9. Mutations (heritable changes in an organism's DNA) can arise from (a) DNA polymerase errors during replication; (b) radiation- or chemical-induced structural alteration of bases; (c) spontaneous hydrolysis of glycosidic bonds in DNA; (d) insertions or deletions due to intercalating agents; (e) error-prone repair processes such as SOS repair and recombination repair; (f) errors in general recombination; and (g) rearrangements mediated by transposons. (Sections 24-4, 24-5C, and 24-6)

10. Direct DNA repair is typically accomplished via a single enzyme that recognizes a damaged or modified nucleotide and restores it to its original state. These enzymes include methyltransferases that remove a methyl group, or DNA photolyases that reverse UV damage. Nucleotide excision repair requires multiple enzymes, including an endonuclease that removes an oligonucleotide containing the damage, a polymerase to replace nucleotides, and DNA ligase to seal the nick. (Sections 25-5A and B)

11. General recombination occurs according to the Holliday model. Aligned strands of homologous DNA are nicked, and the nicked strands cross over to pair with the nearly complementary strand. The crossover point can migrate in either direction by further strand exchange before the break is sealed. Depending on how the four-stranded crossover structure is resolved, the resulting homologous DNA molecules may have exchanged a single segment of DNA or DNA all the way to the end of the chromosome. (Section 25-6A)

12. In *E. coli,* recombination is supported by RecBCD, which unwinds duplex DNA and makes single-strand cuts. RecA catalyzes strand exchange via the formation of a three- or four-stranded DNA helix in an ATP-dependent manner. Topoisomerases maintain the proper level of supercoiling; SSB binds to single-stranded regions; RuvB is an ATP-dependent motor that drives branch migration; RuvA targets RuvB to the Holliday junction, RuvC cuts apart the two recombinant duplexes, and DNA ligase seals the nicks. (Section 24-6A)

13. Transposons mediate genetic recombination by causing the insertion or deletion of a segment of DNA. The movement of a transposon may interrupt a gene that contains the insertion site. The transposon may also carry one or several genes (even from another species) that can be introduced to a new site in the host genome. Segments of DNA between transposons may be inserted or deleted via recombination between the two transposons. (Section 24-6B)

Questions

Overview of DNA Replication

1. What feature of DNA replication, as shown in Figure 24-4, is inconsistent with the known enzymatic properties of DNA polymerases? What observation reconciled this inconsistency?

2. What prevents the reverse reaction of DNA polymerization *in vivo*?

Prokaryotic DNA Replication

3. Describe the roles of the exonuclease activities of Pol I.

4. For each of the enzymes and proteins listed below, match the function or feature related to DNA replication in *E. coli*.

A. DNA polymerase I
B. DNA polymerase II
C. DNA polymerase III
D. DNA ligase
E. DnaB
F. Single-strand binding protein
G. Primase
H. Tus protein
I. DNA gyrase

______ Required for the initiation of DNA synthesis
______ Essential for the condensation of Okazaki fragments
______ A helicase required for unwinding the DNA duplex
______ Involved in the termination of DNA strand replication
______ The enzyme that is strictly involved in DNA replication
______ The enzyme critical for relieving buildup of positive supercoils
______ The enzyme principally involved in DNA repair
______ Prevents the reannealing of DNA in the replication fork
______ Excision and replacement of RNA primers
______ Incapable of nick translation
______ Requires NAD^+ in *E. coli*
______ The most abundant DNA polymerase

5. Radioactive DNA probes can be made by nick translation of linear pieces of DNA. Why is a trace amount of DNase I necessary?

6. From the gut of a slug in your garden you have identified a new circular, double-stranded DNA-containing bacteriophage. Electron micrographs of infected bacteria during production of phage progeny reveal the structures shown below (thick lines represent double-stranded DNA and thin lines represent single-stranded nucleic acids). Alkali treatment of these replicative DNA forms destroys their single-stranded "tails." Bacteria incubated with ^{15}N-nucleosides were infected with this bacteriophage. CsCl equilibrium density centrifugation of DNA reveals two bands of phage DNA. The heavier band contains 100 times more DNA than the lighter band.
   (a) What is the composition of the single-stranded nucleic acid in the structures shown on the opposite page?
   (b) Suggest a mechanism for DNA replication in this bacteriophage.

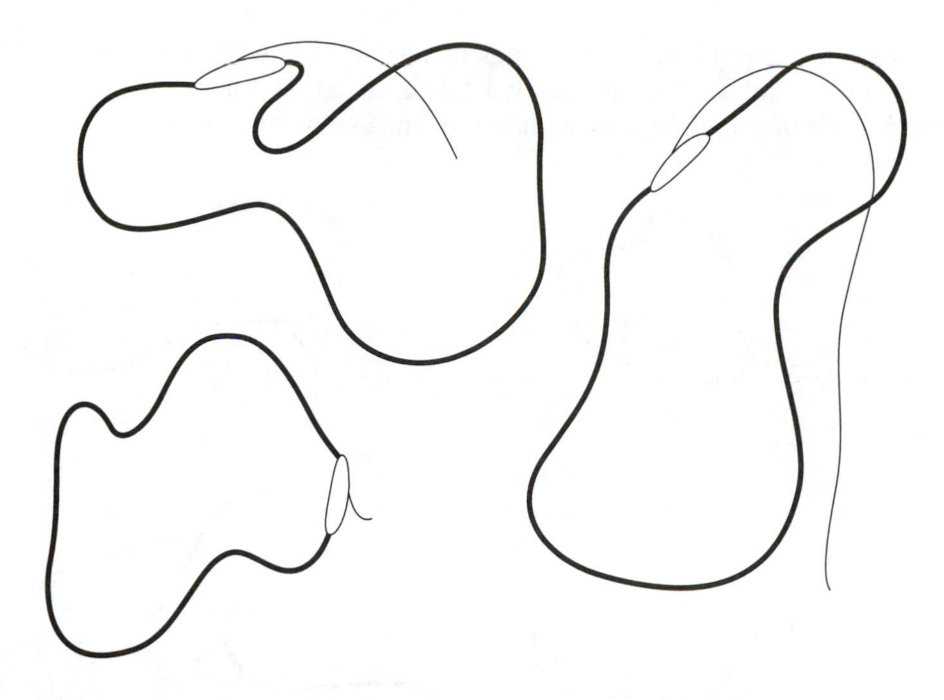

Eukaryotic DNA Replication

7. Which eukaryotic DNA polymerase is most likely responsible for the following activities: (a) Leading-strand synthesis, (b) Okazaki-fragment synthesis, (c) DNA repair.

8. Why does reverse transcription of retroviral DNA require a specific host tRNA for DNA synthesis? What section of the tRNA is most likely to be involved?

9. When reverse transcriptase is used in the oligo(T)-primed synthesis of double-stranded DNA from a specific cellular mRNA, the aggregate length of the resulting double-stranded cDNA is sometimes twice as long as that of the mRNA template. Explain this phenomenon.

10. What mammalian cell type must retain its telomerase activity?

11. Shown below are schematic drawings of various circular DNAs undergoing replication. The thick lines indicate double-stranded DNA, and thin lines indicate single-stranded DNA. Match each of the structures with one of the statements below.

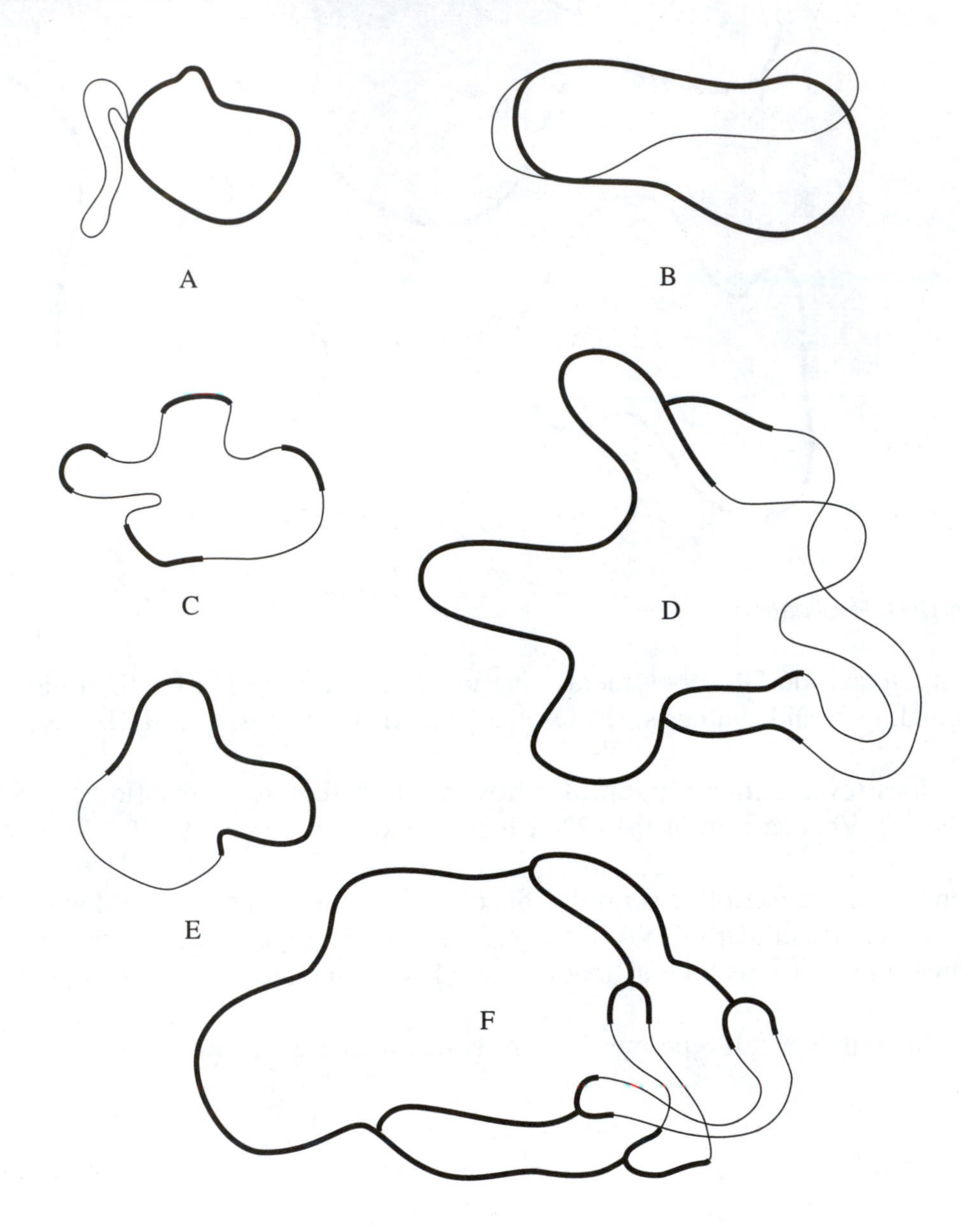

_____ This structure is likely to be found in rapidly dividing bacteria.

_____ This structure corresponds to an intermediate in the synthesis of the replicative form of a bacteriophage.

_____ This structure is consistent with the rolling circle model of DNA replication.

Mutation

12. In the pairs of DNA sequences below, the lower duplex represents a mutagenized daughter duplex. Identify the mutation as a transition, a transversion, an insertion, or a deletion.

(a) 5′ GCCTAGAACCAGTAC 3′
3′ CGGATCTTGGTCATG 5′

5′ GCACTAGAACCAGTA 3′
3′ CGTGATCTTGGTCAT 5′

(b) 5′ CGCATAGCTACTGGAA 3′
3′ GCGTATCGATGACCTT 5′

5′ CGCAGAGCTACTGGAA 3′
3′ GCGTCTCGATGACCTT 5′

13. What kinds of mutations does 5-bromouracil generate?

14. Match the compound on the left with the kind of mutation on the right.

______ Acridine orange	A. Transition
______ Nitrous acid	B. Transversion
______ Ethylnitrosourea	C. Insertion or deletion
______ Dimethyl sulfate	
______ Ethidium bromide	

DNA Repair

15. Match the proteins below with their roles in DNA repair.

A. O^6-methylguanine–DNA methyltransferase
B. UvrABC endonuclease
C. uracil *N*-glycosylase
D. RecA

______ Removes 7-methyladenine residues from damaged DNA
______ Serves as a sink for methyl residues abstracted from O^6-methylguanine residues
______ Removes deaminated cytosine residues
______ Removes damaged DNA via postreplication repair
______ Removes thymine dimers
______ Initiates the SOS response

Recombination

16. Why is transposition a misnomer in describing the operation of bacterial transposons?

17. You transformed a strain of *E. coli* with a plasmid containing a transposon that includes an ampicillin-resistance gene. To recover the plasmid, you pick a colony from a culture plate containing ampicillin. Much to your surprise, the plasmid you recover is twice the size you expected. You then pick 100 more colonies and discover that only one of them has plasmid of the expected size.
    (a) Why is the recovery of plasmid of expected size so low?
    (b) Why can you recover any plasmid of expected size at all?

Answers to Questions

1. The largely symmetric labeling of each replication fork suggests that both strands of DNA are synthesized simultaneously. However, this implies that one strand is synthesized from 5′ to 3′ and the other strand is synthesized from 3′ to 5′. A polymerase that acts in the 3′→5′ direction has never been found. Okazaki demonstrated that one of the new strands is synthesized continuously in the 5′→3′ direction and one strand is synthesized discontinuously in the 5′→3′ direction, after which the DNA fragments are ligated together.

2. The polymerase reaction is irreversible *in vivo* because the pyrophosphate released on incorporation of a nucleotide into a DNA strand is quickly cleaved to 2 P_i.

3. The 3′→5′ exonuclease activity, located near the polymerase active site, recognizes and removes mismatched bases. The 5′→3′ exonuclease site of Pol I binds to DNA at single-strand breaks and initiates nick translation.

4. __G__ Required for the initiation of DNA synthesis
   __D__ Essential for the condensation of Okazaki fragments
   __E__ A helicase required for unwinding the DNA duplex
   __H__ Involved in the termination of DNA strand replication
   __C__ The enzyme that is strictly involved in DNA replication
   __I__ The enzyme critical for relieving buildup of positive supercoils
   __A, B__ The enzyme principally involved in DNA repair
   __F__ Prevents the reannealing of DNA in the replication fork
   __A__ Excision and replacement of RNA primers
   __B, C__ Incapable of nick translation
   __D__ Requires NAD^+ in *E. coli*
   __A__ The most abundant DNA polymerase

5. A trace amount of DNase is necessary to produce single-strand nicks that are the starting point for the 5′→3′ exonuclease and polymerase activities of Pol I.

6. (a) The single-stranded nucleic acid is likely to be composed entirely of RNA, suggesting that replication of the phage genome begins with the synthesis of a "genomic" RNA.

(b) The appearance of only two bands of phage DNA suggests that the "genomic" RNA serves as a template to generate new double-stranded DNA. In this scenario, ^{15}N-nucleosides are incorporated into a DNA strand complementary to the RNA and then into a second DNA strand complementary to the first. Hence, both strands of newly synthesized DNA are labeled with ^{15}N. During infection, when progeny phage are produced, there is much more of the heavy new phage DNA than there is of the original light ^{14}N-containing DNA.

7. (a) DNA polymerase δ
   (b) DNA polymerase α
   (c) DNA polymerases ε and β

8. The host tRNA serves as an RNA primer for the synthesis of the DNA. The 3′ end is the most likely section of the molecule to serve as the primer, since it provides the 3′-OH group to which deoxynucleotides can be attached.

9. The 3′ end of the newly synthesized cDNA strand folds back to prime the synthesis of a complementary DNA strand, presumably on initiation of RNase H activity, which digests away the original mRNA template. This folded DNA then serves as a template for the synthesis of a complementary strand by DNA polymerase.

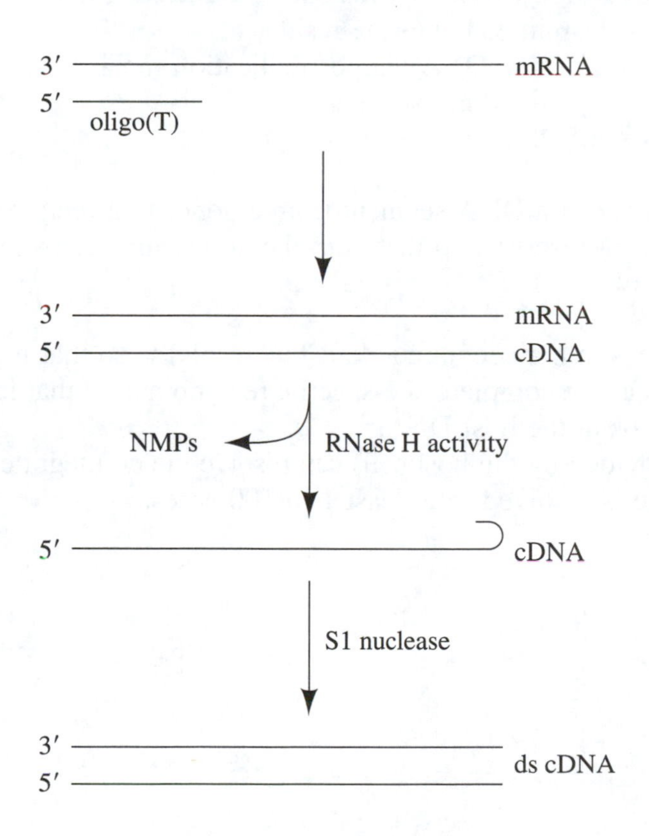

10. The cells that give rise to eggs and sperm must retain telomerase activity in order to produce immortal gametes.

11. __F__ This structure is likely to be found in rapidly dividing bacteria.
__E__ This structure corresponds to an intermediate in the synthesis of the replicative form of a bacteriophage.
__A__ This structure is consistent with the rolling circle model of DNA replication.

12. (a) The mutation is an insertion of an A · T base pair in the third position.
(b) The mutation is a transversion at the fifth position (T · A to G · C).

13. 5-Bromouracil is a base analog that can substitute for thymine and base pair with adenine. However, its enol tautomer, which forms more readily than that of T, can also base pair with guanine, thereby generating a T · A → C · G transition. When 5BU substitutes for cytosine, it generates a C · G → T · A transition.

14. __C__ Acridine orange
__A__ Nitrous acid
__B__ Ethylnitrosourea
__B__ Dimethyl sulfate
__C__ Ethidium bromide

15. __B__ Removes 7-methyladenine residues from damaged DNA
__A__ Serves as a sink for methyl residues abstracted from O^6-methylguanine residues
__C__ Removes deaminated cytosine residues
__D__ Removes damaged DNA via postreplication repair
__B__ Removes thymine dimers
__D__ Initiates the SOS response

16. There is no transfer of a DNA segment from a donor to a recipient; instead, the transposon is replicated during insertion so that both the donor and recipient ultimately contain transposon sequences.

17. (a) The transposon may contain a defective resolvase so that it can integrate into the host DNA but cannot complete site-specific recombination that lets the original plasmid be recovered from the host DNA.
(b) RecA (provided by the host cell) can resolve the cointegrate at a low level, so that the cointegrate is resolved in at least 1 in 100 cases.

Chapter 25 Transcription and RNA Processing

This chapter focuses on the biochemistry of RNA synthesis, from the initiation of transcription through the subsequent modification of the transcription products into biologically functional RNAs. The chapter begins by discussing the structure and biochemical properties of prokaryotic RNA polymerase, comparing and contrasting them with those of DNA polymerase. The discussion then progresses to the three main activities of transcription: template binding and RNA chain initiation, chain elongation, and chain termination. The chapter then examines the remarkably more complex biochemistry of transcription in eukaryotic cells, which involves multiple DNA sequence elements (promoters, enhancers, and silencers) and many proteins (transcription factors). In addition, eukaryotic cells use three different isoforms of RNA polymerase to synthesize different classes of RNA. The chapter concludes with a discussion of posttranscriptional processing, focusing first on the extensive modifications of eukaryotic RNAs. Here you are introduced to some of the minor forms of RNA involved in RNA processing (small nuclear RNAs and guide RNAs). Finally, rRNA and tRNA processing in prokaryotes and eukaryotes is considered.

Essential Concepts

1. Cells contain several types of RNA, mainly ribosomal RNA (rRNA), transfer RNA (tRNA), and messenger RNA (mRNA). rRNA constitutes better than 80% of the total mass of cellular RNA and accounts for two-thirds of the mass of ribosomes. tRNAs are small RNAs that deliver amino acids to the ribosome. mRNAs contain the nucleotide sequences that encode the amino acid sequences of polypeptides. In addition to these three large classes, which account for over 95% of the mass of RNA in eukaryotic cells, there are small nuclear RNAs (snRNAs) involved in mRNA splicing, guide RNAs involved in RNA editing, and other small RNAs.

RNA Polymerase

2. The transcription of DNA into RNA is carried out by RNA polymerases, which contain multiple different subunits. Bacteria contain one RNA polymerase core enzyme, whereas eukaryotes have four or five isoforms. All RNA polymerases catalyze the reaction

$$(\text{RNA})_{n \text{ residues}} + \text{NTP} \rightleftharpoons (\text{RNA})_{n+1 \text{ residues}} + \text{PP}_i$$

The enzyme uses the substrates ATP, CTP, GTP, and UTP, synthesizing an RNA chain in the $5' \rightarrow 3'$ direction. Hydrolysis of the released PP_i makes polymerization irreversible.

3. The RNA polymerase holoenzyme in *E. coli* contains four types of subunits: $\alpha_2\beta\beta'\sigma$. The σ subunit dissociates from the complex on the initiation of transcription, leaving the core enzyme to continue transcription.

4. Many prokaryotic genes exist in contiguous sets of functionally related genes called operons and are transcribed as polycistronic mRNAs. Most protein-encoding genes (structural genes) in eukaryotes, however, are transcribed individually as monocistronic mRNAs.

5. The prokaryotic RNA holoenzyme binds to DNA sequences called promoters, which are "upstream" from the coding sequence of a gene or operon. RNA polymerase initiates transcription about 10 nucleotides "downstream" from a consensus DNA sequence called the Pribnow box.

6. Different genes or operons have different promoter sequences, which allows for differential gene expression in response to environmental needs or developmental changes in the organism. Promoter selectivity is regulated by the σ subunit of RNA polymerase. The rate of transcription of a gene is directly proportional to the rate at which the holoenzyme forms a stable initiation complex.

7. A promoter region can be identified by a technique called footprinting. RNA polymerase and DNA containing the putative promoter are incubated with an alkylating reagent, which results in cleavage of the DNA backbone at the alkylated residues. The products of this reaction can be analyzed by electrophoresis to determine the region of DNA that binds to RNA polymerase.

8. The RNA polymerase core enzyme binds DNA tightly ($K \approx 5 \times 10^{-12}$ M, with a half-life of ~60 min). The RNA polymerase holoenzyme, however, binds DNA relatively weakly ($K \approx 10^{-7}$ M) except at specific promoter sequences recognized by the σ subunit. Hence, the σ subunit allows the holoenzyme to rapidly scan many sites until it finds its promoter. RNA polymerase melts ~11 bp of DNA to form an open complex. Once initiation of an RNA chain has occurred, the σ subunit is jettisoned, and the tight-binding core enzyme elongates the RNA chain with high processivity until a termination site is reached.

9. Transcription termination is relatively imprecise. In prokaryotes, termination occurs in one of two major ways:
   (a) RNA polymerase halts when it encounters a hairpin-forming GC-rich region that is followed by a series of U residues on the DNA sense strand.
   (b) Termination is mediated by rho factor, a hexameric protein that recognizes a nascent RNA and unwinds the DNA–RNA duplex in the open complex, thereby releasing the RNA transcript at a particular site.

Transcription in Eukaryotes

10. Eukaryotes have three distinct isoforms of RNA polymerase:
    (a) RNA polymerase I, located in nucleoli, synthesizes the precursors of most rRNAs.
    (b) RNA polymerase II, located in the nucleoplasm, synthesizes mRNA precursors.
    (c) RNA polymerase III, also located in the nucleoplasm, synthesizes the precursors of 5S rRNA, tRNA, and a variety of small nuclear and cytosolic RNAs.

11. Each RNA polymerase contains up to 12 subunits, three of which are homologous to the α, β, and β′ subunits of prokaryotic RNA polymerase.

12. In eukaryotic promoters, which are much more complex and diverse than prokaryotic promoters, the position and orientation may vary considerably relative to the transcription start site.

13. RNA polymerase I recognizes only one promoter sequence, but this sequence is species specific. In contrast, RNA polymerases II and III recognize many different promoters. Some RNA polymerase III promoters are located within the gene's transcribed sequence.

14. There are two classes of RNA polymerase II promoters, both analogous to prokaryotic promoter sequences:
    (a) Most constitutively expressed genes (housekeeping genes) contain a GC box (GGGCGG or its complement).
    (b) Most genes that are expressed in specific cells at specific times contain an AT-rich region known as the TATA box.

    In addition, there are other upstream sequences that are critical for transcription initiation (e.g., the CCAAT box).

15. DNA sequences upstream of the gene play crucial roles in the regulation of transcription. These sequences include enhancers (which stimulate transcription initiation rates) and silencers (which repress transcription initiation rates). Consequently, the regulatory proteins that bind to these sequences are called activators or repressors, respectively. All these regulatory proteins are considered to be transcription factors.

16. In eukaryotes, at least six proteins (called general transcription factors or GTFs) are required to recruit RNA polymerase to a promoter and initiate transcription.

17. The multiprotein complex that initiates transcription is called the preinitiation complex (PIC). The formation of the PIC begins with the binding of the TATA-binding protein (TBP) to the TATA box of a promoter. Several other GTFs bind in succession, followed by RNA polymerase II and the remaining GTFs.

18. One of the PIC proteins (TFIIH) is an ATP-dependent helicase, which forms the open complex. In addition, TFIIH phosphorylates the largest subunit of RNA polymerase II, an event correlated with the transition of RNA chain initiation to RNA chain elongation. As elongation ensues, various GTFs dissociate from the complex.

19. RNA polymerases I and III use TBP and different sets of GTFs at their promoters.

Posttranscriptional Processing

20. In prokaryotes, primary transcripts are translated without further modification; however, in eukaryotes most primary transcripts must be specifically modified by:
    (a) The addition of nucleotides to the 5′ and 3′ ends and, in rare cases, in the interior of the transcript.
    (b) The exo- and endonucleolytic removal of polynucleotide sequences.
    (c) Modification of specific nucleotide residues.

21. Eukaryotic mRNAs have a cap structure consisting of a 7-methylguanosine residue joined to the 5′ end of the mRNA via a 5′–5′ triphosphate linkage. This cap is added to the growing transcript by guanylyltransferase.

22. Most eukaryotic mRNAs also have a string of adenosine residues, the poly(A) tail, at their 3′ end. The primary transcript is cleaved past an AAUAAA sequence and the poly(A) tail is added by poly(A) polymerase. The poly(A) tail, which ranges from 20 to 250 nt, protects the mature mRNA from ribonuclease activity in cells.

23. Unlike prokaryotic genes, most eukaryotic structural genes give rise to primary transcripts, called heterogeneous RNA (hnRNA), that are much longer than predicted from the size of the polypeptides they encode. These transcripts are processed by splicing, which is the excision of intervening sequences (introns) and the ordered joining of the flanking expressed sequences (exons) to form a mature mRNA.

24. Each splicing reaction involves two transesterification reactions mediated by ribonucleoprotein complexes called spliceosomes. Several small nuclear RNAs (snRNAs) mediate proper alignment of splice junctions within the spliceosome. Splicing begins with the nucleophilic attack by the 2′-OH group of an intron adenosine residue on the intron's 5′-phosphate group. The second transesterification reaction splices the 3′-OH group of the 5′ exon with the 5′ end of the 3′ exon, releasing the intron in the form of a lariat.

25. The DNA sequences of exon–intron junctions are conserved. In particular, the 5′ dinucleotide of an intron is invariably GU and the 3′ dinucleotide is invariably AG.

26. In certain organisms, mRNA undergoes editing in which numerous U residues are inserted and removed to generate a translatable mRNA. Guide RNAs base pair with the immature mRNAs to direct these alterations.

27. In *E. coli,* primary rRNA transcripts contain the sequences for 5S rRNA, 16S rRNA, 23S rRNA, and up to four tRNAs. These RNAs are released by the action of several specific RNases that commence processing while the primary transcript is still being synthesized.

28. Eukaryotic 45S rRNA primary transcripts contain the sequences for 5.8S rRNA, 18S rRNA, and 28S rRNA. Roughly 110 sites are methylated prior to the nucleolytic removal of spacer sequences that separate the three rRNAs.

29. A few eukaryotic rRNAs have introns that are removed by transesterification reactions in which the pre-rRNA acts as a catalyst. Group I introns are removed by a self-splicing reaction that requires a free guanine nucleotide but no protein. Group II introns react via a lariat intermediate and require no free nucleotide.

30. Prokaryotic and eukaryotic transfer RNA is processed by trimming the ends of the primary transcript and modifying specific bases. Many eukaryotic pre-tRNAs also have an intron that is spliced out. In addition, tRNA nucleotidyltransferase adds two C's and an A to the 3′ end of the immature tRNA (prokaryotic pre-tRNA already contains this trinucleotide).

Guide to Study Exercises (text p. 843)

1. If "gene" refers to a sequence of nucleotides in DNA that corresponds to the sequence of amino acids in a polypeptide, then it would not include the sequences that encode tRNA,

rRNA, and other small RNA molecules (i.e., RNAs that are not translated into protein). If "gene" refers to a sequence of DNA that is transcribed as a single unit, then it would include operons, which contain several protein-coding sequences in one RNA transcript. Furthermore, many transcripts are processed by the addition, removal, or covalent modification of nucleotides before they become fully functional or can be translated. In these cases, the "gene" does not reflect the final sequence of the RNA or protein molecule. In addition, many DNA sequences that are essential for the proper expression of a "gene" are located far upstream or downstream of the transcribed region but could be considered part of the gene because of their regulatory function. Finally, in some cases, a single DNA sequence may contain overlapping coding regions so that it is impossible to separate the "genes." (Sections 25-1 and 25-3)

2. The molecular structures of RNA and DNA polymerases are roughly similar: a hand shape with the active site at the bottom of a deep cleft. DNA replication requires that DNA polymerases work in pairs to synthesize two new strands of DNA. RNA synthesis requires only one polymerase to produce a single RNA strand. DNA polymerase uses dATP, dCTP, dGTP, and dTTP as substrates, whereas RNA polymerase uses ATP, CTP, GTP, and UTP. Both enzymes follow the same mechanism: The growing chain's 3′-OH group nucleophilically attacks the α-phosphoryl group of an incoming nucleotide that has been selected by its ability to pair with a template DNA strand. The hydrolysis of the released PP_i makes the polymerization reaction irreversible. DNA polymerase has an error rate of 10^{-8} to 10^{-10}; the RNA polymerase error rate is much higher, $\sim 10^{-4}$. DNA polymerase uses as its template all the organism's DNA so that a complete copy of the genetic information can be made. RNA polymerase uses as its template only a small portion of the genome that is selected through interaction with accessory proteins such as transcription factors. (Sections 24-2 and 25-1)

3. Genes that are arranged in operons can be regulated in a coordinated fashion and can be translated rapidly and simultaneously. This is an advantage when the products of the genes are needed at the same time and/or in similar amounts. This is a disadvantage when the organism actually needs only one of the gene products, because the effort of transcribing, translating, and then degrading the unneeded mRNA and protein is wasteful. (Section 25-1B)

4. A consensus sequence is a nucleotide sequence that represents the average sequence with a particular function. The consensus sequence usually does not occur in such a form in any organism but is an ideal constructed from the frequency with which the four bases appear at each position in the sequence in a number of organisms. (Section 25-1B)

5. The RNA polymerase holoenzyme, which includes the σ subunit, binds to DNA relatively weakly ($K \approx 10^{-7}$ M). This allows the holoenzyme to quickly scan DNA to locate a promoter specific for the σ subunit. Following transcription initiation, the σ subunit is released, leaving the core enzyme. This complex binds more tightly ($K \approx 5 \times 10^{-12}$ M) but nonspecifically to DNA. This allows the enzyme to associate stably with DNA along the length of the gene while it synthesizes RNA. (Section 25-1B)

6. Eukaryotic RNA polymerase I, located in the nucleolus, synthesizes most rRNA precursors. RNA polymerase II occurs in the nucleoplasm and synthesizes mRNA precursors. RNA polymerase III occurs in the nucleoplasm and synthesizes the precursors of tRNA and other small RNAs, including some rRNA. (Section 25-2A)

7. Assembly of the eukaryotic preinitiation complex (PIC) requires at least six general transcription factors (GTFs). First, TBP (a component of TFIID) binds to and distorts the TATA box of a promoter. Next, TFIIA, TFIIB, TFIIE, TFIIF, and TFIIH assemble at the promoter. Their order of addition is essential for recruiting RNA polymerase and preparing the DNA for transcription. Other components of TFIID may bind but are not strictly required to initiate transcription. (Section 25-2C)

8. Eukaryotic mRNA is posttranscriptionally modified by the addition of a 5′ cap and a 3′ poly(A) tail and by the removal of introns and the joining of exons to form a translation-ready mRNA. In rare cases, bases may be added, deleted, or modified through an editing process. Eukaryotic rRNA is processed by methylation of specific sites followed by nucleolytic cleavage and trimming to release individual rRNA molecules from a large primary transcript. In some cases, the rRNAs undergo self-splicing to remove introns. Eukaryotic tRNA posttranscriptional processing includes nucleolytic removal of extra nucleotides, splicing, residue modification, and the addition of CCA to the 3′ ends of tRNAs. (Section 25-3)

9. Splicing in both mRNA and group I introns proceeds by transesterification reactions and therefore does not require an external source of free energy. mRNA splicing requires the spliceosome, a complex of proteins and snRNPs, whereas splicing in group I introns requires only a free guanosine or guanine nucleotide. In mRNA splicing, a 2′,5′-phosphodiester bond forms between an intron adenosine residue and the intron's 5′-terminal phosphate group. This leaves a lariat structure that is eliminated when the 3′-OH group of the 5′ exon displaces the 3′ end of the intron, forming a new 3′,5′-phosphodiester bond with the 5′ end of the 3′ exon. In group I intron splicing, the guanosine 3′-OH group attacks the intron's 5′ end, displacing the 3′-OH group of the 5′ exon and forming a 3′,5′-phosphodiester bond rather than a 2′,5′-phosphodiester bond as in mRNA splicing. The new 3′-OH group of the 5′ exon then attacks the 5′-terminal of the 5′ exon to displace the intron. In a third reaction unique to pre-rRNA processing, the excised intron cyclizes. (Sections 25-3A and B)

Questions

RNA Polymerase

1. The prokaryotic RNA polymerase holoenzyme contains the core enzyme and a σ factor. What is the function of the core enzyme? What are its limitations in terms of enzymatic activity? What is the role of the σ factor?

2. In a binding assay, a bacterial RNA polymerase binds its 70-base promoter 100 times faster when the promoter is part of an 8 kb plasmid than when the promoter is part of a 200-bp linear DNA. What might account for this difference in binding behavior? Assume that the polymerase is incubated with an equal weight of each type of DNA.

3. What evidence suggests that newly transcribed RNA is not wound around its template DNA?

4. You are interested in discerning the role of σ factors in prokaryotic transcription, using a system of purified core polymerase, a DNA template, a σ factor, and labeled nucleotides (^{32}pppN). The incorporation of ^{32}P into RNA at 2 different core enzyme concentrations (10^{-10} M and 10^{-11} M) and constant concentration of σ factor (10^{-12} M) is shown below.
   (a) What does the graph measure?
   (b) What can you conclude about the behavior of σ factor?
   (c) What can you conclude about the behavior of the core enzyme?

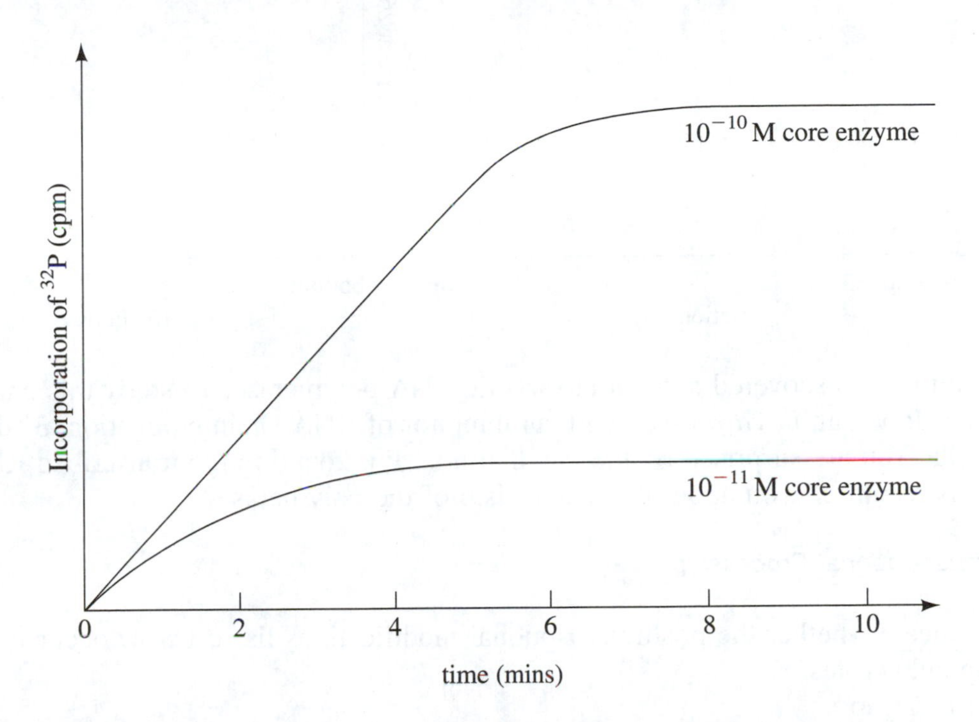

5. What evidence indicates that a GC-rich region of DNA followed by an AT-rich region is sufficient but not necessary to promote RNA chain termination?

Transcription in Eukaryotes

6. Distinguish between enhancers and promoters.

7. To a mammalian cell culture you add varying amounts of α-amanitin. After several hours of treatment you isolate total RNA and analyze it by sucrose density gradient centrifugation. How can varying concentrations of α-amanitin be used to distinguish the classes of eukaryotic RNA polymerases? Draw the profiles of RNA in the sucrose density gradient

following centrifugation. Note that 98% of the total RNA mass is the four major rRNA species.

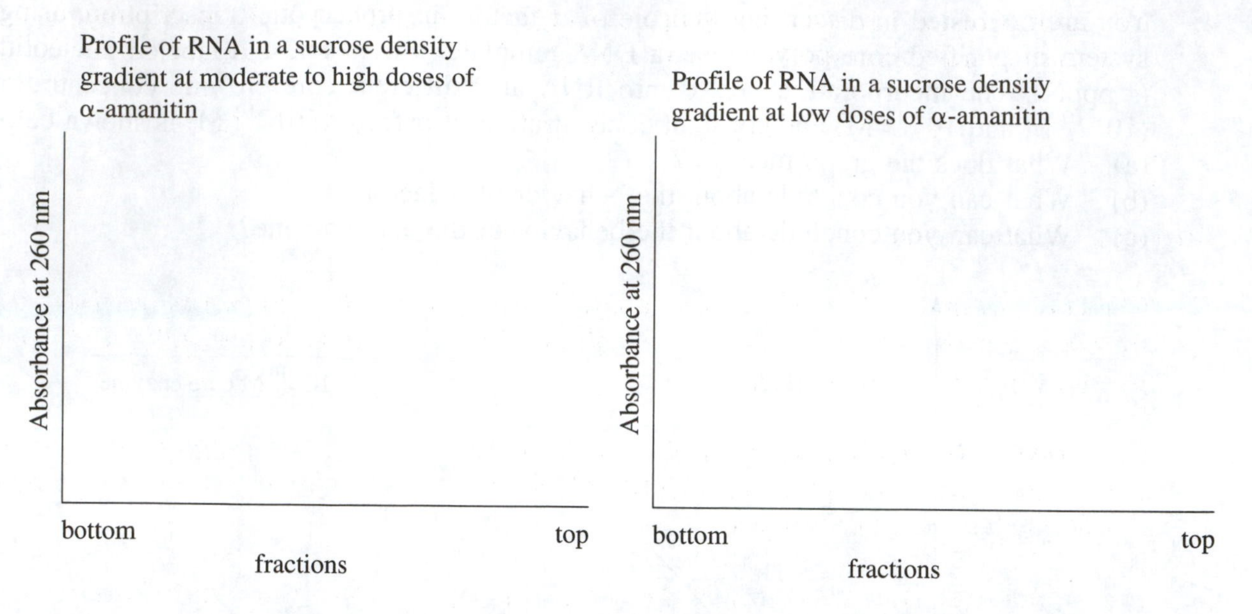

8. You have discovered a novel eukaryotic RNA polymerase. To study this enzyme, you have developed an *in vitro* assay with an inhibitor of RNA chain elongation, 3′-deoxy-5′[α-^{32}P] CTP. You are surprised by the fact that no ^{32}P is found in the transcribed RNA! What does this result tell you about the mechanism of the polymerase?

Posttranscriptional Processing

9. Indicate whether the posttranscriptional modifications listed below occur in prokaryotes or in eukaryotes.
   (a) 5′ Cap
   (b) Polyadenylation
   (c) Methylation of nucleotide residues
   (d) Endonucleolytic cleavage
   (e) Splicing

10. Which of the following is contained in the rRNA cistron of *E. coli*?
    (a) 16S rRNA (b) 23S rRNA (c) a tRNA gene (d) 5S RNA

11. What RNA-processing enzyme complexes contain RNA? In which does RNA participate in catalysis?

12. What kinds of posttranscriptional processing occur in eukaryotic tRNA precursors?

13. Group I introns incubated with increasing concentrations of poly(C) show Michaelis–Menten kinetics for the cleavage of poly(C); that is, in the presence of increasing amounts of poly(C), the rate increases to a maximum. Is this proof that these introns are catalysts? Explain.

14. The mRNA capping reaction involves cleavage of a methylated guanosine triphosphate to release its pyrophosphate. You wish to measure RNA chain initiation rates in isolated nuclei. Which of the isotopes shown below would you use? Explain. What assumptions are necessary to use this experimental strategy?

$$^{32}\text{pppA} \qquad \text{p}^{32}\text{ppA} \qquad \text{pp}^{32}\text{pA}$$

Answers to Questions

1. The core enzyme carries out RNA synthesis but it cannot initiate transcription. The σ factor binds specifically to a promoter; this allows the DNA to melt apart so that RNA polymerase can begin transcribing.

2. The polymerase binds nonspecifically to the plasmid and finds its promoter in a one-dimensional search, which proceeds faster than when the enzyme must locate its promoter through a three-dimensional search.

3. *E. coli* topoisomerase mutants show a buildup of positive or negative supercoils during transcription, which would not occur if RNA polymerase simply followed the helical path of the DNA template. In fact, electron micrographs such as Figure 25-9 indicate that the RNA extends away from the DNA as transcription proceeds.

4. (a) Since radioactivity from ^{32}pppN is only incorporated into the 5′ nucleotide, the graph measures RNA chain initiations.
   (b) Since more initiations occur when more core enzyme is present despite constant σ factor concentration, the σ factor can dissociate from the core enzyme after initiation and bind to a new core enzyme to initiate a new transcript.
   (c) Since the curves flatten at both core enzyme concentrations, the core enzyme does not dissociate from the DNA template (another factor must facilitate dissociation of the core enzyme from the template). Consequently, once all the core enzyme binds to DNA and initiates transcription, no new RNA chain initiations are observed.

5. Termination resulting from the presence of GC- and AT-rich regions does not always occur. In some cases, particularly where these sequences are missing, rho factor–dependent termination occurs.

6. Enhancers are eukaryotic transcriptional control elements where specific proteins bind and thereby promote RNA polymerase binding at the promoter. Enhancers are located at variable distances from the gene and in different orientations. Promoters, in contrast, are located near the transcription start site. A promoter is sufficient for transcription initiation and can operate with or without an enhancer. An enhancer cannot operate without a promoter.

7. At high doses, α-amanitin inhibits RNA polymerases II and III. Therefore, only RNA polymerase I transcripts are produced; these include the transcripts for 28S rRNA, 18S rRNA,

and 5.8S rRNA. At lower doses, α-amanitin inhibits RNA polymerase II but not RNA polymerase III. Thus, the RNA polymerase III–transcribed 5S rRNA also appears.

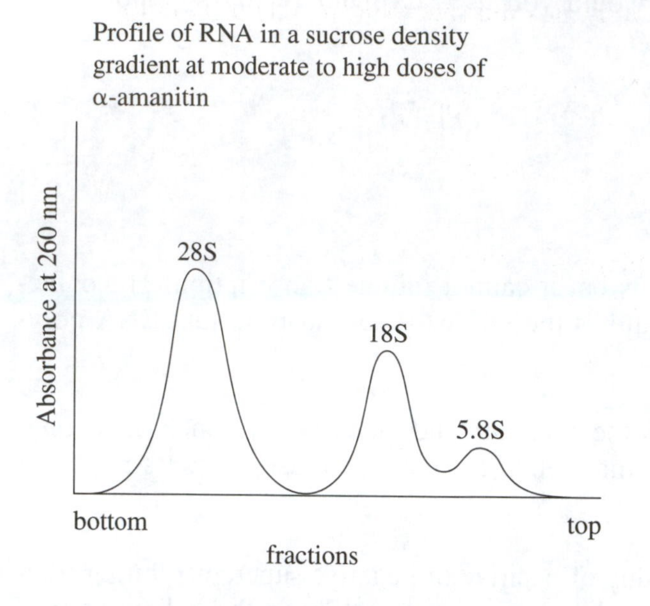

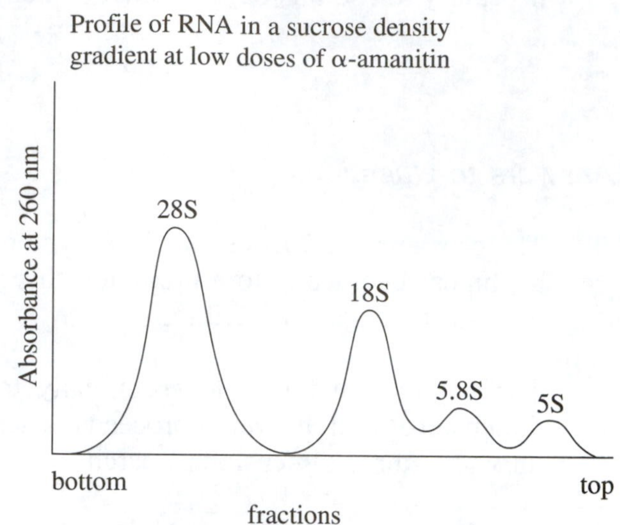

8. Normal RNA chain elongation requires a 3′-OH group. Because the inhibitor lacks this functional group, it should be incorporated into the RNA but prevent further polymerization. The absence of ^{32}P in this experiment suggests that RNA chain elongation may proceed in the 3′→5′ direction. In this case, the RNA chain would grow through nucleophilic attack by the 3′-OH group of the incoming nucleotide, a reaction in which the 3′-deoxy compound cannot participate.

9. (a) 5′ Cap: eukaryotes
   (b) Polyadenylation: eukaryotes
   (c) Methylation of nucleotide residues: eukaryotes and prokaryotes
   (d) Endonucleolytic cleavage: eukaryotes and prokaryotes
   (e) Splicing: eukaryotes

10. All of these elements are contained in the rRNA primary transcript.

11. RNA-processing enzyme complexes that contain RNA include RNase P, snRNPs, and the mRNA-editing machinery. Only the RNase P RNA participates in catalysis.

12. Nucleolytic removal of nucleotides, splicing of two exons, addition of CCA residues at the 3′ end, and chemical modification of certain nucleotides.

13. No. It must be demonstrated that the intron is regenerated in the course of the reaction. It is possible that the intron is irreversibly cleaved in the course of poly(C) cleavage.

14. The β-labeled compound p^{32}ppA should be used because the ^{32}P will appear only in the first incorporated nucleotide (in subsequent residues, the β and γ phosphates are eliminated). If the α-labeled compound were used, all A residues in the mRNA would be labeled. During cap addition, two phosphates of the methylated GTP and one terminal mRNA phosphate are

removed, so only a β-label would remain. The assumptions in this experiment are that the initial residue is A rather than G and that the capping reaction is not rate limiting, so that the rate of appearance of β-labeled nucleotide indicates the rate of RNA chain initiation.

$$
\begin{array}{c}
p^{32}ppApNpNpNpNpN\text{-} \\
\big\downarrow + p\text{-}p\text{-}p\text{-}G \\
Gp^{32}ppApNpNpNpNpN\text{-} \\
+ PP_i + P_i
\end{array}
$$

Chapter 26 Translation

This chapter explores the mechanisms by which messenger RNA directs the synthesis of a polypeptide. The discussion begins with a look at the genetic code, including the arguments and experiments that led to the conclusion that an mRNA nucleotide sequence is decoded as a nonoverlapping series of triplets. Hence, in this coding scheme, there are 64 triplet combinations, or codons, that encode the 20 amino acids. Because each amino acid can be coded for by more than one codon, the genetic code is said to be degenerate. However, whereas some amino acids are encoded by as many as six codons, others are encoded by only one (this phenomenon may reflect the way the genetic code evolved). The discussion then turns to the molecules that "translate" the genetic code: transfer RNA. The structures of the tRNAs are explored, from their modified bases to their unusual secondary and tertiary structures. Each tRNA can covalently attach an amino acid at its 3′ end to form an aminoacyl–tRNA. Two families of aminoacyl–tRNA synthetases, which show surprisingly little sequence similarity, mediate this aminoacylation. Structural information on the mechanisms by which aaRSs recognize their cognate tRNAs is also presented.

The chapter then discusses the "wobble hypothesis," which accounts for the degeneracy of the genetic code at the level of base-pairing interactions between the codons of mRNA and the anticodons of tRNAs. Some organisms have a "21st amino acid," selenocysteine, which is incorporated into certain proteins using a unique tRNA. The structure of the ribosome is discussed next in some detail. The chapter concludes with a description of the mechanism of polypeptide chain initiation, elongation, and termination. Included in this section are discussions of the role of GTP hydrolysis in protein synthesis, translational accuracy, posttranslational modifications, the effect of diphtheria toxin on protein synthesis, and the mechanisms by which antibiotics interfere with translation.

Essential Concepts

1. Translation shares important features with DNA transcription and replication:
   (a) Translation occurs in a large multiprotein machine, which includes accessory factors that regulate the initiation, elongation, and termination of polypeptide synthesis.
   (b) Even though it is amino acids that are polymerized, translation depends on Watson–Crick base pairing between mRNA and tRNA so that amino acids are joined as genetically specified.
   (c) Accuracy is essential, so the translational machinery has proofreading capabilities.
   (d) The process is endergonic and requires the cleavage of "high-energy" phosphoanhydride bonds.

The Genetic Code

2. The DNA sequence encoding a protein is made up of a triplet of nucleotides called a codon. A triplet code allows more than one codon to specify an amino acid; hence, the code is said to be degenerate. For example, leucine is specified by six codons, serine by four, and glutamate by two. Codons that specify the same amino acid are termed synonyms.

3. Transfer RNA deciphers the codon. Each tRNA contains a trinucleotide sequence, called the anticodon, which is complementary to an mRNA codon specifying the amino acid attached to the tRNA.

4. The code is read sequentially and contains punctuation. The sequence AUG (read 5′ to 3′ along the mRNA), or occasionally GUG, at the 5′ end of the mRNA marks the first codon to be translated. The codons UAA, UAG, and UGA are called stop codons because they signal the termination of translation.

5. The arrangement of codons is nonrandom. Most synonyms share the first two nucleotides. Interestingly, codons with different nucleotides in the first position tend to specify chemically similar amino acids. Codons with second position pyrimidines encode mostly hydrophobic amino acids, and those with second position purines encode mostly polar amino acids. This suggests a nonrandom evolution of the genetic code that minimizes the deleterious effects of point mutations (see Box 26-1).

6. The "standard" code is not universal; exceptions have been observed in mitochondria and a few protozoans.

Transfer RNA and Its Aminoacylation

7. Almost all known tRNAs can be arranged in a so-called cloverleaf secondary structure, which has the following key features:
   (a) A 5′-terminal phosphate group.
   (b) Three keyhole-shaped stem-and-loop structures called the D arm, the anticodon arm, and the T or TψC arm. The D arm is named for the modified base dihydrouridine in the stem. The anticodon arm contains the anticodon sequence in a loop. The TψC arm is named for the sequence thymidine–pseudouridine (ψ)–cytidine in the stem.
   (c) A 3′ CCA sequence containing a free 3′ hydroxyl group. This trinucleotide is either genetically specified (in prokaryotes) or enzymatically appended to the immature tRNA (in eukaryotes).
   (d) Between the anticodon arm and the TψC arm is a variable arm whose sequence and length (3–21 nucleotides) varies among tRNAs as well as among species.
   (e) Transfer RNAs have 15 invariant positions (always the same base) and 8 semivariant positions (only a purine or pyrimidine) that occur mostly in the loops.

8. Up to 25% of the bases of tRNA are modified posttranscriptionally. The biological significance of many of these modifications is still mysterious, since they are not required for the structural integrity of tRNA or its proper binding to the ribosome.

9. Transfer RNA has an L-shaped tertiary structure in which structural stability is maintained by extensive base stacking interactions and base pairing within and between the helical stems.

10. Attachment of an amino acid to a tRNA is carried out in a two-reaction process by an aminoacyl–tRNA synthetase (aaRS). The first reaction involves "activation" of the amino acid by its reaction with ATP to form an aminoacyl–adenylate with the release of PP_i. The mixed anhydride then reacts with tRNA to form the "charged" aminoacyl–tRNA with the

release of AMP. The hydrolysis of pyrophosphate (PP_i) ensures that the overall reaction is irreversible.

$$\begin{array}{ll} 1. & \text{Amino acid} + \text{ATP} \rightleftharpoons \text{aminoacyl–AMP} + PP_i \\ 2. & \text{Aminoacyl–AMP} + \text{tRNA} \rightleftharpoons \text{aminoacyl–tRNA} + \text{AMP} \\ \hline & \text{Amino acid} + \text{tRNA} + \text{ATP} \rightleftharpoons \text{aminoacyl–tRNA} + \text{AMP} + PP_i \end{array}$$

11. There are two structurally unrelated classes of aminoacyl–tRNA synthetases (Class I and Class II). These enzymes differ in
    (a) The mechanism by which they recognize their tRNA substrates.
    (b) The initial site of aminoacylation on the tRNA.
    (c) Amino acid specificity.

12. Each set of isoaccepting tRNAs is recognized by a single aminoacyl–tRNA synthetase (aaRS). Isoaccepting tRNAs have different anticodons but are conjugated to the same amino acid. aaRSs appear to recognize the unique structural features of the cognate tRNAs, not specific base sequences. Some aaRSs are capable of proofreading, which occurs at a different site from the aminoacylation active site.

13. During protein synthesis, the proper tRNA is selected only through codon–anticodon interactions; the aminoacyl group is not involved. Many tRNAs bind to two or three of the codons that specify their cognate amino acids, which is due to non-Watson–Crick base pairing between the third codon position and the first anticodon position. For example, U in the first position of the anticodon can base pair with A or G at the third codon position. Some tRNAs have inosine at the first anticodon position, which can base pair with U, C, or A. The degeneracy in binding at this position is called wobble.

Ribosomes

14. Ribosomes carry out the following functions:
    (a) They bind mRNA so that its codons are matched, with high accuracy, to tRNA anticodons.
    (b) Ribosomes contain three specific binding sites: the A site for the incoming aminoacyl–tRNA; the P site for the tRNA to which the growing polypeptide chain is attached; and the E site for the outgoing, uncharged tRNA.
    (c) They mediate interactions of nonribosomal proteins that are required for polypeptide chain initiation, elongation, and termination. Many of these factors use the free energy of GTP hydrolysis to carry out their functions.
    (d) Ribosomes catalyze peptide bond formation between the incoming amino acid and the growing polypeptide chain.
    (e) They are responsible for the translocation of the mRNA so that codons are read sequentially.

15. Prokaryotic ribosomes are enormous ribonucleoprotein machines containing dozens of proteins and several RNA molecules. The *E. coli* ribosome has a sedimentation coefficient of 70S and is composed of two subunits: The 30S subunit contains the 16S rRNA, while the

50S subunit contains the 23S and 5S rRNAs. The tunnel-like opening in the 30S subunit appears to accommodate a single strand of mRNA (without any secondary structure); a similar opening in the 50S subunit may harbor the growing polypeptide chain.

16. The structure of the ribosome has been elucidated through a combination of techniques: standard electron microscopy, immune electron microscopy, X-ray diffraction, NMR, and chemical cross-linking studies. The large rRNA of each subunit provides a structural scaffold for the assembly of the ribosomal proteins, many of which contain a structural motif common in other RNA-binding proteins, called the RNA-recognition motif (RRM).

17. Eukaryotic ribosomes resemble prokaryotic ribosomes in shape and function but are larger and more complex. The 80S ribosome is composed of a 40S small subunit (with 18S rRNA) and a 60S large subunit (with 5S, 5.8S, and 28S rRNAs). The 18S rRNA is homologous to the prokaryotic 16S rRNA, while the 5.8S and the 28S rRNAs are homologous to the prokaryotic 23S rRNA.

18. In *E. coli,* the rate of rRNA synthesis is coupled to the rate of protein synthesis by the stringent response, which is activated when the supply of charged tRNA is limiting. Ribosomes bound to uncharged tRNAs stimulate the synthesis of ppGpp, which reduces the transcription of rRNA and tRNA genes and stimulates the transcription of genes involved in amino acid biosynthesis.

Polypeptide Synthesis

19. Polypeptide synthesis has the following key features:
    (a) Ribosomes read mRNA in the 5′ to 3′ direction.
    (b) Polypeptide synthesis proceeds from the N-terminus to the C-terminus.
    (c) Chain elongation occurs by linking the growing polypeptide chain to the incoming charged tRNA.
    (d) Several ribosomes can bind to a single mRNA to form a complex called a polysome or polyribosome.

20. Polypeptide chain initiation begins with Met–tRNA in eukaryotes and a modified Met–tRNA in prokaryotes, *N*-formylmethionyl–tRNA (fMet–tRNA). In prokaryotes, the AUG start codon is distinguished from other AUG codons by base pairing between the 16S rRNA and a 5′ mRNA sequence called the Shine–Dalgarno sequence. Eukaryotes do not contain an analogous recognition sequence; initiation occurs at the first AUG of a eukaryotic mRNA.

21. Initiation requires soluble proteins called initiation factors. Eukaryotes employ far more initiation factors (eIF-*n*) than do prokaryotes, which have only three factors (IF-1, IF-2, and IF-3).

22. Initiation of translation in *E. coli* occurs in three steps:
    (a) IF-3 binds to the 70S ribosome and promotes its dissociation into the 30S and 50S subunits. IF-1 stimulates this process.
    (b) mRNA, IF-2 in complex with GTP and fMet–tRNA, and perhaps IF-3 bind to the 30S subunit.

(c) IF-3 is released, allowing the 50S subunit to bind to the 30S subunit. This stimulates hydrolysis of the GTP bound to IF-2, resulting in a conformational change in the 30S subunit and the release of IF-1 and IF-2, which can participate in further initiation reactions.

23. Elongation of a nascent polypeptide chain in prokaryotes proceeds as a three-reaction cycle until a termination codon is encountered.
    (a) An incoming aminoacyl–tRNA, in a complex with EF-Tu · GTP, binds to the A site. This interaction is accompanied by the hydrolysis of GTP, which releases P_i and EF-Tu · GDP from the ribosome. EF-Ts replaces GDP, which is then replaced by GTP, thereby regenerating the EF-Tu · GTP complex.
    (b) A peptide bond is formed through the nucleophilic displacement of the P-site tRNA by the amino group of the aminoacyl–tRNA in the A site. Hence, the nascent polypeptide is lengthened at its C-terminus by one amino acid and transferred to the A site, a process called transpeptidation. The 23S rRNA appears to be the catalyst in peptide bond formation.
    (c) The new peptidyl–tRNA is translocated from the A site to the P site with the assistance of EF-G · GTP. Hydrolysis of GTP drives translocation, releasing GDP, P_i, and EF-G from the ribosome. The uncharged tRNA in the P site slides into the E site. EF-G · GTP increases the affinity of the E site for the deacylated tRNA while decreasing the affinity of the A site for this molecule. On the other hand, EF-Tu · GTP increases the affinity of the A site for an incoming aminoacyl–tRNA. Hence, the alternating activities of EF-Tu and EF-G keep the ribosome cycling in a unidirectional manner.

24. Elongation in eukaryotes is similar to that in prokaryotes, except that eEF-1 replaces EF-Tu and EF-Ts, and eEF-2 replaces EF-G.

25. Diphtheria toxin inactivates eEF-2 by transferring an ADP-ribosyl moiety from NAD^+ to a modified His residue in eEF-2.

26. In *E. coli*, two release factors, RF-1 and RF-2, recognize stop codons. Binding of either RF-1 or RF-2 to the ribosome stimulates the transfer of the peptidyl group to water, thereby terminating polypeptide chain elongation. A third factor, RF-3, in complex with GTP, stimulates the binding of RF-1 and RF-2. GTP hydrolysis releases RF-1 and RF-2 from the ribosome. Eukaryotic translational termination is similarly mediated by eRF, which carries out all the activities of RF-1, RF-2, and RF-3.

27. A mutation that converts an aminoacyl-coding ("sense") codon to a stop codon is called a nonsense mutation and results in the premature termination of translation. Such mutations can be rescued by a nonsense suppressor tRNA, which contains an anticodon to a stop codon but carries the amino acid of its wild-type progenitor.

28. Translational accuracy is thought to be mediated by a kinetic proofreading mechanism. In this model, a noncognate aminoacyl–tRNA dissociates from the ribosome faster than does the EF-Tu · GTP complex, which would otherwise commit the ribosome to form a peptide bond between the aminoacyl–tRNA and the peptidyl–tRNA.

29. Posttranslational modifications of newly synthesized proteins are common, especially in eukaryotes. In some cases, limited proteolysis may activate a protein, which is called a proprotein. A preproprotein also contains a signal peptide that is removed in the endoplasmic reticulum, prior to excretion of the protein. Common modifications of proteins include phosphorylation (at Ser, Thr, Tyr, and His residues), glycosylation, and hydroxylation of Pro and Lys residues in collagen.

30. Some commonly used antibiotics interfere with the initiation or elongation steps in prokaryotic translation. Puromycin mimics tyrosyl–tRNA and aborts transpeptidation. Streptomycin inhibits translational initiation. Chloramphenicol inhibits the peptidyl transferase reaction. Tetracycline inhibits aminoacyl–tRNA binding to the ribosome.

Guide to Study Exercises (text p. 885)

1. Replication, transcription, and translation are the three processes that are central to the transmission and expression of biological information. Nucleic acids are essential participants, as are complexes of specific proteins that often operate in concert with nucleic acids. In each case, a polymer is synthesized by a series of regulated reactions that govern how polymerization is initiated and terminated and how the polymer is extended as directed by a template. Various proofreading mechanisms ensure the accuracy of replication, transcription, and translation.

   Notable differences among these processes are the identities of the resulting polymer: Replication produces DNA, transcription produces RNA, and translation produces protein. In addition, replication involves the entire DNA molecule so that a complete copy of the genome can be made, whereas transcription and translation are more selective, involving specific regions of DNA. Replication occurs once in a cell's lifetime, but transcription of a gene may occur many times, and translation of that gene transcript many more times still. (Sections 24-1, 25-1, and 26-4)

2. Codons must consist of three nucleotides because if they consisted of only two, there would be only $4^2 = 16$ possible codons—not enough to encode 20 amino acids. If codons consisted of four nucleotides, there would be $4^4 = 256$ possible codons—an excessive amount, even considering the need for redundancy and for translation start and stop signals. Three-nucleotide codons permit $4^3 = 64$ different combinations—slightly more than are strictly necessary. (Section 26-1)

3. The major structural features of tRNA include a 7-bp acceptor stem, the 3′ end of which is the site of aminoacylation; a stem-and-loop structure called the D arm because it frequently contains the modified base dihydrouridine; the anticodon arm, a 5-bp stem with a loop containing the three-base anticodon; and a stem-and-loop structure called the TψC arm because it includes the sequence TψC (ψ is pseudouridine). In addition, many tRNAs are posttranscriptionally modified. In a fully folded tRNA, the acceptor stem and the TψC arm form one leg of an L, and the D arm and anticodon arm form the other leg. Extensive stacking interactions and base pairing maintain the structure. (Section 26-2A)

4. Both types of aminoacyl–tRNA synthetases first activate an amino acid by reacting it with ATP and then transfer the adenylated amino acid to the 3′-terminal ribose of the appropriate tRNA. The two classes of aaRSs differ in their amino acid specificity and exhibit different conserved sequence motifs. Mechanistic differences between the two classes of enzymes include the site of tRNA aminoacylation: Class I enzymes attach the amino acid to the 2′-OH group, whereas Class II enzymes charge the 3′-OH group. In addition, many Class I enzymes must interact with the anticodon of their target tRNA, whereas several Class II enzymes do not. (Section 26-2B)

5. The accuracy of translation is enhanced by proofreading at two points. First, many aminoacyl–tRNA synthetases have proofreading active sites to help ensure that the proper amino acid is attached to a specific tRNA. Second, the ribosome is thought to mediate a kinetic proofreading process to ensure proper codon–anticodon pairing. In kinetic proofreading, an incorrect aminoacyl–tRNA dissociates from the ribosome before a bound EF-Tu hydrolyzes its bound GTP to GDP, which would otherwise commit the ribosome to form a peptide bond. In contrast, the correct aminoacyl–tRNA remains bound to the ribosome and hence forms a peptide bond. (Sections 26-2B and 26-4D)

6. The ribosome's three tRNA-binding sites are the A (aminoacyl), P (peptidyl), and E (exit) sites. An individual tRNA binds to the three sites in succession during the elongation phase of translation. An aminoacyl–tRNA enters the A site with the assistance of EF-Tu · GTP. Following transpeptidation, in which the polypeptide is transferred to the aminoacyl group, the tRNA moves to the P site with the assistance of EF-G. After it has given up its peptidyl group, it moves to the E site, then dissociates from the ribosome. The tRNAs that occupy the A and P sites maintain contact with the mRNA through their anticodons so that amino acids can be polymerized in the order specified by a gene. (Section 26-4B)

7. The initiation factors promote ribosome dissociation in preparation for a new round of translation. IF-2 (in *E. coli*) or eIF-2 (in eukaryotes) forms a ternary complex with GTP and the initiator tRNA that can then bind to the ribosome, which assembles and begins translating an mRNA.

   Elongation factors include EF-Tu (in *E. coli*) and eEF-1 (in eukaryotes), which, in complex with GTP, deliver an aminoacyl–tRNA to the ribosome. Hydrolysis of the GTP to GDP + P_i makes this binding reaction irreversible. EF-G (in *E. coli*) and eEF-2 (in eukaryotes) facilitate ribosomal translocation in a reaction that also involves GTP hydrolysis. The alternating binding of elongation factors keeps the ribosome moving through transpeptidation and translocation. EF-Ts facilitates the exchange of GDP for GTP by EF-Tu.

   Release factors (RF-1, RF-2, and RF-3 in *E. coli* and eRF in eukaryotes) recognize stop codons and bind to the ribosome to terminate polypeptide synthesis. Again, GTP hydrolysis drives this process. (Sections 26-4A, B, and C)

8. The transpeptidation reaction itself does not need a source of external free energy, but free energy is required for several other phases of ribosomal function. This free energy is supplied in the form of GTP hydrolysis to GDP + P_i, which drives protein conformational changes. These changes occur during the delivery of the initiator tRNA or another aminoacyl–tRNA to the ribosome, during ribosomal translocation, and during translation termination. (Sections 26-4A, B, and C)

Questions

The Genetic Code

1. Why is the genetic code degenerate?

2. How many amino acid homopolymers can be specified by polynucleotides consisting of two alternating nucleotides?

3. A cell extract from a new species of *Amoeba* is used to translate a purified mouse hemoglobin mRNA, but only a short peptide is produced. What is the simplest explanation for this result?

4. Predict the effect on protein structure and function of an A · T to G · C transition in the first codon position for lysine.

5. Identify the polypeptide encoded by the DNA sequence below, in which the lower strand serves as the template for mRNA synthesis.

```
5′ GGACCTATGATCACCTGCTCCCCGAGTGCTGTTTAGGTGGG 3′
3′ CCTGGATACTAGTGGACGAGGGGCTCACGACAAATCCACCC 5′
```

Transfer RNA and Its Aminoacylation

6. What two reactions are carried out by aminoacyl–tRNA synthetases?

7. Which of the following tRNA structural features are always involved in interactions with an aaRS?
   (a) acceptor stem (b) anticodon loop (c) TψC arm

8. A new species of yeast has just one $tRNA^{Ala}$, with the anticodon IGC. What does this suggest about codon usage in this organism?

9. Which of the following amino acids are likely to be linked to only one species of tRNA? What are their anticodons?
   (a) Phe (b) Leu (c) His

Polypeptide Synthesis

10. Reticulocyte (immature red blood cell) lysates can be made devoid of mRNA by ribonuclease treatment followed by inactivation and removal of the ribonuclease. These lysates are now capable of synthesizing protein when mRNA, GTP, and aminoacyl–tRNAs are added. How could you use this system to demonstrate that a protein is synthesized from the N-terminus to the C-terminus?

11. Match each of the functions on the left with the appropriate prokaryotic protein on the right.

______ Binds fMet–tRNA and GTP.	A. RF-1
______ Binds aminoacyl–tRNA and GTP.	B. RF-2
______ Recognizes stop codons.	C. RF-3
______ Binds GTP and stimulates RF-1 and RF-2 binding.	D. IF-1
______ Hydrolyzes GTP to GDP.	E. IF-2
______ Promotes the transfer of peptidyl–tRNA from the A site to the P site.	F. IF-3
______ Inhibits the interaction of the 30S and 50S subunits.	G. EF-Ts
______ Facilitates mRNA binding to the 30S ribosome.	H. EF-Tu
______ Displaces GDP from EF-Tu.	I. EF-G

12. How could you show that the inhibition of protein synthesis in a bacterial cell extract by puromycin is not irreversible?

13. You have isolated a temperature-sensitive mutant *E. coli*. At 34°C, the growth rate is nearly normal but is reduced 10-fold at 37°C. You eventually discover that EF-Tu · GDP binding to the ribosome is two-fold higher in the mutant at 37°C. What aspect of translation is affected by the mutation?

14. Which amino acid substitutions might occur in *E. coli* opal suppressors that arise by single-base mutations?

15. The error rate of translation in some rapidly growing *E. coli* strains is as high as 2 percent. What is the percentage of correctly translated protein in such bacteria (assume the protein has 100 residues)? How can these bacteria do so well?

16. Distinguish between eukaryotic and prokaryotic polypeptide initiation, elongation, and termination.

17. What is the adaptive significance of posttranslational cleavage of proteins such as collagen and various hormones?

Answers to Questions

1. The genetic code, which consists of 64 codons, is degenerate because the codons are used in a redundant manner. Only 21 codons would be strictly required to code for 20 amino acids and a stop codon.

2. The six possible polynucleotides each encode two polypeptides since each has two possible reading frames, yielding 11 different polypeptides [two are poly(Arg)]. Poly(AC) yields poly(Thr) from ACA codons and poly(His) from CAC codons; Poly(AG) yields poly(Arg) from AGA codons and poly(Glu) from GAG codons; Poly(AU) yields poly(Ile) from AUA codons and poly(Tyr) from UAU codons; Poly(CG) yields poly(Arg) from CGC codons and poly(Ala) from GCG codons; Poly(CU) yields poly(Leu) from CUC codons and poly(Ser) from UCU codons; and Poly(GU) yields poly(Val) from GUG codons and poly(Cys) from UGU codons.

3. An *Amoeba* tRNA interprets a mouse sense codon as a stop codon, thereby halting translation of the mouse RNA.

4. An A to G transition would change the lysine AAA and AAG codons to GAA and GAG, both of which encode glutamate. This would result in an acidic residue replacing a basic residue in the protein. If the change were on the protein surface, it might have no effect on protein structure or function, but if the change involved a buried ion pair, a catalytic residue, or some other essential residue, its effect might be significant.

5. The mRNA includes an AUG start codon and a UAG stop codon and has the sequence

   5′ GGACCU**AUG**AUCACCUGCUCCCCGAGUGCUGUU**UAG**GUGGG 3′

   The encoded polypeptide has the sequence

   H_3N^+–Met–Ile–Thr–Cys–Ser–Pro–Ser–Ala–Val–COO^-

6. The first reaction "activates" an amino acid via condensation with ATP to form an aminoacyl–AMP + PP_i. The second reaction condenses this mixed anhydride with its cognate tRNA to form an aminoacyl–tRNA + AMP.

7. Only the acceptor stem is involved in all tRNA–aaRS interactions.

8. This yeast uses Ala codons GCA, GCC, and GCU but not GCG, since I can pair with A, C, or U but not G.

9. Only Phe and His are likely to be linked to one tRNA species. The anticodon of $tRNA^{Phe}$ is GAA, so it can bind both UUU and UUC by wobble pairing (see Table 26-3). Similarly, the anticodon of $tRNA^{His}$ is GUG, so it can bind codons CAC and CAU. $tRNA^{Leu}$ must have at least three anticodons (UAA, IAG, and CAG or UAG) to accommodate the six Leu codons (UUA, UUG, CUU, CUC, CUA, and CUG).

10. Add the mRNA for a known protein to the lysate. After a minute or two, add puromycin to terminate translation prematurely. Analyze the sequences of the resulting peptides to determine whether they correspond to the C- or the N-terminus of the protein. Alternatively, add to the lysate an isotopically labeled amino acid (as an aminoacyl–tRNA) that occurs only near one terminus of the protein and determine whether the label becomes incorporated into the peptides.

11. __E__ Binds fMet–tRNA and GTP.
    __H__ Binds aminoacyl–tRNA and GTP.
    __A, B__ Recognizes stop codons.
    __C__ Binds GTP and stimulates RF-1 and RF-2 binding.
    __C, E, H, I__ Hydrolyzes GTP to GDP.
    __I__ Promotes the transfer of peptidyl–tRNA from the A site to the P site.
    __F__ Inhibits the interaction of the 30S and 50S subunits.
    __F__ Facilitates mRNA binding to the 30S ribosome.
    __G__ Displaces GDP from EF-Tu.

12. Puromycin, which resembles the 3′ end of Tyr–tRNA, competes with aminoacyl–tRNAs for binding to ribosomes. Increasing the concentrations of aminoacyl–tRNAs in the extract should at least partially overcome the effects of puromycin and restore protein synthesis.

13. EF-Tu · GDP dissociation is essential for polypeptide elongation. Hence, in the mutant at 37°C, slower dissociation of the EF-Tu · GDP complex slows translation. The dramatic decrease in growth rate may be a function of kinetic proofreading in which an increase in the time of EF-Tu · GDP binding to the ribosome increases the chance of misincorporating an aminoacyl–tRNA.

14. The anticodon of an opal suppressor tRNA must be UCA. The anticodons that can mutate to UCA via a single-base change are UCC, UCG, UCU, UAA, UGA, UUA, ACA, CCA, and GCA, which correspond to amino acids Arg, Cys, Gly, Leu, Ser, and Trp (UUA corresponds to the stop codon UAA).

15. The probability that any one amino acid is incorporated correctly is 0.98. Hence, the probability that a 100-residue protein is synthesized correctly is 0.98^{100}, or 13.26%! The bacteria survive because many translation errors do not alter protein function. In addition, protein turnover is sufficiently rapid to overcome the effects of abnormal proteins. In fact, many nonfunctional proteins have shorter-than-normal half-lives.

16. Initiation. Eukaryotes lack Shine–Dalgarno binding interactions, and they have many more initiation factors, including an mRNA cap-binding protein.

    Elongation. Eukaryotic eEF-1 carries out the functions of prokaryotic EF-Tu and EF-Ts, while eEF-2 functions similarly to EF-G.

    Termination. Eukaryotic polypeptide termination requires only one factor, eRF, which carries out all the functions of the three prokaryotic RFs.

17. Mature collagen self-assembles into large polymeric structures. Such assembly inside the cell would likely destroy the cell. Hence, the immature procollagen, which is incapable of self-assembly, exits the cell before being proteolytically converted to its mature form. In the case of peptide hormones, a mature hormone present in the cell might find its receptor and activate a signal transduction cascade that is inappropriate for that cell. This is prevented by extracellular activation, via posttranslational proteolysis, of the prohormone.

Chapter 27 Regulation of Gene Expression

This final chapter introduces you to one of the main efforts of modern biochemistry, understanding the regulation of gene expression, which underlies most metabolic functions of cells. In eukaryotes, the regulation of gene expression is more complex than in prokaryotes, largely due to the size of the eukaryotic genome, the numerous levels of packaging of this DNA, and the complexity of the cells and the numerous cell types in eukaryotic organisms. The chapter begins by examining genome organization in prokaryotes and eukaryotes, focusing on gene number, gene clusters, and nontranscribed DNA. The yeast and *E. coli* genomes are used to illustrate the kinds of information provided by the emerging field of genomics. Included in this discussion is an overview of the organization of DNA in eukaryotes with regard to repetitive and unique DNA sequences. Box 27-1 highlights the association of alterations in the number of certain repeated sequences with several neurological diseases, while Box 27-2 describes the use of restriction fragment length polymorphisms in inferring human genealogy.

The regulation of prokaryotic gene expression is discussed next, with attention focused on a few systems that illustrate key features common to the regulation of all prokaryotic genes. The *lac* repressor is an example of a negative regulator; the *trp* operon provides an example of repression and attenuation. Catabolite repression is an example of gene activation. Next, bacteriophage λ serves as a model system in which differing quantitative parameters, such as binding affinity and protein concentration, between two proteins competing for the same binding sites dictate a developmental decision. The story of how Cro protein and the λ repressor control the lytic versus lysogenic life cycle is one of the most elegant in molecular genetics.

The chapter then turns to eukaryotic gene regulation, which can occur at many points between the exposure of promoter sites within chromatin to the translation of the mRNA. One of the most complex cases of gene regulation occurs with the construction and expression of immunoglobulin genes. The chapter closes with an overview of gene expression in the development of the fruit fly, *Drosophila melanogaster*. Here, a set of genes that regulate the expression of other genes orchestrates embryogenesis. *Hox* genes encode a remarkably conserved set of transcription factors that participate in embryonic patterning in *Drosophila* and a wide variety of organisms.

Essential Concepts

1. Gene expression is neither random nor fully preprogrammed. Information in an organism's genome is used in an orderly fashion during development but also responds to changes in the organism's internal or external environment.

Genome Organization

2. There is surprisingly modest correlation between the amount of DNA in an organism's genome (its C value) and its apparent morphological or metabolic complexity. This is known as the C-value paradox.

3. The new discipline of genomics, the study of organisms' genomes, endeavors to elucidate the functions of all genes. Potential protein-coding genes are identified as open reading frames (ORFs). The function of an ORF can sometimes be assigned by comparing its sequence to that of a known gene. Genomics also can explore evolutionary relationships between genes, organelles, and organisms.

4. In general, an organism's genes are distributed randomly throughout its genome; however, some genes are clustered together in both prokaryotes and eukaryotes. Prokaryotic operons include genes related to a specific metabolic function. In eukaryotes, genes for histones and globins occur in clusters. rRNA genes are also clustered in both prokaryotes and eukaryotes.

5. A significant portion of prokaryotic and eukaryotic genomes consists of DNA that is never transcribed. In prokaryotes, these nontranscribed sequences are predominantly promoter and operator sequences adjacent to genes but also include insertion sequences and remnants of integrated bacteriophages. In eukaryotes, nontranscribed sequences include promoters and other regulatory regions; however, the bulk of these sequences are repetitive DNA of unknown function, for example, the *Alu* sequences that appear over 300,000 times in the human genome. Some repetitive DNA is concentrated in the centromeres and telomeres.

6. Several neurological diseases are linked to the expansion of trinucleotide repeats in certain genes. These trinucleotide repeats exhibit some genetic instability: Above a threshold of about 35–50 copies, they multiply with successive generations. The longer the repeat, the more severe the disease and the earlier its age of onset (Box 27-1).

Regulation of Prokaryotic Gene Expression

7. Prokaryotic gene expression is mainly controlled at the level of transcription. Translational control is probably not necessary since mRNAs have half-lives of only a few minutes.

8. Transcription of the *lac* operon is under negative control. The *lac* repressor prevents transcription initiation by binding to the operator. A metabolite of lactose, 1,6-allolactose, acts as an inducer of transcription by binding to the *lac* repressor, which causes a conformational change in the protein that dramatically reduces its affinity for the operator.

9. The presence of glucose prevents the expression of genes involved in the catabolism of other sugars, a phenomenon known as catabolite repression. In the presence of cAMP (which signals the absence of glucose), catabolite gene activator protein (CAP) is a positive regulator of transcription of the *lac* operon and certain other operons. CAP–cAMP binds to the promoter region of the *lac* operon via its two HTH domains and stimulates transcription.

10. Attenuation of transcription occurs in the *E. coli trp* operon, which is also controlled by the *trp* repressor (the repressor binds to the operator when tryptophan, which is a corepressor, is also present). The 5′ end of the *trp* operon mRNA contains a leader sequence that encodes a short polypeptide containing two consecutive Trp residues. When Trp is limiting, a ribosome translating the leader peptide stalls at the Trp codons. Meanwhile, transcription of the leader sequence continues and the mRNA beyond the ribosome forms a hairpin secondary structure (the antiterminator) that allows transcription to proceed into the operon's five genes. When Trp (and therefore Trp–$tRNA^{Trp}$) is present, the ribosome proceeds rapidly through the leader sequence, which allows the formation of an alternative hairpin structure in the mRNA that terminates transcription.

11. *E. coli* bacteriophage λ can follow two possible life cycles after infecting its host: (1) the lytic mode, in which the virus rapidly reproduces and lyses the cell; and (2) the lysogenic mode, in which the genome of the phage integrates into a specific site in the host's genome.

On integration, the virus is referred to as a prophage and the host a lysogen. The prophage can be excised and initiate a lytic cycle in a process called induction.

12. Gene transcription in the lytic mode has three phases:
    (a) During early transcription, RNA polymerase transcribes from two promoters, yielding the *N* gene product (gp*N*) and Cro protein, among others.
    (b) During delayed early transcription, gp*N*, a transcriptional antiterminator, promotes the transcription of mRNA beyond the termination sites of the early transcripts. The longer transcripts encode proteins involved in the excision of the prophage and another transcriptional antiterminator, gp*Q*.
    (c) During late transcription, gp*Q* promotes further transcription to express genes required for packaging new phage particles and lysing the host cell.

13. The lysogenic pathway begins with site-specific recombination between the *att*P locus of the phage and the *att*B locus of *E. coli*. The *int* gene product, integrase, mediates integration. Lysogeny is favored by high concentrations of cII, a product of early transcription. cII promotes transcription of *cI,* which encodes the λ repressor.

14. The genetic switch that determines which life cycle is pursued by the phage takes place at the operator o_R, which has three subsites and is located between the promoters p_{RM} and p_R. Transcription from p_R yields Cro, a repressor of all λ genes. Transcription from p_{RM} yields the λ repressor, which stimulates its own transcription but represses transcription of all other λ genes. λ repressor binds to o_R and maintains lysogeny by maintaining the production of repressor and preventing Cro production. If sufficient Cro binds to o_R, however, transcription of *cI* is repressed, and lytic growth ensues.

15. The λ repressor and Cro contain HTH domains and bind to DNA as homodimers. The λ repressor binds to operator subsites o_{R1} and o_{R2} cooperatively. DNA damage may lead to cleavage of λ repressor between its N-terminal (DNA-binding) and C-terminal (dimerization) domains, which abolishes its cooperative binding to DNA and allows *cro* and other early genes to be transcribed. The resulting Cro protein prevents further λ repressor synthesis, thereby irreversibly setting the lytic cycle in motion.

Regulation of Eukaryotic Gene Expression

16. The regulation of gene expression in eukaryotes can occur at several levels, as outlined below. Keep in mind that no regulatory mechanism operates exclusively as an on–off switch and that several mechanisms may govern expression of a particular gene.
    (a) **Gene Availability.** Genes may be lost through transposition, rearranged through recombination, condensed into transcriptionally silent heterochromatin, or chemically modified in different tissues or at different stages of development.
    (b) **Gene Dosage.** Some genes are amplified, for example, through the formation of polytene chromosomes.
    (c) **Transcriptional Control.** Regulation of the initiation, elongation, termination, or post-transcriptional processing phases of mRNA synthesis directly affects mRNA concentrations. Additional mechanisms for stabilizing and degrading mRNA influence the cellular half-life of the mRNA.

(d) **Translational Control.** Regulation of the initiation, elongation, or termination phases of mRNA translation influences the concentrations of polypeptides. Regulatory mechanisms that stabilize or degrade polypeptides also affect the rates of protein turnover.
(e) **Protein Activation.** Posttranslational processing reactions, such as phosphorylation, acetylation, glycosylation, proteolysis, or assembly into multimers, may govern protein activity.
(f) **Allosteric Effects.** Allosteric effectors that alter enzymatic properties or the assembly of multiprotein structures play a role in regulating gene expression.

17. Most of the DNA in a eukaryotic organism is packaged in a highly condensed form of chromatin called heterochromatin. Transcribed DNA is packaged more loosely as euchromatin and is more accessible to the transcription machinery.

18. For transcription to proceed, nucleosomes must be disassembled to some extent so that regulatory proteins and RNA polymerase can bind to the DNA. Nucleosome structure may be altered by acetylation of histone Lys residues, which reduces the histones' positive charge and thereby disrupts higher order chromatin structure. Histone acetylases are associated with the transcription machinery; histone deacetylases reverse the effects of acetylation, silencing genes by inhibiting transcription.

19. In eukaryotes, transcription factors that stimulate or repress transcription often bind to sites, called enhancers and silencers, respectively, that may be located a considerable distance from the gene's promoter. For example, the rate of transcription of a gene may increase when one or more transcription factors bind to upstream enhancers as well as to the preinitiation complex (PIC) that assembles at the TATA box (or its equivalent). The orientation of the enhancer and its distance from the transcription start site do not seem to matter, provided that the two elements are far enough apart for the DNA to form a loop that allows the transcription factor to bind to the enhancer and the PIC.

20. Many genes are expressed in response to extracellular signals. In some cases, ligand binding to certain cell-surface receptors results in the phosphorylation of cytoplasmic proteins called signal transducers and activators of transcription (STATs). The STATs then dimerize and enter the nucleus, where they bind to DNA to stimulate the transcription of specific genes.

21. Steroid hormones alter gene expression by binding to dimeric intracellular receptors that then migrate to the nucleus, where they bind to enhancers called hormone response elements. The steroid receptors consist of a ligand-binding domain, a DNA-binding domain, and a transcription activation domain.

22. The eukaryotic cell cycle is divided into four phases:
(a) M phase (for mitosis), when mitosis and cell division occur.
(b) G_1 phase (for gap), a variable but long part of the cell cycle.
(c) S phase (for synthesis), when DNA replication occurs.
(d) G_2 phase, a relatively short period of preparation for mitosis.
(e) Most cells enter a G_0 (quiescent) phase after M phase and await a signal to re-enter the cell cycle.

23. Progression through the cell cycle requires the expression of certain genes, including the cyclins (whose levels rise and fall in the cell cycle) and the cyclin-dependent kinases (CDKs). CDKs, which are active only when appropriately phosphorylated and bound to a cyclin, phosphorylate several nuclear proteins, including histone H1 and proteins that disassemble the nucleus and rearrange the cytoskeleton.

24. One of the targets of CDKs is the retinoblastoma protein (Rb), whose phosphorylation inhibits the expression of genes that promote cell proliferation. Rb belongs to a family of proteins called tumor suppressors because their loss leads to uncontrolled cell growth. The tumor suppressor p53 accumulates in response to DNA damage and leads to inhibition of CDKs, thereby arresting cell division to allow DNA repair or, if repair is unsuccessful, triggering cell suicide (called apoptosis).

25. Genetic recombination is a fundamental process in germline cells; however, somatic cells of the immune system undergo a similar process called somatic recombination. The heavy and light chains of immunoglobulins are constructed by joining together different fragments: *L* (leader), *V* (variable), *J* (joining), and *C* (constant) segments in the light chain genes, and *L, V, D* (diversity), *J,* and *C* segments in the heavy chain genes. These gene segments are brought together through recombination of the DNA in immature *B* cells and by selective splicing of the mRNA precursor. Antibody diversity results from the large number of available *V, D, J,* and *C* gene segments, from variation in the crossover point at the *V*/*J* joint, from the addition or removal of nucleotides at the *V*/*D* and *D*/*J* joints, and from somatic hypermutation of the immunoglobulin gene segments.

26. Posttranscriptional regulatory mechanisms include alternative mRNA splicing, which generates tissue-specific isoforms of a gene product.

27. The half-life of an mRNA may depend on internal sequences that influence the rates of deadenylation of the 3′ end and removal of the 5′ cap. Translation itself may decrease the half-life of some mRNAs.

28. Translational control occurs in embryonic cells, which contain stockpiled maternal mRNA. Globin synthesis is also controlled at the level of translation, via heme-controlled repressor, so that the protein is synthesized only when heme is also available.

29. Embryonic development in *Drosophila melanogaster* begins with rapid nuclear divisions in the fertilized egg followed by a cellularization process that leads to formation of a cell monolayer called the blastoderm. During this development, the embryo begins a segmentation process that ultimately specifies the differentiation of adult body structures.

30. A hierarchy of gene interactions orchestrates early patterning of the *Drosophila* embryo:
    (a) Maternal-effect genes define the embryo's anteroposterior and dorsoventral axes, even before embryogenesis begins, via the deposition of mRNA in different portions of the egg. The resulting protein gradients subsequently influence the expression of embryonic genes.

(b) Segmentation genes, which specify the correct number and polarity of embryonic body segments, include gap genes, pair-rule genes, and segment polarity genes. The products of these genes activate and repress other genes, producing sharp stripes of proteins that define the embryonic body segments.

(c) Homeotic selector genes specify segment identity; their mutation changes one body part into another. The developmental fate of a body segment depends on its position in a gradient of homeotic gene products.

31. Many homeotic genes contain a conserved 60-residue polypeptide sequence called the homeodomain or homeobox. Homeodomains occur in many developmentally important genes, called *Hox* genes, in vertebrates and invertebrates. The homeodomain encodes an HTH DNA-binding motif, and *Hox* proteins are transcription factors that regulate the expression of other genes.

Guide to Study Exercises (text p. 930)

1. Eukaryotic genomes are larger than prokaryotic genomes mainly because eukaryotes are more complex than prokaryotes in both morphology and metabolism and hence contain more genes. The greater size of eukaryotic genomes also reflects the presence of unexpressed sequences, many of which are repetitive sequences. Some of this DNA belongs to chromosomal structures (centromeres and telomeres) that are not present in prokaryotes. (Sections 27-1A and C)

2. The functions of many ORFs are unknown because (a) the ORF may encode a protein that does not resemble any previously sequenced protein, and (b) even if an ORF is identified as coding for a certain class of protein (e.g., a membrane transport protein), the protein's fine structure (e.g., its ligand-binding site) and therefore its metabolic function (e.g., what substance it transports) may remain obscure. (Section 27-1A)

3. rRNA and tRNA genes are transcribed but not translated; therefore, they are needed in larger quantities than genes that are transcribed to mRNA. Clustering of rRNA and tRNA genes may increase the efficiency of their transcription. Clustering may also facilitate the balanced production of RNAs that are needed in equal amounts, for example, for the assembly of ribosomes. (Section 27-1B)

4. The *lac* operon is regulated so that expression of its three genes, which are required for metabolizing lactose, can be adjusted according to the availability of lactose and glucose. The *lac* repressor, a negative regulator, binds to the *lac* operon's operator to inhibit transcription unless a lactose derivative binds to the repressor and causes the repressor to dissociate from the operator. Thus, the availability of lactose governs the production of the enzymes that metabolize it. If glucose is present, however, the *lac* operon is repressed, regardless of whether lactose is also present. When glucose is absent, cAMP levels increase and cAMP binds to the catabolite gene activator protein (CAP). The CAP–cAMP complex is a positive regulator of the *lac* operon because it binds to the *lac* promoter and stimulates transcription. (Sections 27-2A and B)

5. Attenuation regulates the expression of genes involved in amino acid biosynthesis by a mechanism that senses the availability of that amino acid. When the amino acid is present, transcription of the operon encoding the biosynthetic enzymes terminates due to the formation of a terminator hairpin in the leader region of the transcript. As a result, the enzymes are not synthesized when they are not needed. When the amino acid is scarce, the terminator hairpin does not form and transcription proceeds through all the genes of the operon so that the required enzymes can be produced. The terminator hairpin is prevented from forming because a ribosome, which begins translating the leader region of the transcript, pauses where the absent amino acid should be incorporated. This pause allows the transcript to lengthen and adopt an alternative hairpin structure that precludes formation of the terminator hairpin. (Section 27-2C)

6. The genetic switch in bacteriophage λ governs the transition from lysogenic to lytic growth. During lysogenic growth, the λ repressor (the product of the *cI* gene) blocks transcription from the promoters p_L and p_R, thereby preventing the synthesis of proteins necessary for lytic growth while stimulating transcription of its own gene. It does so by binding as a dimer to subsites o_{R1} and o_{R2} of the operator o_R.

   Lytic growth often begins with damage to the host's DNA, which activates the SOS response. This leads to proteolysis of λ repressor such that the repressor loses its ability to cooperatively bind to o_{R1} and o_{R2}. The decrease in binding of λ repressor allows transcription from p_R and p_L to proceed, thereby generating the proteins required for lytic growth as well as Cro protein. Cro binds to the o_{R3} subsite of o_R, which prevents further synthesis of λ repressor. At high concentrations, Cro binds to all three o_R subsites and represses all transcription, but at this point, the lytic process has already been irretrievably initiated. (Section 27-2D)

7. The acetylation of the Lys residues in the N-terminal tails of histones in the nucleosome core neutralizes the positive charges of these Lys residues and thereby weakens their affinity for neighboring nucleosomes and possibly for the negatively charged DNA. This disrupts chromatin structure, thereby allowing RNA polymerase to transcribe the DNA. Histone deacetylation restores the positive charges of the Lys residues, promoting the formation of higher order chromatin structure. This renders the DNA in the nucleosome less accessible to the transcription machinery. (Section 27-3A)

8. Transcription efficiency is largely insensitive to the length of the DNA between an enhancer and a promoter because this DNA forms a loop. The size of the DNA loop does not matter, provided that the transcription factors that bind to the enhancer can also interact with the general transcription factors of the preinitiation complex at the promoter. (Section 27-3B)

9. Antibody diversity results from somatic recombination and hypermutation during the development of *B* cells. In somatic recombination, several DNA segments are pieced together to form genes encoding heavy and light chains of immunoglobulins. Because hundreds of different gene segments are available and because the point at which the sequences recombine varies by several nucleotides, billions of unique proteins can potentially be produced. Additional variation arises from the addition or deletion of nucleotides at the recombination joints and from the high rate of mutation in the immunoglobulin gene segments in *B* cell progenitors. (Section 27-3C)

10. During early embryogenesis in *Drosophila,* synchronized nuclear divisions lead to the formation of a blastoderm. Further development includes gastrulation and organogenesis. The body segments that become established during this phase persist into adulthood. Embryonic pattern formation is determined by the expression of maternal genes, whose mRNA is already present in the egg, and later by embryonic genes that are transcribed after about the eleventh round of nuclear division. Maternal-effect genes (which define the embryo's polarity) and segmentation genes (which specify the number and polarity of body segments) influence development through the formation of protein gradients that affect the transcription and translation of other genes. The expression of homeotic genes controls the differentiation of individual body segments. These genes encode transcription factors that contain a common DNA-binding motif called a homeodomain. (Section 27-3E)

Questions

Genome Organization

1. From Figure 27-2, select the two types of organisms that most dramatically illustrate the C-value paradox.

2. Can the *D. melanogaster* histone genes be considered a type of moderately repetitive DNA?

3. How might recombination during meiosis result in the expansion of trinucleotide repeats?

Regulation of Prokaryotic Gene Expression

4. What observation indicates that transcription of the *lac* operon is never completely repressed?

5. You have been handed several strains of *E. coli.* Suggest an easy way to determine how many are lysogens for bacteriophage λ.

6. Match the genes below with their functions.

   A. *cI* B. *cII* C. *N* D. *Q* E. *cro*

   ______ Its product acts as an antiterminator for the transcription of delayed early genes.
   ______ Required for the establishment of lysogeny.
   ______ Binding of its product to subsites of operator o_R commits the phage to lytic growth.
   ______ Its product acts as an antiterminator for the transcription of late genes.
   ______ Required to maintain lysogeny.
   ______ Relative levels of its product dictate the λ life cycle pathway.

7. Delayed early transcription of bacteriophage λ extends from p_L past the *att*P site into the *b* region (see Fig. 27-20). The RNA transcribed from the *b* region forms a hairpin loop that is a substrate for a 3′ exonuclease. How might the *b* region function to promote lytic growth?

8. Taking into account the answer from Question 7 above, how might the *b* region of bacteriophage λ affect the induction of a lysogen? Hint: Examine the gene order around the attachment site after integration (Fig. 27-22).

9. An o_{R2} point mutation does not affect λ repressor binding, but transcription from p_{RM} is severely repressed. Explain.

Regulation of Eukaryotic Gene Expression

10. Match the structural elements on the left with their functions on the right.

______ Promoter	A. Transcription factor binding site
______ Silencer	B. Transcription terminator
______ Operator	C. Protein coding region
______ Attenuator	D. Inducer binding site
______ Enhancer	E. RNA polymerase binding site
______ Structural gene	F. Repressor binding site

11. Which of the following DNA structures are likely to be transcriptionally active: Barr body, euchromatin, heterochromatin, DNase I–sensitive regions, polytene chromosome puffs, and highly methylated DNA?

12. Which phase of the cell cycle is likely to be missing in rapidly dividing embryonic cells of the frog *Xenopus laevis?*

13. Why is a mutated p53 gene present in so many cancers?

14. Inactivation of the eukaryotic proto-oncogene *c-myc* (see p. 682) involves premature termination of its RNA transcripts near the promoter. What feature of this phenomenon is likely to be similar to attenuation in prokaryotes? What feature must be different?

15. The two general mechanisms for immunoglobulin diversity are ____________________ and ____________________.

16. Why is it important that cells other than reticulocytes not synthesize heme-controlled repressor?

17. Match the following genes involved in *Drosophila* development with the statements below.

A. Homeotic genes	C. Gap genes
B. Pair-rule genes	D. Segment polarity genes

______ Gradients of these gene products define the polarity of body segments.
______ A deletion in one of these genes converts an antenna into a leg.
______ The spatial expression of these genes is regulated by the distribution of a maternal mRNA.
______ Mutations in these genes result in deletion of portions of every second body segment.
______ The products of these genes contain helix–turn–helix motifs.

18. How do Bicoid and Nanos proteins determine the differential expression of the *giant, Krüppel*, and *knirps* genes?

19. Do homeotic genes encode morphogens? Explain.

Answers to Questions

1. Some algae (with less than five cell types) have more than 1000 times more DNA than some arthropods (with dozens of different cell types).

2. The *Drosophila* genome contains ~100 copies of the histone genes, which can therefore be classified as moderately repetitive sequences.

3. A recombination event that was not in register would result in two daughter cells with differing numbers of trinucleotide repeats, as diagrammed below for a CAG repeat (only one DNA strand is shown). The longer the repeat, the more out-of-register the recombination is likely to be.

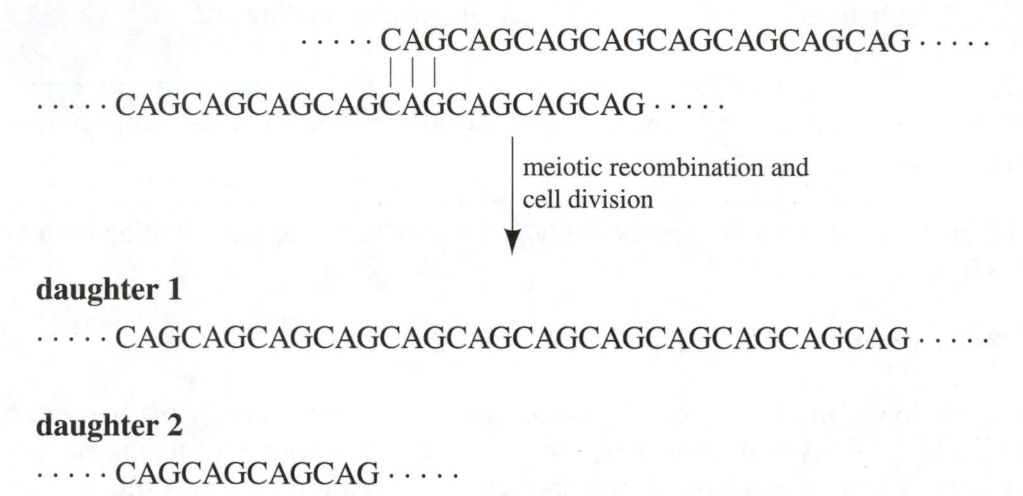

4. The natural inducer of the *lac* operon is 1,6-allolactose, which is produced from lactose by the action of β-galactosidase. The *lac* operon must be transcribed at a low level in order to generate a small amount of β-galactosidase as well as galactoside permease, which allows lactose to enter the cell.

5. Subjecting the bacteria to ultraviolet radiation or another agent that causes DNA damage would result in excision of the prophage in cells that are lysogens.

6. __C__ Its product acts as an antiterminator for the transcription of delayed early genes.
__B__ Required for the establishment of lysogeny.
__E__ Binding of its product to subsites of operator o_R commits the phage to lytic growth.
__D__ Its product acts as an antiterminator for the transcription of late genes.
__A__ Required to maintain lysogeny.
__A, E__ Relative levels of its product dictate the λ life cycle pathway.

7. Exonuclease digestion that begins in the *b* region may proceed "rightward" into the *int* gene, thereby preventing its translation. Without integrase, the phage could not integrate into the host DNA, so only lytic growth would be possible.

8. After integration, the *b* region is located to the right of p_L and the *int* and *xis* genes (see diagram below) and therefore cannot promote the degradation of the transcript from p_L. The resulting integrase and excisionase help excise the prophage from the host DNA.

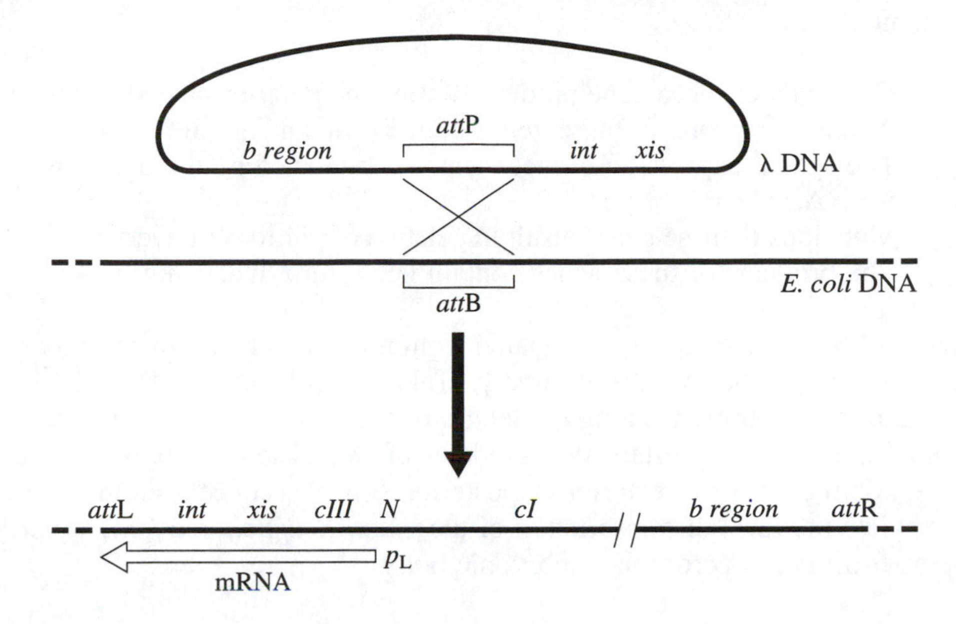

9. The mutation must alter a site where RNA polymerase interacts with the DNA to initiate transcription.

10. __E__ Promoter
 __A__ Silencer
 __F__ Operator
 __B__ Attenuator
 __A__ Enhancer
 __C__ Structural gene

11. The transcriptionally active structures are euchromatin and the polytene chromosome puffs. DNase I–sensitive regions only mark potentially active loci. The Barr body, heterochromatin, and highly methylated DNA are transcriptionally inactive.

12. These cells progress directly from M phase to S phase, essentially skipping G_1 phase. Early frog embryo cells divide about every 20–30 minutes.

13. p53 normally prevents cell division when DNA is damaged. Loss of this function leads to mutations that contribute to uncontrolled cell proliferation.

14. The eukaryotic transcript likely forms a transcription termination structure similar to that in prokaryotic attenuation. The obvious difference is that in eukaryotes, ribosomal binding and translation cannot be involved in transcription termination.

15. Somatic recombination and somatic mutation.

16. Heme-controlled repressor phosphorylates eIF-2 so that eIF-2 · GTP cannot be regenerated. This prevents translation initiation and would inhibit protein synthesis in all cells. The inhibition of protein synthesis does not harm reticulocytes, which synthesize little protein other than hemoglobin.

17. __D__ Gradients of these gene products define the polarity of body segments.
    __A__ A deletion in one of these genes converts an antenna into a leg.
    __C__ The spatial expression of these genes is regulated by the distribution of a maternal mRNA.
    __B__ Mutations in these genes result in deletion of portions of every second body segment.
    __A__ The products of these genes contain helix–turn–helix motifs.

18. Bicoid and Nanos proteins are translated from mRNAs that are maternally deposited in the anterior and posterior poles, respectively, of the unfertilized egg. They are therefore present at different concentrations along the length of the early embryo (see Fig. 27-45). The gradients of these proteins establish a gradient of expression of Hunchback protein that decreases nonlinearly from anterior to posterior. Specific concentrations of Hunchback protein regulate the level of transcription of the *giant, Krüppel,* and *knirps* genes so that these proteins form bands across the embryonic body.

19. Homeotic genes encode transcription factors, which cannot be morphogens because they do not diffuse between cells or between nuclei in a syncytium.

DONALD VOET
University of Pennsylvania

JUDITH G. VOET
Swarthmore College

KINEMAGE EXERCISES

for

FUNDAMENTALS OF BIOCHEMISTRY

INTRODUCTION AND SET-UP

Kinemages are molecular graphics displays that the user can manipulate in real time on a variety of computers including Macintoshes and IBM-compatible PCs. The files contain kinemages for display on Macintosh and PC computers.

The kinemages are provided in 21 Exercises whose file names are "Enn_wxyz.kin" (where nn is the Exercise number and wxyz is a four letter acronym; see the Table-of-Contents). To display the kinemages in an exercise, double click on the icon for the program MAGE. When queried, click on the "Proceed" box. Then, under the "File" pull-down menu, invoke the "Open File..." option and select the Exercise of your choice. In exercises that contain more than one kinemage, you must select these additional kinemages from the "KINEMAGE" menu.

The Kinemage Exercises included here are meant to enrich your study of biochemistry by allowing you to visualize and manipulate various proteins and nucleic acids that occur as figures in the text. The accompanying booklet contains questions and suggestions to guide you through the exercises. These questions and suggestions are also presented in the "TEXT : Kinemages" window of each exercise.

If you have not previously worked with kinemages, we suggest that you first examine E01_Tutr.kin, Teach-Image, a tutorial designed to instruct you in the use and capabilities of MAGE. Follow the instructions in the booklet or in the "TEXT : Kinemages" window to gain facility with using the program.

ACKNOWLEDGEMENTS

The program MAGE was written and generously provided by David C. Richardson, Biochemistry Department, Duke University, Durham, NC 27710, USA. He also wrote and provided the program PREKIN, which we used to help generate the kinemages shown here.

John H. Connor (Terminator Graphics Limited, 1118 Kimball Dr., Durham, NC 27712, USA) skillfully generated preliminary versions of eight of the exercises presented here (see the Table-of-Contents). Kim M. Gernert (4D Projections, 2902 Ode Turner Rd., Hillsborough, NC 27278, USA) did so for Exercise 3, and Thomas LaBean (Department of Biochemistry, Duke University, Durham, NC 27710, USA) did so for Exercise 20. Virginia Indivero (Lecturer, Swarthmore College, Swarthmore, PA 19081) worked through all of the kinemages and provided valuable feedback.

Many of the kinemages presented here are based on atomic coordinate sets that were obtained from the Protein Data Bank (PDB). These coordinate sets are each identified by a 4-character accession code (e.g., 3ABC) that is given at the end of the TEXT file of the corresponding exercise. These coordinates sets, as well as numerous others, can be obtained via the World Wide Web from http://www.rcsb.org/pdb. The files containing the coordinate sets also contain the names of the individuals who determined the structure under consideration, give an outline of how the structure was determined and an indication of its reliability, and provide a list of relevant references.

Donald Voet
Judith G. Voet

TABLE OF CONTENTS

EXERCISE 1 (E01_Tutr.kin): Teach-Image—An Introduction to Kinemages.*

EXERCISE 2 (E02_DNA.kin): DNA—Structure B Form.
Kinemage 1: B-DNA (Section 3-2B; Figure 3-9).
Kinemage 2: The Watson–Crick base pairs (Figure 3-11).

EXERCISE 3 (E03_ScSt.kin): Protein Secondary Structures—Peptide Groups, Helices, Beta Sheets, and Beta Bends.
Kinemage 1: The peptide group (Figs. 6-2 and 6-4).
Kinemage 2: The alpha helix (Fig. 6-7).
Kinemage 3: Beta pleated sheets (Figs. 6-9 and 6-10).
Kinemage 4: Beta bends/reverse turns (Fig. 6-20).
Kinemage 5: Visualization of the secondary structures of Lysozyme.

EXERCISE 4 (E04_FbPr.kin): Fibrous Proteins.*
Kinemage 1: Ribbon diagram of a coiled coil such as occurs in alpha-keratin (Fig. 6-14*b*).
Kinemage 2: The heptad repeat and hydrophobic interactions in coiled coils (Fig. 6-14*a*).
Kinemage 3: Ribbon and spacefilling diagrams of the collagen triple helix (Fig. 6-17).
Kinemage 4: Collagen backbone and the effect of a mutation (Fig. 6-18).

EXERCISE 5 (E05_Cyto.kin): *c*-Type Cytochromes—Structural Comparisons Illustrating Structural Evolution.
Kinemage 1: Cytochromes *c* (Sections 6-2C and 17-2E; Figs. 6-26, 6-31, and 17-14).

EXERCISE 6 (E06_Hemo.kin): Myoglobin and Hemoglobin—Structural Comparisons and Allosteric Interactions.*
Kinemage 1: Deoxymyoglobin (Fig. 7-1).
Kinemage 2: Evolutionary relationship of deoxymyoglobin and the α and β subunits of deoxyhemoglobin (Section 5-4).
Kinemage 3: Conformational changes in hemoglobin on oxygen binding—an overview; BPG binding to deoxyemoglobin (Figs. 7-5, 7-9, 7-10 and 7-15).
Kinemage 4: Conformational changes in hemoglobin's alpha subunit on oxygen binding (Fig. 7-9).
Kinemage 5: Conformational changes at hemoglobin's alpha1–beta2 interface on oxygen binding (Fig. 7-10).

EXERCISE 7 (E07_Sacc.kin): Saccharides—Glucose, Sucrose, Hyaluronic Acid, and a Complex Carbohydrate.*
Kinemage 1: D-Glucopyranose: Section 8-1B; Figs. 8-4 and 8-5.
Kinemage 2: Sucrose: Section 8-2A.
Kinemage 3: Hyaluronic Acid: 8-2D; Fig. 8-12.
Kinemage 4: Structure of a Complex Carbohydrate: Section 8-3C; Fig. 8-16.

EXERCISE 8 (E08_TrMb.kin): Transmembrane Proteins—Bacteriorhodopsin, the Photosynthetic Reaction Center, and Porin.*
Kinemage 1: Bacteriorhodopsin (Section 10-1A; Fig. 10-4).
Kinemage 2: Photosynthetic reaction center (Sections 10-1A and 18-2B; Fig. 10-5).
Kinemage 3: Porin (Sections 10-1A and 10-5B; Figs. 10-6 and 10-33).

EXERCISE 9 (E09_Lyso.kin): HEW Lysozyme—Catalytic Mechanism.
Kinemage 1: Lysozyme (Section 11-4; Figs. 11-15 and 11-17).

EXERCISE 10 (E10_SerP.kin): Serine Proteases—Catalytic Mechanism, Inhibition, and Activation.*
Kinemage 1: Structural Overview of a trypsin/inhibitor complex: (Section 11-5; Fig. 11-23).
Kinemage 2: Evolutionary structural comparisons of trypsin, chymotrypsin, and subtilisin (Fig. 11-25).
Kinemage 3: Chymotrypsin in complex with 2-phenylethyl boronic acid, a transition state analog (Fig. 11-27).
Kinemage 4: Structural comparison of chymotrypsin with chymotrypsinogen; mechanism of activation (Fig. 11-30).

EXERCISE 11 (E11_ATCs.kin): Aspartate Transcarbamoylase (ATCase)—Allosteric Interactions.
Kinemage 1: Quaternary structure of ATCase and its allosteric conformational changes (Section 12-3; Fig. 12-12).
Kinemage 2: Conformational changes caused by the binding of substrate (Fig. 12-13).

EXERCISE 12 (E12_TIM.kin): Triose Phosphate Isomerase (TIM)—Catalytic Mechanism.
Kinemage 1: The eight-stranded alpha/beta barrel (Section 6-2B; Figs. 6-27*c* and 6-28*d*).
Kinemage 2: TIM active site (Section 14-2E; Fig. 14-6).

EXERCISE 13 (E13_PFK.kin): Phosphofructokinase (PFK)—Allosteric Interactions.
Kinemage 1: Conformational changes in a dimeric unit of PFK (Section 14-4A; Figs. 14-22 and 14-24).
Kinemage 2: The major conformational changes in a subunit of PFK (Fig. 14-24).

EXERCISE 14 (E14_Phos.kin): Glycogen Phosphorylase—Allosteric Interactions.
Kinemage 1: Structure of glycogen phosphorylase *a* (Section 15-1A; Fig. 15-3).
Kinemage 2: Conformational differences between T and R states of glycogen phosphorylase *b* (Fig. 15-5).
Kinemage 3: The dimer interface on activation of glycogen phosphorylase.

EXERCISE 15 (E15_cAPK.kin): cAMP-Dependent Protein Kinase (cAPK).
Kinemage 1: The catalytic subunit of cAPK (Section 15-3B; Fig. 15-14).

EXERCISE 16 (E16_CaM.kin): Calmodulin (CaM).*
Kinemage 1: The structure of CaM highlighting its EF-hand motifs (Section 15-3B; Figs. 15-16 and 15-17).
Kinemage 2: The structure of CaM in complex with its target peptide (Fig. 15-18).

EXERCISE 17 (E17_DNA2.kin): DNA—Structures of A, B, and Z forms.
Kinemage 1: Comparison of the structures of A-, B-, and Z-DNAs (Section 23-1; Fig. 23-2).
Kinemage 2: The Watson–Crick base pairs (Section 23-1A; Fig. 23-1).
Kinemage 3: Sugar pucker, 3′-*exo* and 3′-*endo* (Section 23-1B; Fig 23-6).
Kinemage 4: B-DNA (Section 23-1; Fig. 23-2).
Kinemage 5: A-DNA (Section 23-1; Fig. 23-2).
Kinemage 6: Z-DNA (Section 23-1; Fig. 23-2).

EXERCISE 18 (E18_EcoR.kin): *Eco*RI and *Eco*RV Restriction Endonucleases—Complexes with Target DNAs.

Kinemage 1: *Eco*RI–DNA (Fig. 23-30).
Kinemage 2: *Eco*RV–DNA (Fig. 23-31).

EXERCISE 19 (E19_Rprs.kin): 434 Repressor/DNA Interactions.*

Kinemage 1: 434 Repressor—DNA interactions (Fig. 23-33).

EXERCISE 20 (E20_tRNA.kin): $tRNA^{Phe}$—Structural Interactions.@

Kinemage 1: Overview of functional domains (Fig. 26-6).
Kinemage 2: Structural features (Section 26-2A).
Kinemage 3: Tertiary base-pairing interactions (Fig. 26-7).

EXERCISE 21 (E21_GnRS.kin): Glutaminyl–tRNA Synthetase (GlnRS) in Complex with $tRNA^{Gln}$ and ATP.

Kinemage 1: The overall complex (Fig. 26-11).

*Kinemages generated in large part by John H. Connor, Terminator Graphics Limited, 1118 Kimball Dr., Durham, NC 27712, USA.

#Kinemage generated in large part by Kim M. Gernert, 4D Projections, 2902 Ode Turner Rd., Hillsborough, NC 27278, USA.

@Kinemage generated in large part by Thomas LaBean, Department of Biochemistry, Duke University, Durham, NC 27710, USA.

EXERCISE 1. TEACH-IMAGE, AN INTRODUCTION TO KINEMAGES

Welcome to TEACH-IMAGE, an introductory kinemage that is intended to show you some of the many features you will find in MAGE. MAGE is an interactive program that allows you to inspect molecular structures in a more hands-on fashion than is possible with static diagrams and photographs. MAGE renders images in three dimensions and enables you, the reader, to rotate, explore, and even change them.

If you have not already done so, **double click on the icon for the program MAGE. When queried, click on the "Proceed" box. Then, under the "File" pull-down menu, invoke the "Open File..." option and select E01_Tutr.kin.**

There are three windows that are shown in kinemages:

1: The text window, which contains the same text as in the manual.
2: The caption window, the long window at the bottom of the screen.
3: The graphics window, which has a black background and may be behind, in front of, or beside the text window.

These windows can be accessed by clicking directly on them, or by selecting the appropriate window under the "Windows" pulldown menu.

Click on the graphics window to bring it to the front. Click on the box in the upper right corner of the MAGE color graphics bar to make the graphics window fill the screen.

The graphics window presently contains three items (not all immediately visible), the most visible of which is the word "CELL". **Click on the picture window to bring it to the foreground so you can manipulate the features of this kinemage.** Further explanation is provided below but should be used only as an occasional reference as you explore the contents of this kinemage.

Try clicking on and off the boxes on the upper right-hand side. These boxes control the parts of the kinemage that are shown. The top button "*Cell" turns on and off the whole word, CELL. The indented buttons below it turn on and off individual letters when the "*Cell" button is on. The other buttons, "Item #2" and "Item #3", presently appear to have no function, but as you progress, their function will become evident.

In the middle of the panel, just above the dashed line, there are a few more buttons that turn on and off items in any of the previous groups. "Consonant", for instance, turns on and off "C1", "L3" and "L4"; "Vowels" turns on and off "E2".

Click on the ANIMATE button (the bottom box) a few times. It steps between and turns on the groups whose names begin with an asterisk, "*". In this kinemage, the word builds up: "C", to "CE", to "CEL", to "CELL". This feature will be helpful in showing conformation changes of the molecules. Animation may also be invoked by repeatedly hitting the letter "a" on the keyboard. In kinemages with a "2ANIMATE" box, this second animation sequence may instead be invoked by repeatedly hitting "b" on the keyboard.

Click on an atomic position in a kinemage to display identifying information about that atom. Atoms are located at the intersections or ends of the displayed lines, whereas these lines represent the bonds between atoms. Clicking on an atom will normally display, at the bottom left of the screen, the atom's name (which usually indicates its type and position in the molecule), together with its residue name and number. However, in the present kinemage, other information is listed at some of the vertices. **Click on a few spots in the kinemage and look at the lower left hand corner to see these descriptions.** Some points (in this kinemage) simply provide the x, y, and z coordinates of their positions This can also be seen by using the pull down menu series: "Other", "Kluges Menu" and "XYZ point", whereupon the coordinates of picked atoms will be displayed at the top right of the screen. Compare these numbers with the labels at the bottom left. **Consecutively click on any two points to display the distance between them, in Å, at the lower right of the screen.**

You can measure the angle or dihedral angle between points by invoking "measures" under the pulldown menu "Other". A button labeled "measure" will appear near the bottom of the right-hand panel of the window. Make sure it is checked. (You can toggle a button on and off by clicking on it.) **Click on two, three, and then four points.** White lines will appear to connect these points. The red dots which appear are the average points between the last two, last three, and last four points selected. The characters at the bottom of the screen progressively show the distance, angle, and dihedral angle of the last two, three, and four points picked. Click on the "markers" box; one of two markers will show up to highlight the point chosen.

Try rotating the word. To do this, click in the graphics window (holding down the mouse button) and drag the mouse. Clicking on the top one-third portion of the screen controls the rotation in the plane of the monitor (rotation about the z axis; the arrow will have a circle at its tail). Dragging to the right rotates the display clockwise; dragging to the left rotates it counterclockwise. Clicking on the bottom two thirds of the screen controls the rotation in and out of the plane [about the x and y axes; the arrow will have a plus (+) at the end of its tail]. **Click on the lower right side of the graphics window and drag the mouse straight to the left (rotate about the y axis).** This will flip the word "CELL" backwards, putting the "L" on the left and the "C" on the right. **Click on the bottom of the graphics window and drag the mouse straight up (rotate about the x axis).** This will turn the word "CELL" upside-down. **You can also change the orientation of the picture by invoking "View2" under the pulldown menu** "VIEWS". Such set views highlight and show specific aspects of the molecule. You can save a particular view by rotating to that view and invoking "set reader's view" under the "Other" pulldown menu. Change the view and then select "Reader's View under VIEWS to return to the view that you saved. Note that the "Reader's View" will not be saved upon exiting a kinemage. **If you lose track of your orientation, just choose "View 1" to go back to the original view.**

Rotate the molecule until other images become apparent, thus emphasizing the importance of visualizing objects in three dimensions rather than two. **Rotate the molecule until you can see a second green word, "Item #2" ("View 3").** Even in this view, the green word is not very legible because it is too small.

Increase the size of the image by using the ZOOM slide. The ZOOM slide is a vertical control bar on the right of the graphics window. The larger the number, the larger the image. The ZOOM value can be changed by clicking on the arrow at the top or the bottom of the bar (slowest), by clicking in the shaded portion of the bar (medium speed), or by dragging the box (fastest). Changing the ZOOM value to approximately 2.20 sufficiently enlarges the second image to easily read "MOLECULES".

The green word is not at the center of display. Thus, you need to adjust the centering of the molecule. **Adjust the centering by turning on the button "pickcenter" near the bottom of the button list, and then clicking on the point (atom) you want centered.** This repositions the whole image to now rotate about that point. **Click on the first "E" of "MOLECULES". Do not forget to turn off the "pickcenter" option when you are finished** as failing to do so can lead to frustrating recenterings when you least want them. **Now rotate the image and increase its size to be able to read "MOLECULES" clearly (View4).** Clicking parts of the kinemage off, for instance "Cell", may also be useful to unclutter the display area.

Once you have brought the second word close, turn on "zclip" (near the bottom of the button list) and rotate the molecule. Letters will disappear as you do so if the ZSLAB slide is set sufficiently small. This feature is particularly useful when inspecting complex molecules because it unclutters the image. The extent of zclipping can be controlled by the ZSLAB slide and the ZTRAN slide. The ZSLAB slide controls how much of the molecule can be seen (the thickness of the viewing slab). **Go to View5; only "OLECUL" can be seen. Increase ZSLAB to 200 to see the whole word;** decrease ZSLAB to 60 to just see "LEC". The ZTRAN slide controls the position of the molecule in that slab. Use the ZTRANS slide to move the word through the slab. At a ZTRANS value of −90, "MOL" is seen; at 105 "ULE" is seen. Hitting the letters "x" or "z" on the keyboard translates the molecule in and out of the screen (towards +z or −z); hitting the upper case letters "X" or "Z" does the same thing but at ten times the rate. However, these keyboard controls do not affect the value of ZTRANS.

Reset to View5. A third object ("Item #3"), a red word, should now be visible. **Rotate, pickcenter, and zoom so that you can see this last word, "ATOMS",** and you have achieved proficiency at everything necessary to explore kinemages. If you are having any difficulty look under the pulldown menu item VIEWS. The successive views give differing perspectives at various zooms and can help orient you in case you get sidetracked. Does your final image look like the one in View7? Can you make it do so?

More advanced techniques:

Rotatable Bonds

To the far right of the menu panel is a horizontal bar labeled "Rotatable". Reset the VIEW to "View 3" so that you can see the word "CELL" and you are looking nearly down the edge of the word. By clicking in the horizontal bar; you can rotate the upper arm of the "L" about its "bottom line". The bond of the bottom line is highlighted in pink when you turn on the master button "Rotatable". The number just above the bar shows the angle at which the upper arm is now displayed. To measure this dihedral angle (dhr=), invoke "measures" under the "Other" pulldown menu. Then click on the top left of the "E", the bottom left of the "E", the corner of the "L", and then the end of the upper arm of the "L".

Linewidths and Depth Cueing

Under the "Display" pulldown menu, select "onewidth". How does this option affect the word as you rotate it? Again under "Display", deselect "onewidth" and select "thinline" (or hit "t" on the keyboard). This should make all line widths quite thin, an option that may help to clear up a complex molecular image. Turn off "thinline" and turn on "perspective" (or hit "p" on the keyboard) to give exaggerated depth cueing.

Exercise in large part by John H. Connor
Terminator Graphics Limited
1118 Kimball Dr.
Durham, NC 27712
USA

Kinemages:

4D Projections, 2902 Ode Turner Rd., Hillsborough, NC 27278.

EXERCISE 2. DNA STRUCTURE

KINEMAGE 1: B-DNA; Section 3-2B; Figure 3-9.

KINEMAGE 2: The Watson–Crick Base Pairs; Section 3-2A; Figure 3-11.

DNA, the archive of hereditary information, forms double helices whose component strands are complementary and antiparallel. In this exercise, we explore the structure of the Watson–Crick double helix, B-DNA (Fig. 3-9). We then study the structures of the Watson–Crick base pairs (Fig. 3-11).

If you have not already done so, **double click on the icon for the program MAGE. When queried, click on the "Proceed" box. Then, under the "File" pull-down menu, invoke the "Open New File..." option and select E02_DNA.kin.**

Click on the graphics window to bring it to the front. Click on the box in the upper right corner of the MAGE color graphics bar to make the graphics window fill the screen.

Kinemage 1: B-DNA (Section 3-2B; Fig. 3-9).

View1 shows B-DNA with its helix axis vertical and looking down its 2-fold axis of symmetry into its minor groove. All atoms of the 12-bp duplex helix are shown as large balls (with C, N, O, and P atoms white, blue, red, and seagreen) that, for the sake of clarity, are slightly smaller than space-filling size. **Use the "Backbone" and "Bases" buttons under the "Spacefilling" button** to turn the base pairs and the two sugar-phosphate chains of the duplex on and off separately. **Use the "MinorGroove" and "MajorGroove" buttons** to highlight the minor and major grooves (in cyan and yellow). Turning on the "MinorGroove" button highlights, in cyan, an atom that lines the minor groove on each base (atoms C2, O2, N2, and O2 on A, T, G, and C, respectively). Turning on the "MajorGroove" button highlights, in yellow, an atom that lines the major groove on each base (atoms N6, O4, O6, and N4 of A, T, G, and C, respectively) Views2 through 4 are different orientations of the DNA.

Compare View1 (similar to Fig. 3-9) and View2 of B-DNA. Note that its major groove is considerably wider than its minor groove although the two grooves are more or less equally deep. This is particularly evident in View3, a view along the grooves in which the major groove faces left in the center of the DNA and the minor groove faces left near both the top and bottom of the DNA. The different widths of the grooves arise from the asymmetry of the ribose–phosphate groups that comprise their walls. **In View1 or View2, turn off the "Backbone" button under the "Spacefilling" button,** to see that the base pairs form a solid stack in which the bases are in van der Waals contact (the apparent gaps between the bases are due to the less-than-van der Waals radii of the balls representing the atoms).

Turn off the "Spacefilling" button and turn on the upper "SingleStrnd" button to display the path taken by one of the two identical polynucleotide strands of the B-DNA. Note that even the bases in a single strand of B-DNA are well stacked.

Now turn off both the "Spacefilling" and upper "SingleStrnd" buttons and then turn on the "Wireframe" button to display the entire duplex molecule in stick form colored skyblue with its ribose ring oxygen atoms represented by small red balls. The backbone and bases can be individually controlled with the corresponding "Backbone" and "Bases" buttons. The "Top bp" button highlights the top base pair in white.

Turn on the lower (below the dashed line) "SingleStrand" button to highlight one of the sugar–phosphate backbone strands in magenta. **Turn on the associated "Bases" button** to highlight the bases of this strand in gold. **Turn off the "Wireframe" button** to trace the pathway of a single strand of B-DNA. **Then, turn on the upper "SingleStrnd" button,** so the B-DNA is displayed with one of its strands in skeletal form and the other in spacefilling form.

Kinemage 2: The Watson-Crick Base Pairs (Section 3-2A; Fig. 3-11).

View1 shows an A·T Watson–Crick base pair from ideal B-DNA (Fig. 3-11). The C, N, and O atoms of the bases (including ribose atom C1′, the glycosidic carbon atom) are represented by gray, blue, and red balls respectively, that can be turned on and off with the "Base Atoms" button. The bonds of the thymine (T) base are yellow and those of the adenine (A) base are sea-green. The 5′-ribose phosphate groups, which can be turned on and off with the "Ribose Phos." button, are magenta and their ribose ring oxygen atoms are represented by small red balls. The hydrogen bonds through which the bases are paired are represented by dashed white lines that can be turned on and off with the "H bonds" button.

Click the "ANIMATE" button to replace the A·T base pair with a G·C Watson–Crick base pair. The G·C pair is colored identically to the A·T pair except that the bonds of the guanine (G) base are skyblue and those of the cytosine (C) base are orange. Note that the atomic positions of the ribose phosphate groups, including those of the glycosidic carbons (the ribose C atoms that are bonded to the bases), are unchanged by this base pair switch. **View2 shows the bases edgewise.** There would likewise be no change in the atomic positions of the ribose phosphate groups if the A·T and G·C base pairs were replaced by T·A or C·G. Thus, the conformation of a DNA's sugar–phosphate backbone is unaffected by the identities of its Watson–Crick base pairs. This is one of the requirements for building a regular helix whose structure is independent of sequence.

View2 also shows that the two ribose–phosphate groups in each base pair are oriented oppositely, making the DNA's sugar–phosphate backbones antiparallel.

The bases in DNA lie at smaller distances from the helix axis than the sugar–phosphate backbones, so each DNA double helix has two helical grooves running along its periphery (we examined these grooves in KINEMAGE 1). **Identify the atoms of each base pair that point toward the major and minor grooves.** These grooves can be easily identified as follows: **Go back to View1 and turn on the "Glycosid.Bon" button to highlight the glycosidic bonds in red.** The minor groove is the one in which the two glycosidic bonds of the base pair make an angle of less than 180°. The groups that line the minor groove are donated by the edges of the bases that face the opening of this angle. **Turn on the "MinorGroove" button to highlight, in cyan, an atom that lines the minor groove on each base** (atoms C2, O2, N2, and O2 on A, T, G, and C, respectively). Click on the center of an atom to cause its atom type to appear in the lower left corner of the screen. Similarly, the edges of the bases that face the opening of the angle made by the glycosidic bonds that is greater than 180° line the DNA's so-called major groove. **Turn on the "MajorGroove" button to highlight, in yellow, an atom that lines the major groove on each base** (atoms N6, O4, O6, and N4 of A, T, G, and C, respectively).

EXERCISE 3. PROTEIN SECONDARY STRUCTURES

KINEMAGE 1: The Peptide Group: Figs. 6-2 and 6-4.

KINEMAGE 2: The Alpha Helix: Fig. 6-7.

KINEMAGE 3: Beta Pleated Sheets: Figs. 6-9 and 6-10.

KINEMAGE 4: Beta Bends (Reverse Turns): Fig. 6-20.

KINEMAGE 5: Visualization of Secondary Structures in Lysozyme.

A protein's primary structure is defined as the amino acid sequence of its polypeptide chain; its secondary structure is the local spatial arrangement of a polypeptide's backbone (main chain) atoms; its tertiary structure refers to the three-dimensional structure of an entire polypeptide chain; and its quaternary structure is the three-dimensional arrangement of the subunits in a multisubunit protein. In this exercise we examine the structures of the most frequently occurring elements of secondary structures in proteins; those of α helices, β pleated sheets, and β bends (the tertiary and quarternary structures of proteins are the subjects of other exercises). We shall see that, in the case of α helices and β sheets, successive amino acid residues in a polypeptide chain have the same main chain conformations. We begin by examining the conformations that are sterically available to amino acid residues.

If you have not already done so, **double click on the icon for the program MAGE. When queried, click on the "Proceed" box. Then, under the "File" pull-down menu, invoke the "Open New File..." option and select E03_ScSt.kin.**

Click on the graphics window to bring it to the front. Click on the box in the upper right corner of the MAGE color graphics bar to make the graphics window fill the screen.

Kinemage 1: The Peptide Group (Section 6-1A; Figs. 6-2 and 6-4).

View1 shows an amino acid in ball-and-stick representation whose display is controlled by the "AA (1)" button. Its atoms are colored by atom type (N blue, C brown, O red, and H white) and its bonds are white. **Click on the center of an atom** to cause its atom type and its residue name and number to be displayed at the lower left corner of the screen.

The amino acid's sidechain ("R-Group") is initially represented only by its Cβ (Cb) atom shown as a purple ball-and-stick. **Turn on the "Sidechain" button** to see the full sidechain in ball-and-stick form, with its bonds cyan. **Turn on the "sc H's" button** to see the hydrogen atoms of the sidechain.

Turn on the "AA (2)" button, a Lys residue (Lys 2). What happens to one of the carboxyl oxygen atoms of "AA (1)" when it forms a peptide bond with "AA (2)"? Go to View2 (Fig. 6-2). Measure the bond lengths and angles of the peptide and compare your values with those in Fig. 6-2. Distances can be measured by clicking first on one atom and then on the next. The distance between them is then displayed in the lower right hand corner. Angles are measured by invoking the "measures" option under the "Other" pulldown menu. Clicking on three atoms in a row then gives the bond angle.

Turn on the "Planes" button to highlight the plane of the resulting peptide group. **Is the peptide group actually planar?** View3 sights along this plane allowing you to verify that the Cα (Ca), the carbonyl carbon (C), and the carbonyl oxygen (O) of "AA (1)", together with the main chain nitrogen (N), its covalently bonded amide hydrogen, and the Cα of "AA (2)" are indeed coplanar. The sidechain of residue "AA (2)" is also turned on by the "Sidechain" button. Locate the four carbons of its sidechain and the charged nitrogen (Nz) at the end of this sidechain.

Turn on the "AA (3)" button to display the third amino acid of the tripeptide (compare with Fig. 6-3), This is a Leu residue (Leu 3). Again, **what happens to one of the carboxyl oxygens on formation of this peptide bond? Turn on the "Sidechain" button to display the sidechain of "AA (3)".**

Turn on the "Planes" button to highlight the two planar peptide groups that link the three amino acids (compare with Fig. 6-4). Such planar groups are rigid and hence peptide bonds are not free to rotate. However, both main chain bonds flanking the Cα atom of each amino acid residue can rotate. A polypeptide's main chain conformation is therefore specified by the torsion angles (dihedral angles) about the Cα—N bond (ϕ) and the Cα—C bond (ψ) of each of its residues. **Identify the ϕ and ψ bonds by turning on the "PHI" and "PSI" buttons** to highlight, in green, the corresponding bonds of the tripeptide's central Ca atom.

Rotate the molecule to look directly down the Cα—N bond (PHI; View4). Vary the ϕ torsion angle of the peptide by clicking in the top horizontal bar labeled PHI. You can either (1) click on the small arrows at the end of the bar (slow), (2) click in the shaded area of the bar (fast), or (3) drag the slider (small square) within the bar (fastest) to change the torsion angle. The second number above the bar indicates the new torsion angle. **Rotate ϕ with just the "AA (2)" button on,** to see how the amide group rotates about the Cα—N bond. Then **turn on "AA (1)" and "AA (3)"** to see the change you have made in the polypeptide's conformation. Note that the ϕ and ψ sliders only control the ϕ and ψ angles of residue 2; they do not alter the conformations of residues 1 or 3.

Rotate the molecule to look down the Cα—C bond (ψ; View5). With "AA (1)", "AA (2)" and "AA (3)" on, change ψ to rotate the peptide plane about the Cα—C bond. Note the changes you have made to the polypeptide's conformation.

Return to the initial conformation by invoking the "Kluges Menu" under the "Other" pull-down menu and selecting "init angles". In the largely extended conformation of the tripeptide that was initially displayed ($\phi = -128°$, $\psi = 147°$), there are no atom overlaps or contacts. **Turn on the "O 1", "NH 2", "CaH 2", "Cb 2", "O 2" and/or "NH 3" buttons** to display portions of the van der Waals surfaces of the corresponding atoms or groups as dot surfaces that have the same colors as their parent atoms (The van der Waals distances between atoms are the minimum sterically allowed distances between nonbonded atoms). **Are there any overlaps of the dot surfaces around the atoms?** The closest noncovalent contacts that the tripeptide makes in this conformation are between the amide hydrogen of "NH 2" and the carbonyl oxygen "O 2", which are 2.6 Å apart and hence greater than the normally allowed lower limit of 2.4 Å, and between "O 1" and the hydrogen atom of "CaH2", which is 2.3 Å and hence is between the normally allowed limit of
2.4 Å and the outer limit (minimum allowed interatomic distance) of 2.2 Å.

Problem 1: Measure the distance between the amide hydrogen "NH 2" and the carbonyl oxygen "O 2" and between "O 1" and the hydrogen atom of "CaH 2" described in the above paragraph to verify the distances assigned to them. Reminder: By consecutively clicking on any two points, the distance between them, in Å, is given at the lower right of the screen.

There are steric constraints on the torsion angles, ϕ and ψ, that limit the conformational range of a polypeptide backbone. For example, changing both the ϕ and ψ angles to 180°, which puts the polypeptide in its most extended conformation, places several atoms within less than van der Waals distances.

Problem 2: With ϕ and ψ angles at 180°, find the distance between the amide hydrogen of "NH n" and the carbonyl oxygen "O n" and between the carbonyl oxygen "O n−1" and the hydrogen of "CaH n". Are these distances allowed or are they too close?

Change the ϕ and ψ angles to find other conformations that are not allowed due to atomic overlaps. Fig. 6-5 shows a peptide with $\phi = -60°$ and $\psi = 30°$. **Duplicate this illustration by going to View6 and changing ϕ to $-60°$ and ψ to 30°.** There is a steric clash between the carbonyl oxygen "O 1" and the amide hydrogen of "NH 3"; they are separated by only 1.5 Å. Turning on the "O 1" and "NH 3" buttons clearly reveals that the dot surfaces of these atoms intersect and thus indicates that the atoms are in collision.

Problem 3: Place the computer structure in the conformations with the following ϕ and ψ angles and determine whether or not they are sterically allowed by visually examining potential clashes and then measuring the distance between atoms you think may be too close:

#	*ϕ (deg)*	*ψ (deg)*	*Allowed?*	*Steric Clash between which atoms?*
Example	6	55	NO	C(n) and O($n-1$)
1.	0	180		
2.	116	55		
3.	−57	−47		
4.	−119	113		

Consult the Ramachandran diagram (Fig. 6-6) to see if these angles are allowed and if they correspond to any specific secondary structure.

Kinemage 2: The α Helix (Section 6-1B; Fig. 6-7).

View1 shows four turns of a right-handed α helix in ball-and-stick form with its atoms colored according to type (N blue and O red) and its sidechains represented by their Cα—Cβ bonds in purple (Fig. 6-7).

The torsion angles of the residues in α helices lie in a small range (~10°) about $\phi = -57°$ and $\psi = -47°$ (see Fig. 6-6). The 5 residues in the first turn all fall in this range. **Go to View2 for a closeup of the first turn. Turn on the "Planes" button to highlight the peptide groups linking three successive amino acid residues. Which color plane (yellow or green) rotates the ψ bond and which the ϕ bond?** (*Hint:* The planes are connected by Cα whose bond enters one plane to N to form the ϕ bond and the other plane to C to form the ψ bond.) **Determine the ϕ and ψ angles** by invoking the "measures" option under the "Other" pulldown menu. Then click on [C(59) to N(60) to Ca(60) to C(60)] for ϕ (labeled dhr in the kinemage in the lower right hand corner) and [N(60) to Ca(60) to C(60) to N(61)] for ψ. Note their range of variation.

A polypeptide segment whose amino acid residues all have similar ϕ, ψ values forms a secondary structure. The particular ϕ, ψ values of an α helix result in the amide hydrogens and the carbonyl oxygens pointing in opposite directions along the helix axis. The carbonyl oxygen of any residue n points along the helix in the direction of the C-terminal end and towards the amide hydrogen of residue n+4, thereby forming an H-bond (represented by a dashed green line when the "H-Bonds" button is on) with an N−O distance of ~2.8 Å. Thus, the H-bonding requirements of the polypeptide backbone are satisfied in the α helix. **Measure the distances of some of the H-bonds of the α helix. Do they maintain this distance? Are there any unsatisfied H-bond donors or acceptors in the α helix?**

The right-handed α helix has 3.6 residues per turn and a pitch (rise per turn) of 5.4 Å. **Can you confirm the rise and pitch of this α helix visually?**

The R groups of the helix all project backwards down the polypeptide chain (View1) and outward from the helix (View3). View3 is a top view that looks down the helix axis. The backbone and the planes of the polypeptides form an approximately square outline; the side chains point out and spiral around like a pinwheel.

Turn on the "Polar sc" and "Nonpolar sc" buttons. You can see that in this helix the side chains cluster according to their polarities. The nonpolar residues usually cluster on the side of the helix that packs toward the interior of the protein and the polar residues face the solvent. This is called an amphiphilic helix. Helices that are completely buried in proteins or in nonpolar membranes have mostly nonpolar surfaces.

Go to View1 and turn off all of the turns but "Turn 1" and the "Polar sc" and "Non-polar sc" buttons. Turn on the "Potential HB" button. This shows, in light green, the potential H-bonds that can be made if the helix is extended. **Now build the helix by turning on turns 2, 3 and 4 in sequence.** The potential H-bonds turn green as they are fulfilled.

The view of a protein's backbone may be simplified by displaying only its connected Cα atoms (its Cα backbone). **This α helix may be shown in this way, in pinktint, by turning off all of the Turn buttons under the "alpha Helix" button and turning on its "Ca-Ca" button.** A blue ball marks the N terminus of the helix and a red ball marks its C-terminus, thereby indicating the direction of the polypeptide chain.

Kinemage 3: β Pleated Sheets (Section 6-1B); Figs. 6-9 and 6-10.

View1 shows three β strands forming a β sheet. Click the ANIMATE button to switch between parallel and antiparallel β sheets. Turn on the "H-bonds" button and ANIMATE again to observe the difference in H-bond structure for parallel and antiparallel β sheets (Fig. 6-9). **Turn off the "Main Chain" and "H-Bonds" buttons, turn on the "Ca-Ca" button and ANIMATE again.** This displays, in pinktint, the polypeptide segments' Cα chains, with blue balls at their N-termini and red balls at their C-termini. In this form it is easy to see the parallel or antiparallel nature of the strands. **Turn off the "Ca-Ca" button and turn on the "Main Chain" button again before proceeding.**

Now let's dissect the antiparallel β sheet structure. **Animate until the antiparallel β sheet is shown. Turn off strands 3 and 5.** You should now see a single polypeptide chain segment as represented by its main chain with its N and O atoms shown as blue and red balls and with its sidechains represented by its Cα—Cβ bonds in purple. The torsion angles of the 8 central residues of this strand are each in a small range about $\phi = -139°$ and $\psi = 135°$ and hence this strand is nearly fully extended. **Turn on the "Planes" button** to highlight the planes of two successive peptide groups in yellow and green. **Which plane corresponds to rotation about the ϕ angle and which corresponds to rotation about the ψ angle** (see Kinemage 2 for a hint)? **Measure the ϕ and ψ dihedral (torsion) angles around these peptides.** A portion of this strand is displayed in magnified form in View2.

Turn on the "Potential HB" button to see, in greentint, the potential H-bonds that this strand can make. Note that these H-bonds are nearly parallel to each other and are nearly perpendicular to the polypeptide strand (View1 and View2). The "H-Bonds" button does not highlight anything with this single strand, because H-bonding in β sheets occurs between neighboring strands rather than within a single strand.

Now begin building a β sheet by turning on "Strand 5". Is this strand oriented parallel or antiparallel to "Strand 4" (momentarily turning off the "Main Chain" button and turning on the "Ca-Ca" button may help you visualize this)? Can you see that the strands are antiparallel? Turning on the "H-Bond" button reveals that every second residue of each of these two polypeptides forms two H-bonds, highlighted in green, between their backbone NH groups and their carbonyl O atoms. **Now turn on "Strand 3".** It interacts with "Strand 4" similarly to but on its opposite side from "Strand 5", thereby fully satisfying the main chain H-bonding potential of "Strand 4".

Turn off "Strand 5" and "Strand 3" and change to View3. This view is looking edge-on along the peptide planes with the potential H-bonds now pointing in and out of the screen. **In this direction, the β strand has a rippled or pleated appearance, which can also be seen in Fig. 6-10.** Successive side chains of the polypeptide chain extend to opposite sides of the strand, alternately pointing up and down (View3 and View4).

Again turn on "Strand 5" and "Strand 3". The sheet formed by the three strands maintains its pleated appearance with its side chains alternating up and down from the sheet. **Turn on the "Polar sc" and "Nonpolar sc" buttons** to show that the polar (magenta) and nonpolar (gold) side chains cluster on opposite sides of the sheet. This is because the polar (hydrophilic) side of this sheet faces the solvent, whereas the nonpolar (hydrophobic) side faces the interior of the protein (concanavalin A; Fig. 6-25*b*). Fig. 6-11 is a space-filling diagram of the entire 6-stranded β sheet of which the three strands shown here are members.

Now return to View1 and click the "ANIMATE" button. Are the three polypeptide chains shown parallel or antiparallel (this can be seen more easily by temporarily turning off the "Main Chain" and turning on the "Ca-Ca" buttons)? Can you see that the strands are parallel? The hydrogen bonding pattern in this parallel β sheet is similar to that of the antiparallel sheet in that a given residue of the central chain (here "Strand 3") forms two H-bonds with the main chain of a neighboring strand and the next residue does so with the second such strand. However, whereas each residue in the antiparallel sheet forms two H-bonds with a single residue on a neighboring strand, in a parallel sheet, the two hydrogen bonds extending from one residue are made with residue n and residue $n+2$ of the neighboring strand. Consequently, the H-bonds formed by parallel sheets have a more distorted appearance than those of antiparallel sheets. This may account for the somewhat lesser stabilities of parallel sheets relative to those antiparallel sheets.

Viewed edge-on (View3 and View4), the parallel sheet is pleated, as is the antiparallel sheet. You can see that this parallel sheet has a significant right-handed twist; that is, the edges of the sheet curl in the direction of the fingers of a right hand when its thumb is held parallel to one of the polypeptide strands. Indeed, the entire 8-stranded mixed β sheet of carboxypeptidase A, which contains the 3-stranded parallel sheet shown here, has an overall twist of well over 90° (Fig. 6-12). Antiparallel sheets exhibit similar curvatures. Note that the sidechains of parallel sheets, as do those of antiparallel sheets, alternate up and down. In the case of the parallel sheet shown here, you can see that both sides of the sheet are relatively nonpolar. This is because this sheet is located almost entirely within the nonpolar interior of carboxypeptidase A (Fig. 6-12).

Kinemage 4: β Bends (Reverse Turns) (Section 6-1D; Fig. 6-20).

View1 shows four residues in a Type I β bend or reverse turn ("Type I") displayed in ball-and-stick form with its main chain bonds white and its atoms colored according to type (C brown, N blue, H white, and O red). The bonds to the H atoms of the main chain and the side chain Cα—Cβ bonds, which are controlled by the "Hydrogens" and "R-Groups" buttons, are gray and purple. View1 is similar to that of Fig. 6-20.

Click the "H-bonds" button several times to see the H-bond joining the carbonyl oxygen of this reverse turn's first residue (here Ser 159) to the amide hydrogen of its fourth residue (here Ser 162).

Reverse turns include both polar (magenta) as well as nonpolar (gold) sidechains (here controlled by the "Polar sc" and "Nonpolar sc" buttons), even though they almost always occur at protein surfaces. Proline residues commonly occur at position 2 in a β bend because a Pro residue can adopt the ϕ and ψ angles required at this position and probably because its presence bestows rigidity on the structure. Note that residue 2 in this example is Pro 160.

In View1, ANIMATE to convert the Type I β bend to a type II β bend. Type II β bends ("Type II"), differ from Type I β bends by a 180° flip of the peptide unit linking their residues 2 and 3. **Can you see the peptide unit "flip" so as to position the carbonyl oxygen on the opposite side of the turn?** Rotate the structure for different views of this flip.

In Type II β bends, as in Type I β bends, the carbonyl oxygen of residue 1 (here Asn 89) is hydrogen bonded to the amide hydrogen of residue 4 (here Gln 92). However, in Type II β bends, the carbonyl oxygen atom of residue 2 would crowd the Cβ (Cb) of residue 3. Consequently, residue 3 of Type II β bends is usually Gly as it is here (Gly 91). **How does the distance between the sidechain hydrogen of Gly 91 and the carbonyl oxygen of residue 2 (Tyr 90) compare with the C—O van der Waals distance necessary for any other sidechain (2.8 Å)?**

β bends or reverse turns cause the polypeptide chain to abruptly change its direction. They often connect successive strands of an antiparallel β sheet (which is why they are named β bends). However, they can occur in any nonrepetitive structure. **Turn on the "More" buttons to see that the Type I β bend shown here is part of a loop structure (View2) and the Type II β bend joins two approximately perpendicular helices (View3).** In both β bends, their N- and C-terminal extensions are bluetint and pinktint and the hydrogen bonds they form are represented by dashed greentint lines.

Kinemage 5: Visualization of Secondary Structures in Lysozyme from Torsion Angles (ϕ and ψ).

The table below gives the torsion angles, ϕ and ψ, of hen egg white lysozyme for residues 24-73 of this 129-residue protein. Use Figure 6-6 and Kinemage 5 to answer the following questions: (a) What is the secondary structure of residues 26-35? (b) What is the secondary structure of residues 42-53? (c) What is the probable identity of residue 54? (d) What is the secondary structure of residues 69-72?

TORSION ANGLES (phi and psi) FOR RESIDUES 24 TO 73 OF HEN EGG WHITE LYSOZYME

Res. No.	*A. A.*	*ϕ (deg)*	*ψ (deg)*	*Res. No.*	*A. A.*	*ϕ (deg)*	*ψ (deg)*
24	Ser	−60	147	49	Gly	95	−75
25	Leu	−49	−32	50	Ser	−18	138
26	Gly	−67	−34	51	Thr	−131	157
27	Asn	−58	−49	52	Asp	−115	130
28	Trp	−66	−32	53	Tyr	−126	146
29	Val	−82	−36	54	xxx	67	−179
30	Cys	−69	−44	55	Ile	−42	−37
31	Ala	−61	−44	56	Leu	−107	14
32	Ala	−72	−29	57	Gln	35	54
33	Lys	−66	−65	58	Ile	−72	133
34	Phe	−67	−23	59	Asn	−76	153
35	Glu	−81	−51	60	Ser	−93	−3
36	Ser	−126	−8	61	Arg	−83	−19
37	Asn	68	27	62	Trp	−133	−37
38	Phe	79	6	63	Trp	−91	−32
39	Asn	−100	109	64	Cys	−151	143
40	Thr	−70	−18	65	Asn	−85	140
41	Glu	−84	−36	66	Asp	133	8
42	Ala	−30	142	67	Gly	73	−8
43	Thr	−142	150	68	Arg	−135	17
44	Asn	−154	121	69	Thr	−122	83
45	Arg	−91	136	70	Pro	−39	−43

Res. No.	*A. A.*	*φ (deg)*	*ψ (deg)*	*Res. No.*	*A. A.*	*φ (deg)*	*ψ (deg)*
46	Asn	−110	174	71	Gly	−61	−11
47	Thr	−66	−20	72	Ser	−45	122
48	Asp	−96	36	73	Arg	−124	146

Source: Imoto, T., Johnson, L.N., North, A.C.T., Phillips, D.C., and Rupley, J.A., *in* Boyer, P.D. (Ed.), *The Enzymes* (3rd ed.), Vol. 7, pp. 693–695, Academic Press (1972).

EXERCISE 4. FIBROUS PROTEINS

KINEMAGE 1: Ribbon Diagram of a Coiled Coil such as occurs in α-Keratin; Fig. 6-14*b*.

KINEMAGE 2: The Heptad Repeat and Hydrophobic Interactions in Coiled Coils; Fig. 6-14*a*.

KINEMAGE 3: Ribbon and Spacefilling Diagrams of the Collagen Triple Helix; Fig. 6-17.

KINEMAGE 4: Collagen Backbone and the Effect of a Mutation; Fig. 6-18.

Fibrous proteins are, for the most part, characterized by highly repetitive simple sequences. We shall examine here two models for fibrous proteins: a coiled coil of two α helices such as occurs in α-keratin (Fig. 6-14); and that of a trimer that forms a collagen-like triple helix (Figs. 6-17 and 6-18).

If you have not already done so, **double click on the icon for the program MAGE. When queried, click on the "Proceed" box. Then, under the "File" pull-down menu, invoke the "Open File..." option and select E04_FbPr.kin.**

Click on the graphics window to bring it to the front. Click on the box in the upper right corner of the MAGE color graphics bar to make the graphics window fill the screen.

Kinemage 1: Ribbon Diagram of a Coiled Coil; Fig. 6-14b.

KINEMAGE 1 is a ribbon diagram of a coiled coil, similar to Fig. 6-14*b*. One of the segment's two identical chains, "Coil 1", is yellow and the other, "Coil 2", is seagreen. The N- and C-terminal ends of each segment are marked by blue and red balls. You can readily see that the protein forms a parallel coiled coil: its component alpha helices are right-handed but the coiled coil is left-handed. **Look at Views1-3 to see the coiled coil from different angles. Turn Zclip off and on in View3 to see part or all of the structure.** Note that the coiled coil makes only slightly more than 1/2 turn over a distance in which each α helix makes ~13 turns.

Toggle the "ANIMATE" button to see the way in which the two coiled coils wrap around each other. Parallel coiled coils occur in many proteins including α keratin, a fibrous stress-bearing protein occurring in mammalian skin (Figs. 6-14 and 6-15); tropomyosin and myosin, important proteins in muscle (Section 7-3; Fig. 7-26); and the so-called leucine zipper segments that permit the dimerization and hence activation of numerous eukaryotic transcription factors (Section 23-4C). Leucine zippers differ from other 2-helix coiled coils only in that their *d*-residues are almost invariably Leu (Fig. 23-38).

Kinemage 2: The Heptad Repeat and Hydrophobic Interactions in Coiled Coils (Section 6-1C; Fig. 6-14a).

KINEMAGE 2 shows the structure of the coiled coil in atomic detail but shows only the Cβ atoms of the sidechains. View1 shows a centrally located 8-residue segment, "Piece-4" of each subunit, with the covalent bonds of Coil 1 yellowtint, those of Coil 2 greentint, and the Cβ atoms represented by magenta balls. **Click the "a-Residues" and "d-Residues" buttons to see the Cβ atoms of the *a* and *d* residues highlighted in green and gold (Fig. 6-14*a*).**

The amino acid sequence of the polypeptide exhibits a rough heptad repeat, $(a\text{-}b\text{-}c\text{-}d\text{-}e\text{-}f\text{-}g)_n$, with the residues at the *a* and *d* positions having small hydrophobic side chains (Ala, Val, Leu, and Ile). This provides a hydrophobic strip on one face of each α helix that mediates its association with another such helix to form a coiled coil (Fig. 6-14*a*). The slight discrepancy between the 3.6 residues per turn of a normal α helix (Section 6-1B) and the 3.5-residue repeat of the *a*- and *d*-residues causes this hydrophobic strip to wrap about its α helix in a gentle left-handed helix, thereby accounting for the formation of the left-handed coiled coil.

Go to View3 and turn on all seven "Piece-n" buttons to display the coiled coil in side view. Here, the protein is represented by its main chain and Cβ atoms colored as described above. View3 through View8 show the coiled coil in various orientations about its superhelix axis. **Go to View1 and turn off "zclip"** to show the entire coiled coil viewed along its superhelix axis. Note how its *a*- and *d*-residue side chains are packed along the superhelix axis. The closeness of this packing accounts for the presence of only small hydrophobic side chains at these positions; larger hydrophobic side chains, those of Phe and Trp, in these positions would pry the two helices apart and thereby destabilize the coiled coil. **Finally, click on several of the Cβ atoms located about the periphery of the coiled coil** to convince yourself that these residues are almost entirely hydrophilic in character.

Kinemage 3: Ribbon and Spacefilling Diagrams of the Collagen Triple Helix (Section 6-1C; Fig. 6-17).

Collagen, the most abundant protein in vertebrates, is an extracellular protein that comprises the major protein component of such stress-bearing structures as bones, tendons, and ligaments. Collagen is characterized by a distinctive repeating sequence: $(\text{Gly-X-Y})_n$ where X is often Pro, Y is often 5-hydroxyproline (Hyp), and *n* may be >300. This, as we shall see, causes each collagen chain to assume a left-handed helical conformation with 3.3 residues per turn and a pitch (rise per turn) of 10.0 Å. Three such chains associate in parallel to form a right-handed triple helix (Fig. 6-17).

In Kinemages 3 and 4, we study a model compound for naturally occurring collagen, a 30-residue synthetic polypeptide of sequence $(\text{Pro-Hyp-Gly})_4$-Pro-Hyp-Ala-$(\text{Pro-Hyp-Gly})_5$, three chains of which associate to form a collagen-like triple helix of parallel strands that is 87 Å long and ~10 Å in diameter.

View1 shows the triple helical molecule in ribbon form seen perpendicular to its triple helical axis and with its three parallel and identical chains, "Chain 1", "Chain 2", and "Chain 3", colored purple, gold, and white, respectively—similar to Fig. 6-17. **View2 is down the triple helical axis,** a perspective in which this ribbon diagram appears to have a hollow center. However, **click the "2ANIMATE" button** to show the spacefilling form and prove to yourself that the center is not hollow. Return to the ribbon diagram by clicking the "2ANIMATE" button again before continuing.

Go back to View1 and repeatedly click the "ANIMATE" button. This "grows" Strand 1 from its N- to its C-terminus in differently colored 3-residue increments. Note how the molecule's three strands twist around each other and that the triple helix makes one turn every ~7 three-residue repeats.

Repeatedly click the "2ANIMATE" button to alternately display the original ribbon diagram and a spacefilling diagram of the polypeptide chains together with their side chains. The chains of the spacefilling diagram, which are colored identically to those of the ribbon diagram, can be individually turned on and off. Displaying one or two chains as ribbons and the remainder in spacefilling form may better reveal the helical character of the triple helix.

Kinemage 4: Collagen Backbone and the Effect of a Mutation (Section 6-1C; Fig. 6-18; Box 6-1).

This kinemage displays all of the atoms of the collagen model compound $(\text{Pro-Hyp-Gly})_4$-Pro-Hyp-Ala-$(\text{Pro-Hyp-Gly})_5$ in stick form (Fig. 6-18; note that the "essential" Gly residue in this model compound's central triplet is replaced by Ala). **View1 shows the triple helix in side view** with "Chain 1" in pinktint, "Chain 2" in yellowtint, and "Chain 3" in white. The Pro, Hyp, and Ala side chains, which are independently controlled by the corresponding buttons, are green, cyan, and magenta, respectively. **Use View1 and View2, which is down the triple helix axis, to prove to yourself that all Pro and Hyp side chains are on the periphery of the triple helix.** These rigid groups are thought to help stabilize the collagen conformation.

View3 and View4 are side and top views of a segment of the collagen helix in which its three polypeptides all consist of repeating triplets of ideal sequence, $(\text{Gly-Pro-Hyp})_n$. **Go to View3 to see that the three polypeptide chains are staggered in sequence by one residue, that is, a Gly on Chain 1 is at the same level along the triple helix axis as a Hyp on Chain 2 and a Pro on Chain 3. Turn on the "H bonds" button (H bonds are represented by dashed orange lines), to see that this staggered arrangement permits the formation of a hydrogen bond from the Gly main chain NH of Chain 1 to the Pro main chain O on Chain 2** (and likewise from Chain 2 to Chain 3 and from Chain 3 to Chain 1). Since the main chain N atoms of both Pro and Hyp residues lack H atoms, this exhausts the ability of the main chain to donate hydrogen bonds. Although the center of the triple helix appears to be hollow in View4, taking into account the van der Waals radii of its various atoms reveals that the center of the triple helix is, in fact, quite tightly packed. Indeed, the above hydrogen bonds pass very close to the center of the triple helix. This close packing accounts for the absolute requirement for a Gly at every third residue in a functional collagen molecule. Since, as you can see, the Gly $C\alpha$ atoms are near the center of the triple helix, the side chain of any other residue at this position would, as we shall see below, significantly distort and hence destabilize the collagen triple helix.

View5 and View6 show the side and top views of the triple helix segment containing an Ala on each chain instead of a Gly. The effect of replacing the Gly H atom side chain with a methyl group to form Ala, the smallest residue substitution possible, is quite striking. The interior of the collagen triple helix is too crowded to accommodate an Ala side chain without significant distortion. **The triple helix in this region therefore unwinds and expands so that no H-bonds form in this region. The unwinding of the triple helix in the region about the Ala residues is, perhaps, best seen by returning to KINEMAGE 3. You can see that the triple helix is bulged out in the center of View1.** These conformational changes, which disrupt collagen's rope-like structure, are responsible for the symptoms of such human diseases as osteogenesis imperfecta and certain Ehlers-Danlos syndromes (Box 6-1).

The coordinates for the coiled-coil (tropomyosin) were obtained from Xiaoling Xia and Carolyn Cohen, Brandeis University; those for the collagen-like polypeptide were obtained from 1CAG.

Exercise in large part by John H. Connor
Terminator Graphics Limited
1118 Kimball Dr.
Durham, NC 27712
USA

EXERCISE 5. CYTOCHROMES *c*

If you have not already done so, **double click on the icon for the program MAGE. When queried, click on the "Proceed" box. Then, under the "File" pull-down menu, invoke the "Open File..." option and select E05_Cyto.kin.**

Click on the graphics window to bring it to the front. Click on the box in the upper right corner of the MAGE color graphics bar to make the graphics window fill the screen.

Kinemage 1: Cytochromes c: Sections 6-2C and 17-2E; Figs. 6-26, 6-31, and 17-14.

Cytochrome *c* is a small monomeric protein that functions in the mitochondrial electron transport chain to transfer an electron from Complex III (cytochrome *c* reductase) to Complex IV (cytochrome *c* oxidase). It contains a heme group whose Fe atom shuttles between its Fe(II) and Fe(III) states and hence acts as cytochrome *c*'s electron carrier.

View1 shows the reduced [Fe(II)] form of tuna cytochrome *c* in its "standard" orientation (Fig. 6-26) with its Cα chain green and its heme group purple. Although the polypeptide chain is relatively short (103 residues), it surrounds the heme group so as to provide it with a hydrophobic environment (Fig 6-26*b*). It folds into five α helices comprising 53 residues and five reverse turns but contains no β pleated sheets. **Can you find cytochrome *c*'s five α helices and five reverse turns?**

The cytochromes *c*, which are all mitochondrial enzymes, have closely homologous sequences (Table 5-6) and hence nearly identical three-dimensional structures. The *c*-type cytochromes, which occur in prokaryotes, are more distantly related to the eukaryotic cytochromes *c* as evidenced by their considerably more divergent sequences. Nevertheless, these proteins have remarkably similar three-dimensional structures (Fig. 6-31). **Compare two of these structures by displaying View1 and switching on the "c(550)" button.** This shows the X-ray structure of cytochrome c_{550} (with its Ca chain in greentint and its heme group in pinktint), a *c*-type cytochrome from *Paracoccus denitrificans,* a respiring bacterium that can use nitrate as an oxidant. **ANIMATE for another way of comparing the structures of the two proteins.** Note that the two structures are nearly superimposable with the exceptions of several external polypeptide loops on the much longer cytochrome c_{550}. **With the two structures superimposed, measure the distances between a heme atom or Cα atom of the tuna protein and the corresponding heme atom or Cα atom of the *Paracoccus* protein,** (the sequence numbers of the *Paracoccus* residues have the suffix "P"). Note that many of these distances are less than 0.5 Å.

We'll now study the attachments of the heme group to the polypeptide chain in cytochrome *c*. **Switch off the "c(550)" and the "Ca chain", switch on the "Heme Bonds" and go to View3. Manipulate the model** so you can see that the Fe atom (orange sphere) is octahedrally liganded by the four heme pyrrole N atoms (small blue spheres) as well as an imidazole N atom of His 18 (seagreen) and the S atom (small yellow sphere) of Met 80 (gold). In addition, the heme group is linked to the polypeptide chain via thioether bonds between the S atoms of Cys 14 and Cys 17 (yellow) and the two heme ethyl groups (originally the vinyl groups in iron-protoporphyrin IX, Box 17-1).

Go back to View1, switch on the "Ca chain" and the "Lys residue" buttons. The Lys side chains displayed in blue and cyan are those whose Nζ atoms are strongly and less strongly protected from acetylation by acetic anhydride by the presence of cytochrome *c* oxidase or cytochrome *c* reductase (Fig. 17-14). Note that these Lys side chains more or less surround what View2 reveals is the heme group's solvent-exposed edge. It is thought that cytochrome *c* oxidase and cytochrome *c* reductase have complementary rings of Asp and Glu side chains through which they form specific complexes with cytochrome *c*.

The atomic coordinates for tuna cytochrome *c* and *Paracoccus* cytochrome c_{550} were obtained from 5CYT and 1COT, respectively. The two coordinates sets were transformed so as to maximally overlap the two proteins.

EXERCISE 6. MYOGLOBIN AND HEMOGLOBIN

KINEMAGE 1: Deoxymyoglobin: Fig. 7-1.

KINEMAGE 2: Evolutionary Relationship of Deoxymyoglobin and the α and β Subunits of Deoxyhemoglobin: Section 5-4.

KINEMAGE 3: Conformational Changes in Hemoglobin on Oxygen Binding—An Overview; BPG binding to Deoxyemoglobin: Figs. 7-5, 7-9, 7-10, and 7-15.

KINEMAGE 4: Conformational Changes in Hemoglobin's α Subunit on Oxygen Binding: Fig. 7-9.

KINEMAGE 5: Conformational Changes at Hemoglobin's α_1–β_2 Interface on Oxygen Binding: Fig. 7-10.

This exercise consists of 5 Kinemages designed to examine the structures of myoglobin (Mb) and hemoglobin (Hb), focusing on the structural changes in Hb responsible for cooperative O_2 binding. Kinemage 1 studies myoglobin structure, illustrating the folding of its α helices and the binding of the heme. Kinemage 2 compares myoglobin and the two different polypeptide subunits of hemoglobin, the α and β chains, showing the evolutionary homology between all three of these polypeptides. Kinemage 3 gives an overview of the binding of oxygen (O_2) to hemoglobin, visualizing the conformational changes that occur in the protein on interaction of O_2 with the heme groups. BPG binding is also presented in this Kinemage. Kinemages 4 and 5 show specific changes in the interactions of hemoglobin's subunits on binding oxygen.

Hemoglobin and myoglobin are oxygen-binding proteins that evolved from a common ancestor through gene duplication (Fig. 5-20). Myoglobin (Mb) is a monomeric muscle protein that functions to facilitate rapid oxygen diffusion and, in aquatic mammals, to store O_2. Hemoglobin (Hb), in contrast, is an $\alpha_2\beta_2$ heterotetramer that functions to transport O_2 in the blood.

Each Mb molecule binds only one molecule of O_2 via a simple mass action process and hence has a hyperbolic saturation curve (Section 7-1B). Hb, whose α and β subunits are homologous in both sequence and structure, binds up to four O_2 molecules in a cooperative manner and therefore has a sigmoidal saturation function. This occurs because O_2 binding to any of its subunits facilitates a conformational change in that subunit. However, as the symmetry model of allosterism explains (Section 7-2E), Hb's four subunits are so tightly coupled that all four subunits must simultaneously change their conformations to their high O_2-affinity states, whether or not they have bound O_2. Thus, the binding of an O_2 molecule to one subunit of Hb increases the O_2-binding affinity of its other subunits. In this exercise, we examine the structures of Mb and Hb with emphasis on the structural changes in Hb responsible for its cooperative O_2 binding.

If you have not already done so, **double click on the icon for the program MAGE. When queried, click on the "Proceed" box. Then, under the "File" pull-down menu, invoke the "Open File..." option and select E06_Hemo.kin.**

Click on the graphics window to bring it to the front. Click on the box in the upper right corner of the MAGE color graphics bar to make the graphics window fill the screen.

Kinemage 1: Deoxymyoglobin (Section 7-1; Fig. 7-1).

View1 shows a ribbon diagram, in gray, of the "Globin" (protein) portion of sperm whale deoxymyoglobin in which its eight helical segments, A through H, are displayed with two strands. **Toggle the "ANIMATE" button** to sequentially color these helices and their preceding nonhelical segments in rainbow order. You can see that the globin consists mostly of α helices; it has no β sheets and its nonhelical segments mostly serve as links that connect the helices. **Look down the barrel of some of the longer helices. Are they all straight?**

The heme is shown, in pink, in wireframe form with its N, O, and Fe atoms displayed as blue, red, and orange balls. Note how the heme is almost completely enclosed by the globin. **Which few chemical groups of the heme are exposed to the solvent?** (Clicking on atoms displays their identity in the lower left hand corner.) **Can you rationalize this exposure?**

Figs. 6-41 and 7-1 are different representations of Mb as seen from the same direction as View1, the standard view of Mb. **Rotate the protein around** to convince yourself that the globin is approximately disc-shaped with a diameter that is about twice its thickness. **Turn on the "Main Chain" button** to display the polypeptide backbone in white with its N and O atoms represented by blue and red balls. **How closely does the ribbon follow the main chain?**

Click "Animate" until the Mb ribbon is gray. Turn off the "Mb Ribbon" button. What are the orientations of the main chain carbonyl groups and the amide N atoms relative to each other which allow formation of the H-bonds of the α helices (not drawn)?

View2 is a closeup of the heme from the same direction as View1. **Turn on the "HemeLigand" button** to display, in cyan, the sidechain of His 93, the proximal His, liganding the heme's Fe(II) ion (white bond). The Fe(II) is also liganded in a square-planar array by the heme's four pyrrole N atoms and hence has a total of 5 pyramidally arranged ligands. In oxyMb, the reversibly bound O_2 molecule ligands the Fe from the opposite side of the heme as does the proximal His so that the Fe(II) becomes octahedrally coordinated. The Fe atom is not oxidized by its O_2 ligand; it remains in the Fe(II) oxidation state.

Rotate the image about the vertical axis until you see the heme edge-on. Is the heme planar? Note that the Fe atom is displaced towards the proximal His by 0.55 Å from the best plane though the porphyrin ring atoms. In oxyMb, the Fe is only 0.22 Å out of the heme plane and still on the side of the proximal His. We shall see in KINEMAGE 4 that this relatively slight movement of the Fe atom towards the heme plane on oxygenation is an essential element in hemoglobin's mechanism of cooperative oxygen binding.

Kinemage 2: Evolutionary Relationship of Deoxymyoglobin and the α and β Subunits of Deoxyhemoglobin (Section 5-4).

View1 shows a ribbon diagram of human deoxyHb viewed along its 2-fold axis of symmetry as it is in Fig. 7-5*a*. The beta subunits, which lie closest to the viewer in View1, are skyblue with pink hemes and the alpha subunits are bluetint with pinktint hemes. The Fe atoms in all hemes are represented by orange balls, the sidechains of the proximal His residues are cyan, and the bonds through which they coordinate the Fe atoms are white.

The table below provides the % sequence identities among Mb and the two subunits of Hb:

Hbβ-Mb	24%
Hbβ-Hbα	42%
Hbα-Mb	25%
Hbβ-Hbα-Mb	18%

Compare the 3-D structures of Mb, Hbα and Hbβ subunits as follows: **Click the "ANIMATE" button** to display the ribbon diagram of sperm whale deoxyMb (seagreen with a yellowtint heme) oriented so as to best superimpose it on the β_1 subunit of deoxyHb. Now go to **View2** for an enlarged view of deoxyMb and **turn off the "Hb Tetramer" button. Click the "ANIMATE" button again** to replace the deoxyMb with the β_1 subunit of deoxyHb. **Given the 24% sequence identity, are you surprised by the close structural resemblance of these two globins?**

Click the "Hb alpha" button to display the α subunit of deoxyHb oriented so as to superimpose it on the β_1 subunit. The α subunit lacks a D helix. **How does the absence of the D ring affect its structure relative to the Hb β subunit (and Mb)?** Click "ANIMATE" once again to return to deoxyMb.

More Advanced Studies: Symmetry of the Hemoglobin Tetramer

Go back to View1, toggle "ANIMATE" until the β_1 subunit appears, and again turn on the "Hb Tetramer" button, thereby returning to this kinemage's initial display. Now turn on the "Axes" button to display Hb's molecular axis system. The Hb molecule's exact twofold axis (red; seen in View1 as a red dot at the intersection of the two green axes), as we have seen, is perpendicular to the screen and passes through the center of the molecule. This axis relates the two β subunits, which lie nearest the viewer in View1, as well as the two α subunits, which lie farthest from the viewer in View1. Note the large channel that parallels the ~50-Å length of the exact twofold axis. It is $>$10 Å across at its narrowest opening so that, even taking into account presence of the side chains with which it is lined, it must be filled with solvent (Fig. 6-32 is a space-filling representation of Hb as viewed along this solvent-filled channel).

Since the α and β subunits are structurally similar, the Hb molecule also has two axes of twofold pseudosymmetry (green), which are perpendicular to each other and to the tetramer's exact twofold axis. The axis which relates α_1 to β_1 and α_2 to β_2 is horizontal and parallel to the plane of the screen in View1; that which relates α_1 to β_2 and α_2 to β_1 is vertical. These two axes intersect the exact twofold axis at the center (origin) of the tetramer. Hence, although hemoglobin only has exact C_2 symmetry (twofold rotational symmetry), the relatively small although functionally essential structural differences between its α and β subunits give it pseudo-D_2 symmetry (the symmetry of a rectangular prism; symmetry nomenclature is discussed in Section 6-3; Fig. 6-33).

Kinemage 3: Conformational Changes in Hemoglobin on Oxygen Binding—An Overview (Section 7-2C); BPG binding to Hemoglobin (Section 7-2C).

View1 shows human deoxyHb as represented by its Cα chains, viewed down its exact two-fold axis, and with its subunits and hemes colored as in KINEMAGE 2 (β subunits skyblue with pink hemes and α subunits bluetint with pinktint hemes). **Click the "ANIMATE" button** to display the similarly oriented human oxyHb with its α subunits pinktint, its β subunits pink, their associated hemes magenta and purple, and all Fe atoms represented by orange balls. Note that each of the hemes is now in complex with an O_2 molecule. This is more clearly seen in **View2,** a closeup of heme β_1, in which an O_2 molecule (whose atoms are represented by large red balls) is liganding the heme's Fe(II) ion (white bond). **Note that the Fe—O—O bond angle is 159°; that is, the Fe—O—O group is bent rather than linear. What is the Fe—O—O bond angle in the α subunits?** (Angles can be measured by choosing "measures" from the "Other" menu and then clicking on the Fe, O and O atoms in series). In order to see the α heme, you may have to increase the value of ZSLAB in the righthand tool bar and use the "Pickcenter" button to center it in the window.

Return to View1 and turn on the "Axes" button to display the hemoglobin molecule's exact 2-fold axis (red; just a dot in the center of View1), and its two pseudo-twofold axes (green) that relate its α and β subunits. **Click the "ANIMATE" button several times** to see the conformational change that hemoglobin undergoes on oxygenation. The nature of this conformational shift can be more readily understood by going to **View3,** which is View1 as seen from the right but turned 90 clockwise—the same view as in Fig. 7-6. **In View3, turn off the "a2-b2 Dimer" button and then toggle the "ANIMATE" button.** Note that, on oxygenation, the α_1-β_1 dimer ("a1-b1 Dimer") undergoes a largely rigid counterclockwise rotation of approximately 15° about an axis perpendicular to the screen that passes through the α_1 subunit (lower) and intersects the exact twofold axis (red). Repeating the animation with only the "a2-b2 Dimer" displayed shows that, upon oxygenation, this dimer undergoes an identical rotation about the same axis but in the opposite direction from that of the "a1-b1 Dimer". Consequently, the Hb molecule maintains its exact twofold symmetry. **Measure the movement of the Cα atoms most distant from the axis of rotation in going from deoxy to oxyHb** (click on both oxyHb and deoxyHb buttons to superimpose both structures. Click alternately on one atom and then the next to get the distance between them. Distance is displayed in the lower right-hand corner. It is sometimes necessary to click back and forth several times to get the distance to stabilize). Satisfy yourself that the atoms are displaced by around 3 Å in the deoxy to oxy conformational change. We examine the molecular details of this conformational change in KINEMAGES 4 and 5.

BPG Binding to Hemoglobin.

D-2,3-Bisphosphoglycerate (BPG) binds strongly, in 1:1 ratio, to deoxyHb but only weakly to oxyHb (Section 7-2C). Consequently, BPG heterotropically reduces hemoglobin's oxygen-binding affinity as Fig. 7-14 indicates. **You can see how BPG binds to deoxyHb by returning to View1 and turning on all four subunits of deoxyHb.** (Be sure both the "a1-b1 Dimer" and "a2-b2 Dimer" buttons are turned on, and the "OxyHb" button is turned off.) **Turn on the "BPG" button** to display the bound BPG in ball-and-stick form with its bonds green and its O and P atoms represented by red and yellow balls. **Turn on the "BPGinteract" button** to display the positively charged protein groups that interact with the highly anionic BPG. They are shown in wireframe form in cyan, with their N atoms represented by blue balls, and with the hydrogen bonds they make with BPG displayed as dashed white lines. **View4** is a closeup of this interaction analogous to Fig. 7-15.

Note how BPG binds in the upper portion of deoxyHb's central channel to its β side chains. Although BPG is an asymmetric molecule, it binds to deoxyHb in a pseudo-twofold symmetric fashion (the red dot above the C2 atom of BPG is Hb's exact twofold axis seen end-on). Eight basic groups, including the N-terminal amine group of each β chain, surround the anionic BPG, which has a charge of -4 under physiological conditions. Four of the basic groups form hydrogen bonds with the BPG; the others (His 143β) are too far away from the BPG ($>$3.5 Å) to do so. **Measure the distance from each His 143β N atom to the closest BPG phosphate O atoms.** These groups are still sufficiently close to it to form significant electrostatic interactions. This is corroborated by the observation that the γ subunits of human fetal hemoglobin (HbF), which are 73% identical to the β subunits that replace them in human adult hemoglobin (HbA), have a Ser residue in place of His 143β and bind BPG with reduced affinity. [More recent structural data indicate that the interactions between deoxyHb and BPG are somewhat different from those pictured in Fig. 7-15, most notably in that the N-terminal amino group on Val 1 and the sidechain of His 2 on each β subunit have exchanged positions and that the BPG carboxylate group is differently oriented such that it interacts with the β_1 as well as the β_2 subunit.]

Still in View4, turn on the "BPGinteract" button under "OxyHb" and click the "ANIMATE" button. Note that the β subunits in oxyHb have come together by ~4 Å relative to their positions in deoxyHb, thereby making the central channel too narrow to bind BPG. Moreover, some of the basic groups involved in binding BPG in deoxyHb have shifted to new positions in oxyHb (where they are colored seagreen) that are as much as 10-Å distant from their positions in deoxyHb while others have moved closer together. To see this best, **simultaneously turn on the "DeoxyHb" and "OxyHb" buttons and turn off all the subunit buttons.** Note that some of the basic groups that bind BPG in deoxyHb would sterically interfere with it in oxyHb if it occupied the same position in both conformers. **Measure the distance between His 143's in the two β subunits in the deoxy and the oxy states to see how much the site narrows on oxygenation.** To see an alternate view of how oxygenation eliminates Hb's BPG binding site, **go to View5** (which is View3 rotated by 135° about the vertical axis and tipped slightly towards the viewer)**, turn off the "Labels" and "BPG" buttons, and click the "ANIMATE" button several times.**

Kinemage 4: Conformational Changes in Hemoglobin's α Subunit on Oxygen Binding (Section 7-2C; Fig. 7-9).

In this kinemage we examine the conformational changes in the vicinity of the heme group induced by oxygenation of the α subunit of human hemoglobin. The β subunit exhibits similar conformational changes. **View1 shows the α subunit of deoxyHb ("DeoxyHb-a")** in much the same way as subunit α_2 is seen in Fig. 7-5*a*. The subunit is colored as it is in KINEMAGE 3 but with the coordinate bonds to the Fe(II) ion all in greentint, the sidechains of the proximal and distal His residues both cyan, and the other sidechains that are in contact with the heme colored gold if they are nonpolar and purple if they are polar. The carboxylate groups of the heme's two propionate sidechains have been deleted for the sake of clarity; they will reappear in KINEMAGE 5. **Click the "ANIMATE" button to display the similarly oriented α subunit of oxyHb ("OxyHb-a")**, also colored as it is in KINEMAGE 3, but with its coordinate bonds to the Fe(II) ion green, and with the sidechains that directly contact the heme colored as they are in deoxyHb. **View both states simultaneously by turning on both the "DeoxyHb-a" and the "OxyHb-a" buttons.** This is useful when observing changes in conformation and measuring distance changes.

Go to View2, a closeup of the heme group that resembles Fig. 7-3. Toggle the "ANIMATE" button to compare the conformations of the heme group and its surrounding protein in the deoxy and oxy states. View3 and View4 are an overview and a closeup from the vantage of Fig. 7-9. Note that in the deoxy state the porphyrin ring is slightly domed towards the proximal His while in the oxy state the Fe is in the center of the planar ring. **Measure the distance that the Fe atom moves in going from the deoxy to the oxy state.** (To do this, display "DeoxyHb-a" and "OxyHb-a" simultaneously and click alternately on the two Fe atoms several times until the distance (lower right of screen) becomes constant). **Measure the Fe—N(porphyrin) bond lengths in the deoxy and oxy states. How does this change in bond length facilitate the change in geometry on O_2 binding? How much movement in the proximal His (His 87) is caused by this change in Heme geometry?** This movement evidently triggers the T state (deoxy) → R state (oxy) conformational change. Before we trace how this occurs, **note that the O_2 in oxyHb is H-bonded to the distal His (His 58)**, which occupies very nearly the same position as it does in deoxyHb.

Go to View2 or View4 and turn on the "HemeContac" button. This highlights atom Ce1 of His 87 by a cyan ball as well as the contact between this atom and heme atom N a by a dotted white line. **Toggle the "ANIMATE" button** in View2 and/or View4 to observe the movement on oxygenation of His 87 and the F-helix to which it is attached. **How does the F-helix move on oxygenation? Does any other part of the molecule move as much?**

The β subunit of Hb exhibits a nearly identical collection of linked movements on oxygenation. It is these movements that, according to the Perutz model (Section 7-2C), induce the quaternary conformational shift from the T state conformation of deoxyHb to the R state conformation of oxyHb.

Finally, note the relative sizes of the O_2 molecule and the Hb α subunit and consider the magnitude of the conformational changes that the former induces in the latter.

Kinemage 5: Conformational Changes at Hemoglobin's α_1-β_2 Interface on Oxygen Binding (Section 7-2C).

View1 shows deoxyHb (Fig. 7-5) viewed and colored as in KINEMAGES 3 and 4 but, for the sake of clarity, with only its α_1 and β_2 subunits displayed. **Click the "ANIMATE" button** for the corresponding view of oxyHb. The sidechains whose interactions we examine in this kinemage are yellowtint on the β_2 subunit and seagreen on the α_1 subunit.

Views of the α_1 and β_2 subunits contacts can be controlled by two buttons: (1) **The "Switch" button** controls the visibility of the switch region side chains. It has alternative sets of contacts in the T and R states; and (2) **The "Salt Bridges"** button controls the visibility of the salt bridges and H-bonds formed by the C-terminal side chains that stabilize the T state (Fig. 7-11).

The switch region, whose side chains are displayed under the control of the "Switch" button, consists of the C helix (residues 38-42) and residue 94 (the first residue of the G helix) of alpha1 and the FG corner (residues 97-102) of β_2. **Toggle the "ANIMATE" button in View2 or View3 to observe that the relative motions of opposing groups of the switch region are on the order of 5 Å (measure the movement and see if you agree).** Hence, as you can see, Hb's T and R states have different sets of intersubunit contacts. **In View3, find His 97β_2.** Confirm that in deoxyHb, as Figs. 7-5*a* and 7-10*a* indicate, His 97β_2 is near Thr 41α_1, whereas in oxyHb, it is near Thr 38α_1, one turn back along the α_1 chain's C helix. Note that, in both conformations, knobs on the β_2 subunit are inserted into grooves on the α_1 subunit. Any intermediate position would be sterically strained because it would bring His 97β_2 and Thr 41α_1 too close together.

Observe the two alternative intersubunit hydrogen bonds in the switch region: **Find the H-bond between Asp 99β_2 and Tyr 42α_1 in the T (deoxy) state Hb. Watch it break in going to the R (oxy) state on animation. Can you find the alternate H-bond that forms in the R state? What amino acid residues does it connect? Predict what would happen to the relative stabilities of the T and R states if the amino acid at 102β were mutated to Ala.**

What stabilizes deoxyHb in the T state conformation? In View4, turn off the "Switch" button, turn on the "Salt bridges" masterbutton, and display deoxyHb. In deoxy Hemoglobin, what are the H-bonding interactions of the C-terminal residue of the α_1 subunit, Arg 141 (lower left)? How much movement does Arg 141 undergo in changing from the deoxy to the oxy state? What happens to the H-bonding when it goes to the oxy state? (Did you notice in KINEMAGE 4 that in addition to the F helix movement on oxygenation, the C-terminus of the displayed α subunit also moved a great deal? Now you see what this movement has done.)

Study Fig. 7-11 to help identify the other H-bonding interactions that disappear in going from deoxy to oxy Hemoglobin.

View5 in deoxyHb shows the C-terminal residue of the β_2 subunit, His 146, participating in H-bonds and salt bridges. Identify the residues participating with it in these interactions (See Fig. 7-11 for help). **Click the "ANIMATE" button** to reveal the β_2

C-terminal residue conformational change on shifting from the deoxy to the oxy state. **Explain from these observations why the T state is more stable than the R state in the absence of O_2 and why the first O_2 binds Hb with lower affinity than the last O_2.**

The Bohr effect (Hb's release of H^+ ion on binding O_2; Section 7-2C) results from p*K* changes in various residues induced by the changes in their electrostatic environments that occur in going from the T state to the R state. **Identify sidechains whose p*K*'s might be expected to change on going from deoxy to oxy Hb. Will the change in p*K* result in absorption or release of H^+ on going from deoxy to oxy Hb?**

Once you have investigated the interactions displayed in KINEMAGE 5, you should go back to KINEMAGE 3 to examine the switch region in the context of the entire hemoglobin molecule (when the "Axes" button is on, these regions occur along the vertical green axis in View1 and the horizontal green axis in View3 of KINEMAGE 3). The sidechains involved in these interactions are not included in KINEMAGE 3, but the region is, nevertheless, reasonably apparent, and the overall context of the rearrangements is clearer than in KINEMAGE 5.

The coordinates for sperm whale deoxymyoglobin, human deoxyhemoglobin, and human oxyhemoglobin were obtained from 1MBD, 2HHB, and 1HHO, respectively. The coordinates for BPG and the sidechains with which it interacts in human oxyhemoglobin were obtained from Guy Dodson, University of York, York, U.K.

Exercise in large part by John H. Connor
Terminator Graphics Limited
1118 Kimball Dr.
Durham, NC 27712
USA

EXERCISE 7. SACCHARIDES

KINEMAGE 1: D-Glucopyranose: Section 8-1B; Figs. 8-4 and 8-5.

KINEMAGE 2: Sucrose: Section 8-2A.

KINEMAGE 3: Hyaluronic Acid: 8-2D; Fig. 8-12.

KINEMAGE 4: Structure of a Complex Carbohydrate: Section 8-3C; Fig. 8-16.

Sugars or saccharides are one of the basic building blocks of biological molecules and are used as energy fuels by almost all cellular organisms. Many saccharides have the empirical formula $(CH_2O)_n$ and hence are known as carbohydrates. This simple empirical formula, however, masks their structural and conformational complexity. In this exercise we explore the structures of four saccharides of increasing complexity, those of the monosaccharide D-glucopyranose, the disaccharide sucrose, the linear polysaccharide hyaluronate, and a branched-chain carbohydrate.

If you have not already done so, **double click on the icon for the program MAGE. When queried, click on the "Proceed" box. Then, under the "File" pull-down menu, invoke the "Open File..." option and select E07_Sacc.kin.**

Click on the graphics window to bring it to the front. Click on the box in the upper right corner of the MAGE color graphics bar to make the graphics window fill the screen.

Kinemage 1: D-Glucopyranose (Section 8-1B; Figs. 8-4 and 8-5).

View1 shows a side view of β-D-glucopyranose ("b-Anomer") in the axial chair conformation in which the nonhydrogen substituents of the pyranose ring are all in axial positions (Fig. 8-5, right). The glucose molecule is shown in ball-and-stick form in which C atoms are green, O atoms are red, and H atoms are represented by small bluetint balls.

The conformation shown ("Axial Chair") is sterically unfavorable. **Measure the distances of the close approaches between axial atoms on the same side of the ring** and compare them to the 2.6- and 2.7-Å outer limit van der Waals O—O and C—O distances. **Now click the "ANIMATE" button to show the equatorial chair conformation of β-D-glucose** in which the ring's nonhydrogen substituents are all in the equatorial position ("Eq. Chair"; left panel of Fig. 8-5) and hence not in contact. Here the ring's H atom substituents are all in axial positions but, since the normally allowed H—H distance is 2.0 Å, you can convince yourself that "Eq. Chair" is an unstrained conformation of β-D-glucopyranose.

Now click the "2ANIMATE" to view α-D-glucopyranose ("a-Anomer"; Fig. 8-4). This molecule differs from "b-Anomer" only in the configuration about its anomeric carbon atom, C1. **Clicking "2ANIMATE" again displays β-D-glycopyranose in its "Eq. Chair" conformation.** The conformation of most polysaccharides is greatly affected by anomeric differences since nearly all adjacent sugar units in polysaccharides are linked by glycosidic bonds, that is, by bonds to atom C1. For example, the only difference between starch and cellulose is the anomeric configuration of their glucosyl residues. (NOTE: Clicking "ANIMATE" when "a-Anomer" is displayed always displays "b-Anomer" in its "Axial Chair" conformation.)

View2 shows a view of the molecule similar to that in the right panel of Fig. 8-4 and View3 shows it more or less in top view.

Kinemage 2: Sucrose (Section 8-2A).

The disaccharide sucrose [O-α-D-glucopyranosyl-(1 → 2)-β-D-fructofuranoside; common table sugar), the most abundant disaccharide in nature, occurs in plants where it functions as a transport form of carbohydrates. Sucrose consists of glucopyranose and fructofuranose in $\alpha_1 \rightarrow \beta_2$ covalent linkage (Section 8-2A). Thus, sucrose is one of the few saccharides in which two anomeric carbon atoms are in glycosidic linkage to each other via a bridging O atom.

View1, a side view, shows sucrose in ball-and-stick form with its atoms colored according to type (C green, O red, and H small bluetint balls). Only H atoms that are direct ring substituents are displayed. The glucose residue is shown in its stable equatorial chair conformation ("Eq. Chair" of KINEMAGE 1). **View2 is a top view of the glucopyranose residue and View3 is a top view of the fructofuranose residue.** Fructofuranose contains a 5-membered ring, which is conformationally more constrained than is glucopyranose's 6-membered ring but is by no means inflexible (see Fig. 23-6 for illustrations of two different conformers of β-D-ribofuranose). Thus, taking into account the relatively free rotation about each of sucrose's two glycosidic bonds, it is clear that this molecule has considerable conformational flexibility.

Kinemage 3: Hyaluronic Acid (Section 8-2D; 8-12).

Hyaluronic acid consists of 250 to 25,000 β(1 → 4)-linked disaccharide units that consist of D-glucuronic acid and *N*-acetyl-D-glucosamine in β(1 → 3) linkage (Fig. 8-12). The carboxylate group on each of hyaluronic acid's glucuronate residues is ionized at physiological pH's. The viscoelastic properties that the resulting closely spaced negative charges confer on hyaluronic acid (really hyaluronate) suit this important glycosaminoglycan to its function as a major constituent of ground substance (where it is a component of proteoglycans; Section 8-3A), synovial fluid (the lubricating fluid of joints), and the vitreous humor of the eye.

The structure of calcium hyaluronate is shown here as represented by four consecutive disaccharide units in ball-and-stick form with its atoms colored according to type [C green, N blue, O red, H represented by small bluetint balls, and calcium ions (Ca^{2+}) represented by large greentint balls]. The bonds of the *N*-acetyl-D-glucosamine ("NAG") residues are cyan and those of the D-glucuronate ("GCU") residues are hotpink. H atoms substituent to O atoms are not shown.

View1 is a side view of the polysaccharide. Ca^{2+} hyaluronate forms an elongated left-handed helix with three disaccharide units per turn of length (pitch) 28.3 Å. **Rotate the molecule about the vertical axis to see the helical conformation. Repeatedly click the "ANIMATE" button to "grow" the hyaluronate helix in one disaccharide increments.** The four disaccharide units displayed here form 1-1/3 turns of the helix. Note that the bottom disaccharide unit in View1 has the same orientation as the top disaccharide unit and that corresponding atoms of these two disaccharides are 28.3 Å apart. This conformation is maintained, in part, by hydrogen bonds ("H bonds") between neighboring residues (dashed white lines).

View2 is a close-up of a disaccharide repeating unit of hyaluronate (Fig. 8-12).

Kinemage 4: Structure of a Complex Carbohydrate (Section 8-3C; Fig. 8-16).

KINEMAGE 4 displays an octasaccharide found bound to the bovine protein galectin-1. (Galectin-1 binds terminal galactose residues that have $\beta(1 \rightarrow 4)$ linkages.) This kinemage demonstrates that even a few sugars that are linked with different branching patterns can form a quite complex molecule.

View1 shows the entire, largely extended, octasaccharide in wireframe form with its O and N atoms represented by red and blue balls. The octasaccharide consists of three *N*-acetyl-D-glucosamine residues ("NAG"; cyan bonds), two D-galactose residues ("Gal"; pinktint bonds), and three D-mannose residues ("Man"; yellowtint bonds). **Can you identify the linkages between these saccharide units?** Turning on the "Labels" button will enable you to check your answers.

Go through Views2–6 to isolate each type of residue.

The coordinates for Ca^{2+} hyaluronate were obtained from 4HYA. Those for the octasaccharide in KINEMAGE 4 were obtained from 3SLA.

Exercise in large part by John H. Connor
Terminator Graphics Limited
1118 Kimball Dr.
Durham, NC 27712
USA

EXERCISE 8. TRANSMEMBRANE PROTEINS

KINEMAGE 1: Bacteriorhodopsin: Section 10-1A; Fig. 10-4.

KINEMAGE 2: Photosynthetic Reaction Center: Sections 10-1A and 18-2B; Fig. 10-5.

KINEMAGE 3. Porin: Sections 10-1A and 10-5B; Fig. 10-6 and 10-33.

Biological membranes consist of lipid bilayers in complex with proteins (Section 10-1). Membrane-associated proteins are classified as being either integral proteins, which are tightly bound to membranes via hydrophobic interactions, lipid-linked proteins, which are covalently attached to lipids that anchor them to the membrane, or peripheral proteins, which are but loosely associated with membranes. Integral proteins partially or completely penetrate the lipid bilayer with which they are associated, and are therefore also called transmembrane proteins.

In this exercise, we consider the structures of three transmembrane proteins, those of bacteriorhodopsin, the photosynthetic reaction center, and porin. These constitute the first integral proteins whose structures were determined to near-atomic resolution.

If you have not already done so, **double click on the icon for the program MAGE. When queried, click on the "Proceed" box. Then, under the "File" pull-down menu, invoke the "Open File..." option and select E08_TrMb.kin.**

Click on the graphics window to bring it to the front. Click on the box in the upper right corner of the MAGE color graphics bar to make the graphics window fill the screen.

Kinemage 1: Bacteriorhodopsin (Section 10-1A; Fig. 10-4).

Bacteriorhodopsin, a homotrimer from the halophilic bacterium *Halobacter halobium,* is the only protein component of that bacterium's purple membrane, ~0.5 micron wide patches that form on its surface when O_2 is scarce. Bacteriorhodopsin functions as a light-driven proton pump that ejects protons from the bacterial cytoplasm so as to power the synthesis of ATP in the same manner as does oxidative phosphorylation (Sections 17-2C and 17-3).

View1 shows a subunit of bacteriorhodopsin in its maximum cross section as viewed from within the plane of the membrane. The bacterial cytoplasm is above and the extracellular medium is below (Fig. 10-4). The polypeptide, which is represented by its Cα chain, forms seven transmembrane helices, A through G (skyblue; controlled by the "Helices" button), connected by nonhelical loops (white; controlled by the "Loops" button). The conformations of all these loops but those connecting helices D to E and F to G are tentative due to the poorly visualized electron density in these regions of the structure.

Repeatedly click the "ANIMATE" button to "grow" the Cα chain, from N- to C-terminus (A through G), one helix with its following loop per click. Alternatively, each helix + loop segment may be individually highlighted by turning on the corresponding "Helix" button.

View2 displays the bacteriorhodopsin subunit as seen from the left in View1, whereas View3 shows it as seen from the cytoplasm (the top in View1). Note that the helices, which are all approximately the same length (32-39 Å; 22–26 residues), are successively antiparallel, extend approximately perpendicularly to the lipid bilayer, and are arranged in clockwise order, A through G, in View3 so as to give the subunit an elongated cross section in the plane of the bilayer. Also, as is best seen in View3, the seven transmembrane helices are not precisely parallel but gently twist together to form a short segment of a left-handed coiled coil. Such 7-helix bundles are common structural motifs in transmembrane receptors, particularly those that interact with heterotrimeric G-proteins (Section 21-3B).

Go to View4 and turn on the "Trimer" button to display Bacteriorhodopsin's quaternary structure. View4 is along the trimer's threefold axis of rotation, which is perpendicular to the plane of the bilayer and, as you can see, is centrally located in a large channel that runs the length of the trimer and is presumably filled with bilayer and solvent. View4 reveals that the outer helices of each subunit (helices E, F, G, and A) are inclined with respect to the trimer's threefold axis of rotation. Note that the only close contacts between subunits are between Helix B and its flanking loops in one subunit and Helix D and its flanking loops in the clockwise-adjoining subunit in View4.

Return to View1, turn off the "Trimer" button, and turn on the "ChargedRes" button. This marks bacteriorhodopsin's charged residues (Arg, Lys, Asp, and Glu) by magenta balls at their Cα positions. By rotating the molecule around you can see that all of its 28 charged residues are located either at the ends of the helices, on their connecting loops, or within the subunit's central hydrophilic proton-pumping channel (see below). Thus, these charged residues are all in contact with the lipids' polar head groups and/or the solvent. None of them are on the exterior surfaces of the transmembrane helices' central segments, the segments that penetrate the nonpolar central region of the lipid bilayer and hence must themselves be highly nonpolar. Note also that α helices completely satisfy the hydrogen bonding potentials of their main chains (Section 6-1B and KINEMAGE EXERCISE 3-2), thereby further contributing to the nonpolar nature of transmembrane helices.

Bacteriorhodopsin has a retinal prosthetic group (gold) that is linked through the formation of a Schiff base (imine; hotpink) with the Nζ atom of Lys 216 (seagreen). Retinal's noncyclic portion contains four conjugated double bonds that, as displayed here, all have the *trans* form. However, the absorption of light isomerizes (photoisomerizes) the double bond nearest the Schiff base to yield 13-*cis*-retinal, which subsequently reconverts to the all-*trans* form. It is these isomerizations that power bacteriorhodopsin's proton pump.

In View1, turn on only the "Brhodopsin" and "Retinal" buttons and the "Helices", "Loops", and "ChargedRes" buttons. Then, turn on the "H+ Pump" button to display the side chains of several charged residues lining the subunit's central hydrophilic channel (Asp residues 85, 96, 115, and 212 in red, and Arg 82 in seagreen), whose conservative mutations (e.g., Asp $\rightarrow$ Asn) each significantly reduce or eliminate bacteriorhodopsin's proton pumping capacity. All of these charged groups but Asp 96 are shown in closeup view in View5. It is thought that these acid–base groups, and in particular those of Asp 85 and Asp 96, function to relay protons from the cytoplasm to the N atom (blue ball) of the Schiff base linking the retinal to Lys 216 and from there to the cell's exterior. This process is driven by the positional and p*K* changes of these groups induced by retinal's photoisomerization to its 13-*cis* form and its subsequent reconversion to its all-*trans* form.

Kinemage 2: Photosynthetic Reaction Center (Sections 10-1A and 18-2B; Fig. 10-5).

In purple photosynthetic bacteria, light energy is converted to chemical energy in a process mediated by the photosynthetic reaction center. The photosynthetic reaction center from *Rhodopseudomonas (Rps.) viridis,* which consists of four subunits in complex with four hemes, four chlorophyll molecules, four other prosthetic groups, and a nonheme Fe atom, was the first transmembrane protein whose X-ray structure was determined at atomic resolution. **View1 shows the "front" of this protein as viewed, as in Fig. 10-5*a*, from within the bacterial plasma membrane in which it is embedded.** The external face of the membrane is above and its cytoplasmic face is below. The protein is represented by its Cα chains with its 4-heme *c*-type cytochrome subunit (336 residues) colored green, and its so-called H (259 residues), M (323 residues), and L (273 residues) subunits colored magenta, cyan, and orange, respectively. The display of these subunits is collectively controlled by the "Ca Chain" button and individually controlled by their corresponding buttons. In many photosynthetic bacteria, the photosynthetic reaction center lacks a bound cytochrome subunit. For example, as Fig. 18-8 indicates, the photosynthetic reaction center of *Rhodobacter (Rb.) sphaeroides* consists only of the H, M, and L subunits but is otherwise nearly identical in structure to the *Rps. viridis* photosynthetic reaction center.

The photosynthetic reaction center's transmembrane segment consists of 11 nearly parallel alpha helices, one from the H subunit and five each from the M and L subunits. **Switch on the "Sidechains" button to see the hydrophobic character of these helices.** This displays the Cα

atoms of the protein's Ile and Phe residues as green and yellow spheres (controlled by the "Ile + Phe" button); those of its positively charged residues, "Arg + Lys", as blue spheres; and those of its negatively charged residues, "Asp + Glu", as red spheres—all of which can also be displayed by subunit through the use of the corresponding buttons. **The distribution of these sidechains may be more clearly seen by momentarily switching off the "Ca Chain" button and/or going to View2.** Note the complete absence of charged residues in the protein's helical region. Indeed, as Fig. 10-5*b* reveals, even the main chain N and O atoms, all of which participate in intrahelical hydrogen bonds, are far less exposed on the surface of this region than they are on the other parts of the protein's surface. This nonpolar column constitutes the protein's transmembrane region. The remaining surface regions of the protein, which are exposed to the aqueous environments of the cytoplasm and the extracellular medium, are normally polar as is indicated by their abundance of charged residues and scarcity of Ile and Phe residues.

View3, which is View2 as seen from the right, indicates that the transmembrane region of the protein is far narrower in this dimension than it is in the other and that the helices are inclined with respect to each other. This is seen more clearly in View4, a top view of only the protein's transmembrane region (temporarily turning off the "zclip" button displays the entire protein). View4 also reveals, again more clearly than View2 or View3, that the transmembrane helices of the M (cyan) and L (orange) subunits are each arranged in interlocking arcs and that they are related to each other by a pseudo-twofold axis of rotation which is perpendicular to the plane of the membrane in which the protein is embedded. In fact, the M and L subunits have homologous sequences.

The Remaining Portion of KINEMAGE 2 Accompanies Section 18-2B.

How does the photosynthetic reaction center convert light energy to chemical energy? **Go to View1 and turn on the "Chromophrs" button to display the protein's chromophoric (light-absorbing) groups—which are more clearly seen when the "Ca Chains" button is off.** Starting with the *c*-type cytochrome, its four bound heme groups are shown in wireframe form colored, from top to bottom, pink, hotpink, magenta, and purple with their Fe atoms each represented by an orange ball. You can see that neighboring hemes approach each other as closely as ~8 Å. The hemes of cytochromes are redox active (Box 17-1); they can alternate between their Fe(II) and Fe(III) states and, hence, appear to form a conduit for the transfer electrons to the nearby (~10 Å in closest approach to the bottom heme) bacteriochlorophyll *a* molecules (see below).

The chromophores associated with the M and L subunits are more clearly seen in View2 (note that the H subunit does not contact any chromophore). In *Rps. viridis,* these chromophores consist of two bacteriochlorophyll *a* molecules (BChl *a*; Fig. 18-2) in close association to form a so-called "Special pair" (green with their bound Mg^{2+} ions represented by bluetint balls); two single "BChl a" molecules (seagreen); two bacteriopheophytin *a* molecules ("BPheo a"; greentint; a pheophytin is a chlorophyll whose Mg^{2+} ion has been replaced by two protons); a menaquinone molecule ("Menaquin."; yellow; structural formula on p. 634); a nonheme Fe(II) ion ("Fe"; large orange ball); and a "Ubiquinone" molecule (CoQ; Fig. 17-10*b*; gold). The ring systems of these molecules (but not their phytyl side chains) are related by the same pseudo-twofold axis of rotation that relates the L and M subunits that bind them. In some photosynthetic bacteria, including *Rb. sphaeroides,* the menaquinone is replaced by a second ubiquinone.

The primary event in photosynthesis is the absorption of a photon by the special pair. **View5 shows that, in this assembly, the ring systems of its two component BChl *a* molecules are very nearly parallel and in close apposition** such that their Mg^{2+} ions are separated by 7.6 Å and some of the atoms in the two ring systems are less than 4 Å apart. The special pair, which acts electronically as a single molecule, is often referred to as P870 since its major longwave absorption band peaks at 870 nm (in species which contain bacteriochlorophyll *b,* the corresponding absorption maximum is 960 nm and hence such a special pair is referred to as P960). **Turning on the "Ligands" button in View5 displays, in skyblue, the His sidechains liganding each Mg^{2+} ion in the protein's four BChl *a* molecules.** The BPheo *a* molecules lack such a ligand, which accounts for their lack of a bound Mg^{2+} ion.

Click the ANIMATE button to see the sequence of electron transfers (photooxidations) mediated by the photosynthetic reaction center in response to photon absorption. Here, a molecule in its excited state is highlighted in red as in Fig. 18-9. **The first click** shows the photosynthetic reaction center immediately after it has absorbed a photon ("0 s"): Only P870 is excited. However, within 3 picoseconds **(2nd click; "3 pico-s")**, P870 has transferred an electron to the Bpheo *a* molecule on the right side of the protein. There is no spectroscopic evidence that this electron ever resides on the BChl *a* molecule between the P870 and this BPheo *a.* This BChl *a* molecule has therefore been dubbed the accessory BChl *a,* although it almost certainly influences the electron transfer process (but in a manner that is poorly understood).

By 200 picoseconds after photon absorption (**3rd click; "200 pico-s"**), the excited electron has migrated to the menaquinone (also known as Q_A), which thereby assumes its semiquinone ($Q_A{}^-$) oxidation state (quinones are among the few organic molecules that form stable free radicals). All of these electron transfers are to progressively lower energy states (Fig. 18-10) and, consequently, this process is essentially irreversible.

The remainder of the photosynthetic electron-transfer process occurs relatively slowly. By 100 microseconds after photon absorption (**4th click; "100 micro-s"**), the electron has migrated to the ubiquinone (also known as Q_B), which thereby assumes its semiquinone oxidation state ($Q_B{}^-$). The electron does not linger on the Fe(II) ion between the menaquinone and the ubiquinone. In fact, the presence of this ion is not required for electron transfer to proceed at a physiologically supportable rate. Most likely, the Fe(II) ion functions to fine tune the photosynthetic reaction center's electronic properties. Turning on the "Ligands" button and going to View6 (a magnified portion of View1 but with the top of the protein tilted towards the viewer) reveals that this Fe(II) ion is liganded by four His side chains (skyblue) and both carboxyl oxygens of a Glu side chain (red) arranged in what could be taken to be a highly distorted octahedon.

For the reaction to proceed further, the now positively charged (photooxidized) P870 must re-acquire its ejected electron. This occurs via a process described below. Once the P870 has returned to its ground state, it can absorb another photon, thereby initiating a new sequence of electron transfers. Now, however, the $Q_B{}^-$ (ubisemiquinone) is further oxidized to its $Q_B{}^{2-}$ form, which eventually takes up two protons from the cytoplasmic side of the membrane to yield ubiquinol (Q_BH_2). The ubquinol, which is relatively loosely bound to the protein, dissociates into the membrane and is replaced by a ubiquinone molecule from the membrane-bound pool, thereby returning the photosynthetic reaction center to its original state **(5th click of the "ANIMATE" button).** Thus, ubiquinone acts as a molecular transducer that utilizes the energy acquired in two consecutive photon absorptions to power a two-electron chemical reduction.

The electrons taken up by the ubiquinone are eventually returned to the photooxidized special pair so as to complete the electronic circuit. This process, as is discussed in Section 18-2B, is species dependent and involves the membrane-bound pool of ubiquinone molecules and a series of cytochromes, including the transmembrane protein complex cytochrome bc_1 and the peripheral membrane protein cytochrome c_2 (which function similarly to ubiquinone, Complex III, and cytochrome *c* in mitochondria; Section 17-2). *In Rps. viridis,* the final link in this electron-transfer chain is the 4-heme *c*-type cytochrome that is bound to the extracellular face of the photosynthetic reaction center. Thus, in photosynthetic bacteria, photosynthesis results in no net oxidation. Rather, the electron transfer process is accompanied by the transport of the protons taken up by Q_BH_2 on the cytoplasmic side of the plasma membrane to the external medium. The discharge of the resulting proton gradient drives the synthesis of ATP in a process that closely resembles oxidative phosphorylation (Section 17-3B) but is known as photophosphorylation.

Finally, you should note that electrons are transferred almost entirely along the right branch of the chain of chromophores as seen in View2, even though the left branch appears to be structurally equivalent. Evidently, small structural differences and/or differences in the properties of nearby sidechains greatly favor the flow of electrons through the right-side BPheo *a*. Thus, the function of the left-side BPheo *a* and its neighboring "accessory" BChl *a,* if any, are unknown.

Kinemage 3: Porin (Sections 10-1A and 10-5B; Figs. 10-6 and 10-33).

Porins are transmembrane proteins that serve as macromolecular sieves in gram-negative bacteria, mitochondria, and chloroplasts. They are mounted in the outer membranes of these bacteria/organelles where they permit the influx of solutes smaller than ~600 D, such as nutrients, while preventing the efflux of larger molecules such as proteins. The known structures of porins from several bacterial species are all closely similar despite the low sequence similarities among some of them.

View1 shows a subunit of the homotrimeric OmpF porin of *E. coli* in greentint as represented by its Cα chain (Fig. 10-6). Its upper, "rough" surface faces the external medium and its lower, "smooth" surface faces the periplasmic/intermembrane space (recall that gram-negative bacteria, mitochondria, and chloroplasts are all enclosed by a double membrane; Fig. 8-14*b*).

Porin differs from other transmembrane proteins of known structure in that it has no transmembrane helices. Rather, it consists almost entirely of a 16-stranded antiparallel β sheet that curls around in the usual right-handed manner (Section 6-1B) to form a hollow barrel-like structure. **Turn on the "Barrel" button to highlight, in cyan, the subunit's β strands and, in yellow, its longest loop, which acts to limit the inner diameter of the barrel.** Note that the polypeptide chain closes on itself such that the N-terminal amino group (blue ball on the lower right of View1) forms a hydrogen bonded salt bridge (dashed white line) with the C-terminal carboxyl group (whose oxygens are represented by red balls). Thus, an OmpF porin subunit consists of a pseudo-cyclic polypeptide. The main chain hydrogen bonding capacity of the β strands is thereby fully utilized, which is, in part, responsible for this β barrel's ability to penetrate a membrane (see below)—as is likewise the case with transmembrane helices.

View2 and View3 are views into the barrel from the top and bottom of View1. Can you see that the OmpF porin barrel is ~20 Å in diameter but is constricted to a diameter of ~7 Å at is narrowest point by the yellow loop? A vertical cross section through the center of the OmpF porin barrel showing its solvent-accessible surface and thereby the extent of this constriction is presented in Fig. 10-33.

Porin's side chain distribution is characteristic of transmembrane proteins. **Turn on the "Outside" button" and then the "Tyr" button.** This displays, in magenta, the Tyr sidechains that festoon the outside of the porin barrel with their phenolic O atoms represented by red balls. These sidechains form two belts near the top and bottom of the barrel. In each of these belts, the bonds to the phenolic O atoms point away from the barrel's midline so as to delimit the protein's ~25-Å-high nonpolar transmembrane band (see also Fig. 10-6*c*). **Turn on the "Ile + Phe" button to display Ile and Phe sidechains in wireframe form in green and gold.** Where are the majority of these nonpolar residues located? **Turn on the "Arg + Lys" and the "Asp + Glu" buttons to display these positively and negatively charged sidechains in blue and red.** Are these residues located in the transmembrane band? Note that the majority of these charged residues are distributed in a ~120° arc around the upper side of the barrel in View1. The reasons for this will become apparent below.

Go to View2 and/or View3, turn off the "Outside" button, and turn on the "Inside" button. This displays the above-named sidechains that line the inside of the barrel. **Is the inside of barrel hydrophobic or hydrophilic?** Can you see that it is lined with charged and polar sidechains (the latter being represented by those of Tyr) that extend into the barrel, whereas there are only four Phe residues and no Ile residues (Ile has the highest hydropathy of all protein sidechains, Table 7-5) that are exposed inside the barrel? This accounts for OmpF porin's ability to permit the passage of polar solutes (Section 10-5B).

Go to View4 (which is related to View1 by a ~180° rotation about the barrel axis), turn off the "Inside" button, and turn on the "Bilayer" button. This displays, in wire frame form, a theoretically generated model of a lipid bilayer composed of phosphatidylcholine (Table 9-2) in complex with OmpF porin in which the polar phosphocholine head groups are purple and the nonpolar diacylglycerol residues are orange. View1 shows these interactions from the opposite side of the barrel, whereas View2 and View3 reveal that the bilayer contacts that portion of the porin barrel that does not participate in intersubunit contacts (see below). **Turning off the "Tails" button for clarity and turning on the "Arg + Lys" and "Asp + Glu" buttons together with the "Outside" button reveals how the above-described ~120° swath of charged residues at the upper left of View1 (upper right of View4) is proposed to interact with the head groups of the bilayer's upper leaflet.** The positively charged choline residues in the upper leaflet of the bilayer associate with the Asp and Glu residues extending mostly from above the bilayer and the Arg and Lys residues, which are located at approximately the same level as the head groups, interact with the bilayer's negatively charged phosphate groups. **Turning on the "Tyr" button reveals that the polar ends of the Tyr sidechains extend towards the head groups in both leaflets** from within the bilayer's nonpolar region.

View5, a closeup view of the upper left region in View1, displays the lipid-protein interactions in this model. Turning on the "Tyr" button and the "Outside" button displays how the lipid bilayer is thought to interact with porin's transmembrane region.

Now, turn off the "Tyr" and the "Bilayer" button, go to View6, and turn on the "Trimer" button. This displays the trimer as viewed along its threefold axis of rotation with the trimer's two previously undisplayed OmpF porin subunits colored seagreen and skyblue. The orientation of the initially displayed subunit is similar to that in View2. **By turning on the "Arg + Lys" and "Asp + Glu" buttons together with the "Outside" master button, you can see that the above described ~120° band of charged sidechains about the top of the barrel are, as might have been expected, exposed on the surface of the trimer.** Momentarily turning on the "Bilayer" button reveals how the trimer is thought to be surrounded by the lipid bilayer (only that portion of the bilayer in contact with the initially shown subunit is displayed). Note that, near the trimer's threefold axis, a loop near the N-terminal of each subunit extends partially across the clockwise-adjacent subunit to help bind the trimer together. View7 is an oblique view of the trimer showing how the β sheet on one subunit abuts the β sheet from its clockwise-adjacent neighbor.

The atomic coordinates for bacteriorhodopsin were obtained from Richard Henderson and Nikolaus Grigorieff of the MRC Laboratory of Molecular Biology, Cambridge, U. K.; those for the photosynthetic reaction center from *Rps. viridis* and *E. coli* OmpF porin were obtained from 1PRC and 1OPF. The coordinates for the phosphatidylcholine lipid bilayer in complex with OmpF porin were generated by Eric Jakobsson and coworkers of the University of Illinois.

Exercise in large part by John H. Connor
Terminator Graphics Limited
1118 Kimball Dr.
Durham, NC 27712
USA

EXERCISE 9. LYSOZYME

Kinemage 1: Lysozyme: Section 11-4; Figs. 11-15 and 11-17.

Hen egg white (HEW) lysozyme is a small (129-residue) monomeric enzyme that, like all lysozymes, catalyzes the hydrolysis of the C1—O1 bond linking an *N*-acetylmuramic acid (NAM) residue to an *N*-acetylglucosamine (NAG) residue in the beta (1 $\rightarrow$ 4)-linked poly(NAM–NAG) that forms the backbone of bacterial cell wall peptidoglycans (Figs. 8-15 and 11-14). The enzyme likewise hydrolyzes poly(NAG).

If you have not already done so, **double click on the icon for the program MAGE. When queried, click on the "Proceed" box. Then, under the "File" pull-down menu, invoke the "Open File..." option and select E09_Lyso.kin.**

Click on the graphics window to bring it to the front. Click on the box in the upper right corner of the MAGE color graphics bar to make the graphics window fill the screen.

View1 shows the Cα backbone of lysozyme and a NAG_6 substrate bound at its active site as in Fig. 11-15. The protein backbone is white and its catalytically important side chains, those of Glu 35 and Asp 52, are hotpink and pink, respectively, with their side chain oxygen atoms represented by red balls. The NAG_6 substrate is seagreen with its *N*-acetyl groups greentint and its N and O atoms represented by small blue and red balls. Only the A, B, and C residues of NAG_6 are experimentally observed; their coordinates were obtained from an X-ray structure of NAG-NAG-NAG (NAG_3) in complex with lysozyme. In contrast, the positions of the D, E, and F residues were obtained through model building and hence, these saccharides are represented by dashed lines. These model-built residues can be turned on and off with the "Model" button when "NAG6" is on.

Rotate the protein to see the long and deep cleft in which the NAG_6 binds.

Click the ANIMATE button to see how a different inhibitor, NAM-NAG-NAM, binds to lysozyme (MGM; mostly yellow with its *N*-acetyl groups and the lactyl groups of its muramic acid residues yellowtint and its N and O atoms represented by small blue and red balls). This procedure also gives you a sense of the reaction catalyzed by lysozyme. **Superimpose NAG_6 and MGM by turning on both the "MGM" and "NAG6" buttons.** Note that MGM and the experimentally observed NAG_3 portion of NAG_6 are not identically bound to the enzyme; only the rings in the B and C positions overlap, as can be seen better in View2. **Measure a few of the distances between corresponding B and C ring atoms of NAG_6 and MGM to ascertain just how closely these positions overlap.**

Turn on View2 with the "MGM" button on and the "NAG6" and "Lysozyme" buttons off. The MGM D ring's C1—O1 bond (green) is the bond that lysozyme would cleave in a true substrate, that is, if D ring atom O1 was also linked to atom C4 of a NAG residue bound in lysozyme's E site. The Phillips mechanism postulates that lysozyme sterically distorts its bound D ring to a half-chair conformation such that D ring atoms C1, C2, C5, and O5 are coplanar (Fig. 11-17). Such a conformation closely resembles that of the oxonium ion transition state through which the bond cleavage reaction is proposed to occur (Fig. 11-19). The X-ray structure of the MGM complex of lysozyme shows that the above D ring atoms (whose connecting bonds are gold) are, in fact, nearly coplanar. **"Pickcenter" on one of the D ring atoms (and then turn off "pickcenter") and rotate the MGM trisaccharide so as to convince yourself that its D ring's C1, C2, C5, and O5 atoms are coplanar. How closely does the NAG_6 D ring (which was model-built 25 years before the X-ray structure of the MGM-lysozyme complex was determined) correspond to the D ring of MGM?**

Go to View 3, switch "Lysozyme" and "MGM" on, and "NAG6" and "CaBackbone" off. The Phillips mechanism postulates that Glu 35's carboxyl group is a proton donor in the bond cleavage reaction (general acid catalysis), whereas Asp 52's carboxylate group acts to electrostatically stabilize the hydrolytic reaction's oxonium ion transition state (electrostatic catalysis). **Invoke "measures" under the "Other" pulldown menu and measure the angle between atoms Cd of Glu 35, Oe1 of Glu 35, and O1 of the D ring. Measure the distance between the latter two atoms (which are linked by the dashed yellowtint line). How well do this angle and distance compare with the ideal geometry for an O—H···O hydrogen bond (2.8 Å and 120°)?** Convince yourself that the closest approach that Glu 35 makes to the D ring's C1 and O1 atoms is that between Oe1 Glu 35 and O1 mgm D. Likewise, **find the closest approach between the carboxylate O atoms of Asp 52 and the D ring's C1 and O1 atoms. Is this distance small enough to permit the formation of a covalent intermediate during the catalytic reaction?**

Go to View4 (Fig. 11-17), switch "Lysozyme" and "MGM" off, and "NAG6" and "sidechains" on. Can you find the H-bond made by the NH of Val 109 with O6 of the D ring? Mutation of Val 109 to Pro, which lacks the NH group to make such a hydrogen bond, inactivates the enzyme. Why?

The HEW lysozyme and MGM coordinates were obtained from the X-ray structure of their complex as determined by Natalie Strynadka and Michael James, University of Alberta, Edmonton, Alberta, Canada. The X-ray-based coordinates of the A, B, and C rings of NAG_6 were taken from 1HEW, whereas the model-built coordinates of its D, E, and F rings were provided by Louise Johnson, Oxford University, Oxford, U. K.

EXERCISE 10. SERINE PROTEASES

KINEMAGE 1: Structural Overview of a Trypsin/Inhibitor complex: Section 11-5; Fig. 11-23.

KINEMAGE 2: Evolutionary Structural Comparisons of Trypsin, Chymotrypsin, and Subtilisin: Fig. 11-25.

KINEMAGE 3: Chymotrypsin in Complex with 2-Phenylethyl Boronic Acid, a transition state analog: Fig. 11-27.

KINEMAGE 4: Structural Comparison of Chymotrypsin with Chymotrypsinogen; mechanism of activation: Fig. 11-30.

The serine proteases are peptide-hydrolyzing enzymes that are characterized by the presence of a highly reactive and catalytically essential serine residue (Section 11-5). They have been extensively studied over more than four decades by a variety of techniques including X-ray, kinetic, enzymological, and more recently, genetic engineering methods, and hence comprise perhaps the most well characterized and mechanistically best understood group of enzymes. In this exercise, we examine the structures of several serine proteases, namely bovine trypsin, bovine chymotrypsin, subtilisin, and bovine chymotrypsinogen, in order to further understand their structures, evolutionary relationships, and catalytic mechanism.

If you have not already done so, **double click on the icon for the program MAGE. When queried, click on the "Proceed" box. Then, under the "File" pull-down menu, invoke the "Open File..." option and select E10_SerP.kin.**

Click on the graphics window to bring it to the front. Click on the box in the upper right corner of the MAGE color graphics bar to make the graphics window fill the screen.

Kinemage 1: Bovine Trypsin (Section 11-5B, Fig. 11-23).

In this kinemage we examine the structure of bovine trypsin in complex with a polypeptide inhibitor.

View1 shows trypsin in ribbon form in its "standard" orientation (Fig. 11-23). The catalytic triad, controlled by the "Cat Triad" button, consists of His 57, Asp 102, and Ser 195 (shown in cyan, red, and magenta, respectively), the latter being the enzyme's catalytically essential Ser residue. **Click on the "Inhibitor" box to display a Lys-containing inhibitor peptide (a fragment of bovine pancreatic trypsin inhibitor) bound to the active site.** This inhibitor is on the way to forming a tetrahedral intermediate on the enzyme's surface. **Click on the "Spec.Pocket" button to display the interactions of the inhibitor's Lys residue with the protein.**

Go to View2 for a closeup of the active site. Click the "scissile bond" button on and off to highlight in cyan the substrate bond to be broken. Click on the "covalent bond" button to show the covalent bond that forms between the active site serine and a substrate during reaction. Switch on the "OxyanionHole" button. This highlights the main chain of residues 192 through 196 in green and shows the main chain N atoms of Gly 193 and Ser 195 as small blue spheres. These two atoms form hydrogen bonds to and thereby stabilize the enzymatic reaction's tetrahedral intermediate (Section 11-3F and Fig. 11-27). **Rotate the molecule around to see this interaction better.**

Kinemage 2: Structural Comparisons of Trypsin, Chymotrypsin, and Subtilisin (Section 11-5B; Fig. 11-25).

The pancreas synthesizes three closely related serine proteases that it secretes into the small intestine (Table 5-5): trypsin, which is specific for basic groups (Arg and Lys) preceding the scissile (to be cleaved) peptide bond; chymotrypsin, which is specific for residues with bulky nonpolar side chains (Phe, Trp, and Tyr) in this position; and elastase, which cleaves past small nonpolar side chains (Ala, Gly, Ser, and Val). In this kinemage we examine the structures of trypsin and chymotrypsin and then compare these structures with that of the unrelated bacterial serine protease subtilisin.

View1 shows trypsin colored skyblue. **Click "ANIMATE" once** to show chymotrypsin oriented similarly in bluetint. Now **turn on the "Trypsin" button** to display the superimposed structures of chymotrypsin and trypsin. **How closely do these two structures resemble each other?** (Section 11-5D).

Go to View2 to study the superimposed active sites. How well do the catalytic groups of trypsin and chymotrypsin superimpose?

Turn on the "OxyanionHole" button. This highlights the main chain of residues 192 through 196 in green and shows the main chain N atoms of Gly 193 and Ser 195 as small blue spheres. These two atoms, as we shall see in KINEMAGES 3 through 5, form hydrogen bonds to and thereby stabilize the enzymatic reaction's tetrahedral intermediate (Section 11-3F and Fig. 11-27). **How good is the overlap of the residues enclosing the so-called oxyanion hole in chymotrypsin and trypsin?**

Turn on the "Spec.Pocket" button. This highlights, in purple, the three main chain segments that form the enzyme's specificity pocket, the pocket that binds the side chain of the residue that precedes the scissile peptide bond. Trypsin binds positively charged side chains (Arg and Lys) in this pocket. Residue 189, which occurs at the bottom of the pocket in trypsin, is Asp (side chain in red). In chymotrypsin, which binds bulky nonpolar residues (Phe, Trp, and Tyr) in its specificity pocket, residue 189 is Ser (side chain in seagreen). **How well do the corresponding main chain segments forming the specificity pockets of the two enzymes overlap? How well do the side chains of Asp 189 in trypsin and Ser 189 in chymotrypsin overlap?** It is these side chains, as we shall see in KINEMAGES 4 and 5, that are, in part, responsible for the different substrate specificities of the pancreatic serine proteases.

The pancreatic serine proteases are homologous in sequence and structure with each other and with numerous other serine proteases such as many of the proteins that participate in blood clotting, including thrombin (Box 11-5). However, several serine proteases have been characterized that have no discernible sequence homology with the pancreatic serine proteases. One such protein is subtilisin, a protease produced by *Bacillus subtilus,* whose side chain specificity is similar to but considerably more permissive than that of chymotrypsin. **Go back to View1 and click the "ANIMATE" button.** This shows the Cα backbone of subtilisin in yellowint. **Simultaneously display the chymotrypsin backbone to reveal that these two proteins have little if any discernible structural homology.** The same, of course, is true with trypsin and subtilisin. (See Fig. 11-25 for sequence comparison of chymotrypsin and subtilisin).

Despite the lack of structural resemblance between subtilisin and the pancreatic serine proteases, the active sites of these proteins are remarkably similar. **Compare the active sites of subtilisin and chymotrypsin by going to View2, switching on the "ActiveSite" buttons of both chymotrypsin and subtilisin, switching off their polypeptide backbones, and turning off all buttons but "Cat. Triad".** Subtilisin has a catalytic triad composed of the side chains of its Ser 221, His 64, and Asp 32 (colored magenta, cyan, and red, respectively), that are arranged almost identically to chymotrypsin's (and trypsin's) catalytic triad. Indeed, subtilisin, in this kinemage, was oriented by superimposing its catalytic triad on that of chymotrypsin. **Turn on the "OxyanionHole" button and compare the positions of subtilisin's and chymotrypsin's oxyanion holes. What can you say about them?** (See Fig. 11-27 for an illustration of chymotrypsin's active site).

Turn on the "Spec.Pocket" button to display, in magenta, the two subtilisin peptide segments forming this protein's specificity pocket. What can you say about the structural resemblance between these polypeptide segments and those forming the specificity pockets of chymotrypsin and trypsin? The pancreatic serine proteases and subtilisin are related by convergent evolution: Both have essentially the same catalytic mechanism but little other structural resemblance.

At least one other serine protease, wheat germ serine carboxypeptidase II, has a catalytic triad but is otherwise unrelated, in structure and sequence, to both the pancreatic serine proteases and to subtilisin (Fig. 11-25). In addition, numerous esterases, including several lipases (Section 19-1) and acetylcholinesterase (Box 11-4), have catalytic mechanisms resembling that of the serine proteases. In acetylcholinesterase, which hydrolyzes the neurotransmitter acetylcholine, the catalytic triad contains a Glu rather than an Asp side chain.

Kinemage 3: Chymotrypsin in Complex with 2-Phenylethyl Boronic Acid, a Transition State Analog (Section 11-5C; Fig. 11-27).

The X-ray structures of numerous serine proteases in their complexes with a great variety of inhibitors have been determined. In this kinemage we examine the structure of bovine chymotrypsin in complex with its inhibitor 2-phenylethyl boronic acid (PBA). Since electron deficient boronic acids easily form tetrahedral adducts and chymotrypsin preferentially binds bulky nonpolar groups such as a phenyl ring in its specificity pocket, PBA specifically binds to chymotrypsin as a transition state analog.

View1 shows the by now familiar structure of bovine chymotrypsin in its "standard" orientation. Switch on the "Active site" button and the "OxyanionHole" and "Spec.Pocket" buttons to show the catalytic triad, the main chain segment lining the oxyanion hole, and the elements forming the specificity pocket, all colored as in previous kinemages. Now go to View2 to view the enzyme's active site region and PBA in complex with chymotrypsin. The PBA is colored orange with its B and O atoms represented by greentint and red spheres.

The PBA is well along its reaction coordinate to form a covalent bond with the side chain O atom of Ser 195. **Measure the length of this bond, shown in yellow and controlled by the "CovalentBnd" button. Compare this length to the 1.39-Å lengths of the PBA's two other B—O bonds and the ~3.2-Å van der Waals distance between B and O atoms. What is the geometry of the bonds about the Boron?** (This can be quantitatively established by measuring the bond angles about the boron atom by invoking the "measures" option under the "Other" pull-down menu). **What interactions do you note between PBA and the oxyanion hole [the main chain N atoms of both Gly 193 and Ser 195 (Fig. 11-27*b*)]? Finally, what is the location of the phenyl ring with respect to the specificity pocket (keep in mind that the phenyl ring has covalently attached H atoms that are not displayed here)?**

Kinemage 4: Structural Comparison of Chymotrypsin with Chymotrypsinogen (Section 11-5D; Fig. 11-30).

Chymotrypsin is synthesized in the pancreas as chymotrypsinogen. It is activated in the small intestine, as is diagrammed in Fig. 11-30, by tryptic cleavage between its Arg 15 and Ile 16 residues. This catalytically active so-called π-chymotrypsin nevertheless undergoes further autolytic cleavage by the excision of its Ser 14–Arg 15 and Thr 147– Asn 148 dipeptides to yield the catalytically active α-chymotrypsin, which we refer to as simply chymotrypsin. Chymotrypsin therefore consists of three polypeptide chains although, since it is synthesized as a single polypeptide chain, it is not considered to consist of subunits. In this kinemage, we compare the structures of chymotrypsin and chymotrypsinogen to see how the initial tryptic cleavage activates chymotrypsinogen.

View1 shows chymotrypsin in its standard orientation represented by its Cα backbone and colored with its Peptide 1 (residues 1-13 but with residues 12-13 not visualized here) gold, Peptide 2 (residues 16-146) skyblue, and Peptide 3 (residues 149-245) pink. **Click "ANIMATE" to instead display chymotrypsinogen in the same orientation with its corresponding polypeptide segments yellowtint, bluetint, and pinktint, respectively. To directly compare the two proteins, turn on both the "Chymotryp." and "Chym-ogen" buttons. They can be further compared in View2,** which displays the proteins as seen approximately from the right in View1. **Can you find places where the two structures deviate?**

Go to View3, turn on both the "Chymotryp." and "Chym-ogen" buttons, turn off all the "Peptide" buttons, and turn on the "Cat. Triad" button. Note how closely the catalytic

groups of chymotrypsin and chymotrypsinogen are superimposed. Why is chymotrypsinogen so much less catalytically active than is chymotrypsin?

Turn on the "OxyanionHole" master button. This displays the peptide segment that binds the oxyanion in the catalytic reaction's transition state in green (as usual) for chymotrypsin and in seagreen for chymotrypsinogen. **How well are these corresponding segments superimposed starting at around the main chain N atom of Gly 193 to the main chain N of Ser 195 (blue balls)? Can your observations explain the decreased enzymatic activity of chymotrypsinogen?**

Now go to View4 and turn on the "Spec.Pocket" button to display the three polypeptide segments forming the specificity pocket: those of chymotrypsin are purple and those of chymotrypsinogen are magenta. How well do these specificity pockets superimpose? Can the difference in chymotrypsinogen's specificity pocket account for its decreased capability of binding substrate?

What is the structural basis of the foregoing conformational shifts? Turn off the "Chym-ogen" button and turn on "ScissionSite" button. This displays, in gold, the main chain of Ile 16, the newly liberated N-terminus in the tryptic activation of chymotrypsinogen, together with the side chain of Asp 194, in pink in chymotrypsinogen and in red in chymotrypsin. These two oppositely charged groups, as we saw in previous kinemages, form a hydrogen bonded salt bridge (dashed white line). **Turn on the "Chym-ogen" button to display the position of the main chain N atom of Ile 16 (orange) in chymotrypsinogen. Measure the distance moved by this group in going from chymotrypsinogen to chymotrypsin. Turn on both "Peptide 2" buttons and go to View1 to see the movement of Ile 16 from the exterior of chymotrypsinogen to the interior of chymotrypsin** (where the lesser dielectric constant presumably strengthens the Ile 16–Asp 194 salt bridge). Also note that the side chain of Asp 194 has a quite different conformation in the two proteins. **What effect does the salt bridge between Ile 16 and Asp 194 in chymotrypsin (which cannot form in chymotrypsinogen) have on the stability of chymotrypsin's catalytically active conformation?**

To help answer this question, go to View5 and turn on both the "Chymotryp." and "Chym-ogen" Peptide 1 buttons. Be sure the "Cat. Triad", "Oxyanion Hole" and "Scission Site" buttons are all on, and press the ANIMATE button several times. What is happening to Gly 193 in the Oxyanion Hole? To help identify Gly 193, turn on the "Labels" button. Gly 193 is greentint in chymotrypsinogen and green in chymotrypsin. **Measure the distance moved by the N atom of Gly 193 in going from chymotrypsinogen to chymotrypsin. What might this movement mean to the structure of the oxyanion hole?**

The coordinates for bovine chymotrypsin in KINEMAGES 1, 2, and 4 were obtained from 4CHA; those for bovine trypsin in KINEMAGE 2 were obtained from 3PTN (transformed so as to superimpose its Cα backbone on that of chymotrypsin); those for subtilisin in KINEMAGE 2 were obtained from 1CSE (transformed so as to superimpose its catalytic triad on that of chymotrypsin); those for bovine chymotrypsin in its complex with 2-phenylethane boronic acid in KINEMAGE 3 were obtained from 6CHA; and those for bovine chymotrypsinogen in KINEMAGE 4 were obtained from 1CHG (transformed so as to superimpose its Cα backbone on that of chymotrypsin).

Exercise in part by John H. Connor
Terminator Graphics Limited
1118 Kimball Dr.
Durham, NC 27712
USA

EXERCISE 11. ASPARTATE TRANSCARBAMOYLASE: ALLOSTERIC INTERACTIONS

KINEMAGE 1: Quaternary Structure of ATCase and Its Allosteric Conformational Changes: Section 12-3; Fig. 12-12.

KINEMAGE 2: Conformational Changes Caused by the Binding of Substrate: Fig. 12-13.

Aspartate transcarbamoylase (ATCase) is a multisubunit enzyme that catalyzes the reaction of carbamoyl phosphate with aspartate to yield *N*-carbamoyl aspartate and P_i (Section 12-3). This irreversible reaction constitutes the first committed step in the synthesis of pyrimidines (Section 22-2A). Since either a shortage or an excess of pyrimidines (cytosine and thymine) relative to purines (adenine and guanine) greatly increases the error rate in DNA synthesis, it is essential that the rate of pyrimidine synthesis be closely regulated to precisely satisfy the needs of the cell. In *E. coli,* this occurs through the control of the catalytic activity of ATCase: The enzyme is inhibited by CTP (Fig. 12-10), a product of the pyrimidine synthesis pathway (Fig. 12-11 and Section 22-2C), and is activated by ATP, whose high concentration indicates that purines are in excess. Both of these metabolites bind to a site on the enzyme different from its catalytic site, a so-called allosteric site, and hence ATCase is an allosteric enzyme. Indeed, ATCase closely follows the symmetry model of allosterism (Section 7-2E).

In the symmetry (or MWC) model of allosterism, a symmetric multisubunit enzyme has (at least) two conformational states: a less catalytically active state known as its T state and a more catalytically active state known as its R state. The conformational changes between the T and R states, nevertheless, preserve the symmetry of the enzyme (hence the name "symmetry model") so that, in this model, states in which the enzyme's various chemically equivalent subunits have different conformations are forbidden. Substrate(s) preferentially binds to the R state, thereby stabilizing the protein in this conformation and increasing the protein's affinity for substrate. This cooperative process is known as a positive homotropic effect. Thus both aspartate and carbamoyl phosphate are positive homotropic effectors of ATCase. By the same token, activators preferentially bind to an allosteric site on the R-state enzyme, thereby also increasing its affinity for substrate (a positive heterotropic effect), whereas inhibitors preferentially bind to an allosteric site on the T-state enzyme, thereby decreasing its affinity for substrate (a negative heterotropic effect). Hence, ATP and CTP are positive and negative heterotropic effectors of ATCase.

In this exercise we examine and compare the quaternary and tertiary structures of *E. coli* ATCase in its T and R states. We also inspect how the unreactive bisubstrate analog *N*-(phosphoacetyl)-L-aspartate (PALA; p. 345) binds to ATCase's active site.

If you have not already done so, **double click on the icon for the program MAGE. When queried, click on the "Proceed" box. Then, under the "File" pull-down menu, invoke the "Open File..." option and select E11_ATCs.kin.**

Click on the graphics window to bring it to the front. Click on the box in the upper right corner of the MAGE color graphics bar to make the graphics window fill the screen.

Kinemage 1: Quaternary Structure of ATCase and Its Allosteric Conformational Changes (Section 12-3; Fig. 12-12).

******************************* A Cautionary Note *******************************

ATCase is a very large protein. Thus, as you may have already noticed, MAGE takes considerably longer than usual to display this kinemage as well as to change the display. In particular, any attempt to rotate the entire protein must be done in small increments to get predictable results.

****

E. coli ATCase is a 310-kD enzyme that consists of six copies each of two types of subunits: a 310-residue catalytic (*c*) subunit and a 158-residue regulatory (*r*) subunit.

View1 shows the entire T state ATCase molecule as viewed in Fig. 12-12*a*. The three subunits of the upper catalytic trimer (by custom, subunits c_1, c_2, and c_3, starting on the left and going clockwise) are red and those of the lower catalytic trimer (c_4, c_5, and c_6, which are respectively below c_1, c_2, and c_3) are cyan. **Click on the "Cat. Trimer" button to change the colors of c_2 and c_3 to pink and pinktint for better identification.** Each regulatory subunit is in direct contact with only one catalytic subunit and is assigned the same subscript (e.g., r_1 contacts c_1). The three upper regulatory subunits in View1 (r_1, r_2, and r_3) are green and the lower three (r_4, r_5, and r_6) are gold. The three regulatory dimers are each composed of a green and a gold subunit (r_1 with r_6, r_2 with r_4, and r_3 with r_5). **Click on the "Reg. Dimer" button to change the subunits of one of the dimers to greentint and yellow for better identification.** Isolated catalytic trimers (c_3) are fully active and non-cooperative. The regulatory dimers interact with the catalytic trimers to form the allosteric c_6r_6 holoenzyme.

The asymmetric portion of a protein, which in ATCase consists of one catalytic subunit and its associated regulatory subunit (e.g., c_1 and r_1), is referred to as a protomer. By this terminology (Section 6-3), which we use here, ATCase is a hexamer of protomers (although, since it consists of 12° subunits of two types, it may be alternatively described as being a heterododecamer). **View2 shows ATCase from the side as viewed in Fig. 12-12*b*.**

Return to View1 and click the "ANIMATE" button to show the enzyme in its R state (toggling "ANIMATE" alternately displays ATCase's R and T states). Note that the T to R transition rotates the upper catalytic trimer counterclockwise relative to the lower catalytic trimer by about 12° such that they more completely eclipse each other when viewed from the top.

Go to View2 for a more dramatic view of the allosteric changes in ATCase. Here it can be seen that the T to R transition causes the catalytic trimers to separate along the 3-fold axis by around 11 Å . This quaternary transition is accompanied by a major structural reorganization of the contacting residues in vertically aligned subunits (e.g., c_3 and c_6). **Switch off all of the buttons except the "c3,c6" button and click on ANIMATE a few times to see this structural reorganization (Fig. 12-13).**

Go to View3 and ANIMATE to see that the regulatory dimers rotate clockwise about 15° and separate by about 4 Å in going from the T to the R state.

To see how the T-state protein is constructed, again turn on View1, turn off the "Hexamer" button, and turn on the "Protomer" button. This displays the c_1-r_1 protomer. **Now turn off "Protomer" and turn on "Dimer".** This adds the c_6-r_6 protomer to the structure. Note that r_1 contacts r_6 via a β pleated sheet.

Turn off "Dimer" and turn on "Tetramer". This adds the c_2-r_2-r_4-c_4 dimer to the display. Your perception of depth here might be enhanced by temporarily switching off "onewidth" under the "Display" pulldown menu. **Now, turn off "Tetramer" and turn on "Hexamer" to add the c_3-r_3-r_5-c_5 dimer.** Note that the three dimers are arranged in a manner reminiscent of the overlapping flaps of a closed cardboard carton with the first dimer over the second over the third over the first.

To see how the subunits of a catalytic trimer contact each other, go to View2, turn off the "Hexamer" and turn on the "c3,c6" button. ANIMATE to see how these contacts change in going from T to R state. Note that these contacts appear to be rather tenuous. We shall examine these contacts in more detail in KINEMAGE 2.

Kinemage 2: Tertiary and Quaternary Conformational Changes Caused by the Binding of Substrate (Fig. 12-13).

View1 shows the c_3 and c_6 vertically interacting subunits of ATCase in the T state oriented as in Fig 12-13. The CP domain is skyblue and the Asp domain is greentint in c_6. c_3 is red. **Identify the CP and Asp domains by turning on and off the "CP site" and "Asp site" buttons.** In c_6, the CP binding site is outlined by yellow balls and the Asp binding site is outlined by gold balls. **ANIMATE to see the structural changes in going from the unbound T state to the PALA-bound R-state.** PALA is a bisubstrate analog whose N, O, and P atoms are represented by blue, red, and gold balls. It is only shown in the c_6 subunit. **Go to Views2 and 3 for an overview and a closeup of the PALA site.** Note that the phosphoryl group of PALA binds to the CP site and the carboxyl groups bind to the Asp site. The two sites come closer together as a result of the T to R state conversion, as illustrated in Fig. 12-13.

Go to View4 and highlight the c_3-c_6 interface with the "c3-c6Interfa" button. Click the "Int SideCh." button to see the intersubunit interactions of the amino acid sidechains that undergo a conformational change in the T to R transition in yellow for c_3 and in orange for c_6. The hydrogen bonds linking these groups are represented by dashed white lines. **ANIMATE in View4 with the "Int SideCh." button checked to see the intersubunit H-bonds that break in going from the T state to the R state, reminiscent of hemoglobin.**

The atomic coordinates for T-state ATCase in complex with ATP were obtained from 4AT1; those for R-state ATCase in complex with PALA were obtained from 8ATC; and those for CTP and the T-state ATCase residues that differ significantly in conformation from those in 4AT1 were obtained from 1RAI.

EXERCISE 12. TRIOSE PHOSPHATE ISOMERASE

KINEMAGE 1: The Eight-Stranded α/β Barrel: Section 6-2B; Figs. 6-27*c* and 6-28*d*.

KINEMAGE 2: TIM Active Site: Section 14-2E; Fig. 14-6.

Triose phosphate isomerase (TIM) is a glycolytic enzyme that catalyzes the interconversion of glyceraldehyde-3-phosphate (GAP) to dihydroxyacetone phosphate (DHAP) via an enediol (or enediolate) intermediate (Section 14-2E). TIM is a dimer of identical subunits that each contain an eight-stranded α/β barrel (Figs. 6-27*c* and 6-28*d*) as a dominant motif.

If you have not already done so, **double click on the icon for the program MAGE. When queried, click on the "Proceed" box. Then, under the "File" pull-down menu, invoke the "Open File..." option and select E12_TIM.kin.**

Click on the graphics window to bring it to the front. Click on the box in the upper right corner of the MAGE color graphics bar to make the graphics window fill the screen.

Kinemage 1: The Eight-Stranded α/β Barrel (Section 6-2B).

KINEMAGE 1 displays the backbone conformation of yeast TIM in complex with the inhibitor 2-phosphoglycolate (PGC; p. 391). **View1 shows the enzyme displayed with the top of its α/β barrel facing the viewer.** (The "top" of the barrel is the edge toward which the C-terminal ends of the β strands point). The enzyme is colored with the parallel β strands of the α/β barrel ("sheets") yellow, their α helices ("helices") orange, and the remaining parts of the protein including the loops linking successive α helices and β strands of the α/β barrel ("loops"), greentint. The PGC is drawn in ball-and-stick form with C green, O red, P gold, and bonds white.

View2 shows the α/β barrel structure from the side. Turn off the "helices" and "loops" buttons to better see the barrel structure of the β sheet. (Visualization might be improved here by turning off "onewidth" under the "Display" pulldown menu. This gives better depth perception.) The eight β strands form a parallel β sheet that is twisted in a right-handed manner (the normal direction of β sheet twist; Section 6-1B and 6-2C) so as to close on itself, thereby forming a β barrel. This β barrel is encircled by a kind of barrel made of eight parallel α helices that are largely antiparallel to the β strands. **Turn on "helices" and "loops" to see that the β strands and α helices alternate along the polypeptide strand** so that the α/β barrel can be considered to be constructed of a series of overlapping β–α–β motifs (Fig. 6-28). **Turn on the "Labels" button to identify the N and C termini of the polypeptide chain and trace the chain to convince yourself that the β strands are all parallel and that the active site is at the end of the barrel containing the C-terminal ends of the β strands (the top).**

Kinemage 2: TIM Active Site (Section 14-2E; Fig. 14-6).

View1 shows the active site of TIM complexed with the inhibitor 2-phosphoglycolate (PGC; p. 391) as it is seen in Fig. 14-6. The interconversion of GAP and DHAP involves an enediolate intermediate (p. 391). PGC is an enediolate mimic and thus a transition state analog. Three protein side chains that participate in TIM's catalytic mechanism are displayed, those of Lys 12 (blue), His 95 (purple), and Glu 165 (red).

View2 is a magnified view of the active site. Find the carboxylate group of Glu 165, which has been implicated as the general base that abstracts the C2 proton from GAP and replaces it at C1 in forming DHAP. Is it properly positioned between the C1 and C2 atoms of PGC to do so? Note that the N_ε atom of His 195, which functions as a general acid to protonate O1 of GAP and then as a general base to abstract a proton from O2 of the resulting enediol(ate) intermediate, is hydrogen bonded (dashed purple line) to O1 of PGC. Note also that the cationic N_ζ atom of Lys 12, which is thought to electrostatically stabilize the developing negative charges in the transition states of the TIM reaction, is also hydrogen bonded to O1 of PGC (dashed blue line).

Go back to View1 and ANIMATE to see the "flexible loop" open and close over the active site (cyan, "Flx Loop Dn" and magenta, "Flx Loop Up"). The closure of this loop apparently stabilizes the enediol(ate)-like transition state, and holds the enediol(ate) intermediate in a conformation that greatly favors product formation over the elimination of the phosphate group to yield the toxic methyl glyoxal (p. 392).

Go to View3 and ANIMATE for a side view of the TIM barrel that nicely shows the movement of the "loop".

The atomic coordinates for the yeast TIM-PGC structure were obtained from 2YPI; those of the flexible loop in yeast TIM alone were obtained from 1YPI transformed to superimpose on the 2YPI structure.

EXERCISE 13. ALLOSTERISM IN PHOSPHOFRUCTOKINASE

KINEMAGE 1: Conformational Changes in a Dimeric Unit of PFK: Section 14-4A; Figs. 14-22 and 14-24.

KINEMAGE 2: The Major Conformational Changes in a Subunit of PFK: Fig. 14-24.

Phosphofructokinase (PFK) is a glycolytic enzyme that catalyzes the transfer of a phosphoryl group from ATP to fructose-6-phosphate (F6P) to yield ADP and fructose-1,6-bisphosphate (FBP) (Section 14-2C). The PFK reaction is strongly exergonic (irreversible) under physiological conditions (Table 14-1) and hence is one of the glycolytic pathway's rate-determining steps. In most organisms/tissues, PFK is the glycolytic pathway's major flux-controlling enzyme; its activity is controlled by the concentrations of an unusually large number of metabolites including ATP, ADP, AMP, and fructose-2,6-bisphosphate (F2,6P; Section 15-4C).

The symmetry model of allosterism (Section 7-2E) requires that an oligomeric (multisubunit) protein maintain its molecular symmetry in undergoing a transition from its high-activity R state to its low-activity T state. Hence, this transition must be concerted, with no intermediate states. The X-ray and enzymological evidence indicates that PFK, a tetramer of identical subunits, is an allosteric enzyme that follows the symmetry model. Hence, the binding of one molecule of its substrate F6P, which binds to the R state enzyme with high affinity but to the T state enzyme with low affinity, causes PFK to take up the R state, which in turn, increases the binding affinity of the enzyme for additional F6P (a homotropic effect). Activators, such as ADP and AMP, bind to so-called allosteric sites, binding sites distinct from the active site, where they likewise facilitate the formation of the R state and hence activate the enzyme (a heterotropic effect; ADP, being a product of the PFK reaction, also binds at the enzyme's active site). Similarly, inhibitors such as PEP bind to allosteric sites (which in the case of PFK overlap the activating allosteric sites) where they promote the formation of the T state, thereby inhibiting the enzyme.

This exercise consists of two kinemages that illustrate some of the allosterically induced conformational changes that occur in PFK from *Bacillus stearothermophilus.*

If you have not already done so, **double click on the icon for the program MAGE. When queried, click on the "Proceed" box. Then, under the "File" pull-down menu, invoke the "Open File..." option and select E13_PFK.kin.**

Click on the graphics window to bring it to the front. Click on the box in the upper right corner of the MAGE color graphics bar to make the graphics window fill the screen.

Kinemage 1: Conformational Changes in a Dimeric Unit of PFK (Section 14-4A; Figs. 14-22 and 14-24).

PFK from *B. stearothermophilus* is a tetramer of identical 320-residue subunits. This kinemage shows the same two subunits of the tetramer as does Fig. 14-22. **View1 shows the two subunits in their R state conformation** as represented by their Cα backbones with Subunit 1 in pinktint and Subunit 2 in pink. Two side chains in each subunit are shown, those of Glu 161 (red) and Arg 162 (cyan), which are important participants in PFK's allosterically facilitated conformational change (see below). An F6P (hotpink) and an ADP (green; "ADP-active") are bound in the active site of each subunit. An additional ADP (yellow; "ADP-allo") is bound in a separate so-called allosteric site of each subunit. The ADPs each have an associated Mg^{2+}, which is represented here by a ball of the same color as the ADP to which it binds.

Click the "ANIMATE" button to switch the dimer between its R and T states. In its T state, Subunit 1 is bluetint and Subunit 2 is skyblue. The side chains of Glu 161 and Arg 162 in both subunits are red and cyan as before (only the Cα and Cβ atoms of the Arg 162 side chain in Subunit 1 are observed in the X-ray structure of the T state; those of Subunit 2 are all observed). The T state enzyme binds the inhibitor 2-phosphoglycolate (gold; "PGC"; p. 391), a nonphysiological analog of the glycolytic intermediate phosphoenolpyruvate (PEP; Section 16-2J). Note that the binding site of PGC in the T state overlaps the allosteric binding site of ADP in the R state ("ADP-allo") and hence their binding is mutually exclusive. The T state active sites, which do not contain F6P, are marked by "ghost" F6Ps (gray; "F6P site"), which have the same positions as do the F6Ps in the R state enzyme.

View2 is a closeup of the upper portion of View1 showing both the active site and the allosteric site in this region. Note that the active site is located at the interface between two subunits and that the allosteric site interacts directly with the adjacent active site. **Compare the R state and T state conformations by displaying both at once or clicking on "ANIMATE". Can you identify the Mg^{2+} ion associated with each of the ADPs bound to the enzyme in the R state? Which ADP atoms coordinate these Mg^{2+} ions?**

The phosphate group of PGC binds to the allosteric site in the T state in very nearly the same position that the β phosphate group of "ADP-allo" binds to the R state allosteric site; both phosphate groups bind to the side chains of the same three Arg residues (not shown).

In the high-activity R state, the positively charged side chain of Arg 162 forms a hydrogen bonded salt bridge with the negatively charged 6-phosphate group of F6P (white dashed lines), an interaction which presumably stabilizes the R state relative to the T state and is therefore in part responsible for F6P's homotropic effect.

Kinemage 2: The Major Conformational Changes in a Subunit of PFK (Section 14-4A; Fig. 14-24).

KINEMAGE 2 corresponds to Fig. 14-24 and shows those segments near the allosteric site (residues 53-60 on the right side of the Fig. 14-24 are not shown here). As in KINEMAGE 1, the polypeptide is represented by its Cα chain with R state Subunits 1 and 2 in redtint and pink, and T state Subunits 1 and 2 in bluetint and skyblue. **KINEMAGE 2 comes up in the R state** showing the phosphate group of F6P (hotpink) bound in the enzyme's active site in a hydrogen bonded salt bridge (dashed white lines) with the side chain of Arg 162 (cyan). An ADP (yellow; "ADP-allo") occupies the adjacent allosteric site. **Click once on "ANIMATE" to switch to the T state.** This replaces the ADP in the R state allosteric site with the inhibitor and PEP analog PGC (gold). F6P no longer occupies the active site but its position in the R state is indicated by the "ghost" F6P (gray; "F6P site").

What happens to the central polypeptide helical segment (residues 149-164) in the R to T transition? What does this do to the relative positions of the negatively charged Glu 161 and the positively charged Arg 162? Click on "F6P site". What influence would the presence of the carboxylate group of Glu 161 have on the phosphate group of F6P were it to bind in the active site? Does this explain, at least in part, why T state PFK has low affinity for F6P? Go to View2 for a closeup of the F6P-sidechain interactions.

The atomic coordinates for R state PFK were obtained from 4PFK; those for T state PFK were obtained from Philip Evans, MRC Laboratory of Molecular Biology, Cambridge, U. K.

EXERCISE 14. GLYCOGEN PHOSPHORYLASE

KINEMAGE 1: Structure of Glycogen Phosphorylase *a:* Section 15-1A; Fig. 15-3.

KINEMAGE 2: Conformational differences between T and R states of Glycogen Phosphorylase *b:* Fig. 15-5.

KINEMAGE 3: The Dimer Interface on Activation of Glycogen Phosphorylase.

Glycogen phosphorylase is a dimer of identical 842-residue subunits that catalyzes the rate determining step of glycogen breakdown: the phosphorolytic excision of a glucose residue from a nonreducing end of glycogen to yield glucose-1-phosphate (G1P). The enzyme does so via an oxonium ion intermediate in which the phosphate group of its pyridoxal phosphate (PLP) cofactor functions as an acid–base catalyst (Fig. 15-4). The catalytic activity of glycogen phosphorylase is regulated both by allosteric interactions that closely follow the symmetry model of allosterism (Section 7-2E) and by covalent modification/demodification. The modification and demodification reactions, which are catalyzed by the enzymes phosphorylase kinase and phosphoprotein phosphatase-1 (Fig. 15-12), interconvert two forms of glycogen phosphorylase (Fig. 15-13): Phosphorylase *b,* which is covalently unmodified, tends to assume its catalytically inactive T conformation, and allosterically responds to the inhibitors ATP and glucose-6-phosphate (G6P) and the activator AMP; and phosphorylase *a,* which has a phosphoryl group esterified to its Ser 14 hydroxyl group, tends more to assume its catalytically active R conformation, and allosterically responds only to the inhibitor glucose. In Kinemage 1 we examine the structure of glycogen phosphorylase *a* (Fig. 15-3). In Kinemage 2 we study the allosteric interactions and conformational changes that occur in going from T state to R state glycogen phosphorylase *b* (Fig. 15-5). Kinemage 3 shows the changes in the dimer interface on activation.

If you have not already done so, **double click on the icon for the program MAGE. When queried, click on the "Proceed" box. Then, under the "File" pull-down menu, invoke the "Open File..." option and select E14_Phos.kin.**

Click on the graphics window to bring it to the front. Click on the box in the upper right corner of the MAGE color graphics bar to make the graphics window fill the screen.

Kinemage 1: Structure of Glycogen Phosphorylase a (Section 15-1A; Fig. 15-3).

View1 displays a homodimer of rabbit muscle glycogen phosphorylase in its R state, much as it is depicted in Fig. 15-3*a*. In the enzyme's "Subunit 1", the N-terminal domain's interface subdomain ("interface") is pinktint, its glycogen-binding subdomain ("glyc-bind.") is hotpink, and the C-terminal domain ("C-terminal") is orange, whereas the corresponding portions of "Subunit 2" are bluetint, skyblue, and seagreen, respectively. In both subunits, the serine phosphate side chain at position 14 ("Ser14-P") is green, the nearby AMP bound at the allosteric activator site ("AMP-allo") is gold, the glucose bound at the active site is purple, and the enzyme's bound PLP cofactor is cyan. Note that the catalytic site is ~70 Å from the catalytic site of the symmetry related subunit, and is ~45 Å and ~35 Å from the "AMP-allo" and "Ser14-P" of the same subunit. The glycogen fragments bound at each subunit's glycogen storage site ("Glycogen"), a 3-residue and a 5-residue unbranched alpha (1 → 4)-linked glucose chain, are white.

In View2, the dimeric protein in View1 is seen from the right. Turn on the "Labels" buttons and turn the "Subunit 2" button on and off several times to study the relatively tenuous contacts between subunits. Identify each subunit's projecting tower helix (residues 261-274). Identify the N-terminal regions of the two subunits, which include their Ser14-P residues.

The interactions between subunits in these two areas are crucial to the T ↔ R interconversion of this allosteric enzyme. We examine these interactions in greater detail in KINEMAGE 3.

Examine an individual subunit to become familiar with the various binding sites. Turn off the "Subunit 2" button, turn on the "Labels" button under "Subunit 1", and select View3. Click on and off the buttons corresponding to each of the sites to further identify them. Note the position of the N terminal polypeptide segment containing Ser 14. This segment undergoes a strikingly large conformational change on activation. We further consider this conformational change in KINEMAGES 2 and 3.

Kinemage 2: Conformational Differences between T and R States of Glycogen Phosphorylase b (Section 15-1A; Fig. 15-5).

KINEMAGE 2 compares the conformations of a glycogen phosphorylase *b* subunit in its T and R states. **View1 displays a T state phosphorylase *b* subunit viewed similarly to the left side of Fig. 15-5.** The subunit is represented by its Cα diagram, mostly in magenta, but with its N-terminal segment (residues 11-21; residues 1-10 are disordered) seagreen, its N-terminal helix (residues 21-38) cyan, its tower helix (residues 261-274) blue, and its loop residues 281-286, which we refer to here as the 280s loop, white. (NOTE: We did not see this 280s loop in Kinemage 1 because it is only seen in the T state of the enzyme. In the R state, the 280s loop is disordered).

Click the "ANIMATE" button to display a subunit of R state glycogen phosphorylase *b* in the same orientation as the T state subunit (similar to the right side of Fig. 15-5). The Cα backbone of the R state enzyme is mostly pink but with its N-terminal segment, N-terminal helix, tower helix, and 280s loop the same colors as they are in the T state enzyme. (NOTE: In the R state enzyme, residue 11 is also disordered (invisible in the X-ray structure) as are most of the 280s loop (residues 282-286) and the subunit's C-terminal segment (residues 838-842).)

Groups that are initially displayed in spacefilling form may be shown in wireframe form by turning off the "Spacefilling" button. Turn their buttons on and off to identify these individual groups. **Repeatedly click the "ANIMATE" button to compare the conformations of the T and R state subunits. This comparison can also be made by simultaneously turning on the "T State" and "R State" buttons. Identify Ser 14 and measure the distance it moves in going from the T to R state.** This large conformational change is shown from a different perspective in View2.

Go back to View1 and, with the "R State" button on, turn on the "PO4-allo" button. This displays, in yellow, an activating sulfate ion that has bound close to the position that is occupied by the phosphoryl group of Ser 14P in phosphorylase *a* (recall that sulfate and phosphate are both tetrahedral anions; sulfate ion was present in the solution from which the enzyme crystallized). The "PO4-act" button displays, also in yellow, a sulfate ion that is bound in the enzyme's active site near the PLP's phosphate group. This is presumably the same position occupied by the substrate phosphate ion.

View3 is a closeup of the enzyme's active site region. Turn off both "Cα chain" buttons and make sure that the T State button is on and the R State button is off. Look at the positions of the residues of the 280s loop and Arg 569. Turn off the "Spacefilling" button to better see the side-chains. Now click the "ANIMATE" button to change the structure to the R state. What are the positions of the side chains of Asp 283 and Asn 284 in the T state relative to the positions of Arg 569 and the substrate phosphate ion in the R state? Do you agree that Asp 283 and Asn 284 would sterically and electrostatically interfere with a substrate phosphate ion bound in the position it occupies in the R state? Turn on the "Spacefilling" button and click the "ANIMATE" button several times. This may help visualize the interference of Asp 283 and Asn 284 with a substrate phosphate ion. This steric and electrostatic interference, at least in part, explains the lack of catalytic activity of the T state enzyme. In the R state enzyme, the 280s loop is disordered and presumably no longer occludes the active site. Moreover, the side chain of Arg 569 has swung around to a position from which it forms a hydrogen bonded salt bridge to the substrate phosphate ion (dashed white line), thereby stabilizing its binding of this phosphate ion to the enzyme.

View4 displays the tower helix. Turn on both "Cα chain" buttons. Toggle the "ANIMATE" button to show the conformational changes this chain segment undergoes on switching between the T and R states. The tower helix (residues 261-274) is directly connected to the 280s loop; its change of conformation apparently motivates the change of conformation at the active site. We further examine the allosteric conformational changes of the tower helix and other groups at the dimer interface in KINEMAGE 3.

Kinemage 3: The Dimer Interface on Activation of Glycogen Phosphorylase (Section 15-1A).

This kinemage compares the conformations of the R and T states of glycogen phosphorylase in the vicinity of its dimer interface so as to identify the interactions responsible for the allosteric communication of conformational changes between subunits.

View1 shows glycogen phosphorylase *b* in its T state ("Phos b-T"). The protein is represented by its Cα backbone with the segment containing the N-terminal three helices of Subunit 1 ("Sub1-Nterm"; residues 5-80; orange) and a segment that includes the tower helix and the 280s loop ("Sub1-Tower"; residues 265-310; gold). The corresponding segments of Subunit 2 are seagreen and greentint. The side chains of "Ser14" and the AMPs bound at the allosteric effector sites ("AMP-allo") are initially shown in spacefilling form colored according to atom type but, when the "Spacefilling" button is off, they are shown in wireframe form colored pinktint and bluetint.

Click the "ANIMATE" button to display the same polypeptide segments of glycogen phosphorylase *a* in its R state ("Phos a-R"). Here, "Sub1-Nterm" and "Sub1-Tower" are magenta and pinktint, whereas "Sub2-Nterm" and "Sub2-Tower" are skyblue and cyan. The sidechains of the Ser 14-phosphate residues ("Ser14-P"), "AMP-allo", and the active site-bound glucose molecules are all shown in spacefilling form colored according to atom type (C green, N blue, O red, and P yellow). Turning off the "Spacefilling" button displays these groups in wireframe form with the Ser14-P sidechains red, the AMPs blue, and the glucose molecules green.

Click the "ANIMATE" button several times to observe the differences in T- and R-states. Compare View1 and View2 and rotate the image to see that the tower helices form an antiparallel association whose angle and extent of contact changes on going from T- to R-state. Similarly, on going from T- to R-state, the N-terminal helix of each subunit moves to contact helix 2 and the loop connecting it to helix 3 of the opposite subunit at the back of View1 and on the right of View2.

Go to View3 for a closeup of the region about Ser 14 and "AMP-allo" of Subunit 1. ANIMATE several times to observe the changes in the interactions of T- and R-states.

Go back to View2 and ANIMATE to observe that the conformational changes at the N-terminal and Tower regions involve a ~10° rotation of one subunit relative to the other in going from T- to R-state. This causes the residues at the dimer interface to shift by several Å, a motion that, as we saw in KINEMAGE 2, catalytically activates the enzyme. Note that the 280s loop is not disordered in R-state phosphorylase *a* as it is in R-state phosphorylase *b* (KINEMAGE 2).

The coordinates for R-state glycogen phosphorylase *a* in KINEMAGES 1 and 3 were provided by Stephen Sprang, University of Texas Southwest Medical Center. The coordinates for R-state glycogen phosphorylase *b* in KINEMAGE 2 were obtained from 7GPB and those of T-state glycogen phosphorylase *b* in KINEMAGES 2 and 3 were obtained from 8GPB and transformed to superimpose on the R state enzyme in these kinemages.

EXERCISE 15. THE CATALYTIC SUBUNIT OF cAMP-DEPENDENT PROTEIN KINASE (cAPK)

KINEMAGE 1: The Catalytic Subunit of cAPK: Section 15-3B; Fig. 15-14.

cAMP-dependent protein kinase (cAPK) catalyzes the phosphorylation, by ATP, of the Ser and Thr residues of its target sequence, Arg-Arg-X-Ser/Thr-Y, where the hydroxyl group of the Ser/Thr residue is the site of phosphorylation, X is any small residue, and Y is a large hydrophobic residue. The phosphorylation modifies the activities of the numerous cellular proteins that contain this target sequence. Thus, for example, cAPK activates phosphorylase kinase but inactivates glycogen synthase (Fig. 15-20).

In the absence of adenosine-3′,5′-cyclic monophosphate (cAMP), cAPK is a catalytically inactive heterotetramer, R_2C_2, in which R is its regulatory subunit and C is its catalytic subunit. When the cAMP concentration is sufficiently high, two molecules of cAMP bind to each regulatory subunit resulting in the release of catalytically active C subunits (Fig. 15-20; *top*). Thus, an increase in the cAMP concentration, which generally results from the hormonal stimulation of a cell, causes the phosphorylation and consequently the activation or deactivation of a variety of cellular proteins.

If you have not already done so, **double click on the icon for the program MAGE. When queried, click on the "Proceed" box. Then, under the "File" pull-down menu, invoke the "Open File..." option and select E15_cAPK.kin.**

Click on the graphics window to bring it to the front. Click on the box in the upper right corner of the MAGE color graphics bar to make the graphics window fill the screen.

Kinemage 1: The Catalytic Subunit of cAMP-Dependent Protein Kinase (Section 15-3B; Fig. 15-14).

View1 resembles Fig. 15-14, and shows the 350-residue catalytic subunit of cAPK in complex with Mg–ATP and a 20-residue segment of a naturally occurring protein kinase inhibitor. The inhibitor contains the sequence Arg-Arg-Asn-Ala-Ile, a nonphosphorylatable analog of cAPK's target sequence, Arg-Arg-X-Ser/Thr-Y (in which the normally phosphorylated Ser/Thr is replaced by Ala). **Click on the "Inhibitor" button to turn the inhibitor on and off. Click on the "Target seq." button to turn on and off the small cyan balls marking this target sequence.** The protein and its peptide inhibitor are represented by their Cα backbones in purple and gold, respectively.

Click on the "Conserved" button to identify the Cα atoms of the residues that are most highly conserved in the many known protein kinases (marked by small white balls). The ATP is represented in space-filling form colored according to atom type (C, green; N, blue; O, red; and P, yellow) controlled by the "Spacefilling" button or in wireframe form colored pink and controlled by the "Wireframe" button. The two magnesium ions that are associated with the ATP's triphosphate group are represented by skyblue balls and controlled by the "Mg2+" button. The inhibitor peptide, the ATP, and the magnesium ions are also collectively controlled by the "Substrate" button. (Note that View1 resembles Fig. 27-35, which represents the structure of the related protein kinase CDK2, as well as Fig. 15-14).

Turn off and on the "Substrate" button to clearly see that the enzyme is bilobal with a deep cleft between the lobes that is occupied by the Mg–ATP and the inhibitor peptide segment. Note how the C subunit's conserved residues are largely clustered around the Mg–ATP.

Now go to View2 for a closer look at the enzyme's catalytic site. Turn on the "Substrate" and the "Catalysis" buttons. Can you find the ATP's γ phosphorus atom (highlighted in gold), the atom that is transferred to the substrate's Ser/Thr hydroxyl group? (*Hint:* Use the "gamma P" button.) **Can you find the target hydroxyl group to which the γ phosphate group would be transferred in an actual substrate?** It is marked with a red ball. Also displayed, in cyan with O atoms represented by red balls, is the side chain of the C subunit's conserved Asp 166, a group positioned to act as a general base in the removal of a proton from the target Ser/Thr hydroxyl group.

Coordinates were obtained from 1ATP.

EXERCISE 16. CALMODULIN

KINEMAGE 1: The structure of CaM highlighting its EF-hand motifs: Section 15-3B; Figs. 15-16 and 15-17.

KINEMAGE 2: The structure of CaM in complex with its target peptide: Fig. 15-18.

Calmodulin (CaM) provides an excellent example of calcium binding by proteins that sense the concentration of calcium. Many such proteins contain what are known as EF-hands in which the loop binds a calcium ion. They are named for the E and F helices of the calcium-binding protein parvalbumin, the helical components of the first EF-hand motif of known X-ray structure, together with the conformational resemblance of this motif to the index finger and thumb of a right hand (Fig. 15-17).

Calmodulin consists of four EF-hands that are separated by short connecting segments. These EF-hands bind calcium cooperatively, thereby making calmodulin a sensitive indicator of the concentration of free calcium in the cell. Calmodulin's calcium-induced conformational change affects the activities of a large number of proteins (Section 15-3B, for example). This EXERCISE contains two kinemages that illustrate calmodulin's structural repetitiveness, stability, and flexibility.

If you have not already done so, **double click on the icon for the program MAGE. When queried, click on the "Proceed" box. Then, under the "File" pull-down menu, invoke the "Open File..." option and select E16_CaM.kin.**

Click on the graphics window to bring it to the front. Click on the box in the upper right corner of the MAGE color graphics bar to make the graphics window fill the screen.

Kinemage 1: The Structure of CaM Highlighting Its EF-hand Motifs (Section 15-3B; Figs. 15-16 and 15-17).

View1 shows the Cα backbone of calmodulin (CaM) oriented as in Fig. 15-16 with its four EF-hands colored similarly to those in Fig. 15-17. CaM has a bilobal structure, that is, it is dumbbell-shaped. Thus, its structurally similar two lobes (domains), which each contain two EF-hands, are separated by a long central helix whose central connecting region is colored magenta. CaM's four bound Ca^{2+} ions are represented by yellow spheres. **Press the "ANIMATE"**

button several times to see how the EF hands are added, together with the central helix, to form the molecule.

Rotate the molecule to assure yourself that all of the EF-hands have similar structures. Look down the long helix (View2) to see how the two sets of EF-hands extend to opposite sides of the helix.

Go to View3 and turn off "EF-Hand 2" and "Bend Helix". This displays an EF-hand motif in the orientation shown in Fig. 15-17. **See if you can repeat this for the other EF-hands.**

Kinemage 2: The Structure of CaM in Complex with Its Target Peptide (Fig. 15-18).

The crystal structure of mammalian CaM in complex with its target peptide [here a synthetic peptide whose sequence is based on that of chicken smooth muscle myosin light chain kinase (MLCK)] is quite different from that of free CaM. Part of the long central helix unwinds to form a more flexible structure, thereby allowing both lobes of the structure to come together, much like a set of jaws, to enclose the peptide.

View1, which resembles Fig. 15-18*a*, shows a view of CaM bound to the MLCK peptide. Here, the N-terminal lobe is on the right, and the C-terminal lobe is on the left. Note the close proximity of the two ends of the CaM polypeptide.

View2, which resembles Fig. 15-18*b*, looks down the helix of the MLCK peptide of the CaM–MLCK complex. CaM almost entirely engulfs the MLCK peptide. **Rotate the molecule and see how little of the peptide is exposed to solvent. Does this suggest a mechanism for activation of CaM-dependent proteins?**

In View3, use the ANIMATE button to visualize the conformational changes that CaM presumably undergoes on binding its target peptide. Click on the button "Bend Region" to highlight that part of the helix that changes conformation when CaM binds peptide. View3, View4, and View5 provide different views of this conformational change. It is unusually large and occurs via the uncommon mechanism of helix unwinding.

The coordinates for the X-ray structure of rat testes calmodulin were obtained from 3CLN. The coordinates for the X-ray structure of mammalian calmodulin in complex with chicken smooth muscle myosin light chain kinase were obtained from 1CDL.

Exercise in large part by John H. Connor
Terminator Graphics Limited
1118 Kimball Dr.
Durham, NC 27712
USA

EXERCISE 17. A-, B- AND Z-DNA STRUCTURES

KINEMAGE 1: Comparison of the Structures of A-, B-, and Z-DNAs: Section 23-1; Fig. 23-2.

KINEMAGE 2: The Watson-Crick Base Pairs: Section 23-1A; Fig. 23-1.

KINEMAGE 3: Sugar Pucker, 3′-*Exo* and 3′-*Endo:* Section 23-1B; Fig 23-6.

KINEMAGE 4: B-DNA: Section 23-1; Fig. 23-2.

KINEMAGE 5: A-DNA: Section 23-1; Fig. 23-2.

KINEMAGE 6: Z-DNA: Section 23-1; Fig. 23-2.

DNA, the archive of hereditary information, forms double helices whose component strands are complementary and antiparallel. There are three major conformers of DNA (Section 23-1): B-DNA, the native form, which is stable at high humidities; A-DNA, which is stable at lower humidities; and Z-DNA, which consists of sequences of alternating purine and pyrimidine nucleotides and is stable at high salt concentrations. In this exercise, we explore the structures of these DNAs. We begin by comparing all three molecules, we then study the structures of the Watson–Crick base pairs and how sugar pucker influences the conformation of DNA's sugar phosphate backbone, and finally we examine the structure of each of the three DNA conformers in detail.

If you have not already done so, **double click on the icon for the program MAGE. When queried, click on the "Proceed" box. Then, under the "File" pull-down menu, invoke the "Open File..." option and select E17_DNA2.kin.**

Click on the graphics window to bring it to the front. Click on the box in the upper right corner of the MAGE color graphics bar to make the graphics window fill the screen.

Kinemage 1: Comparison of the Structures of A-, B-, and Z-DNAs (Section 23-1; Fig. 23-2).

In this kinemage, the three DNA conformers are shown side-by-side for comparison. The A- and B-DNAs are shown in their ideal (canonical) conformations whose structural parameters are given in Table 23-1. Both of them consist of a pair of self-complementary 12-nucleotide strands of sequence d(GCGCAATTGCGC). The Z-DNA, which is also an idealized structure, consists of a pair of self-complementary 12 nucleotide strands of sequence d(CGCGCGCGCGCG).

View1 shows, from left to right, A-, B-, and Z-DNAs as viewed with their helix axes vertical (similar to Fig. 23-2a). All three DNAs are skyblue with their ribose ring oxygen atoms (O4′ or O4*) represented by small red balls. Note that the three DNAs in this kinemage move together as a rigid unit so that, in some orientations, to view one DNA molecule without overlap, you will have to turn off the others. However, we shall inspect each of these DNAs in greater detail in KINEMAGES 4, 5, and 6.

Temporarily turn off "onewidth" under the "Display" pulldown menu and set "ztran" to −70 (lower, far right slider) to get better depth perception. Then determine the handedness of each of the helices. To do this, decide which of your hands, when wrapped around the helix in the direction in which it is rising, results in the thumb pointing up. **Can you see that A- and B-DNAs are right-handed double helices, whereas Z-DNA is a left-handed double helix? Turn on the "Backbone 2" button to color the strands of the helices differently. Also determine the pitch (rise per turn), diameter and base tilt of each of the helices and compare your results with the data in Table 23-1. Can you verify that, whereas the planes of the base pairs in B- and Z-DNAs are nearly perpendicular to their helix axes, those of A-DNA are tipped by 20° from this orientation?** To highlight the bases, turn on the "Bases" button.

In View2, all three DNAs are viewed, side-by-side, from the top (similar to Fig. 23-2*c*). This view is perhaps more clearly seen by turning on "thinline" under the "Display" pulldown menu. **Turn on the "Top bp" button to highlight the top base pair of each DNA in white** with its connecting hydrogen bonds represented by dashed lines. Also **turn on the "Backbone 2" and "Bases" buttons.** In each type of DNA, the sugar–phosphate backbone is wrapped around the periphery of its helix with the bases extending from them towards the helix center. However, in A-DNA, the sugar–phosphate backbone, whose ribose rings are oriented approximately perpendicular to radial lines extending from the helix axis, forms a thin layer around the helix' periphery, whereas in B-DNA, whose ribose rings are oriented more nearly radially, the sugar–phosphate layer is thicker.

From the positions of the ribose ring oxygen atoms (red balls) throughout each helix it is apparent that in both A- and B-DNA each nucleotide has an equivalent position within the helix. (In A-DNA the ribose ring oxygens are all around the outer perimeter while in B-DNA they are all somewhat inside). This is not the case, however, with Z-DNA. It has two rings (in View2) of ribose ring oxygen atoms: The outer one is formed by the purine nucleotides (here all G's) and the inner one is formed by the pyrimidine nucleotides (here all C's). This is because the Z-DNA ribose rings have different sugar pucker depending on whether they are attached to a purine or pyrimidine base (Table 23-1; we shall examine these ribose ring pucker conformations in KINEMAGE 3). **Go back to View1, and turn on the "Curves" button. This draws imaginary lines joining successive ribose ring oxygen atoms on one strand of A-, B-, and Z-DNAs. Can you see that both A- and B-DNAs have relatively smooth curves while Z-DNA has a zigzag shape?** The "Z" in Z-DNA is due to the alternating conformations of the nucleotides in a Z-DNA strand. The repeating unit of A- and B-DNAs is therefore a nucleotide, whereas that in Z-DNA is a dinucleotide or, alternatively, a base pair.

Go back to View2 and look at the centers of the three helices. Can you see that the helix axis passes through the center of each base pair in B-DNA and near the top edge of each base pair in Z-DNA? This makes these two helices have solid cores. In contrast, the bases of A-DNA lie so far from the helix axis that this DNA has a hollow 6-Å-diameter core, wide enough to allow the passage of water molecules. Thus, as we shall see in KINEMAGE 5, A-DNA can be described as a base-paired ribbon wrapped around a cylinder.

Kinemage 2: The Watson-Crick Base Pairs [Section 23-1A; Fig. 23-1; Also review Section 3-2A (This KINEMAGE is also presented in EXERCISE 2)].

View1 shows an A·T Watson–Crick base pair from ideal B-DNA (Fig. 23-1). The C, N, and O atoms of the bases (including ribose atom C1′, the glycosidic carbon atom) are represented by gray, blue, and red balls, respectively, that can be turned on and off with the "Base Atoms" button. The bonds of the thymine (T) base are yellow and those of the adenine (A) base are seagreen. The 5′-ribose phosphate groups, which can be turned on and off with the "Ribose Phos." button, are magenta and their ribose ring oxygen atoms are represented by small red balls. The hydrogen bonds through which the bases are paired are represented by dashed white lines that can be turned on and off with the "H bonds" button.

Click the "ANIMATE" button to replace the A·T base pair with a G·C Watson–Crick base pair. The G·C pair is colored identically to the A·T pair except that the bonds of the guanine (G) base are skyblue and those of the cytosine (C) base are orange. Note that the atomic positions of the ribose phosphate groups, including those of the glycosidic carbons (the ribose C atoms that are bonded to the bases), are unchanged by this base pair switch. **View2 shows the bases edgewise.** There would likewise be no change in the atomic positions of the ribose phosphate groups if the A·T and G·C base pairs were replaced by T·A or C·G. Thus, the conformation of a DNA's sugar–phosphate backbone is unaffected by the identities of its Watson–Crick base pairs. This is one of the requirements for building a regular helix whose structure is independent of sequence.

View2 also shows that the two ribose–phosphate groups in each base pair are oriented oppositely, making the DNA's sugar–phosphate backbones antiparallel.

The bases in DNA lie at smaller distances from the helix axis than the sugar–phosphate backbones, so each DNA double helix has two helical grooves running along its periphery. These grooves can be easily identified as follows: **Go back to View1 and turn on the "Glycosid.Bonds" button to highlight the glycosidic bonds in red.** The minor groove is the one in which the two glycosidic bonds of the base pair make an angle of less than 180°. The groups that line the minor groove are donated by the edges of the bases that face the opening of this angle. **Turn on**

the "MinorGroove" button to highlight, in cyan, an atom that lines the minor groove on each base (atoms C2, O2, N2, and O2 on A, T, G, and C, respectively). Similarly, the edges of the bases that face the opening of the angle made by the glycosidic bonds that is greater than 180° line the DNA's so-called major groove. **Turn on the "MajorGroove" button to highlight, in yellow, an atom that lines the major groove on each base** (atoms N6, O4, O6, and N4 of A, T, G, and C, respectively).

*Kinemage 3: Sugar Pucker, 3′-*Exo *and 3′-*Endo *(Section 23-1B; Fig. 23-6).*

The interior bond angles in a ribose ring are all approximately equal to the tetrahedral angle (109.5°) and, since the interior angles of a regular pentagon are all approximately this value (108°), it might naively be expected that the ribose ring is planar. However, the ring substituents are eclipsed when the ring is planar. To reduce this energetically prohibitive crowding, the ribose ring assumes a puckered conformation, with one or two of the ring C atoms out of the plane of the ring (Fig. 23-6). In the majority of cases, only one of the ribose ring atoms departs markedly from the plane of the other four atoms. If the out-of-plane atom is displaced to the same side of the ribose ring as its atom C5′, the ring is said to have the *endo* conformation, whereas if the out-of-plane atom is displaced to the opposite side of the ring from C5′, the ring is said to have the *exo* conformation. The most common ribose ring conformations are C3′-*endo,* which occurs in A-DNA, and C3′-*exo* and C2′-*endo,* which are almost identical and occur in B-DNA.

View1 shows deoxyadenosine-3′,5′-diphosphate in the C3′-*exo* (similar to C2′-*endo*) conformation seen in ideal B-DNA (Fig. 23-6*b*). This molecule is shown in ball-and-stick form with its C, N, O, and P atoms gray, blue, red, and yellow, respectively, and its covalent bonds white with the exception that its out-of-plane ribose ring atom, here C3′, together with its connecting bonds, is cyan. **Go to View2, to see the ribose ring edge on** such that atom C1′ eclipses atom O4′ and atom C2′ eclipses atom C4′. Thus, these four atoms are coplanar (if this were not the case, no view would be possible in which there were two eclipsed ring atoms). **Do you agree that atom C3′ is on the opposite side of the C4′–O4′–C1′–C2′ plane from atom C5′, that is, that the ring has the C3′-*exo* conformation?**

Now go back to View1 and click the "ANIMATE" button. This displays the deoxyadenosine-3′,5′-diphosphate in the C3′-*endo* conformation it has in ideal A-DNA. **With "C3′-endo" selected, go to View3 to see the ribose ring edge on in the C3′-*endo* conformation. Can you see that its atom C3′ lies on the same side of the C4′–O4′–C1′–C2′ plane as does atom C5′?**

Go back to View1 and ANIMATE to compare the C3′-*exo* and C3′-*endo* conformers. Measure the distances between the P3′ and P5′ atoms in the 3′-*exo* conformation and in the 3′-*endo* conformation and compare them with those in Fig. 23-6. **Click on both the C3′-*exo* and C3′-*endo* buttons to superimpose the two conformations and look at the positions of the adenine bases. While both are still anti (Fig. 23-5), there is a significant difference in the torsion angle about the glycosidic bond. This difference is part of the reason that, whereas the planes of the base pairs in B- and Z-DNAs are nearly perpendicular to their helix axes, those of A-DNA are tipped by 20° from this orientation.**

Kinemage 4: B-DNA (Section 23-1; Fig. 23-2. see also Section 3-2A).

In the remaining three kinemages of this exercise, we further examine the conformations of B-, A-, and Z-DNAs, the same molecules we compared in KINEMAGE 1. In this KINEMAGE we examine the structure of B-DNA (also presented in EXERCISE 2). **View1 shows B-DNA with its helix axis vertical and looking down its 2-fold axis of symmetry into its minor groove.** All atoms of the 12-bp duplex helix are shown as large balls (with C, N, O, and P atoms white, blue, red, and seagreen) that, for the sake of clarity, are slightly smaller than space-filling size. **Use the "Backbone" and "Bases" buttons under the "Spacefilling" button to turn the base pairs and the two sugar–phosphate chains of the duplex on and off separately. Use the "MinorGroove" and "MajorGroove" buttons to highlight the minor and major grooves (in cyan and yellow).** Turning on the "MinorGroove" button highlights, in cyan, an atom that lines the minor groove on each base (atoms C2, O2, N2, and O2 on A, T, G, and C, respectively). Turning on the "MajorGroove" button highlights, in yellow, an atom that lines the major groove on each base (atoms N6, O4, O6, and N4 of A, T, G, and C, respectively). View2 looks towards the DNA's major groove, View3 shows the DNA molecule with its 2-fold axis of symmetry horizontal and with its helix tilted about this axis so as to view along the DNAs' major and minor grooves, and View4 shows the DNA as viewed along its helix axis.

Compare View1 (similar to Fig. 23-2*b*) and View2 of B-DNA. Note that its major groove is considerably wider than its minor groove although the two grooves are more or less equally deep. This is particularly evident in View3, a view along the grooves in which the major groove faces left in the center of the DNA and the minor groove faces left near both the top and bottom of the DNA. The different widths of the grooves arise from the asymmetry of the ribose-phosphate groups that comprise their walls. **In View1 or View2, turn off the "Backbone" button under the "Spacefilling" button,** to see that the base pairs form a solid stack in which the bases are in van der Waals contact (the apparent gaps between the bases are due to the less-than-van der Waals radii of the balls representing the atoms).

Turn off the "Spacefilling" button and turn on the upper "SingleStrnd" button to display the path taken by one of the two identical polynucleotide strands of the B-DNA. Only the sugar–phosphate backbone of this strand is initially shown but its bases can also be displayed by turning on its corresponding "Bases" button.

Now turn off both the "Spacefilling" and upper "SingleStrnd" buttons and then turn on the "Wireframe" button to display the entire duplex molecule in stick form colored skyblue with its ribose ring oxygen atoms represented by small red balls. The backbone and bases can be individually controlled with the corresponding "Backbone" and "Bases" buttons. The "Top bp" button highlights the top base pair in white.

Go to View5 to see one of the Purine nucleotides close up. Can you tell what the sugar pucker is (Kinemage 3; Fig. 23-6)? Can you tell whether the base is syn or anti (Fig. 23-5)? Compare your identifications with Table 23-1. Go to View6 to see one of the Pyrimidine nucleotides up close. Can you tell what the sugar pucker is (Kinemage 3; Fig. 23-6)? Can you tell whether the base is syn or anti (Fig. 23-5)? Compare your identifications with Table 23-1.

Go back to View1, and turn on the "Curves" button. This draws imaginary lines joining successive ribose ring oxygen atoms on one strand of the B-DNA. Can you see that B-DNA has a relatively smooth curve? The same is true for A-DNA but not for Z-DNA, as you saw in KINEMAGE 1 and will see again in KINEMAGES 5 and 6.

Turn on the lower (below the dashed line) "SingleStrnd" button to highlight one of the sugar–phosphate backbone strands in magenta. **Turn on the associated "Bases" button** to highlight the bases of this strand in gold. **Turn off the "Wireframe" button** to trace the pathway of a single strand of B-DNA. **Then, turn on the upper "SingleStrnd" button,** so the B-DNA is displayed with one of its strands in skeletal form and the other in spacefilling form.

Finally, you should keep in mind that this kinemage shows an ideal structure whose component nucleotides all have exactly the same conformations. In the X-ray structures of B-DNAs (e.g., Fig. 23-2), the conformations of each nucleotide deviate, occasionally to a surprisingly large degree, from their ideal values. It appears that these deviations are sequence-dependent.

Kinemage 5: A-DNA (Section 23-1; Fig. 23-2).

View1 shows ideal A-DNA in spacefilling form with its helix axis vertical and facing its minor groove. Can you see that A-DNA has a conformation very different from that of B-DNA? Turn on the "MajorGroove" and "MinorGroove" buttons to help identify them. Its minor groove is wide, although not so wide as B-DNA's major groove. View2, facing the major groove, shows that the major groove is quite narrow. Yet, as View3 reveals, the major groove (which opens to the left in the center of the helix) is extremely deep: In contrast, the minor groove (which opens to the left near the top and bottom of the helix) is so shallow as to hardly be a groove at all.

Go to View4 and in the Spacefilling mode, highlight the major groove in yellow by turning on the "MajorGroove" button. Can you see that the base atoms of the major groove actually line A-DNA's hollow central cavity? This emphasizes the previously discussed concept that A-DNA can be considered to be a base paired ribbon wound around a cylindrical cavity.

Go to View5, turn off "Spacefilling" and turn on "Wireframe" to see one of the purine nucleotides close up. Can you tell what the sugar pucker is (Kinemage 3; Fig. 23-6)? Can you tell whether the base is syn or anti (Fig. 23-5)? Compare your identifications with Table 23-1. Go to View6 to see one of the pyrimidine nucleotides up close. Can you tell what the sugar pucker is (Kinemage 3; Fig. 23-6)? Can you tell whether the base is syn or anti (Fig. 23-5)? Again, compare your identifications with Table 23-1.

Go back to View1, and turn on the "Curves" button. This draws imaginary lines joining successive ribose ring oxygen atoms on one strand of the A-DNA. Can you see that A-DNA has a relatively smooth curve? The same is true for B-DNA but not for Z-DNA, as you saw in KINEMAGES 1 and 4 and will see again in KINEMAGE 6.

Kinemage 6: Z-DNA (Section 23-1; Fig. 23-2).

The last kinemage of this exercise displays ideal Z-DNA. View1, View2, and View3 show that this left-handed double helical molecule has a particularly narrow but quite deep minor groove (facing left in View3 in the middle of the molecule) but a very shallow major groove. **Turn on the "MajorGroove" and "MinorGroove" buttons to help identify them.**

Turn off "Spacefilling", turn on "Wireframe" and go to View5 to see one of the purine nucleotides close up. Can you tell what the sugar pucker is (Kinemage 3; Fig. 23-6)? Can you tell whether the base is syn or anti (Fig. 23-5)? Compare your identifications with Table 23-1. Go to View6 to see one of the pyrimidine nucleotides up close. Can you tell what the sugar pucker is (Kinemage 3; Fig. 23-6)? Can you tell whether the base is syn or anti (Fig. 23-5)? Compare your identifications with Table 23-1.

Go back to View1, and turn on the "Curves" button. This draws imaginary lines joining successive ribose ring oxygen atoms on one strand of the Z-DNA. Can you see that Z-DNA has a zigzag shape? The "Z" in Z-DNA is due to the zigzag (alternating) conformations of the nucleotides in a Z-DNA strand. The repeating unit of Z-DNA is a dinucleotide or, alternatively, a base pair.

The atomic coordinates for A- and B-DNAs were generated by the program MacImdad® (Molecular Applications Group) using its default parameters for ideal A- and B-DNAs. The atomic coordinates for Z-DNA were obtained from 3ZNA, an idealized model based on a crystal structure of d(CGCGCG).

EXERCISE 18. RESTRICTION ENDONUCLEASES

KINEMAGE 1: *Eco*RI-DNA: Fig. 23-30.

KINEMAGE 2: *Eco*RV-DNA: Fig. 23-31.

Restriction endonucleases are bacterial enzymes that recognize a specific sequence of four to eight bases in double-stranded DNA and cleave both strands of the duplex. They do so as part of a restriction–modification system that also contains a modification methylase that methylates a specific base in the same recognition sequence. The restriction endonuclease only cleaves DNA whose recognition sequence is unmethylated. Hence, when a DNA whose recognition sequence is unmethylated (e.g., DNA from a virus) is introduced into a bacterium, the DNA is rapidly cleaved by the bacterium's restriction endonuclease. The bacterium's newly replicated DNA is hemimethylated (the parent strand is methylated and the daughter strand is unmethylated), which is sufficient to prevent the action of the restriction endonuclease. The daughter strand is eventually methylated by the relatively slow acting modification methylase, thereby also protecting the progeny of this DNA from the restriction endonuclease.

Type II restriction endonucleases have additional properties; they cleave DNA within their recognition sequences and they occur on different proteins from their corresponding modification methylases. Type II restriction endonucleases (hereinafter called just restriction endonucleases) have therefore become indispensable laboratory tools in molecular biology: They are the "scalpels" that are used to cleave DNAs at sequence-specific sites. Moreover, since most restriction endonucleases recognize sequences that are palindromic (have twofold symmetry; e.g., see Table 3-3) and many of them cleave the two single strands comprising the duplex at sites that are symmetrically staggered about the center of the palindrome, the resulting restriction fragments have complementary (sticky) ends. This facilitates the insertion of restriction fragments into restriction cuts made by the corresponding restriction endonuclease.

In this exercise, we examine the structures of two restriction endonucleases, *Eco*RI and *Eco*RV, in their complexes with duplex DNA segments containing their target sequences.

If you have not already done so, **double click on the icon for the program MAGE. When queried, click on the "Proceed" box. Then, under the "File" pull-down menu, invoke the "Open File..." option and select E18_EcoR.kin.**

Click on the graphics window to bring it to the front. Click on the box in the upper right corner of the MAGE color graphics bar to make the graphics window fill the screen.

Kinemage 1: EcoRI·DNA *(Section 23-4A; Fig. 23-30).*

View1 shows *Eco*RI restriction endonuclease in complex with a 12-bp self-complementary DNA of sequence tcgcGAATTCgcg (recognition sequence in upper case letters). View1 is similar to Fig. 23-30*a*. The dimeric protein is represented by its Cα chain with one subunit red and the other blue. The duplex DNA is displayed in spacefilling form with its sugar-phosphate chains (as represented by its C1*, C4*, and P atoms; pinktint and bluetint), its palindromic recognition sequence bases green, and its other bases greentint. The DNA may be alternatively displayed in wireframe form ("Spacefilling" button off and the "Wireframe" button on) with its sugar–phosphate chains pinktint and bluetint and its bases white.

Identify the DNA's minor groove by turning on the "MinorGroov" button, which highlights in cyan a specific atom that faces the minor groove on each base. **Identify the major groove by turning on the "MajorGroov" button** to highlight atoms in the major groove yellow.

View2 and View3 are views of the complex into the minor and major grooves. Note that DNA binds with its palindromic sequence in the symmetric cleft between the two subunits. Indeed, nearly all bacterial proteins that recognize specific DNA sequences (e.g., see EXERCISE 19), as well as many eukaryotic proteins, have at least a twofold axis of symmetry as well as palindromic recognition sites.

Identify the DNA groove (major or minor) in which *Eco*RI binds. Identify the pair of helices, one from each subunit, that reach into the groove. These helices may be highlighted with the "helices" button. Turn off the "EcoRI" button to see them more clearly. Turn the "EcoRI" button back on and go to View4. Can you see how the protein grips its recognition sequence in a two-armed embrace?

Turn off the "EcoRI", "Spacefilling", "MinorGroov", and "MajorGroov" buttons and turn on the "Wireframe" button. Is the DNA in normal B-form? Look at the DNA's central two A·T base pairs. Are they perpendicular to the helix axis or are they "kinked"? View5 is a closeup of these A·T base pairs edgewise. You can clearly see that the dihedral angle between the planes of the central base pairs (their inter-base pair roll angle) is ~50° opening towards the minor groove, thereby unstacking the central base pairs. This unwinds the duplex helix by 28° and widens the major groove in this region by 3.5 Å, which facilitates the access of the *Eco*RI groups that make sequence-specific contacts with the recognition sequence. By rotating View5 about the twofold (horizontal) axis until the other base pairs in the DNA are viewed edgewise, you can see why the duplex helix appears, at first glance, to be straight: There is a compensating bend at the base pair step on each side of the central base pair.

Turn on and off the "sidechains" button to identify amino acid sidechains interacting with the DNA. What amino acid residue is interacting with the bases? What bases are involved? What kinds of interaction do you see? Are the interactions sequence-specific? That is, would the interaction be possible if the base in question were replaced with a different base?

Use Views6 and 7 to examine two different sugar puckers in the distorted DNA. Looking at the ribose ring nearest the center of the screen in each view, which view shows C3′-*exo* (C2′-*endo*) and which view shows C3′-*endo*? Examine the bases to determine which are tilted and which perpendicular to the helix axis.

Kinemage 2: EcoRV·DNA *(Section 23-4A; Fig. 23-31).*

*Eco*RV is a homodimeric restriction endonuclease that binds to the 6-bp palindromic sequence 5′-GATATC-3′, which it cleaves on each strand at its central T and A bases to yield blunt-ended restriction fragments. This kinemage depicts the structure of *Eco*RV in its complex with a 10-bp, duplex, largely B-type, palindromic DNA of single strand sequence 5′-ggGATATCcc-3′, where upper case letters represent the enzyme's recognition sequence. The dimeric protein is represented by its Cα chain with one subunit red and the other blue. The duplex DNA is displayed in spacefilling form with its sugar-phosphate chains (as represented by its C1*, C4*, and P atoms) pinktint and bluetint, its palindromic recognition sequence bases green, and its other bases greentint. The DNA may be alternatively displayed in wireframe form ("Spacefilling" button off and the "Wireframe" button on) with its sugar-phosphate chains pinktint and bluetint and its bases white.

View1 shows the complex as in Fig. 23-31*b*. The enzyme is represented by its Cα chain and the DNA shown in spacefilling form. **Turn on the "MinorGroov" and "MajorGroov" buttons. Which grooves does the protein interact with? Use Views2 and 3 to help visualize. Compare the structures of *Eco*RI and *Eco*RV.**

Specific contacts are made with the DNA from a loop from each subunit as seen in View2. Turn on and off the "loops" button to highlight them.

Turn off the protein subunits and spacefilling modes and turn on the side chains and wire-frame form. Using Views5 and 6, examine the DNA bend and the DNA–protein interactions. Are the bases tilted or perpendicular to the helix axis? Does the DNA resemble the conformation of B-DNA or A-DNA?

Turn on and off the "Sidechains" button to see sequence-specific contacts of amino acid sidechains with the DNA. These sidechains are shown in magenta. In which groove are these sidechains interacting? Identify the sidechains and the functional groups participating in these interactions. Which base atoms are interacting? Are the interactions sequence specific? That is, would the interaction be possible if the base in question were replaced with a different base?

Use Views7 and 8 to examine two different sugar puckers in the distorted DNA. Looking at the ribose ring nearest the center of the screen in each view, which view shows C3′-*exo* (C2′-*endo*) and which view shows C3′-*endo*?

What general conclusions can you draw about the differences and similarities of DNA binding by *Eco*RI and *Eco*RV?

The atomic coordinates for the *Eco*RI·DNA and *Eco*RV·DNA complexes were obtained from 1R1E and 4RVE.

EXERCISE 19. 434 REPRESSOR/DNA INTERACTIONS

KINEMAGE 1: 434 Repressor-DNA Interactions: Fig. 23-33.

Gene expression in both prokaryotes and eukaryotes is regulated largely through the control of transcriptional initiation. This occurs via the binding of proteins to specific sequences on the DNA at sites associated with the RNA polymerase's transcriptional initiation site.

In prokaryotes, the control of transcriptional initiation is a relatively simple process (Section 27-2). It occurs mainly through the binding of a protein (a repressor) to a specific DNA segment (an operator) that overlaps RNA polymerase's initial DNA binding site (a promoter). This prevents RNA polymerase from binding to the promoter in a way that can initiate transcription. Under appropriate conditions (e.g., when a particular metabolite, known as an inducer, binds to the repressor), the repressor is released from the operator, thereby permitting RNA polymerase to initiate transcription. For some prokaryotic genes, the control of transcriptional initiation requires the binding of several identical copies of the corresponding repressor to two or more relatively closely spaced DNA sites and/or the binding of additional proteins known as activators [e.g., catabolite gene activator protein (CAP); Section 27-2B].

The transcriptional initiation of eukaryotic genes is a far more complicated process (Section 27-3). It requires the assembly of a complex multiprotein particle on a gene's promoter as well as the binding of several additional proteins known as transcription factors to specific DNA sites that may be as distant as several thousand bp from the promoter.

In this exercise we examine the structure of a prokaryotic transcriptional control protein that binds to a specific sequence of DNA, the 434 repressor (Fig. 23-32). 434 is a bacteriophage (bacterial virus) that infects *E. coli.* It is closely related to the more widely studied bacteriophage λ that likewise infects *E. coli.*

If you have not already done so, **double click on the icon for the program MAGE. When queried, click on the "Proceed" box. Then, under the "File" pull-down menu, invoke the "Open File..." option and select E19_Rprs.kin.**

Click on the graphics window to bring it to the front. Click on the box in the upper right corner of the MAGE color graphics bar to make the graphics window fill the screen.

Kinemage 1: 434 Repressor–DNA Interactions (Section 23-4B; Fig. 23-32).

The duplex DNA segment to which 434 repressor binds in the structure examined in this exercise has the single-strand sequence 5′-tatACAAGAAAGTTTGTact-3′, where upper case letters represent the 14-bp sequence of the 434 phage's so-called OR1 operator (see Section 27-3D for a discussion of specific bacteriophage operator sites). Note that only the outer four base pairs at the ends of this operator sequence are palindromic. A variation in any of these base pairs reduces the affinity with which 434 repressor binds operator by ~100-fold.

The 434 repressor, as are most prokaryotic DNA-binding proteins of known structure, is a homodimer. Each subunit consists of an N-terminal domain, which binds to the target DNA, and a C-terminal domain, which mediates subunit dimerization. The complex shown here contains only the repressor's 69-residue N-terminal domain (however, residues 64-69 are not visible in the X-ray structure) and hence is referred to as R1-69. Note that in solution by itself, R1-69 is monomeric although it binds to its target DNA as a dimer.

View1 (Fig. 23-32) shows, in wireframe form, the protein–DNA complex of 434 repressor and the double helix formed from the above DNA sequence and its complementary strand. The ribose–phosphate "Backbone" of the strand whose sequence is given above is yellowtint and oriented with its 5′ end below. The complementary strand is white. The "Bases" of the lower and upper halves of the OR1 operator sequence (7 bp each) are magenta and purple and are controlled by the "Half-Site 1" and "Half-Site 2" buttons. The remaining bases at the ends of the DNA are gray. The orientation of View1 and the nucleotide numbering are consistent with that in Fig. 23-32 [e.g., 5L is the fifth nucleotide (nt) from the 5′ end of "Strand 1" of the left half of the operator sequence, 5′L is its Watson–Crick partner, 5R is the fifth nt from the 5′ end of "Strand 2", and 5′R is its Watson–Crick partner; negative numbers are assigned to nucleotides outside the operator sequence and decrease outwards from the center].

The R1-69 subunits that bind to the DNA's L and R halves are represented by their main chains in pinktint and bluetint. In View1, the homodimer is oriented to the right of the DNA with the L subunit below the R subunit. **Can you see that the two monomers are binding in successive major grooves? Rotate** the complex around from View1 to verify that each protein subunit consists of a bundle of five helices linked by turns of various lengths. **Can you follow the direction of these helices?** (*Hint:* Remember that the carbonyl groups pointing out from the backbone chains of the α helix all point toward the C-terminal end of the chain.)

Turn on the "HTH Motif" button to highlight, in hotpink and skyblue, the helix–turn–helix (HTH) motifs of the L and R subunits, so-called because each consists of two alpha helices (helices 2 and 3) inclined at a characteristic angle (~120°) and connected by a turn. Note how the C-terminal helices of the HTH motifs, the so-called recognition helices (both of which are viewed here nearly end-on), fit neatly into adjacent turns of the DNA's major groove where, as we shall see below, each makes sequence-specific contacts with the bases of the operator as well as non-sequence-specific contacts with the DNA ribose–phosphate backbone. Numerous prokaryotic DNA-binding proteins similarly bind in the major grooves of their target DNAs via

HTH motifs (e.g., CAP, Fig. 27-14; and the *trp* repressor, Fig. 23-33). Other than that, however, most of these proteins exhibit little structural similarity. The HTH motif, which has only been observed in DNA-binding proteins, is called a "motif" because it only assumes its characteristic conformation when it is part of a larger protein.

Go to View2 to display the L half of the complex in expanded view (Fig. 23-32*b*). Rotate the protein so as to sight down the axis of the recognition helix (helix 3) and note again how it nestles parallel to the major groove.

Turn on the "H-bonds" button to display the sidechains of the repressor and the hydrogen bonds in which they participate. The specific H-bonds made to the bases are shown as dashed pink lines with the N and O atoms participating in the H-bonds represented by blue and red balls. The five such H-bonds all involve three Gln residues (Q28L, Q29L, and Q33L) projecting from the recognition helix. They respectively interact with adenine 1L [A(1L)] on strand 1 and G(2′L) and T(4′L) on strand 2.

The nonspecific H-bonds that the sidechains and backbone of the repressor make with several phosphate oxygen atoms on each strand of the DNA are shown in green. Although these are not sequence-specific H-bonds, they are nevertheless important in facilitating the binding of 434 repressor to its operator DNA. Note that one of these H-bonds is mediated by a bridging water molecule (large orange ball). It is becoming increasingly evident that such water-mediated interactions are important structural determinants in proteins and nucleic acids.

View3 is the analogous view of the right (R) half of the complex showing the corresponding protein sidechains and DNA bases. Since the DNA is only partially palindromic, the interactions that this half makes with the DNA are not exact duplicates of the L half. **Can you find some of the differences?**

R1-69 does not contact the central base pairs of OR1 (bases 5, 6, and 7 of both halves) in the major groove. However, can you see that the side chains of both Arg 43L and Arg 43R extend into the minor groove (middle of View2, top of View3, and bottom of View4)? There the side chain of Arg 43L makes water-bridged hydrogen bonds with both A(7L) and A(7′R).

Hydrogen bonds, although important for specificity of binding, are not significant contributors of conformational stability in aqueous solution (Section 6-4A). Thus, the complex between 434 repressor and its operator must be stabilized by hydrophobic, electrostatic, and van der Waals forces which, in turn, requires that these two macromolecules have closely matching conformations in their complex. The structure of the R1-69-DNA complex shows that this is, in fact, the case. Thus, for example, the positively charged Arg 43 sidechains appear to be positioned to electrostatically stabilize the conformation of the anionic DNA (see below), a hypothesis that is supported by the observation that changing Arg 43 to Ala decreases the binding affinity of 434 repressor for its target DNA by over 200-fold.

Let us now examine how the DNA distorts in response to R1-69 binding. **Go to View4 and ANIMATE.** This will alternate the DNA conformation found in the protein–DNA complex with the outline of B-DNA. Note that the two consecutive major grooves of B-DNA are too far apart to both interact well with the repressor.

Turn on the "BaseHbonds" button and turn off "H-bonds". The binding of the repressor to the DNA causes the center of the DNA to bend along an arc of radius 65 Å to bring these grooves closer together. **Can you find the base pairs that are distorted in the repressor-DNA complex?** They form new, bifurcated H-bonds with the base pairs above and below them, marked in orange. It seems likely that the formation of these bifurcated hydrogen bonds in the DNA sequence contributes to reducing its free energy of bending.

The atomic coordinates for the 434 R1-69 in complex with the 20-bp OR1-containing DNA were obtained from 2OR1.

Exercise in large part by John H. Connor
Terminator Graphics Limited
1118 Kimball Dr.
Durham, NC 27712
USA

EXERCISE 20. tRNA^{Phe}

KINEMAGE 1: Overview of Functional Domains: Fig. 26-6.

KINEMAGE 2: Structural Features: Section 26-2A.

KINEMAGE 3: Tertiary Base-Pairing Interactions: Fig. 26-7.

In ribosomal polypeptide synthesis, transfer RNA (tRNA) molecules function as the adapters that translate the base sequence of messenger RNAs to the corresponding amino acid sequence (Figs. 3-16 and 26-3). A tRNA that is covalently linked to its cognate amino acid (by a process described in EXERCISE 21 and Section 26-2B) base pairs via its anticodon to a codon on an mRNA that specifies that amino acid according to the genetic code (Table 26-1). The ribosome then catalyzes the transfer of the amino acid residue to the C-terminal end of the growing polypeptide chain (Fig. 3-16).

All cells as well as mitochondria and chloroplasts produce numerous distinct, yet similar, tRNA species. The "standard" genetic code (Table 26-1) has 61 codons that specify amino acids which, according to the wobble hypothesis (Section 26-2C), can be translated by a minimum of 32 tRNAs—including the tRNA required to initiate ribosomal polypeptide synthesis. Most cells, however, contain more than this minimal set with some mammalian cells having more than 150 tRNAs.

Most tRNAs are around 76 nucleotides (nt) in length, but tRNAs as short as 60 nt and as long as 95 nt have been sequenced. Up to 25% of the bases in mature tRNAs are enzymatically modified forms of the A, U, G, and C in the initially transcribed forms of these tRNAs (Figure 26-5 and Section 26-2A). In this exercise, we examine the structure of yeast phenylalanine tRNA (tRNA^{Phe}), the first polynucleotide to have its X-ray structure determined (Fig. 26-6).

If you have not already done so, **double click on the icon for the program MAGE. When queried, click on the "Proceed" box. Then, under the "File" pull-down menu, invoke the "Open File..." option and select E20_tRNA.kin.**

Click on the graphics window to bring it to the front. Click on the box in the upper right corner of the MAGE color graphics bar to make the graphics window fill the screen.

Kinemage 1: Overview of Functional Domains (Section 26-2A).

This kinemage shows a ribbon diagram of yeast tRNA^{Phe}. **View1 shows the L-shaped tRNA** in the same orientation as in Figure 26-6. Nearly all tRNAs have similar secondary structures, the so-called cloverleaf form (Fig. 26-4 for tRNAs in general and Fig. 26-6*a* for yeast tRNA^{Phe} in particular). They have several features in common:

1. **A 5′-terminal phosphate group (green).**
2. **The acceptor stem (yellow).** A 7-bp stem, which includes the 5′-terminal nucleotide. The amino acid residue carried by the tRNA is appended to its 3′-terminal OH group (see below).
3. **The D arm (white).** A 4-bp stem ending in a loop that frequently contains the modified nucleotide dihydrouridine (D) and hence is known as the D arm.
4. **The anticodon arm.** A 5-bp stem ending in a loop (greentint) that contains the tRNA's anticodon (seagreen), the triplet of bases that is complementary to the codon specifying the tRNA.
5. **The variable arm (orange),** so-called because, among the various tRNAs, it consists of from 3 to 21 nt and may contain a stem with as many as 7 bp. In yeast tRNA^{Phe}, it consists of 5 nt.
6. **The TψC arm (cyan).** A 5-bp stem ending in a loop that usually contains the sequence TψC [where ψ is the abbreviation used here for the modified nucleotide pseudouridine (Fig. 26-5)].
7. **The 3′-terminal (CCA) end.** The 3′-terminal ends of all tRNAs invariably consist of the sequence CCA-3′ and end with a free OH group (red) to which the tRNA's cognate amino acid residue is appended to form an aminoacyl–tRNA (Fig. 26-8). The enzymes that catalyze the aminoacylation reaction, the aminoacyl–tRNA synthetases, and their complexes with their corresponding tRNAs, are the subject of Section 26-2B and EXERCISE 21.

Toggle the "ANIMATE" button to observe how the tRNA's tertiary structure is built up from its secondary structural elements. Note that the D arm, the anticodon arm, and the variable loop stack together to form a pseudo-continuous double helix (vertical in View1). The same is true of the acceptor stem and the TψC arm (horizontal in View1). Also note that the anticodon (residues 34-36, seagreen) is exposed at the bottom of the anticodon arm where it can readily interact with the codon. Thus, in the L-shaped tRNA molecule, the anticodon is ~75 Å distant from tRNA's 3′ end to which the corresponding amino acid is appended.

View2 is the view from the back of View1, View3 is from the right in View1 along its horizontal helix, and View4 is from the bottom of View1 along its vertical helix. **Run through the animation sequence in each view to observe the build up of the structure.**

The structure of yeast tRNA^{Phe} closely resembles other known tRNA structures. Their major conformational differences arise from apparent flexibility in the anticodon loop and the CCA terminus, as well as a hinge-like mobility between the two legs of the L.

Kinemage 2: Structural Features (Section 26-2A).

View1 shows the front view of the L-shaped tRNA molecule as represented, in gray, by its "Pseudo" backbone (line segments joining successive C4′ and P atoms as well as between C1′ and C4′ atoms of the same nucleotide) and its bases in wireframe form with A pink, U skyblue, G seagreen, C yellow, and modified bases purple. The sugar–phosphate backbone can alternatively be viewed in wireframe form in white by turning on the "RibosePhos" button (and turning off the "Pseudo" button). **Examine the distribution of the various bases in the structure by turning off all the base subgroups, then turning them on one at a time, and then two at a time. In particular, note the correlation between the distribution of A with that of U and of G with that of C.**

With all the bases displayed, inspect Views1–4 and also rotate the molecule around. Is the helical conformation of the two double helical segments A-type or B-type (Fig. 23-2 and EXERCISE 17)? Decide for yourself by seeing whether the base pairs are tilted or perpendicular to the helix axis, whether the helices are solid or have a hole down the middle, and whether the major and minor grooves are narrow, wide, deep, or shallow (See Table 23-1).

The tRNA is held in its compact native conformation by an elaborate network of intramolecular interactions. These include Watson–Crick base pairs, tertiary base pairing interactions that we shall examine in KINEMAGE 3, and stacking interactions. **Locate examples of each of these types of interactions. Can you find any bases that are stacked even though they are not hydrogen bonded to other bases?**

Examine some of the modified bases. Turn off the "A", "U", "G", and "C" buttons, turn on the "Modified" button, and go to each of the Views listed below. Compare the modified base structure that you see there with Fig. 26-5.

View	*Modified Base*
View5	N^2,N^2-dimethyl-G
View6	wybutosine (Y), a hypermodified base
View7	Pseudouridine (ψ) and 5-methyl-C
View8	Hoogsteen-paired ribothymidine (T) and 1-methyl-A

Notes: In View5, momentarily turn on the "G" button to compare the structures of these modified bases to that of a nearby G. View8 shows the Hoogsteen-paired ribothymidine (T) 54 and 1-methyl-A 58 (Hoogsteen pairing is discussed in Section 23-2B; Fig. 23-19*b*). This Hoogsteen pairing is further examined in KINEMAGE 3.

Kinemage 3: Tertiary Base Pairing Interactions (Fig. 26-7).

View1 shows the "front" view of yeast tRNA^{Phe} with its pseudo-backbone in gray. Toggle the "ANIMATE" button to show the tertiary base pairings of Fig. 26-7 in the order of their nucleotide sequence. Tertiary base pairs are defined as base pairs that tie together a tRNA's secondary structural elements.

Here, the bases under consideration are shown in wireframe form with A pink, U skyblue, G seagreen, C yellow, and modified bases purple; their N and O atoms are highlighted with blue and red balls; and hydrogen bonds, which are controlled by the "H Bonds" button, are represented by red lines.

View the tertiary pairings in closeup by going to a particular view and then turning on the buttons below the indicated view, one at a time. Note particularly the 3-way pairings, the non-Watson–Crick pairings, and the presence of modified bases.

View	*Tertiary Nucleotides (Fig. 26-7)*
View2	(A) 4-69
View3	(B) 3-22-46; 15-48
View4	(C) 9-12-23; 10-25-45; 26-44
View5	(D) 18-55; 54-58
View6	(E) 19-56

The Hoogsteen-paired ribothymidine (T) 54 and 1-methyl-A 58 seen in KINEMAGE 2 are displayed in View5 (D) by turning on the "54 & 58" button (recall that in a Hoogsteen base pair, the purine's atom N7 hydrogen bonds to the pyrimidine; Section 23-2B). **Compare this interaction with the Hoogsteen pairing drawn in Fig. 23-19*b*. How do these pairs differ? Do any of the tRNA's other non-Watson-Crick pairings consist of A + U or G + C?**

The coordinates for this kinemage were obtained from 1TRA.

Kinemage largely by Thomas LaBean
Department of Biochemistry
Duke University
Durham, NC 27710
USA

EXERCISE 21. GLUTAMINYL–tRNA SYNTHETASE IN COMPLEX WITH $tRNA^{Gln}$ AND ATP

KINEMAGE 1: The Overall Complex: Fig. 26-11.

Aminoacyl–tRNA synthetases (aaRSs) function to append amino acids to the 3′-terminal ribose residues of tRNAs to yield the aminoacyl–tRNAs that ribosomes use to synthesize polypeptides (Section 26-2B). Since ribosomes ignore the identity of the aminoacyl component of a bound aminoacyl–tRNA, correct aminoacylation is as important for accurate polypeptide synthesis as is the correct ribosomal selection of the mRNA-specified aminoacyl–tRNA.

The tRNA charging reaction occurs in two steps: (1) The aaRS-catalyzed reaction of its cognate amino acid with ATP to yield aminoacyl–AMP and PP_i, a reaction that for most aaRSs can occur in the absence of tRNA; and (2) the transfer of the aminoacyl group to the 3′-terminal 2′- or 3′-OH group of the enzyme's cognate tRNA to yield the corresponding aminoacyl–tRNA and AMP. Cells must each contain at least 20 aaRSs, one for each of the 20 "standard" amino acids.

Although nearly all tRNAs have quite similar structures and physical properties (Section 26-2B), the various aaRSs in a given cell have little sequence similarity to one another. Nevertheless, the aaRSs are classified into two families of ten members each: (1) Class I aaRSs, which have polypeptide segments of consensus sequences His-Ile-Gly-His (HIGH) and Lys-Met-Ser-Lys-Ser (KMSKS) and aminoacylate their tRNAs' 3′-terminal 2′-OH group; and (2) Class II aaRSs, which have three different segments with homologous sequences and aminoacylate their tRNAs' 3′-terminal 3′-OH group.

Accurate translation requires that the various aaRSs aminoacylate their cognate tRNAs and reject all others. How the various aaRSs recognize only their cognate tRNAs appears to be largely idiosyncratic. For example, some aaRSs recognize their cognate tRNA according to the sequence of its anticodon whereas others do not even bind anticodons. Nevertheless, most of the identity elements recognized by aaRSs are clustered at the tRNAs' acceptor stems and at their anticodon loops (Fig. 26-9).

In this exercise, we examine the X-ray structure of *E. coli* glutaminyl–tRNA synthetase (GlnRS), a Class I aaRS, in its complex with $tRNA^{Gln}$ and ATP. This was the first known structure of an aaRS–tRNA complex.

If you have not already done so, **double click on the icon for the program MAGE. When queried, click on the "Proceed" box. Then, under the "File" pull-down menu, invoke the "Open File..." option and select E21_GnRS.kin.**

Click on the graphics window to bring it to the front. Click on the box in the upper right corner of the MAGE color graphics bar to make the graphics window fill the screen.

Kinemage 1: The Overall Complex (Section 26-2B; Fig. 26-11).

View1 shows only the tRNA of the GlnRS–$tRNA^{Gln}$-ATP complex from the same direction as does Fig. 26-11. The ribose–phosphate backbone of the 76-nt $tRNA^{Gln}$ (here portrayed by the bonds joining successive C4′ and P atoms and with C4′ also linked to C1′) is green and its bases are pink. **Can you see that in most respects $tRNA^{Gln}$ closely resembles other tRNAs of known structure; e.g., yeast $tRNA^{Phe}$ (Fig. 26-6 and EXERCISE 20)? Rotate the molecule around to sight down the axes of the two helices.** Note the hollow central portions of these helices characteristic of the expected A-RNA conformation (KINEMAGE 20; Section 23-1A). Views2 and 3 sight down these axes. **Switch on the "Anticodon" and "Acceptor" buttons to identify these tRNA segments** by coloring their bases orange and all atoms of their ribose–phosphate chain yellow. **Do they have the same conformations as in Fig. 26-6 and EXERCISE 20?**

Now return to View1 and switch on the "GlnRS" button. GlnRS is represented in cyan by its Cα backbone. **On which portions of the RNA molecule are there the most interactions with the protein?**

Views2 and 3 show the tRNA's anticodon region (nts C34, U35, and G36) and the acceptor stem region. Can you see that both of these regions interact strongly with the protein? In View2, note that the ribose–phosphate chain of the tRNA's 3′-terminal segment curls around in a hairpin turn rather than continuing the acceptor stem helix as occurs in uncomplexed tRNAs, e.g., in $tRNA^{Phe}$ (EXERCISE 20). The 3′-end of the tRNA plunges into a deep pocket in the GlnRS, which also binds this enzyme's substrates, ATP and glutamine. **Switch on the "ATP" button to see ATP in red.** The 2′-OH group of A76 (red ball when the "Acceptor" button is on), the site at which this Class I tRNA is aminoacylated, is in close proximity to the α phosphate of the bound ATP (dashed white line). In the aminoacylation reaction catalyzed by the enzyme, an aminoacyl group (which is not present in this structure but forms an amino-acyl–AMP at the α-phosphoryl group of ATP during the reaction) is transferred to this 2′-OH.

In View1, turn on the "DinucFold-1" and "DinucFold-2" buttons. The Class I aaRSs of known structure, GlnRS, TyrRS, and MetRS, each contain a dinucleotide-binding (Rossmann) fold associated with its bound ATP. This fold was first observed in a variety of dehydrogenases (Section 6-2B; Figs. 6-29 and 6-30) but was later discovered to also occur in kinases and GTP-binding proteins. The first part of the dinucleotide fold in GlnRS, residues 25 to 100 ("DinucFold-1"; magenta and yellow), forms a β–α–β–α–β motif that makes many of GlnRS's major contacts with ATP. The second part of this dinucleotide-binding fold, residues 210 to 260 ("DinucFold-2", purple and yellowtint), forms an α–β–α–β motif which makes sequence-specific contacts with the tRNA's acceptor stem. Together, the two parts of the dinucleotide-binding fold form a 5-stranded parallel β sheet (yellow and yellowtint). This dinucleotide-binding fold is closely superimposable with the dinucleotide-binding folds of TyrRS and MetRS. **Turn off all buttons except "ATP" and "DinucFold-1" and "DinucFold-2" and go to View6. Compare this view with Figs. 6-29 and 6-30. Can you see how the two parts of the fold come together to make the 5-stranded β sheet? Does the ATP interact with the N-terminal or C-terminal ends of the β sheets? Turn on the "$tRNA^{Gln}$" button to see how the tRNA also interacts with the dinucleotide fold.**

In View7, turn on the "HIGH" and "KMSKS" buttons and turn off all other buttons except "ATP". This displays the corresponding Class I aaRS conserved sequence segments with their main chains in yellowtint and their side chains in gold. Both of these conserved segments contact the ATP (the N_ζ of Lys 270 forms a hydrogen bonded salt bridge with ATP's α phosphate group; dashed white line) and interact with each other but neither directly interacts with the tRNA.

The atomic coordinates for this exercise were obtained from 1GTR.